Fluorocarbon Resins

1971

Dr. M.W. Ranney

RXtSA

Thirty-Five Dollars

NOYES DATA CORPORATION
Noyes Building
Park Ridge, New Jersey 07656, U.S.A.

FOREWORD

The detailed, descriptive information in this Chemical Process Review is based on U.S. Patents since 1960 relating to the production of fluorocarbon polymers.

This book serves a double purpose in that it supplies detailed technical information and can be used as a guide to the U.S. Patent literature in this field. By indicating all the information that is significant, and eliminating legalistic terminology, this book presents an advanced industrially oriented review of fluorocarbon polymer technology.

The U.S. Patent literature is the largest and most comprehensive collection of technical information in the world. There is more practical, commercial, timely process information assembled here than is available from any other source. The technical information obtained from a patent is extremely reliable and comprehensive; sufficient information must be included to avoid rejection for "insufficient disclosure".

The patent literature covers a substantial amount of information not available in the journal literature. The patent literature is a prime source of basic commercially utilizable information. This information is overlooked by those who rely primarily on the periodical journal literature. It is realized that there is a lag between a patent application on a new process development and granting of a patent, but it is felt that this may roughly parallel or even anticipate the lag in putting that development into commercial practice.

Many of these patents are being utilized commercially. Whether used or not, they offer opportunities for technological transfer. Also, a major purpose of this book is to describe the number of technical possibilities available, which may open up profitable areas of research and development.

These publications are bound in paper in order to close the time gap between "manuscript" and "completed book". Industrial technology is progressing so rapidly that hard cover books do not always reflect the latest developments in a particular field, due to the longer time required to produce a hard cover book.

The Table of Contents is organized in such a way as to serve as a subject index. Other indexes by company, inventor, and patent number help in providing easily obtainable information.

CONTENTS AND SUBJECT INDEX

Contents and Subject Index

Contents and Subject Index

INTRODUCTION

Generally, as an outgrowth of research efforts directed towards the use of fluorocarbon compounds as refrigerants, the polymerization of tetrafluoroethylene was discovered in the thirties. Fluorocarbon polymers, resistant to oxidation and stable at high temperatures, were developed during World War II for use in high corrosive environments as encountered in the production of Uranium 235. Polymers of polytetrafluoroethylene, known commercially as PTFE, have seen the greatest development. As a result of the strong carbon-fluorine and carbon-carbon bonds, close packing is achieved and the polymer molecules are substantially unbranched stiff chains in which the fluorine atoms are closely packed around the carbon backbone. In general, PTFE is a highly crystalline material having a molecular weight lying in the range 500,000 to 5,000,000.

The first commercial production was described in 1946 and appeared on the market in the United States as Teflon (Du Pont) and in England as Fluon (I.C.I.). PTFE is also now produced in France as Soreflon (Rhone-Poulenc), in Germany as Hostaflon (Hoechst), in Italy as Algoflan (Montecatini), and in Japan as Tetraflon (Nitto), and as Polyflon, Daiflon and Daifoil (Osaka Kinsoku Kogyo). Many grades of PTFE are now available, the essential properties being largely dependent on whether the bead (granular), or dispersion (from latex) polymer is used. Kel-F (3M), a polymer of chlorotrifluoroethylene, appeared on the market in about 1958. Kel-F can be fabricated more easily than PTFE, and while generally maintaining the chemical resistance of PTFE, it is nevertheless somewhat less resistant than PTFE to high temperatures.

Polyhexafluoropropylene and copolymers with tetrafluoroethylene have achieved some commercial significance largely through the research efforts of Du Pont. Pennwalt has developed Kynar, a polyvinylidene fluoride based product, which has good heat stability and chemical resistance. Polyvinyl fluoride, sold in film form as Tedlar (Du Pont), is highly resistant to outdoor weathering and affords excellent flexibility under a variety of conditions.

The development of fluorine-containing rubbers was carried out with considerable aid from the U.S. Army and Air Force during the fifties. Kel-F (3M) elastomers, typically a copolymer of chlorotrifluoroethylene and vinylidene fluoride, are amine cured in the presence of zinc oxide and lead phosphate stabilizer. Viton (Du Pont) and Fluorel (M.W. Kellogg, 3M) are copolymers of perfluoropropylene and vinylidene fluoride (typically 30/70). Generally introduced in 1959, these fluoroelastomers, when fully cured, are unaffected by ozone and have excellent resistance to nitric and hydrofluoric acids at normal temperatures.

Fluoroacrylates and fluorinated nitrosoelastomers have also been the subject of considerable research. The fluoroacrylates, for example, poly-(1,1-dihydroperfluorobutyl acrylate) have not achieved commercial significance as they are generally poorer in solvent resistance than the Viton elastomers. Developed by 3M, the nitroso rubbers, copolymers of trifluoronitroso methane and tetrafluoroethylene, are still extremely expensive and have largely been produced by the Reaction Motors Division of Thiokol under government contract. They are completely nonflammable in pure oxygen and exceptionally resistant to oxidation, solvents, acids and fuels.

Most fluoroelastomers are used to make seals and gaskets for use in hot-liquid systems, aircraft and missiles. The overall volume is still small, but they are sustaining a high rate of growth.

POLYTETRAFLUOROETHYLENE

GENERAL PROCESSES — DISPERSION

Polymerization in the Presence of Hydrogen or Methane as Stabilizing Agents

In a process described by K.C. Brinker and M.I. Bro; U.S. Patent 2,965,595, December 20, 1960 and U.S. Patent 3,066,122, November 27, 1962; both assigned to E.I. du Pont de Nemours and Company, tetrafluoroethylene is polymerized in an aqueous medium in the presence of 0.01 to 10 mol percent of the monomer of a compound of the class consisting of hydrogen, methane, ethane and saturated fluorinated hydrocarbons having not more than 2 carbon atoms and at least one hydrogen atom, where the quantity of the additive employed is from 0.01 to 0.05 mol percent when the compound contains six hydrogen atoms, from 0.01 to 0.5 mol percent when the compound employed contains less than six and more than three hydrogen atoms, from 0.01 to 2.5 mol percent when the compound employed contains less than four and more than one hydrogen atom, and from 0.01 to 10 mol percent when the compound employed contains one hydrogen atom.

It was discovered that the above described compounds, if added to the polymerization of tetrafluoroethylene in quantities within the limits set forth, substantially reduce the toxicity of the resulting polymer at temperatures above the sintering temperature, increase the thermal stability of the polymer, and, when used in a dispersion process, increase the stability of the resulting polytetrafluoroethylene dispersion against coagulation.

It is believed that under the conditions of this process, hydrogen, methane, ethane and fluorinated hydrocarbons having not more than two carbon atoms and at least one hydrogen such as monofluoromethane, difluoromethane, trifluoromethane and difluoroethane, cap the reactive ends of essentially high molecular weight polymer chains, thereby forming a slightly modified polytetrafluoroethylene which is improved in thermal stability. In the polymerization to form aqueous dispersions of polytetrafluoroethylene, it is believed that the stabilizing agents aid in the formation of smaller colloidal particles and thus enhance the stability of the dispersion against coagulation.

The modified polymers produced by this process have the same outstanding physical, mechanical, electrical and chemical properties of the unmodified polymers and are not to be compared to the low molecular weight brittle and waxy polymers obtained by the prior art polymerization of tetrafluoroethylene in the presence of organic compounds. Although it may be said that the additives of this process are also chain-transfer or chain-terminating agents, it is not a suppression of growth of the polymer chain which comes into play in the process, but rather the improvement obtained from the formation of a bond between the additive and the tetrafluoroethylene polymer chain. The compounds employed in this process are therefore considered to be stabilizing agents rather than chain-transfer agents. This process is illustrated by the following examples.

Example 1: Into a stainless steel vessel having a capacity of one gallon was charged 1,600 grams of deoxygenated water, 1.6 grams of disuccinic acid peroxide, 8 grams of ammonium perfluorocaprylate, 100 ml. of a commercially available mineral oil, and 5 parts per million of iron based on the weight of the water. The reaction vessel was evacuated and heated to approximately 85°C.

Specified amounts of hydrogen listed in the table below were injected into the reaction mixture. The vessel was then pressured with tetrafluoroethylene to 400 lbs./sq. in. and polymerization was initiated by agitating the reaction mixture. Pressure and temperature were maintained during polymerization, which was continued until the desired solids content was reached. The reaction mixture was cooled to room temperature and excess monomer was removed. The reaction mixture was then separated from the mineral oil and filtered. The solids content of the dispersion was measured by the specific gravity of the dispersion. The stability of the dispersion was determined by agitating the dispersion at 500 revolutions per minute and measuring the time required for coagulation. The coagulated polymer was washed and dried.

Polytetrafluoroethylene

The polymer was molded into 1" x 3" sheets weighing 25 grams, which were then sintered at 350°C., for a period of 30 minutes in an air oven. The specific gravity of the polymer was determined from the sintered sheets. The sintered sheets were heated to 350°C. in a stream of humid air which was then passed through water in which the hydrolyzable fluorine compounds were adsorbed.

The fluoride ion concentration of the water was determined by standard titration methods. From the fluoride ion concentration the fluoride evolution of the polymer was determined. Although the fluoride evolution does not necessarily measure the quantity of all degradation products formed from the heated polymer sample, i.e., only those which are hydrolyzable, a measure of relative degradation is obtained. The fluoride evolution is further a relative measure of the toxic gases released, the exact composition of which is not known.

Run No.	Stabilizer	Stabilizer Quantity in Mol Percent of Monomer	Polymerization Rate, g./l./hr.[1]	Solids Content, %	Stability, min.	Fluoride Evolution, mg./hr.	Specific Gravity
1	None	-	790	36.4	3-5	1.3	2.2389
2	H_2	0.15	650	36.5	11	0.68	2.2492
3	H_2	0.30	658	33.8	13	0.32	2.2391
4	H_2	0.45	419	35.6	>40	0.48	2.2603
5	H_2	1.2	331	34.0	>20	0.44	2.2646
6	H_2	2.4	206	33.2	>40	0.43	2.2710

[1]g./l./hr. = grams of polymer per liter of medium in one hour.

From the above listed data it can be seen that both the stability of the dispersion is increased and the fluoride evolution at 350°C. is decreased by the addition of hydrogen. However, the data also shows the decrease in the rate of polymerization and the increase in specific gravity. The latter is an indication of decreasing molecular weight. Certain small decreases in molecular weight are desirable, as molding properties are improved by such decreases. However, further decreases in molecular weight will result in brittle polymers, with inferior mechanical properties.

Thus both these measurements indicate the necessity to confine the chain-terminating agent to within certain limits so as not to adversely affect the properties of the polymer nor inhibit the polymerization critically. The correlation of the fluoride evolution to the toxicity of the gases released at high temperatures is shown by the following experiments. Samples of run #1 and run #6 were heated to 350°C. over moist air (50% relative humidity) and passed continuously through a bell jar containing two white rats. The experiment was continued until both rats had died or until it was established that the lethal does of toxic gases could not be reached. The following results were obtained:

Run No.	Stabilizer	Stabilizer Quantity in Mol Percent of Monomer	Fluoride Evolution, mg./hr.	Weight of Sample, grams	Exposure Time, hrs.	Mortality Ratio
1	None	-	1.3	0.8	2	2/2
6	H_2	2.4	0.43	10.6	6	0/2

The results indicate that, although the weight of the sample containing no chain-terminating agent was less than one-tenth of the hydrogen capped polymer, the lethal doses of toxic gases had not been reached in after 6 hours in the case of the hydrogen capped polymer, whereas in the uncapped polymer the lethal doses had been reached after two hours, killing both rats.

Example 2: Into a stainless steel vessel having a capacity of one gallon was charged 1,600 ml. of deoxygenated water, 1.6 gram of disuccinic acid peroxide, 8 grams of ammonium perfluorocaprylate, 100 ml. of Kaydol, a commercially available mineral oil, and 5 parts per million of iron based on the weight of the water. The reaction vessel was evacuated and heated to a temperature indicated in the table below.

Run No.	Stabilizer	Stabilizer Quantity in Mol Percent of Monomer	Reaction Temperature °C.	Polymerization Rate, g./l./hr.	Solids Content, %	Coagulum, %	Stability
1	None	-	85-90	275	38.8	0.7	2 min., 45 sec.
2	None	-	85-88	428	24.8	0.64	4 min.
3	CH_4	0.410	85	98	25.7	0	36 min., 20 sec.
4	CH_4	0.037	85	312	28.3	0	8 min., 30 sec.
5	CH_4	0.029	85	365	32.0	0	6 min.
6	CH_4	0.020	85-87	270	37.0	0	6 min., 15 sec.
7	CH_4	0.0008	86	301	35.8	0	6 min.

Polytetrafluoroethylene

Specified amounts of methane listed in the table were injected into the reaction vessel. The vessel was then pressured with tetrafluoroethylene to 400 lbs./sq. in. and polymerization was initiated by agitating the reaction mixture. Pressure and temperature were maintained during the polymerization, which was continued until a dispersion of the desired solids was reached.

The reaction mixture was cooled to room temperature and excess monomer was removed. The reaction mixture was then separated from the mineral oil and filtered and percentage of coagulum determined. The solids content of the dispersion was measured by the specific gravity of the dispersion. The stability of the dispersion was determined by agitating a sample of the dispersion at 500 revolutions per minute and measuring the time required until coagulation occurred. As can be seen from the tabulated results, the stability of tetrafluoroethylene dispersion is significantly increased and co-agulum formation is prevented. However, the addition of excess quantities of methane will tend to inhibit the polymerization.

Acidic Reaction Media

M.I. Bro and R.C. Schreyer; U.S. Patent 3,032,543; May 1, 1962; assigned to E.I. du Pont de Nemours and Company have found that tetrafluoroethylene can be polymerized to high molecular weight polymers in the presence of specific organic acids, e.g., formic and acetic acid, and liquid inorganic acids. As a result of this discovery, it is possible to lower the polymerization temperatures to below the critical temperature of tetrafluoroethylene, i.e., 30°C., and particularly to below 0°C. As a result of the lower polymerization temperatures in the process, the reactivity of the monomer and the growing polymer chain is decreased and thus allows the polymerization of tetrafluoroethylene to high molecular weight polymer in the presence of compounds which otherwise would act as telomerizing agents in the polymerization. The process is illustrated by the following examples.

Example 1: Into a 320 ml. platinum-lined autoclave was charged 150 ml. of 96% sulfuric acid and 0.15 gram of sodium bisulfite. The reaction was cooled to -10°C. and charged with 36 grams of tetrafluoroethylene. The reaction mixture was agitated at a temperature of -8° to -10°C. for a period of one hour. During this period 0.5 gram of potassium bromate in 35 ml. of water was injected into the reaction vessel in 7 ml. portions at 5 minute intervals. The polymerization pressure dropped from 250 to 40 psi at the end of one hour. The resulting reaction mixture was filtered and the isolated polymer was washed with water and dried in a vacuum oven. The dried polytetrafluoroethylene weighed 25.5 grams and the polymer could be molded into tough dense sheets by preforming at 2,000 psi and sintering at a temperature of 380°C.

Example 2: Into a 320 ml. platinum-lined autoclave was charged 75 ml. of formic acid, 75 ml. of water and 0.1 gram of sodium bisulfite. The reaction mixture was cooled to -2°C., evacuated and charged with 60 grams of tetrafluoro-ethylene. The vessel was agitated at a temperature of -2° to 0°C. for a period of one hour. During this period, 0.5 gram of sodium chlorate in 35 ml. of water was injected into the reaction vessel in 7 ml. portions at 5 minute intervals. The polymerization pressure dropped from 280 to 80 psi during this period. The resulting reaction mixture was removed from the vessel and filtered. The solid polymer isolated was washed with additional water and dried in a vacuum oven. On drying there was obtained 55.8 grams of high molecular weight polytetrafluoroethylene which could be pressure-molded and free-sintered at temperatures above 350°C. into tough sheets.

Delayed Addition of Dispersing Agent

J.E. Duddington and S. Sherratt; U.S. Patent 3,009,892; November 21, 1961; assigned to Imperial Chemical Industries Limited, England, describe a process for polymerizing tetrafluoroethylene in an aqueous medium to obtain a colloidal polymer dispersion which comprises initiating the polymerization of the tetrafluoroethylene in the presence of water containing a catalytic amount of a water-soluble catalyst and, only after the polymerization has started but before the amount of polymer formed exceeds 10% by weight of the aqueous phase plus polymer, adding an anionic dispersing agent, and continuing the polymerization reaction. In order to keep the amount of coagulum formed as low as possible, it is preferred to add the dispersing agent before the amount of polymer formed exceeds 7% by weight of the aqueous phase plus polymer.

The amount of coagulum formed during the process depends on a number of factors, including the rate at which the system is stirred. In the preferred way of working, the polymerization is effected in the presence of a saturated hydrocarbon having more than 12 carbon atoms which is liquid under the polymerization conditions, e.g., octadecane, eicosane, tetradecane, cetane and paraffin waxes having melting points below the temperature of polymerization. Such hydrocarbons are normally added to the aqueous medium before initiation of the polymerization in proportions of between 0.1 and 12% by weight of the water. These hydrocarbons are efficient stabilizers against coagulation of the polymer and allow thorough agitation of the polymerization system.

Suitable classes of anionic dispersing agents which may be used are (a) water-soluble salts of sulfuric acid esters of fatty alcohols, i.e., alcohols corresponding to fatty acids of animal and vegetable fats and oils, and soaps, (b) water-soluble salts of aromatic sulfonic acids, (c) water-soluble salts of polyfluorocarboxylic acids having the formula $X(CF_2)_nCOOH$, where X is hydrogen, chlorine or fluorine and n is 6 to 20, and (d) water-soluble salts of polyfluorochlorocarboxylic acids of the formula Cl(CF$_2$—CFCl)$_n$CF$_2$COOH, where n is 2 to 6. On account of their cheapness and availability particularly suitable dispersing agents for use in this process comprise monoethanolamine lauryl sulfate, triethanolamine lauryl sulfate, ammonium lauryl sulfate and sodium lauryl sulfate. Dispersed polytetrafluoroethylene particles having a mean particle size of 0.2 micron and above have been obtained by this process. The following examples illustrate the process.

Example 1: A stainless steel autoclave fitted with a stirrer was charged with 4,000 parts of distilled water, 2.66 parts of disuccinic acid peroxide and 20 parts of eicosane after which the vessel was evacuated to remove oxygen and then pressurized with tetrafluoroethylene gas until the pressure gauge on the autoclave read 15 lbs./sq. in. The reaction medium was heated with stirring to a temperature of 70°C. and gaseous tetrafluoroethylene was introduced to 300 lbs. per square inch pressure. When the pressure dropped to 290 lbs./sq. in., a further quantity of tetrafluoroethylene was introduced until the pressure returned to 300 lbs./sq. in.

This procedure was repeated throughout the reaction. When 200 parts of polymer had been formed, as indicated by a total pressure drop of 80 lbs./sq. in., a solution of 1 part of ammonium perfluorooctanoate dissolved in 100 parts of distilled water was injected into the reaction mixture. After a total polymerization time of 2 hours an aqueous colloidal dispersion of polytetrafluoroethylene containing 10.4% by weight of the dispersed polymer was obtained. The mean particle size of the dispersed particles was 0.25 micron. The amount of coagulated polymer present was 2.0% of the combined weight of the colloidal polymer in dispersion and the coagulum. The rate of the reaction measured by the space/time yield of dispersed polymer was 56.5 grams/liter/hour.

Example 2: The process of Example 1 was repeated using 1 part of sodium lauryl sulfate instead of ammonium perfluorooctanoate as dispersing agent. After two hours and 50 minutes a colloidal dispersion containing 8.4% by weight of dispersed polytetrafluoroethylene was obtained without the formation of coagulum. The rate of the reaction was 32.5 grams per liter per hour. The mean particle size of the dispersed particles was 0.23 micron.

Example 3: The above process was repeated using 1 part of triethanolamine lauryl sulfate as dispersing agent. After three hours a colloidal dispersion containing 11.3% by weight of dispersed polytetrafluoroethylene was obtained together with 4.5% of coagulum based on the combined weight of dispersed polymer and coagulum. The rate of the reaction was 40.0 grams/liter/hour. The mean particle size of the dispersed particles was 0.20 micron.

Programmed Addition of Dispersing Agent

J.O. Punderson; U.S. Patent 3,391,099; July 2, 1968; assigned to E.I. du Pont de Nemours and Company describes a polymerization process which comprises the programming of dispersing agent addition in a manner such that the concentration of dispersing agent present at the start of nucleation is a finite, very low value, but definitely not zero, followed by increasing the concentration of dispersing agent subsequent to the nucleation period to a level sufficient to prevent substantial coagulation as higher levels of solids content are achieved.

The process may variously be termed as split addition, multistage addition, or programmed addition of dispersing agent as contrasted to previous processes such as the conventional initial addition of dispersing agent or the delayed addition process in which no dispersing agent is present during all or most of the nucleation period. The method consists in carrying out a nucleation process in which the number of particles nucleated is precisely controlled by the presence of a definitely established concentration of dispersing agent, the concentration being substantially lower than that which would be needed to stabilize the dispersion against subsequent coagulation at high solids concentrations, and in following the nucleation period with a growth period carried out at higher dispersing agent concentrations.

One of the features of this process is the discovery that an unexpectedly small amount of dispersing agent is needed during the nucleation period of the polymerization to bring the nucleation under control of the dispersing agent. In a run typical of prior art practice, in which 0.15% of APFC dispersing agent is initially charged to the reactor and no further addition of dispersing agent is made during the polymerization, the average particle size obtained at the 35% solids level is 0.227 micron. (APFC stands for ammonium perfluorocaprylate, also known as ammonium perfluorooctanoate; AFC stands for ammonium ω-hydrohexadecafluorononanoate, and AHT stands for ammonium 3,6-dioxa-2,5-di-(trifluoromethyl)undecafluorononanoate).

The space-time yield (STY) in such a run is about 355 grams/liter-hour. Likewise, by the prior art procedure of delaying addition of dispersing agent until after about 9% solids have been formed, adding 0.15% of APFC and continuing the polymerization to 35% solids, an average particle size of greater than 0.4 micron is obtained, and reproducibility of this high value in repeated runs under nominally the same conditions is poor. The STY is below 200 grams/liter-hour.

By operating the process as described here, a dispersing agent concentration of only 0.015% of APFC during the nucleation period and adding additional dispersing agent to make a total of 0.15% of APFC after about 9% of solids has been formed, an average particle size of 0.236 micron is obtained at the 35% solids level (STY = 330). This very modest increase in particle size compared with the 0.227 micron particle size obtained when 0.15% dispersing agent was present during nucleation indicates that the nucleation process is still under control of the dispersing agent, even though its concentration is only one-tenth of the value normally used. Repeating the above split-addition polymerization but reducing the amount of APFC to 0.006% during the nucleation period and adding additional APFC to make a total of 0.15% after about 9% solids has been formed gives a product with an average particle size of 0.277 micron after 35% solids has been achieved (STY = 265). The sharp increase in particle size resulting from the small shift in dispersing agent concentration from 0.015 to 0.006% during nucleation indicates that a sharp zone of transition is being entered between the dispersing agent controlled nucleation of the process and the uncontrolled nucleation of the delayed addition process.

The particle size obtained using 0.006% APFC during nucleation is closer, however, to that obtained using 0.015% APFC during nucleation than to that obtained when no dispersing agent is used, indicating that the nucleation is still largely under control of the dispersing agent. Thus, with APFC dispersing agent in this particular reaction system, a modest control of particle size is available by shifting dispersing agent concentration during nucleation down through the range from 0.15 to 0.015%, and further control is available in the range from 0.015 to 0.006%.

Because of the low concentration of dispersing agent required during the nucleation period, it has been found possible to employ very low concentrations of substantially hydrocarbon-based dispersing agents containing a plurality of hydrogen atoms per molecule and relatively few or no fluorine or other halogen atoms per molecule for this first stage of the polymerization.

Another significant feature of this process is the discovery that with some dispersing agents, it is possible to shift particle size downward from its normal value, instead of upward, by controlled nucleation at low levels of dispersing agent. For instance, when using conventional polymerization with 0.15% of AHT dispersing agent added initially gives a 0.285 micron particle size at 35% solids (STY = 380), whereas the delayed addition process using AHT gives a particle size in excess of 0.4 micron (STY below 200). By the use of the split addition procedure of this process, using a 0.015% dispersing agent concentration during nucleation and a total of 0.15% after 6% of solids is formed, particle size of 0.238 micron at the 35% solids level (STY = 305) is obtained. Using 0.006% of AHT during nucleation gives a further reduction to 0.203 micron particle size (STY = 335). Thus, it is possible in this system to reduce particle size in a smooth and controlled manner over a wide range.

In addition to the control that is available by varying the amount of dispersing agent initially added, further control of particle size can be exercised by varying the time at which further addition is made. This effect is illustrated below in which 0.015% of AHT is added initially and 0.135% of AHT was added subsequently at various times in a number of runs; the particle diameters obtained at 35% solids are as reported below.

% Solids at Time of Second Addition	Particle Diameter in Microns at 35% Solids Level by Light Transmission Measurements	Space-Time Yield, grams/liter-hour
0	0.285	380
4	0.260	354
7	0.252	314
13	0.244	303
18	0.243	345
22.5	* 0.243	–

* Approximately 1.2% coagulum, dry basis.

It can be seen that continuous control over the range of 0.285 to 0.243 micron at the 35% solids level can be readily obtained by this technique. The following examples are illustrative of this process.

Example 1: A horizontally disposed, water-steam jacketed, cylindrical, stainless steel 6,200 ml. autoclave, having a cage-type agitator running the length of the autoclave, and having a water capacity of 6,200 grams is charged with 200 grams of paraffin wax (MP 58°C.), 0.0065 gram of electrolytically reduced iron powder, and 0.20 gram of ammonium 3,6-dioxa-2,5-di-(trifluoromethyl)-undecafluorononanoate (AHT). The autoclave is then evacuated and 2,450 grams of distilled and deoxygenated water is drawn in. The autoclave is then evacuated and 2,450 grams of distilled and deoxygenated water is drawn in. The autoclave is heated to 70°C., at which time the vapor space is again evacuated and then filled with purified TFE at a pressure of 25 pounds per square inch gauge.

Then a solution of 1.623 grams of disuccinic acid peroxide in 750 grams of distilled and deoxygenated water is added and the system is agitated and further heated to 85°C. The pressure of tetrafluoroethylene is then increased to 400 psig, and the supply of TFE is then shut off. As soon as a 10 psi drop in pressure is noted (with temperature remaining constant) indicating a commencing of reaction (commonly termed as "kick-off"), the pressure is restored to 400 psig by addition of TFE and automatically maintained at this value as reaction proceeded at 85°C.

A continuous record is kept of the amount of TFE fed to the autoclave. When 255 grams of TFE has been reacted, an additional 4.7 grams of AHT as a 10% aqueous solution is pumped into the autoclave. The valve in the line through

Polytetrafluoroethylene

which TFE is automatically admitted to the autoclave is left open until a precalculated amount of TFE has been fed, at which time the valve is closed. This precalculation is arranged such that the total amount of TFE fed to the autoclave is sufficient, after the pressure in the autoclave has, by continued reaction at 85°C., decreased to 175 psig and the remaining TFE vented off, to provide a nominal solids content of 35% in the dispersion of polymer and water forming the liquid phase in the autoclave.

The solids content as determined by hydrometer measurement in this experiment is 34.7%. Essentially no coagulum is formed during the polymerization. The rate of polymerization (space-time yield) is 335 grams per liter-hour, based on the measured time elapsed between "kick-off" and the time at which the pressure in the autoclave reached 175 psig. The average particle size as measured by light-transmission measurement is 0.232 micron, while the particle size corresponding to the maximum of the weight distribution of size as measured in analytical ultracentrifuge is 0.221 micron.

A portion of the dispersion product, after cooling and removal of the supernatant solid wax, is diluted, treated with dilute aqueous ammonium hydroxide to a pH level of 9, and coagulated by the procedure of U.S. Patent 2,593,583. The resin possessed a specific gravity of 2.209 when measured according to the standardized procedure given in ASTM D-1457-56T.

Example 2: A control run is made similar to the run of Example 1 except that all of the dispersing agent is added initially; i.e., 4.9 grams of ammonium 3,6-dioxa-2,5-di-(trifluoromethyl)-undecafluorononanoate is charged initially to the same autoclave, and no further quantity is added thereafter. Otherwise the same procedures, conditions, and quantities of other ingredients are those employed in Example 1 are used, the total quantity of water present being 3,250 grams. The polymerization is carried out to 35.0% solids.

The rate of polymerization is 347 grams per liter-hour, and the particle size as measured by light transmission is 0.285 micron. Ultracentrifuge measurements indicate a diameter corresponding to the peak of the weight distribution curve of 0.278 micron. The resin possessed a specific gravity of 2.216 as determined by ASTM D-1457-56T. Comparison of the results of the two examples above shows that the split addition procedure of Example 1 allows the particle size to be shifted downward while a relatively rapid polymerization rate and ability to polymerize to high solids concentration are maintained.

Example 3: The experiment described in Example 1 is repeated with the exception that in the second addition of dispersing agent, made in this case after 350 grams of TFE has reacted, the agent added is 30 grams of chlorendic acid as contained in a 30% by weight stirred aqueous mixture held at 100°C. until injected into the reactor. The polymerization is continued to a solids content of 34.6%. The rate of polymerization is 342 grams/liter-hour. Essentially no coagulum is formed during the polymerization and the average particle diameter as measured by light transmission is 0.253 micron.

Ultracentrifuge measurements indicate a most probable particle diameter of 0.227 micron. The resin possessed a specific gravity of 2.203 as determined by ASTM D-1457-56T. The sintered chip on which the specific gravity measurement is made appeared to be white and free of contamination. Similar results are obtained in runs in which either 1.0 gram of APFC or 1.0 gram of AFC is substituted for 0.2 gram of AHT as nucleating agent and chlorendic acid is subsequently added in the manner described above.

Example 4: The procedure of Example 1 is repeated, except that 0.20 gram of ammonium perfluorooctanoate is substituted in place of the 0.20 gram of AHT originally charged to the autoclave. When 350 grams of TFE has reacted, 4.7 grams of ammonium perfluorooctanoate (instead of 4.7 grams of AHT) as contained in a 10% aqueous solution is introduced into the autoclave, and the polymerization is continued to a solids content of 34.8%. Essentially no coagulum is formed during the polymerization, and the average particle size of the product as measured by light transmission is 0.277 micron. The resin has a specific gravity of 2.204 as determined by ASTM D-1457-56T. The rate of polymerization is 265 grams per liter-hour.

Use of Reaction Modifiers, Methanol and Hexafluoropropylene

A.J. Cardinal, W.L. Edens and J.W. van Dyk; U.S. Patent 3,142,665; July 28, 1964; assigned to E.I. du Pont de Nemours and Company describe high molecular weight dispersion resins obtained by polymerizing tetrafluoroethylene in an aqueous medium consisting essentially of water, initiator, and dispersing agent, distinguished in that the medium also contains, at least during the polymerization of the final 30% of the tetrafluoroethylene polymerized, a modifier effective to maintain the overall rate of polymerization at least 5% below that obtaining for the polymerization of an equal quantity of tetrafluoroethylene in an identical reaction medium continuously saturated with tetrafluoroethylene at the same temperature and pressure in the absence of the modifier.

The modifier consists of one or more members of the group consisting of water-soluble nonpolymerizable chain transfer agents containing at least one covalently bound nonmetallic monovalent atom other than fluorine, perfluoroalkyl-trifluoroethylenes of 3 to 10 carbon atoms, and oxyperfluoroalkyl-trifluoroethylenes of 3 to 10 carbon atoms. The resins have not

7

only the advantageous characteristics of earlier high molecular weight dispersion resins, but also improved paste extrudability. In specific aspects they are further characterized by a narrow particle size distribution such that $\Delta d\frac{1}{2}/d_{av}$, the ratio, to average particle diameter, of the spread in particle diameters at the half-peak concentrations of particle sizes by weight, is less than 0.40; by a distribution of molecular weight along the particle radii such that the ratio of molecular weight of the shell half to the molecular weight of the core half of the particles is less than 3.5; or by the presence of very small amounts, in the range of 0.01 to 0.3 weight percent, of combined perfluoroalkyl- or oxyperfluoroalkyl-trifluoroethylene in at least the outer 30 weight percent of the resin particles as determined by infrared measurement; or by a combination of two or more such features. In preferred aspects they are capable of being paste extruded at reduction ratios of at least 1,600 to 1.

Examples of effective chain transfer agents include methanol, hydrogen, methane, ethane, propane, propylene, carbon tetrachloride, dichlorotetrafluoroethane, bromoform, acetone and propionic acid. The use of chain transfer agents as sole modifiers in accordance with this process provides resins of exceptionally good thermal stability. Perfluoroalkyl- and oxyperfluoroalkyl-trifluoroethylenes of 3 to 10 carbon atoms may be used as modifiers in this process either alone to provide resins having outstanding flexural properties, or together with a chain transfer agent to provide resins which are paste extrudable at very high reduction ratios, on the order of 10,000/1 or more. The lower carbon trifluoroethylenes are preferred for optimum heat aging and paste extrudability properties; hexafluoropropylene is especially preferred for optimum sintering characteristics.

The proportions or modifier employed will vary widely in accordance with the activity of the particular modifier chosen, and also with the temperature and pressure. With normally gaseous modifiers the minimum amounts required to obtain effective rate limiting capacity ordinarily range from 0.1 part per million with a highly active modifier such as propane to 10,000 parts per million for a very mildly active modifier such as pentafluoroethane, based on the weight of monomer charged. Correspondingly, with normally liquid modifiers, these amounts ordinarily are in the range of 10 to 5,000 parts per million, based on the weight of the aqueous medium. However, amounts larger than these minima, in instances up to 30,000 parts per million or more based on monomer, may also be employed.

The modifiers may effectively be incorporated into the reaction mixture when about 70% of the total quantity of the tetrafluoroethylene to be polymerized has reacted, or at any earlier stage of the reaction. Presence of modifiers throughout the polymerization reaction favors a narrow particle size distribution, and therefore somewhat better paste extrudability. Presence of modifier only during polymerization of the final 30% of tetrafluoroethylene on the other hand favors a higher overall reaction rate. Except as effective modifier is present during this final stage of the reaction, however, the benefits of this process are not obtained. It is believed that the presence of the modifier during this stage counterbalances the tendency for the growing particles to develop very long chain, highly crystalline shells.

The process is illustrated by means of the following examples. In the examples and comparisons all parts and percentages are by weight except as otherwise noted. In the examples, APS is ammonium persulfate, KPS is potassium persulfate, DSP represents disuccinic acid peroxide, $(HOOCCH_2CH_2CO)_2O_2$, used together with 2 parts per million of powdered iron based on the water charged; C_8APFC is a mixture of 8-carbon ammonium perfluorocarboxylate predominantly comprising ammonium perfluorocapyrlate; C_9AFC is the ammonium salt of 9-H hexadecafluorononanoic acid; HFP is hexafluoropropylene; and PPTE is n-perfluoropropoxy-trifluoroethylene, $CF_3CF_2CF_2OCF{=}CF_2$. In all of the examples, the resins comprised at least 98 weight percent combined tetrafluoroethylene, showed specific melt viscosities of greater than 1×10^9 poises as measured at 380°C. and predominantly comprised spheroidal particles 0.05 to 0.5 micron in diameter, with a d_{av} in the range of 0.12 to 0.35 micron.

Except as otherwise specified, the values given for average dispersion particle size are uncorrected results obtained by light-transmission analysis based on an assumed value of 0.020 cc/g. for the refractive index increment of colloidal particles of polytetrafluoroethylene dispersed in an aqueous medium at 25°C. The values so reported differ from the average particle sizes determined from electron microscope photographs or by ultracentrifuge analysis by about −20 to +30% depending on the degree to which the actual refractive index increment is different from 0.020 cc/g. and on the degree of agglomeration of the colloidal particles in the dispersion.

Examples 1 to 12: A horizontally disposed, water-steam jacketed, cylindrical stainless steel autoclave, having a paddlewheel agitator running the length of the autoclave, and having a length-to-diameter ratio of 10 to 1 and a water capacity of 3,900 parts, is evacuated, charged with 1,500 parts of demineralized deoxygenated water, and with desired concentrations of dispersing agent and initiator based on the weight of the water. The liquid modifier is charged in the indicated concentration in percent by weight, based on water, to the aqueous phase and is present through the reaction.

The charge is then pressured with about 2 atmospheres of tetrafluoroethylene, stirred at 125 rpm to keep the aqueous phase saturated with tetrafluoroethylene, heated to desired reaction temperature, and further pressured to 28.2 atmospheres absolute with tetrafluoroethylene. Stirring and temperature are then maintained until reaction commences, as evidenced by a drop in pressure, and then further maintained, while continuously maintaining pressure at 28.2 atmospheres absolute with additional tetrafluoroethylene, until a dispersion of desired solids content is obtained. The resulting dispersion is discharged and cooled, after which supernatant solid wax is removed, and the residue is diluted and

Polytetrafluoroethylene

coagulated by the procedure of U.S. Patent 2,593,583. Samples of the products are lubricated and extruded to determine standard paste extrusion performance (EP). Results are summarized in the table below.

Example No.	Initiator Kind	%	Reaction Temp. °C.	Dispersing Agent Kind	%	% Modifier*	% Wax	Overall Reaction Rate, g./l./hr.	% Solids in Dispersion	d_{av}	SSG	Extrusion Performance Extrusion Pressure, kg./cm.^{2}	Extrudate Quality
1	APS	0.048	50	C$_8$APFC	0.15	0.01	6.3	337	35.2	0.161	2.218	660	6
2	APS	0.006	60	C$_8$APFC	0.15	0.01	6.3	290	35.0	0.136	2.220	670	9
3	APS	0.012	60	C$_8$APFC	0.15	0.01	6.3	396	40.5	0.130	2.221	730	6
4	APS	0.012	60	C$_8$APFC	0.15	0.009	6.3	480	41.2	0.141	2.220	625	9
5	APS	0.012	60	C$_8$APFC	0.15	0.02	6.3	205	36.9	0.145	2.234	610	8
6	APS	0.024	60	C$_8$APFC	0.15	0.01	6.3	585	34.5	0.157	2.221	705	7
7	APS	0.006	60	C$_9$AFC	0.15	0.009	6.3	550	39.7	0.185	2.206	465	9
8	APS	0.006	60	C$_9$AFC	0.15	0.011	6.3	515	40.6	0.184	2.220	400	9
9	APS	0.009	60	C$_9$AFC	0.15	0.010	6.3	510	40.3	0.179	2.212	420	10
10	APS	0.006	70	C$_9$AFC	0.15	0.009	6.3	770	40.5	0.230	2.229	330	9
11	DSP	0.1	85	C$_8$APFC	0.214	0.045	6.3	435	30.8	0.156	2.224	595	8
12	DSP	0.1	85	C$_8$APFC	0.214	0.09	6.3	367	30.8	0.152	2.231	650	8

*Runs 1 to 10, methanol modifier. Runs 11 and 12, propionic acid modifier.

In comparison runs under identical conditions in the absence of modifier, it was found that identical quantities of high molecular weight dispersion resins were obtained at rates of from 1.2 to 2 times as great as in the examples, but that in each instance, the resins obtained yielded only fractured extrudates in the extrusion performance test.

Examples 13 to 23: The procedure of Examples 1 to 12 is repeated except that in these cases HFP modifier was premixed with the tetrafluoroethylene to be charged, in the concentration indicated in weight percent based on tetrafluoroethylene, and thus continuously charged to the reaction. Also, in Examples 17, 18, and 19, the concentrations of tetrafluoroethylene in the aqueous media were continuously maintained slightly below the saturation point. The results are summarized below.

Example No.	Initiator Kind	%	Reaction Temp. °C.	Dispersing Agent Kind	%	% Modifier*	% Wax	Overall Reaction Rate, g./l./hr.	% Solids in Dispersion	d_{av}	SSG	Extrusion Performance Extrusion Pressure, kg./cm.^{2}	Extrudate Quality
13	DSP	0.1	85	C$_8$APFC	0.15	0.15	6.3	523	35	0.170	2.205	455	8
14	DSP	0.1	85	C$_8$APFC	0.15	0.5	6.3	475	35	0.188	2.208	335	6
15	DSP	0.1	85	C$_8$APFC	0.15	0.75	6.3	441	35	0.160	2.202	455	10
16	DSP	0.1	85	C$_8$APFC	0.15	†0.75	6.3	350	35	0.160	2.209	365	9
17	KPS	0.005	85	C$_9$AFC	0.15	0.5	6.3	1,440	35	0.173	2.211	425	10
18	KPS	0.005	85	C$_9$AFC	0.15	0.1	6.3	1,430	35	0.158	2.232	505	10
19	KPS	0.005	85	C$_9$AFC	0.15	0.9	6.3	1,310	35	0.143	2.204	600	10
20	KPS	0.003	70	C$_9$AFC	0.15	0.5	6.3	658	35	0.156	2.176	540	7
21	KPS	0.003	70	C$_9$AFC	0.15	0.1	6.3	690	35	0.178	2.177	540	6
22	KPS	0.003	70	C$_9$AFC	0.15	0.9	6.3	503	35	0.158	2.184	515	10
23	DSP	0.1	85	C$_8$APFC	0.75	0.9	6.3	522	35	0.143	2.194	985	8

*Hexafluoropropylene (HFP).
†Approximately.

In comparison runs under identical conditions in the absence of modifier, it was found that except under the conditions of Examples 17, 18, and 19, identical quantities of high molecular weight dispersion resins were obtained at rates of from 1.2 to 2 times as great as in the examples. Under the conditions of Examples 17, 18 and 19, the rates were substantially unchanged in the absence of modifiers. In all comparison runs, the products yielded only fractured extrudates in the extrusion performance test. In further comparison runs under the identical conditions of Examples 17, 18, and 19 in which the aqueous charges were maintained continuously saturated with tetrafluoroethylene, reaction rates 1.2 to 1.5 times greater were obtained in the absence of modifiers, and these products also fractured in the extrusion performance test. Infrared analyses of the products of the examples indicated combined hexafluoropropylene contents of less than 0.2 weight percent. The specific melt viscosities of the hexafluoropropylene-modified resins of Examples 14 to 16 ranged from about 3×10^{10} to 6×10^{10} poises at 380°C.; for resins made under identical conditions in the absence of modifier, the specific melt viscosity was about 10×10^{10} poises. The lower melt viscosity aids in the improvement of the degree of sintering obtained during a fixed thermal treatment.

Low Pressure Process Employing Copper Accelerator

R.H. Halliwell; U.S. Patent 3,110,704; November 12, 1963; assigned to E.I. du Pont de Nemours and Company describes a process for the polymerization of tetrafluoroethylene in which the tendency for adhesions to form in the polymerization reactor is minimized. It has been found that this can be achieved by a process which comprises contacting tetrafluoroethylene at a temperature in the range of 30° to 80°C. with an aqueous inorganic redox initiator system and a copper accelerator, under a pressure in the range of 0.6 to 4 atmospheres absolute. The following examples illustrate this process.

Example 1: An electrically stirred, water-steam jacketed, stainless steel autoclave of 850 parts water capacity is evacuated; charged with 595 parts of deoxygenated distilled water containing, per million parts, 5,000 parts borax, 100 parts each of ammonium persulfate, sodium bisulfite, and ammonium perfluorocaprylate, and 0.4 part copper as copper sulfate; heated to 55°C.; pressured with tetrafluoroethylene to 1.3 atmospheres and stirred at an agitator speed of 600 rpm. Polymerization commences shortly after the tetrafluoroethylene is added. The reaction is continued for about 3 hours, during which time stirring is continued at 600 rpm; temperature is held at 55° to 60°C.; pressure is held at 1.3 to 1.7 atmospheres by addition of tetrafluoroethylene; after the solids content in the autoclave reaches about 25% a mixture of polymer and water is intermittently withdrawn through a draw-off valve at the bottom of the autoclave; and make-up quantities of deoxygenated water containing 100 parts per million each of ammonium persulfate and sodium bisulfite are intermittently added so as to maintain the liquid level in the autoclave.

It is observed that the power required to maintain stirrer speed remains substantially constant throughout the run. At the conclusion of the run, the autoclave is drained, flushed once with water, and found to be free of polymer adhesions. There is obtained after drying a total of 326 parts of fine, sandy, free-flowing granular polymer corresponding to a space-time yield of 300 grams/liter/hour. The polymer so obtained readily passes a 0.25 inch mesh sieve, is of high molecular weight as determined by specific gravity measurements on a molded chip, and shows commercially acceptable powder flow properties.

In a series of comparable runs in the same equipment at pressures of 10 to 15 atmospheres via batch methods without copper and without ammonium perfluorocaprylate, the power requirement to maintain stirrer speed is approximately doubled after about 200 parts of polymer is produced. The polymer obtained contains on the average about 10% of stringy, agglomerated or compacted particles which fail to pass a 0.25 inch mesh screen. The instantaneous reaction rate during polymerization is in the range of 600 to 800 grams/liter/hour. However, because it is necessary at frequent intervals to stop the reaction and clean the reactor by hand so as to remove polymer adhesions, the overall maximum production rate by this procedure is only about 200 grams/liter/hour.

Example 2: A water-steam jacketed, glass reactor, of 2 parts water capacity and equipped with a magnetically actuated stainless steel stirrer, is evacuated; charged with 1.2 parts of deoxygenated distilled water containing, per million parts, 5,000 parts borax, 100 parts ammonium persulfate and 100 parts sodium bisulfite; heated to 42°C.; pressured from a metered continuous source to 1.1 atmospheres with tetrafluoroethylene; and agitated at a stirrer speed of 600 rpm. Polymerization commences shortly after pressuring with tetrafluoroethylene. The reaction is continued while maintaining temperature at 42° to 44°C., pressure at 1.1 atmospheres, and stirring rate at 600 rpm, and produces polymer at a steady rate of about 8.5 grams/liter/hour, as determined by tetrafluoroethylene take-up.

Addition of copper (as sulfate) to the reaction mixture in amount of 0.025 part per million parts of liquid rapidly increases reaction rate to 13 grams/liter/hour. Addition of a further like amount of copper further increases the rate to 24 grams per liter per hour. The polymer obtained is sandy, free-flowing, and free of adhesions. In continued running, the accelerating effect of the copper diminishes, apparently due to plating out on the stirrer. Optimum rate, however, is maintained by addition of copper at the rate of 0.1 part per hour per million parts of liquid. Copper present in markedly higher amounts, however, functions as an inhibitor, and maximum benefit is obtained at copper concentrations in the range of 0.02 to 2 parts per million parts of the liquid medium.

Example 3: The procedure of Example 2 is repeated to establish an initial reaction rate of about 8 grams/liter/hour; thereafter there is added ammonium perfluorocaprylate in amount of 100 parts, per million parts of liquid. The rate slowly increases to 25 grams/liter/hour. Following establishment of the 25 grams/liter/hour rate, there is added copper sulfate to provide copper in amount of 0.02 part per million parts of liquid and the rate increases rapidly to 44 grams/liter/hour. In companion experiments, acceleration similar to that obtained with ammonium perfluorocaprylate is also obtained with other highly halogenated dispersing agents which are substantially inert chemically under the reaction conditions.

Ammonium Persulfate–Silver Salts as Catalyst

M. Ragazzini, D. Carcano, C. Garbuglio and A. Doria; U.S. Patent 3,193,543; July 6, 1965; assigned to Edison, Italy describe a process which comprises polymerizing tetrafluoroethylene under conditions which allow the polymerization to be rapidly carried out and easily controlled in the presence of a catalytic system constituted of an inorganic peroxide and of a water-soluble silver salt. The following example illustrates this process.

Example: In a stainless steel V4A autoclave of 5 liters, equipped with a propeller stirrer and a cooling jacket, after air removing by suction and nitrogen washing, 1,500 grams distilled and deaerated water containing dissolved 2 grams of borax and 1,500 grams tetrafluoroethylene are introduced in sequence. Temperature is maintained at about 5°C. and then, by means of two metering pumps, 250 ml. aqueous solution containing 0.45 gram $(NH_4)_2S_2O_8$ and 250 ml. aqueous solution containing 0.027 gram $AgNO_3$ are introduced simultaneously into the autoclave through two different valves. The following ratios are thus achieved:

$$\frac{(NH_4)_2S_2O_8 \text{ mols}}{AgNO_3 \text{ mols}} = 12.4; \qquad \frac{(NH_4)_2S_2O_8 \text{ gram}}{C_2F_4 \text{ gram}} = 3 \times 10^{-4}; \qquad \frac{H_2O \text{ gram}}{C_2F_4 \text{ gram}} = 1.33$$

In this way the catalyst is formed in the presence of the monomer to be polymerized. Stirring in autoclave is started; the temperature is maintained at about 5°C., by allowing cold water to circulate in the autoclave jacket. After 6 hours, 1,400 grams of white polymer are discharged which shows an apparent density of about 0.3 gram/milliliter. By operating at the same temperature, but introducing the tetrafluoroethylene gradually over a period of time so as to maintain a pressure of about 6 atmospheres, thus avoiding to have the monomer in liquid state, the polymerization is more easily controlled, the polymer will be sufficiently subdivided, easily dischargeable and washable. The apparent density of the granular polymer is 0.40 g./ml. In addition, the polymer is fluent and, different from that obtained when polymerizing with a liquid, does not show any tendency to clay. After processing, the polymers show a tensile strength value of 200 kg. per cm.^{2} and the elongation at break is 400%. All physical determinations were carried out in accordance with ASTM norms.

Aqueous Telomer Dispersions

L.Q. Green and R.W. Moses; U.S. Patent 3,105,824; October 1, 1963; assigned to E.I. du Pont de Nemours and Company describe a stable dispersion in water of a fluorocarbon telomer having wax-like characteristics obtained by reacting tetrafluoroethylene in the presence of methylcyclohexane and 1,1,2-trichlorotrifluoroethane. More specifically, this process is directed to an aqueous dispersion of a fluorocarbon telomer stabilized with a water-soluble surface active agent, the telomer being produced by reacting, at 75° to 200°C., one mol of tetrafluoroethylene in the presence of from 2 to 3 mols of 1,1,2-trichlorotrifluoroethane, from 0.01 to 0.1 mol of an active telogen, and from 0.05 to 3% by weight of an organic peroxide catalyst based upon the tetrafluoroethylene.

This telomer dispersed in 1,1,2-trichlorotrifluoroethane is obtained by use of telomerization techniques, as described, for example, in U.S. Patent 2,540,088. In general, the telomerization is carried out by first charging an autoclave or other pressure vessel with an active telogen, with trichlorotrifluoroethane, with a peroxide catalyst and then introducing tetrafluoroethylene gas under pressure or by passing it into the cooled reactor. The charged reaction vessel is then heated to a temperature between 75° and 200°C. and the reaction allowed to proceed. Pressures will be generated between 300 and 600 psig and as the reaction nears completion, the pressure within the system will be observed to drop.

In preparing this tetrafluoroethylene telomer dispersed in 1,1,2-trichlorotrifluoroethane, it is necessary to carefully control the amounts of tetrafluoroethylene, trichlorotrifluoroethane and active telogen. For each mol of tetrafluoroethylene, it is necessary to have present, in the reaction mass, 2 to 3 mols of trichlorotrifluoroethane, and, from 0.01 to 0.10 mol of active telogen. If less than 0.01 mol of active telogen is used, the product is of higher molecular weight and is less wax-like, approaching, as the telogen is decreased, polytetrafluoroethylene itself.

If much above 0.10 mol active telogen is used per mol of tetrafluoroethylene, the molecular weight becomes too low and the properties of the product progress from wax-like to grease-like to liquid as the amount of active telogen increases. On the other hand, if much more than 3 mols of trichlorotrifluoroethane is used per mol of tetrafluoroethylene, the dispersion is too dilute for practical purposes. If less than about 2.5 mols of the trichlorotrifluoroethane is used, the viscosity of the resultant product is very high, resulting in poor heat transfer during preparation.

The preferred telomer is that prepared by reacting tetrafluoroethylene in the presence of 1,1,2-trichlorotrifluoroethane, methylcyclohexane, and di-tert-butyl peroxide. By way of illustrating how the telomer may be prepared the following procedure is given. A clean, dry 10-gallon stainless steel, steam-jacketed autoclave, equipped with a cooling coil, anchor-type agitator, and intake and discharge tubes, is flushed with nitrogen and filled with a solution of 1,1,2-trichlorotrifluoroethane containing 0.76% by weight of methylcyclohexane and 0.28% by weight of di-tert-butyl peroxide. The take-off valve is set for 600 psig and the temperature raised to 160°C. The above 1,1,2-trichlorotrifluoroethane solution is then fed to the autoclave at a rate of 68.7 lbs. per hour. At the same time tetrafluoroethylene under a pressure of 650 to 750 psig is introduced into the autoclave at a rate of 20 lbs. per hour. When a steady reaction state is reached a dispersion of a tetrafluoroethylene telomer dispersed in 1,1,2-trichlorotrifluoroethane at a solids concentration of about 20% is obtained. The surface active agent used to stabilize the aqueous dispersion of the tetrafluoroethylene telomer may be of the anionic, cationic, nonionic, or amphoteric type. It is added to the dispersion of the telomer in the 1,1,2-trichlorotrifluoroethane solvent in which it is prepared. From about 0.05 to 2 parts, preferably from 0.3 to one part, of surface active agent per part of telomer may be employed. Representative examples illustrating this process follow.

Example 1: Anionic Dispersing Agent — A three-liter round bottom glass flask fitted with a half-moon stirrer, dropping funnel, and a downward condenser was charged with 1,250 grams of a dispersion of tetrafluoroethylene telomer prepared as described above and dispersed in 1,1,2-trichlorotrifluoroethane to have a 16% solids content. To this telomer dispersion was added 166 grams of the morpholine salt of oleic acid. Under mild agitation and with heat from an oil bath about one-half (510 g.) of the trichlorotrifluoroethane was removed.

Then 640 grams of water was slowly added through the dropping funnel while continuing the distillation of the organic solvent. When the vapor temperature reached 62°C. the heating was discontinued. The resulting product was a viscous milky-white liquid having dispersed therein the tetrafluoroethylene telomer to give a solids content of 20%. As a component of a textile printing ink the telomer dispersion provides a ready release of the ink from the printing screens.

Example 2: Mixed Anionic Dispersing Agents — The reaction flask of Example 1 was charged with 1,500 grams of a 20% dispersion of the tetrafluoroethylene telomer in 1,1,2-trichlorotrifluoroethane and 150 grams of an aqueous solution of a sodium alkyl (C_{12}) benzenesulfonate containing 30% active ingredient. The flask was heated in an oil bath, and 750 grams of 1,1,2-trichlorotrifluoroethane distilled off. Then 125 grams of water was slowly added to give a viscous grease-like material.

Upon the addition of 60 grams of sodium stearate the viscous mass thinned to a fluid. At this point an additional 365 grams of water were added as the heating was continued until a total of 1,200 grams of 1,1,2-trichlorotrifluoroethane was removed. The resulting product was a finely divided dispersion of the telomer having a solids content of 31% and containing none of the trichlorotrifluoroethane as free solvent. When diluted with water to provide 0.5% telomer dispersion, the dilute suspension remained stable for several days with only slight settling. The mildest agitation, such as obtained by merely inverting the container, served to redisperse any separated solid materials.

By employing the product telomer dispersion as a lubricant, a beryllium-copper alloy wire can be directly drawn smoothly, uniformly, and with little or no interruption by breakage. In the absence of the subject lubricant and employing a conventional wire-drawing lubricant, the alloy wire is first coated with copper to improve the wire drawing operation and the quality of the resultant wire and the coating is subsequently removed before the wire is used. Without the copper coating with a conventional lubricant, excessive breakage and scarring of the wire and excessive wear of the die orifice take place, and the drawn wire is nonuniform in gauge.

Example 3: Cationic Dispersing Agent — To the reaction flask of Example 1 was charged 3,280 grams of a 9.2% dispersion of the tetrafluoroethylene telomer in 1,1,2-trichlorotrifluoroethane and 180 grams of the oleic acid diester of N,N,N',N'-tetrakis-(2-hydroxypropyl) ethylenediamine singly quaternized with dimethyl sulfate. With heating the flask in an oil bath 1,500 grams of the trichlorotrifluoroethane were distilled and collected. Then 750 grams of water were added slowly while continuing the removal of the trichlorotrifluoroethane until a total amount of 2,980 grams was collected. The resulting product was a smooth, white paste containing 24.3% of telomer particles having a positive electrical charge. The product telomer dispersion provides a valuable lubricant for the sewing of cotton fabric; the needle damage to the fabric encountered in high speed sewing operations is markedly reduced.

Example 4: Nonionic Dispersing Agent — To the reaction flask used in the previous examples was added 2,500 grams of a 20% dispersion of the tetrafluoroethylene telomer in 1,1,2-trichlorofluoroethane and 250 grams of a condensation product of ethylene oxide and isooctyl phenol. The flask and its contents were then heated on an oil bath until 1,300 grams of the trichlorotrifluoroethane were removed. At this stage the slow, dropwise addition of 250 grams of water was begun while continuing the heating until a vapor temperature of 60°C. was reached. During this final distillation the remainder of the solvent and 165 grams of water were removed. The resulting product was a white paste that comprised 60% telomer dispersed in water with particles that were electrically neutral and thereby rendered compatible with either anionic or cationic surface active agents which might be added to the aqueous telomer dispersion or with which the telomer dispersion might come into contact.

Coagulation of Latex by Aeration

In a process described by M.B. Black III; U.S. Patent 3,464,964; September 2, 1969; assigned to Pennsalt Chemicals Corporation a polytetrafluoroethylene latex is coagulated without the use of mechanically induced shear by passing inert gas in an upwardly flowing direction through the aqueous polymer latex to cause the polymer particles to coalesce and coagulate to permit their ready separation from the aqueous phase. Referring to Figure 1.1a, 1 represents a coagulation tower, 2 is an inlet line for aqueous polymer latex from a polymerizer discharge valve, 3 is an inlet line for filtered air

FIGURE 1.1: COAGULATION OF POLYTETRAFLUOROETHYLENE LATEX

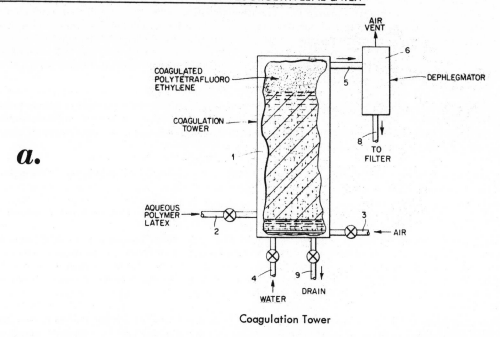

Coagulation Tower

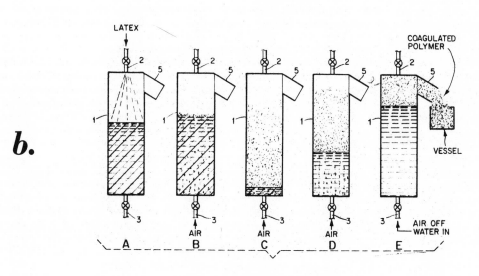

Phases of Coagulation During Aeration

Source: M.B. Black III; U.S. Patent 3,464,964; September 2, 1969

from an air supply source, 4 is an inlet line for adding deionized water which is used for diluting the latex and flushing the tower, 5 is a discharge line from the coagulation tower through which the coagulated polytetrafluoroethylene flows along with exhaust air and the water in which the coagulant is suspended, 6 is a dephlegmator tower with an air vent 7 and a discharge line 8, the latter leading to a filter. At the bottom of coagulation tower 1 is a water drain line 9.

The aqueous polymer latex is introduced through line 2 into the coagulation tower 1, and sufficient deionized water is introduced through line 4 to dilute the latex to a desired dilution. Purified air at a constant volume and pressure is introduced continually at ambient temperature and pressure through line 3. The latex begins to coagulate under the agitation provided by the air, and as finite particles form, the particles coalesce into larger particles as the fine particles come into contact with each other and are occluded by a film of air which floats the particles to the top of the tower 1. The particles accumulate at the top of tower 1 until they are sufficient in number to overflow through line 5 into the dephlegmator 6. The excess air vents through line 7. The coagulated polymer is recovered through line 8 and sent to a filter.

Figure 1.1b shows the phase changes which occur during the coagulation of polytetrafluoroethylene latex according to this process. In Figure 1.1b, A represents a tower 1 into which aqueous polytetrafluoroethylene latex has been introduced through line 2. Tower B shows the latex at about the time aeration is begun by the introduction of air through line 3. Tower C shows the aerated latex about six minutes after aeration has begun, with a gel stage forming. Tower D shows coagulum rising upward in the tower, leaving substantially clear water at the bottom of the tower. Tower E shows the coagulum being floated out of the top of the tower through line 5 by addition of water through line 3, after the air has been turned off.

In contrast to methods employing mechanical agitation, this method imparts a minimum of shear and turbulence to the latex mass. Although the upward flow of air provides some agitation of the latex mass, it is to be observed that agitation is not the sole function of the air. Besides providing a moderate amount of turbulent agitation, the air bubbles, as they pass up the coagulation tower 1, have an unexpected effect which causes the colloidal particles of polytetrafluoroethylene in the latex to coalesce and to form distinct particles during the passage of the air and latex up the column. The rate of ascent of the air up the column essentially must be slow enough to permit the coagulum to form prior to or during the ascent. This process is illustrated by the following examples.

Example 1:   A vertical cylinder measuring 4 inches in inner diameter and about 6 feet in height was equipped with an air inlet at the bottom of the cylinder. Polytetrafluoroethylene latex, prepared substantially by the method described in U.S. Patent 2,593,583, was poured into the cylinder to a height of about 2 feet. The latex, which originally contained 25% solids, was then diluted with deionized water to form a dilute latex containing 10% solids. Filtered air was introduced continuously into the bottom of the cylinder at a rate of 15 to 30 liters per minute at a head pressure of 1.5 psig. At the end of ten minutes, substantially all the latex had coagulated and floated to the surface of the water layer, which was now clear, thus indicating that coagulation of the latex was complete. The coagulated polytetrafluoroethylene particles were skimmed off the water surface and then were dried at 120°C. for 10 hours. Particle size was found to be d_{50} = 560 microns. The resin was paste extruded at a pressure of 1,700 to 2,200 psig, as a resin-naphtha mixture containing about 18% by weight of naphtha as an extrusion aid, at a reduction ratio of 144:1, with no defects. Reduction ratio is defined as R = $(D_b/D_a)^2$, where D_b = the diameter of the extrusion cylinder and D_a = the diameter of the orifice. See G.R. Snelling and J.F. Lontz, Journal of Applied Polymer Science, volume III, 9, pages 257-265 (1960).

Example 2:   The cylinder of Example 1 was filled to a three-foot level with similar latex which was dilute from 25 to 15% solid content. Air was introduced continuously at 15 to 30 liters per minute at a head pressure of 2 pounds per square inch. Coagulation took place in ten minutes. The coagulated resin was recovered, filtered and dried at 120°C. for ten hours. The particle size was found to be d_{50} = 850 microns. Resin-naphtha mixture, made from the resin wetted with about 18% by weight of naphtha, was extruded at a pressure between 1,700 and 2,270 pounds per square inch at a reduction ratio of 144:1, with no defects.

Example 3:   A significant difference has been found to exist in the extrusion pressure of air-coagulated and of motorized-agitator-coagulated particles of polytetrafluoroethylene resin from 10% solids content latices. The air-coagulated resin has been found to extrude at pressures of 1,000 to 2,500 psig lower than the comparable resin made by the motor-agitated method. The data are compared in the table below. Each resin was paste extruded at a reduction ratio of 975:1.

Method	PTFE Latex, ml.	Air Flow, cfm	Agitator, rpm	Coagulation Time, min.	% Solids in Filtrate After Coagulation	Extrusion Pressure, psig
Air	4,000	20	–	22	0.5	11,000
Air	4,000	20	–	14	<0.1	9,500
Motorized-agitator	2,000	–	100	55	1.0	>12,000
Motorized-agitator	2,000	–	100	20	0.5	>12,000

GENERAL PROCESSES — SOLVENT

Perfluorinated Liquid Polymerization Media

M.I. Bro; U.S. Patent 2,952,669; September 13, 1960; assigned to E.I. du Pont de Nemours and Company has found that in the presence of perfluorinated liquid, aliphatic compounds such as perfluorocyclobutane, perfluoromethyl cyclohexane, perfluorokerosene, perfluorotributyl amine, etc., and a catalyst such as a peroxygen compound or an azo compound, high molecular weight perfluorocarbon polymers can be obtained under polymerization conditions used in other processes of obtaining perfluorocarbon polymers.

Due to the greatly increased solubility of perfluorocarbon monomers, under polymer-forming conditions, such as tetrafluoroethylene, hexafluoropropylene, etc., as compared to their solubility in other polymerization media such as water, polymerization reactions may be carried out at faster rates and lower temperatures and yet result in better yields. Some copolymerizations of tetrafluoroethylene with other perfluorinated compounds, such as perfluorocyclobutene can be polymerized with relative ease to tough solids using this process. It has also been discovered that the perfluorinated liquids, emulsified in water with the aid of a dispersing agent used as media in the polymerization of perfluorocarbon polymers will give rise to polymer formation within the perfluorinated liquid part of the emulsion without causing polymerization in the water phase of the emulsion. This was shown by the fact that on breaking the emulsion the water phase could be separated from the liquid perfluorinated compound with only traces of the polymer contained in the water and most of the polymer contained in the perfluorinated liquid phase. Thus the water acts merely as a heat transfer medium. The advantages gained by this phase of this process refer to the physical nature of the resulting polymer and not to its inherent structure.

The saturated perfluorinated liquids used as polymerization media in this process are physically adsorbed on the polymer in the polymerization phase to give a wet spongy solid, but are easily recovered by distillation at reduced pressure, leaving the solid polymer behind, which has in essence the same properties as a polymer of the same monomers made by other processes. For reasons of fast and complete removal of the saturated perfluorinated liquid medium from the perfluorocarbon polymer, it is preferred to use the more volatile perfluorinated liquids such as perfluorocyclobutane, perfluoromethylcyclohexane, perfluorodimethylcyclohexane, and perfluorokerosenes.

Some of these compounds may be prepared by pyrolysis of polytetrafluoroethylene or tetrafluoroethylene as described in U.S. Patent 2,384,821 or U.S. Patent 2,404,374. In using these perfluorosolvents as media for polymerizing perfluorocarbon monomers great care has to be taken that the compounds are pure. Impurities in the solvents will result in the formation of telomerized polymers, not desired in this process. Upon recovery of the perfluorinated medium from the polymer medium mixture the medium may be reused without further treatment. As a matter of fact, trace impurities are "polymerized out," so that on each polymerization the perfluorinated liquid medium attains a higher degree of purity.

The catalysts and initiators that can be used in this process are in general peroxygen compounds and azo compounds such as described in U.S. Patent 2,559,630. The most common medium for the polymerization of perfluorocarbon monomers has been water. This medium however has the disadvantage of limited solubility of organic initiators and catalysts as well as the possibility of chemical attack of the medium on the catalyst, under polymer-forming conditions thus reducing or destroying its activity.

Thus the use of inert organic solvents as used in this process will increase the effectiveness of organic catalysts. This is especially well illustrated by use of perfluorinated peroxygen compounds as initiators. These compounds are very reactive catalysts, but unstable in the presence of water above 0°C. The use of a perfluorinated saturated liquid as the polymerization medium makes the polymerization of tetrafluoroethylene at much lower temperatures and pressures possible. These conditions are below the critical pressures and temperatures of tetrafluoroethylene, so that polymerization may occur in a liquid monomer stage having the medium and the catalyst, a feat generally not accomplished in other processes of preparing tetrafluoroethylene polymer. The following examples illustrate this process.

Example 1:   In a pressure-resistant stainless steel vessel having a capacity of 330 milliliters were placed 130 grams of perfluoromethylcyclohexane containing 0.011 gram of α,α'-azodiisobutyronitrile. The vessel was closed, cooled to -70°C. and evacuated. The vessel and contents were warmed to 75°C. and purified tetrafluoroethylene was added through a valve in the head of the vessel until the pressure in the vessel had built up to 300 to 350 lbs./sq. in. The vessel and contents were then agitated maintaining pressure and temperature.

The pressure in the vessel was maintained by continued addition of purified tetrafluoroethylene. The reaction was continued for 90 minutes. The vessel was then cooled and excess monomer vented off. The polytetrafluoroethylene-perfluoromethylcyclohexane mixture was removed from the stainless steel vessel and placed in a glass container. On evacuation the glass container was heated on a steam bath until all of the solvent had been removed. The yield of the polymer was 75 grams. The polymer could be compression molded at 380°C. into tough films. For comparable results using water as a polymerization medium in similar equipment at higher pressures, 400 to 500 lbs./sq. in. and longer times, 16 hours, yields of the polymer were only 10 to 15 grams.

Example 2:   Example 1 was repeated with the exception of using 250 grams of perfluorocyclobutane instead of 130 grams of perfluoromethylcyclohexane. Similar to Example 1, 28 grams of tetrafluoroethylene polymer was obtained.

Example 3:   In a pressure-resistant stainless steel vessel cooled to -70°C. having a capacity of 330 parts of water were placed 105 grams of perfluorokerosene and 0.064 gram of diheptafluorobutyryl peroxide. The vessel was closed and evacuated. The vessel was then warmed to 15°C. and tetrafluoroethylene was added until a pressure of 100 to 150 lbs. per sq. in. was obtained. The vessel was agitated maintaining temperature and pressure substantially constant. After 10 minutes of agitation the reaction was stopped. Excess monomer was vented off and the polymer-medium mixture removed from the vessel. The perfluorokerosene was recovered from the polymeric tetrafluoroethylene by distillation at reduced pressures. A yield of 45.5 grams of tetrafluoroethylene polymer was obtained. Compression molded, sintered samples were found to have a stiffness of 60,000 lbs./sq. in., a tensile strength of 2,100 lbs./sq. in. and an elongation of 200%.

Example 4: In a pressure-resistant stainless steel vessel having a capacity of 330 milliliters was placed 105 grams of perfluorodimethylcyclohexane having therein dissolved 0.25 gram of α,α'-azodiisobutyronitrile. The vessel was closed and cooled to -70°C. and evacuated. Through a valve in the head of the vessel 10 grams of hexafluoropropylene were added to the reaction mixture. The vessel and contents were then warmed to 75°C. and purified tetrafluoroethylene was added through the said valve until pressure in the vessel had built up to 375 lbs./sq. in. The vessel and contents were then agitated maintaining pressure and temperature.

The reaction was stopped after 120 minutes. The vessel was cooled and excess monomer removed. The mixture of perfluorodimethylcyclohexane and a polymer was removed from the stainless steel vessel and placed in a glass container. On evacuation the glass container was heated on a steam bath until substantially all of the solvent had been removed. The yield of the solid, white copolymer was 32.4 grams. The melting point range of 314° to 319°C. indicated the formation of tetrafluoroethylene-hexafluoropropylene copolymer. Tetrafluoroethylene polymer has a melting point range of 327° to 330°C. The copolymer was compression molded at 360°C. into tough films.

Fluorinated Peracids for In Situ Formation

M.I Bro, R.J. Convery and R.C. Schreyer; U.S. Patent 2,988,542; June 13, 1961; assigned to E.I. du Pont de Nemours and Company describe a process which comprises polymerizing fluorinated monomers in substantially completely fluorinated solvents in the presence of a catalyst formed in situ by the reaction of a substantially fluorinated acid with hydrogen peroxide. The catalysts employed in this process are peroxygenated acids having the general formula, R_fCOOOH, which are formed by the reaction of hydrogen peroxide and a fluorinated acid according to the following equation:

$$R_fCOOH + H_2O_2 \rightleftharpoons R_fCOOOH + H_2O \tag{1}$$

or by the reaction of a fluorinatated acid anhydride with hydrogen peroxide according to the following equation:

$$(R_fCO)_2O + H_2O_2 \rightleftharpoons R_fCOOOH + R_fCOOH \tag{2}$$

where R_f is a perfluoroalkyl or an omega-hydroperfluoroalkyl radical. Since the reaction of the fluorinated acid with hydrogen peroxide is an equilibrium reaction, it is clear that the presence of water is preferably avoided and that it is consequently preferred to react the hydrogen peroxide with the acid anhydride. The reaction of the acid or the anhydride with the hydrogen peroxide is very rapid and thus the catalyst employed in this process may be formed in situ at polymerization conditions.

Hydrogen peroxide is generally obtained in aqueous solutions. Although from a standpoint of completion of reaction, it would be preferable to employ very concentrated solutions of hydrogen peroxide, i.e., 90%, such solutions are not as safe as the more dilute solutions of hydrogen peroxide, i.e., 30%. The excess water in the polymerization system which would result from the use of more dilute solutions can be consumed by an excess of anhydride. Since the hydrogen peroxide is employed in catalytic quantities the quantity of the acid anhydride required to react with the water formed or present is small.

Complete dehydration is of course necessary only when it is desired to obtain the maximum catalytic activity. It is possible to employ the acid and an aqueous solution of hydrogen peroxide and still obtain an active catalyst. It was found that unless in the equilibrium Equation 1 the water concentration is in excess of 50 mol percent of the acid concentration, the equilibrium is sufficiently pushed to the right to form catalytic quantities of the peracid. Hydrogen peroxide employed by itself in the absence of the acid or anhydride will not give rise to significant polymerization rates and leads to the formation of low molecular weight products.

The acid and acid anhydrides employed in this process as catalyst components comprise perfluorinated acids such as trifluoroacetic acid, pentafluoropropionic acid, perfluorobutyric acid, perfluorohexanoic acid, perfluorononanoic acid, etc.; and omega-hydroperfluorinated acids such as deca-fluorohexanoic acid, dodecafluorooctanoic acid, hexadeca-fluorononanoic acid, eicosofluoroundecanoic acid, etc. The anhydrides employed are the anhydrides of the above described acids. This process is illustrated by the following examples.

Example 1: Into a 185 ml. stainless steel autoclave was charged under nitrogen 50 ml. of perfluorodimethylcyclobutane, 5 grams of trifluoroacetic anhydride, and 0.25 ml. of a 0.3% aqueous hydrogen peroxide solution. The nitrogen was replaced with tetrafluoroethylene and additional tetrafluoroethylene was added until a pressure of 400 psig was obtained. The reaction mixture was agitated at room temperature while maintaining the pressure for a period of 30 minutes. The resulting mixture was filtered, and the solid product obtained was washed and dried at 100°C. in a vacuum oven for 15 hours. The yield of polytetrafluoroethylene was 20.2 grams. The polytetrafluoroethylene powder could be compression molded into chips which on heating to a sintering temperature above 327°C. were tough and flexible.

Example 2: Into a 185 milliliter stainless steel autoclave was charged, under nitrogen at 0°C., 50 milliliters of perfluorodimethylcyclobutane, 3 grams trifluoroacetic anhydride, 0.25 gram of a 30% aqueous hydrogen peroxide solution. The mixture was warmed to room temperature and pressured with tetrafluoroethylene to a pressure of 400 psig. The reaction mixture was heated to 75°C. and maintained at that temperature for 10 minutes with agitation. The reaction vessel was then cooled and the reaction products removed. On washing, filtering and drying there was obtained 63 grams of solid polytetrafluoroethylene which could be molded into tough chips. The example was repeated except that instead of the 3 grams trifluoroacetic anhydride and 0.25 gram of the hydrogen peroxide solution there was employed only one gram of a 30% hydrogen peroxide solution. There was obtained 1.5 grams of low molecular weight polytetrafluoroethylene.

Example 3: The procedure of Example 2 was repeated except that the catalyst comprised 10 grams of trifluoroacetic acid and 0.25 gram of a 3% hydrogen peroxide solution. There was obtained 34.5 grams of solid polytetrafluoroethylene.

The advantages of this process are particularly the high catalytic activity of the substantially fluorinated peracids which allows rapid polymerizations at ordinary temperatures and low pressures, and the high degree of safety which is attained by being able to keep the catalyst-forming reagents separately and form the catalyst in situ. Additional advantages are the fact that no metallic reagents need be used which can contaminate the polymer, the fact that the catalyst is soluble in inert solvents which are also solvents for the monomers and that the catalyst system can be employed in bulk polymerizations.

Metal Fluorides and Nitrogen Trifluoride as Catalysts

C.G. Krespan; U.S. Patent 2,938,889; May 31, 1960; assigned to E.I. du Pont de Nemours and Company describes a process of preparing fluorocarbon polymers which comprises bringing in intimate contact at a temperature above 0°C. tetrafluoroethylene, or mixtures with a perfluoroolefin having a terminal difluoromethylene, $=CF_2$ group, the mixtures comprising at least 20 mol percent of tetrafluoroethylene, with catalytic amounts of a fluoride of a metal having an oxidation potential greater than that of mercury, the metal in the fluoride being in its highest valency state.

Suitable catalysts include chromium trifluoride, manganese trifluoride, cerium tetrafluoride, lead tetrafluoride, bismuth pentafluoride, cobalt trifluoride, and silver difluoride. As little as 0.1 mol of catalyst per 100 mols of fluoroolefin is sufficient, although much more can be used if desired, e.g., up to 50% on a molar basis. Preferably, there is used between 0.5 and 5 mols of metal fluoride per 100 mols of the fluoroolefin, or mixture of fluoroolefins, being polymerized.

With the more active catalysts, e.g., cobalt trifluoride or lead tetrafluoride, the polymerization reaction proceeds at temperatures as low as 0°C. and is even exothermic. Preferably, the reaction is carried out at temperatures at least as high as the ambient temperature, e.g., 15° to 25°C. The maximum temperature is not critical, provided it is below the decomposition point of the fluoroolefin polymer, but it is in general unnecessary to exceed 200°C., and the preferred temperature range is that between 15° and 150°C. Under such conditions, substantial conversions are obtained within one to eight hours.

The tetrafluoroethylene polymers or copolymers obtainable by this process are tough solids of high molecular weight above 200,000 and generally in the range of 500,000 to 2,000,000, and excellent physical properties, including good thermal stability, which permit them to be used in the various applications for which these polymers are suitable when prepared by conventional means. The following examples illustrate this process.

Example 1: Into an agitated pressure vessel of about 80 cc capacity containing 28.3 g. of lead tetrafluoride and 15 cc of arsenic trifluoride as the solvent was pressured 40 grams of tetrafluoroethylene in portions of 4 grams over a period of 5 hours at room temperature. The vessel was agitated at room temperature (18° to 23°C.) for an additional hour. The internal pressure during this period did not exceed 350 lbs./sq. in. After evaporation of the arsenic trifluoride, the remaining solid product was extracted with acetic acid, then with dilute hydrochloric acid and with dilute nitric acid. The product was then pulverized in a blending mixer and extracted again with acetic acid-nitric acid and acetic acid-hydrochloric acid mixtures. The remaining solid (36.5 grams, 91% conversion) was polytetrafluoroethylene which was further identified by its x-ray diffraction pattern. The polymer was found by emission spectroscopy to contain lead at the 500 to 2,500 ppm level.

Example 2: A mixture of 14.6 grams of silver difluoride, 15 cc of arsenic trifluoride and 24 grams of tetrafluoroethylene was agitated in a pressure vessel at a temperature of 66°C. under autogenous pressure. The resulting polytetrafluoroethylene, isolated as in Example 1, except that the extractions with hydrochloric acid were omitted, weighed 21.5 grams (90% conversion). Emission spectroscopy indicated the presence of 0.08 to 0.5% of silver. The molecular weight of this polymer was about 1,600,000 when determined by measuring the specific gravity of a sintered chip.

Example 3: A copolymer was prepared by heating a mixture of 7.3 grams of silver difluoride, 15 cc of arsenic trifluoride, 15.1 grams of tetrafluoroethylene and 34 grams of hexafluoropropene for 10 hours at 70° to 150°C. in a sealed vessel under autogenous pressure. After treatment of the reaction mixture as in Example 2, there was obtained 12.5 grams of a solid product having an inherent viscosity at 380°C. of 5.36×10^5 poises under a 5 kg. load. Tough films of this product were pressed at 345°C. The infrared spectrum of one of such films showed the presence of $-CF_3$ and $-CF=$ groups, establishing the identity of the material as a tetrafluoroethylene-hexafluoropropene copolymer. This process provides a convenient method for preparing polytetrafluoroethylene and certain copolymers in excellent yields and under relatively low temperature and pressure conditions.

C.S. Cleaver; U.S. Patent 2,963,468; December 6, 1960; assigned to E.I. du Pont de Nemours and Company describes the use of nitrogen fluorides (N_2F_2, NF_3) as catalysts for the polymerization of unsaturated fluorocarbons. The following example illustrates this process.

Example: A 7.1 gram charge of nitrogen trifluoride was placed in an 80 ml. stainless steel pressure reactor and heated to 250°C. At this temperature, 9.8 g. of tetrafluoroethylene was injected. The temperature and pressure were maintained for one hour at 250°C. and 1,250 lbs./sq. in. respectively. There was obtained white polymeric tetrafluoroethylene and volatile products.

COPOLYMERS WITH HEXAFLUOROPROPYLENE

Control of Process Variables

A process described by M.J. Couture, D.L. Schindler and R.B. Weiser; U.S. Patent 3,132,124; May 5, 1964; assigned to E.I. du Pont de Nemours and Company involves conducting the polymerization of the mixed comonomers, tetrafluoroethylene and hexafluoropropylene, at 95° to 138°C. in an aqueous system in the presence of a dispersing agent and using a free radical polymerization initiator, with the vapor space density of the mixed comonomers being maintained at 0.18 to 0.30 g./cc. The following examples are given to illustrate this process.

Example 1: A cylindrical, horizontally disposed, water-jacketed, paddle-stirred, stainless steel reactor having a length to diameter ratio of about 1.5 and a water capacity of 80.7 parts is evacuated, charged with 46 parts of demineralized water and purged of gases by warming the charge and evacuating the reactor free space. The degasified charge is heated to 95°C., pressured to 390 psig with deoxygenated hexafluoropropylene, made 2.9×10^{-4} molal with respect to potassium persulfate by rapid addition of freshly prepared 0.011 molal solution of potassium persulfate in demineralized water, and then stirred for 15 minutes at 95°C. At the end of the 15 minutes, the reactor is pressured to 650 psig with deoxygenated tetrafluoroethylene so as to achieve a mixture of comonomers which is 75 weight percent hexafluoropropylene and 25 weight percent tetrafluoroethylene.

A freshly prepared 0.007 molal solution of potassium persulfate is injected at the rate of 0.0437 part per minute so that the rate of active radical generation is maintained at about 2.14×10^{-5} mols per minute per liter of solution. The stirring of the reactor contents at 95°C. and the addition of potassium persulfate are continued for 100 minutes after the 650 psig total pressure is attained; during this period the pressure is maintained constant by the continuous addition of tetrafluoroethylene. At the end of 100 minutes the stirring is stopped, the vapor in the reactor is sampled, the reactor is vented and its residual contents is discharged.

There is obtained an aqueous dispersion containing 7.3 parts (about 15 weight percent) of resinous polymeric product. The sample of the vapor space taken from the reactor at the end of 100 minutes is immediately analyzed by infrared techniques and found to contain 75 weight percent hexafluoropropylene. The aqueous dispersion is coagulated by stirring to obtain a particulate coagulum which is filtered from the liquid, washed with distilled water and dried. This material has a specific melt viscosity of 7.5×10^4 poises, and a specific IR ratio of 3.5.

Example 2: A cylindrical, horizontally disposed, water-jacketed, paddle-stirred, stainless steel reactor having a length to diameter ratio of about 1.5 and a water capacity of 80.7 parts is evacuated, charged with 46 parts of demineralized water and purged of gases by warming the charge and evacuating the reactor free space. The contents of the reactor is heated to 120°C. and the reactor is first pressurized to 390 psig with deoxygenated hexafluoropropylene and then with deoxygenated tetrafluoroethylene to 600 psig total pressure so that the composition of the vapor space is 74 weight percent hexafluoropropylene and 26 weight percent tetrafluoroethylene; for a period of 15 minutes, a freshly prepared solution of 0.04 molal potassium persulfate is injected at the rate of 0.175 part per minute so that at the end of 15 minutes the calculated concentration of undecomposed persulfate is 7.85×10^{-5}.

At the end of the 15 minutes the rate of potassium persulfate addition is altered by switching to the injection of a freshly prepared 0.0096 molal solution of potassium persulfate at a rate of 0.0437 part per minute so that the rate of active radical generation is maintained at about 1.24×10^{-5} mols per minute per liter of solution. The stirring of the reactor contents at 120°C. and the addition of potassium persulfate are continued for 100 minutes after the 600 psig total pressure is attained; during this period the pressure is maintained constant by the continuous addition of tetrafluoroethylene. At the end of 100 minutes the stirring is stopped, the vapor in the reactor is sampled, the reactor is vented and its residual contents are discharged. There is obtained an aqueous dispersion containing 18 parts (about 37 weight percent) of resinous polymeric product. The sample of vapor taken from the reactor at the end of 100 minutes is immediately analyzed by infrared techniques and found to contain 74 weight percent hexafluoropropylene.

Examples 3 to 10:   Except where the conditions are noted as being different in the Example Summary, Examples 3 to 10 have been carried out using the procedure of Example 2.

Summary of Examples

No.	Percent dispersing agent	HFP precharge (psig)	Vapor density	Percent HFP	Initiator Scheduling	Polmn. initiator sol. concentration (molal)	Radical gen. rate ($\times 10^{-5}$) moles/min./liter	Dispersion solids content	Spec. melt visc. (poises $\times 10^{-4}$)	Spec. I.R.	Rel. Polmn. rate
1....	0	390	0.235	75	Regular.........	0.007	2.14	7.3	7.5	3.5	1.0
2....	0	390	0.23	74	High Initial...	0.0096	1.24	18	7.5	3.5	1.2
3....	0.1	390	0.23	74	......do.........	0.02	2.6	18	7.5	3.5	2.4
4....	0.2	390	0.23	74	......do.........	0.031	4.03	18	7.5	3.5	3.7
5....	0.2	390	0.23	74	......do.........	0.02	2.6	18	7.5	4.0	2.4
6....	0.2	255	0.19	52.5	......do.........	0.02	2.6	18	7.5	3.5	2.4
7....	0.1	255	0.19	52.5	......do.........	0.01	1.3	18	7.5	3.5	1.2
8....	0.1	418	0.24	77.5	......do.........	0.0125	1.63	18	7.5	3.5	1.5
9....	0.1	390	0.23	74	Low Initial...	0.00174	0.285	9.4	120	3.6	0.8
10...	0.1	390	0.23	74	High Initial...	0.0037	0.605	9.4	120	3.8	1.7

Difluoromethyl End Capping

A process described by R.C. Schreyer; U.S. Patent 3,085,083; April 9, 1963; assigned to E.I. du Pont de Nemours and Company involves stabilizing fluorocarbon polymers through end capping. This process comprises treating a fluorocarbon polymer, obtained through polymerization of a fluoroolefin having the general formula, $CF_2=CFY$, where Y is of the class consisting of fluorine, perfluoroalkyl and omega-hydroperfluoroalkyl radicals, with a peroxide catalyst, with water, preferably in the presence of inorganic compounds having a pH of at least 7, such as stable bases, basic salts and neutral salts, at a temperature of 200° to 400°C., and recovering a fluorocarbon polymer having at least half of all the end groups in the form of difluoromethyl groups.

It was discovered that the carboxylate end groups in the fluorocarbon polymer chain are the principal cause of the instability of fluorocarbon polymer found to occur at melt fabrication temperatures. Carboxylate end groups in the fluorocarbon polymer are formed when the polymerization is initiated through a peroxide catalyst, or when the polymerization is terminated through the formation of a vinyl bond which is subsequently oxidized. The initiation with a peroxide catalyst gives rise to the formation of an oxygen-carbon difluoride bond. This group is unstable and will hydrolyze in the presence of even the smallest traces of water to form a carboxylic acid group, also referred to herein as a carboxylate group.

When this group is heated to the fabrication temperature, it decomposes, carbon dioxide is released, and a vinyl bond is formed in the polymer chain. This vinyl bond, at the elevated temperatures, may react further to either attach to an existing polymer chain, and thereby increase the melt viscosity, or to add oxygen to form an acid fluoride group (—COF), which, in turn, then can be hydrolyzed to form the carboxylate end group again. Since it is extremely difficult to eliminate all oxygen and moisture from the environment of the polymer, most polymerizations even being carried out in an aqueous phase, these reactions become a repeating cycle resulting in the increase in melt viscosity of the polymer and in the build-up of volatile components, such as CO_2, COF_2 and HF, in the polymer.

Since the polymerization is initiated by the peroxide catalysts, it is clear that at least half of the polymer chain end groups have the carboxylate structure. The amount, however, can be greater than half in that the vinyl end group, resulting from termination of the polymerization, can also be converted into a carboxylate end group, and that that termination can also occur through molecular combination.

It was further found that the degradation resulting from the formation of carboxylate groups can be avoided by treating the polymer at elevated temperatures with water. Treatment with water at elevated temperatures causes decarboxylation, but is also accompanied by the formation of a substantial number of highly stable —CF_2H groups. In order to achieve the formation of the stable —CF_2H group, it is necessary that the carboxylate end group be in ionic form. The formation of the —CF_2H group occurs when the polymer is treated with water alone at the temperatures described, but the substantial conversion of carboxylate end groups is slow.

The formation of these stable end groups is increased by the addition of bases, neutral or basic salts to either the aqueous phase or to the polymer. It is believed that the increased rate of formation of stable —CF_2H end groups through the addition of these compounds is due to the effect of these compounds on the ionization of the carboxylate end groups leading to the formation of the carboxylate anion.

In order to cause the reaction of the carboxylate anion with the water, the environment in which the polymer is reacted should contain at least 2% by weight of the environment of water. The environment may be gaseous in nature or liquid. Thus, it is possible to heat the water-polymer mixture in the form of aqueous slurry or to treat the polymer with steam; the concentration of the water may be reduced to the point where the polymer is actually treated with humidified air, as long as the humidity is greater than 0.02 lb. of water per lb. of air.

Polytetrafluoroethylene

In general, two techniques may be employed to carry out the end capping of the perfluorocarbon polymers. These two processes can be characterized as slurry capping and vapor capping. In the slurry capping process the polymer is mixed with sufficient water to create an aqueous slurry and is then heated, under sufficient pressure to maintain the aqueous phase in the liquid form, to the reaction temperatures above 200°C. The salt or the base, if employed, is dissolved in the liquid phase. In the vapor capping process, the base or salt, if employed, is mixed with the polymer, the polymer is then heated to the reaction temperature and moisture is passed over and through the polymer by means of a carrier gas such as air. Instead of moisture in a carrier, steam may also be employed.

The rate at which the polymer is capped depends on the reaction conditions employed. Thus, higher temperatures tend to increase the capping rate. Similarly, increasing the quantity of the added salt or base increases the rate of the capping reaction; also a high concentration of the aqueous phase in the environment will increase the rate of the $-CF_2H$ group formation. The completion of the reaction, i.e., the transformation of substantially all of the carboxylate end groups to $-CF_2H$ end groups can be determined readily by infrared techniques. This process is illustrated by the following examples.

Example 1: Into a 320 ml. stainless steel autoclave was charged 75 grams of wet polymer fluff, obtained from the co-polymerization of tetrafluoroethylene and hexafluoropropylene with a potassium persulfate catalyst in an aqueous medium, the fluff containing 25 grams of a copolymer of tetrafluoroethylene and hexafluoropropylene, the hexafluoropropylene content of the copolymer being 14 to 16 weight percent. To the fluff was added 100 ml. of a 28% aqueous ammonia solution. The autoclave was heated to 250°C. under autogenous pressure and agitated at that temperature for a period of two hours. The resulting polymer and part of the original polymer were dried in a vacuum oven at a temperature of 250°C. for a period of 18 hours. The resulting products are compared in the following table.

Property	Ammonia-Treated Polymer	Untreated Polymer
End group analysis by infrared/10^6 carbon atoms:		
$-CO_2H$ (monomeric)	0	177
$-CO_2H$ (dimeric)	1	212
$-COF$	0	-
$-CF=CF_2$	0	-
$-CF_2H$	380	0
Specific melt viscosity (poises):		
Original	3.2×10^4	39×10^4
After 1 hr. exposure in air at 380°C.	3.2×10^4	150×10^4
Volatiles index	45	110

This table shows the increased stability both in respect to melt viscosity and in respect to volatiles obtained by the treatment with aqueous ammonia. The example was repeated with 1% solution of aqueous ammonia. Infrared analysis showed substantially complete removal of the carboxylate end groups.

Example 2: Into a 320 ml. stainless steel autoclave was charged 25 grams of a copolymer of tetrafluoroethylene and hexafluoropropylene, containing 14 to 16 weight percent of hexafluoropropylene, in the form of 75 grams of wet fluff, one gram of sodium hydroxide and 100 ml. of water. The autoclave was heated to 200°C. under autogenous pressure and agitated at that temperature for a period of one hour. Infrared spectrographic analysis of the resulting dried, end capped copolymer showed the disappearance of all of the carboxylate end groups, the original concentration of which was 177 carboxylate end groups in the monomeric form and 212 carboxylate end groups in the dimeric form per 10^6 carbon atoms in the polymer. The specific melt viscosity of the end capped copolymer remained constant on exposure to air at 380°C. for a period of one hour, whereas the specific melt viscosity of the uncapped copolymer increased fourfold.

Example 3: Into a 320 ml. stainless steel autoclave was charged 100 ml. of water and 25 grams of the copolymer of tetrafluoroethylene and hexafluoropropylene described in Example 2, in the form of 75 grams of wet fluff to which 500 ppm of sodium sulfate had been added. The copolymer was heated for one hour at 250°C. under autogenous pressure with agitation. The experiment was repeated and the polymer was heated for 8 hours. The following results were obtained.

	Infrared End Groups/10^6 Carbon Atoms				
	$-CO_2H$ Monomeric	$-CO_2H$ Dimeric	$-COF$	$-CF=CF_2$	$-CF_2H$
Untreated copolymer	177	212	-	-	-
Copolymer after 1 hr. end capping	84	29	1	32	172
Copolymer after 8 hr. end capping	0	18	0	4	296

Polytetrafluoroethylene

High Energy Ionizing Radiation Improves Melt Flow Properties

G.H. Bowers III; U.S. Patent 3,116,226; December 31, 1963; assigned to E.I. du Pont de Nemours and Company has found that cross-linked polymers of increased molecular weight, compared to the base polymers, can be obtained from fluorocarbon copolymers of the class consisting of copolymers of tetrafluoroethylene and fluoroolefins having the general structure:

$$CF_2=CXC_nF_{2n}Y \qquad \text{and} \qquad CF=CF-C_{n+1}F_{2(n+1)}$$

where n is an interger of 1 and above, X is of the class consisting of fluorine and perfluoroalkyl radicals, and Y is of the class consisting of hydrogen and fluorine, by a process which comprises subjecting the fluorocarbon polymer to high energy, ionizing radiation at a temperature above the glassy state transition temperature of the polymer, but below the thermal decomposition temperature of the polymer.

Contrary to the behavior of fluorocarbon homopolymers, such as polytetrafluoroethylene and polyhexafluoropropylene, it was found that fluorocarbon copolymers of tetrafluoroethylene and such comonomers as hexafluoropropylene, become cross-linked when subjected to high energy, ionizing radiation. However, it is essential that the irradiation of the fluorocarbon copolymer be carried out at the temperatures indicated. Irradiation of the copolymer below its glassy state transition temperature results in degradation. The copolymers which undergo this cross-linking reaction are copolymers of tetrafluoroethylene and substituted, terminally unsaturated perfluoroolefins or omega-hydrofluoroolefins.

A convenient method for practicing the process is to conduct the fluorocarbon copolymer in the form of shaped articles such as film, sheet, fiber, coated wire, hollow tube, or other shape, at a constant rate continuously through a beam of high energy, ionizing radiation. In most instances, the window of the radiation source is maintained at about 10 cm. from the specimen undergoing treatment, but this distance is not critical. The radiation energy imparted to the copolymer is measured in rads which corresponds to 100 ergs per gram of the copolymer. Employing a 2 mev electron beam from a Van de Graaff accelerator at an intensity of 250 microamperes and at a distance of 10 centimeters from the window, a sample will be exposed to 10^6 rads, if each point in the polymer is subjected to an exposure time of one second. A one second exposure can, of course, be obtained by coordinating the rate of exposure with the width of the beam. The total radiation of such a beam will be 12.5 watt-sec./cm.^{2}. This exposure of 12.5 watt-sec./cm.^{2} is also defined for the purposes of the following examples as a "standard pass."

By the term "specific melt viscosity," as used here, is meant the apparent melt viscosity as measured at 380°C. under a shear stress of 6.5 pounds per square inch. The values herein referred to are determined by using a melt indexer of the type described in American Society of Testing Materials, test D-1238-52T, modified for corrosion-resistance to embody a cylinder and orifice of Ampco aluminum bronze and a piston weighing 10 grams having a Stellite cobalt-chromium-tungsten alloy tip. The copolymer, nonirradiated or irradiated, is charged to the 0.375 inch ID cylinder which is held at 380°±0.5°C., allowed to come to an equilibrium temperature during 5 minutes, and extruded through the 0.0825 inch diameter, 0.315 inch long orifice under a piston loading of 5,000 grams. The specific melt viscosity in poises is calculated as 53,150 divided by the observed extrusion rate in grams per minute. This process is illustrated by the following examples.

Example 1:   A copolymer of tetrafluoroethylene and hexafluoropropylene containing 11.2 mol percent of hexafluoropropylene was compression molded into 6 inch by 6 inch sheets, 120 mils in thickness. The sheets were, one at a time, placed into an oven comprising a steel container having an open top heated from the bottom by a temperature-controlled hot plate. The top was covered with 1 mil aluminum foil and 1/4 inch Fiberglas insulation. The air in the oven was preheated to the desired temperature prior to entry into the oven. The fluorocarbon polymer sheet was given 15 minutes to come to the temperature of the oven, maintained at 300°C., and the assembly was then placed on a moving table and passed under the high energy electron beam provided by a Van de Graaff electron generator operating at 2 million electron volts and 250 microamperes current. The sample was subjected to the radiation indicated in the table below. The specific melt viscosity of the irradiated sample was measured by the method described above and compared to the specific melt viscosity of an untreated sample.

Sample	Radiation Dose in Rads	Specific Melt Viscosity in Poises x 10^{-4}
A	0	7.7
B	1×10^4	7.5
C	1×10^5	10.8
D	1×10^6	201
E	1×10^7	No flow

It should be noted that the reproducibility of the melt viscosity is ±5% and thus irradiation of the copolymer with 10^4 rads did not significantly change the structure of the polymer. The polymer, subjected to 10^7 rads, could be deformed at temperatures above the melting point of the linear polymer but recovered its original shape, characteristic of a cross-linked thermoplastic polymer wherein the cross-linked molecules are of high molecular weight.

Example 2:   Using the procedure described in Example 1, a copolymer of tetrafluoroethylene and hexafluoropropylene containing 11.2 mol percent of hexafluoropropylene was irradiated at the temperature and in the atmosphere indicated in the table below. The specific melt viscosity of the irradiated sample was measured and is compared to that of the starting material. The room temperature yield strength and yield elongation of the irradiated copolymer is also compared. Yield strength and elongation were determined by ASTM D-1457-56T.

Sample	Irradiation Temperature, °C.	Irradiation Dose, Rads x 10^{-6}	Irradiation Atmosphere	Specific Melt Viscosity, Poises x 10^{-4}	Yield Strength, psi	Yield Elongation, %
A	-	-	-	7.2	1,930	8.8
B	25	1	air	7.4	1,930	9.3
C	150	1	air	15	1,980	8.8
D	200	1	air	20	1,910	7.1
E	250	1	air	46	1,920	9.2
F	300	1	air	62	1,950	8.6
G	350	1	air	56	1,940	8.5
H	250	3	air	no flow	2,120	6.6
I	250	6	air	no flow	2,180	6.5
J	250	0.1	nitrogen	8.4	1,940	9.3
K	250	1	nitrogen	62.9	1,950	8.9
L	250	10	nitrogen	no flow	2,226	6.2
M	250	100	nitrogen	no flow	3,430	5.2

The table shows the effect of radiation temperature on the degree of cross-linking obtained with the same amount of irradiation, and, furthermore, shows the effect of irradiation beyond the no-flow stage indicating additional improvement. As can be seen, the mechanical properties of the copolymer are improved by irradiation. The effect becomes increasingly pronounced when the same properties are tested at elevated temperature.

Solid Amorphous Copolymers

H.S. Eleuterio; U.S. Patent 3,062,793; November 6, 1962; assigned to E.I. du Pont de Nemours and Company describes a process for preparing normally solid, normally amorphous interpolymers of hexafluoropropylene and tetrafluoroethylene. The interpolymers may be prepared by heating a mixture of 70 to 99 weight percent of hexafluoropropylene and 1 to 30 weight percent complementally of tetrafluoroethylene at a temperature in the range of 70° to 350°C. under a pressure of at least 2,000 atmospheres, preferably 2,000 to 10,000 atmospheres, in the presence of a polymerization initiator. This process is described by the following illustrative examples.

Example 1:   A stainless steel vessel of 120 parts water capacity was charged with 85 parts of perfluorodimethylcyclobutane, 0.15 part of bis(trifluoromethylthio)mercury, and 2 parts of tetrafluoroethylene. The charged was pressurized with hexafluoropropylene to fill the vessel with liquid at room temperature, and confined and heated at 200°C. for 4 hours. A pressure of about 3,000 atmospheres prevailed during the heating period. Upon cooling the charge and flashing off the unreacted monomer and solvent, there was obtained 5 parts of hexafluoropropylene/tetrafluoroethylene interpolymer.

A 2 mil thick film molded from the interpolymer at 250°C. and cooled to room temperature was limp, transparent, and completely amorphous as determined by x-ray. The film showed a specific IR ratio of about 17. The specific IR ratios reported are those determined by the procedure of Belgian Patent 560,454. The intensity of absorption of the film in the infrared at a wavelength of 10.18 microns, relative to the intensity of absorption of a similar film of a homopolymer of hexafluoropropylene prepared under the conditions of this example except that no tetrafluoroethylene was charged was 60%. The polymer was soluble in perfluorodimethylcyclobutane, and manifested an inherent viscosity of 0.2 as measured at 25°C. and 0.5% concentration (grams per 100 ml. of solution) in that solvent, calculated according to the usual equation:

$$\eta_{inh} = \frac{\ln V \text{ solution}/V \text{ solvent}}{0.5}$$

where η_{inh} is inherent viscosity, V solution is the viscosity of the solution at 25°C., and V solvent is the viscosity of the solvent at 25°C. In contrast to the foregoing an otherwise similar run in which the tetrafluoroethylene charge was 25 parts yielded an interpolymer having a specific IR ratio of 10.5, showing a crystallinity of 15% as determined by comparing the areas under the crystalline and amorphous peaks from x-ray analysis.

Example 2: The procedure of Example 1 was repeated except that the amount of tetrafluoroethylene charged was 6 parts and instead of the bis(perfluorodimethylthio)mercury there was charged 0.1 part of cobalt trifluoride. There was obtained 15 parts of interpolymer having a specific IR ratio of about 22, and an absorption relative to polyhexafluoropropylene at 10.18 microns of 90%. The interpolymer film was stiff, transparent, and completely amorphous. It was soluble in perfluorodimethylcyclobutane and showed an inherent viscosity of 0.4 as determined at 25°C. and 0.5% concentration in that solvent.

Similar results are obtained in a run similar except that the charge contains 20 parts oxygen per million parts of monomer charged, and no cobalt trifluoride. In contrast to the foregoing a charge of 0.068 part of bis(trichloroacetyl)peroxide, 32.2 parts of hexafluoropropylene and 3.6 parts of tetrafluoroethylene was charged at low temperature to polymerization vessel of 80 parts water capacity, warmed to –16°C., and maintained at that temperature under autogenous pressure (on the order of 2 atmospheres) until reaction ceased. There was obtained a solid interpolymer of hexafluoropropylene and tetrafluoroethylene which upon being pressed into film at 350°C. and allowed to cool in air to room temperature was brittle and manifested a crystallinity of about 50% as determined by x-ray analysis.

The amorphous interpolymers of this process are useful in most of the applications known for perfluorocarbon resins, and have in addition the feature of ready solubility in perfluorocarbon solvents, from which they may be cast into films or protective coatings, or spun into fibers. They may also be fabricated by molding and extrusion techniques at elevated temperatures. In line with their amorphous nature, articles made from the polymer show relatively slow steady changes in physical properties with change in temperature, in contrast to the behavior of crystalline polymers which tend to show sharp changes in physical properties with change in temperature.

Bis(Trifluoromethyl) Disulfide

M.I. Bro; U.S. Patent 3,023,196; February 27, 1962; assigned to E.I. du Pont de Nemours and Company describes a process for preparing perfluorocarbon copolymers which involves polymerizing mixtures of monoperfluoroolefins of 2 to 9 carbon atoms having a terminal perfluoromethylene group which mixtures contain 5 to 50 weight percent of tetrafluoroethylene (TFE).

The process comprises subjecting such mixtures to a temperature of at least 150°C. in the presence of an initiator of the general formula $[X(CF_2)_nYS]_2Z$, where X is hydrogen or halogen; Y is a single bond or methylene, Z is a single bond or mercury, and n is 1 to 12. In the preferred case, where the pressure is at least 200 atmospheres, the reaction proceeds at rates which are controllable, yet markedly faster than those previously achieved. Bis(trifluoromethyl) disulfide and bis(trifluoromethylthio) mercury are especially preferred for the preparation of high molecular weight resinous perfluorocarbon interpolymers.

This process is described by means of the following examples The term "specific melt viscosity" as used refers to the apparent melt viscosity as measured at 380°C. under a shear stress of 6.5 pounds per square inch. The term "specific IR ratio" refers to the quotient of the absorbance of light at a wavelength of 10.18 microns divided by the absorbance of light at a wavelength of 4.25 microns of a film of 2 mil thickness prepared by compression molding a sample of product at 350°C. and quench cooling the molten molding in ice water.

Indications from material balances and controlled decomposition studies are that the specific IR ratio, multiplied by 4.5, is equal to the weight percent of combined hexafluoropropylene in a TFE/hexafluoropropylene copolymer. The term "MIT flex life" refers to the number of flexes through an angle of 300° at 23°C. under a tension of 1,500 pounds per square inch that are withstood without failure by 5 mil thick film of polymer, prepared by shaping the product in the molten state, as measured on a Tinius-Olson folding endurance tester of the type developed at the Massachusetts Institute of Technology.

Example 1: A stainless steel tube of about 85 parts water capacity is flushed with nitrogen, evacuated, and charged with about 40 parts of perfluorodimethylcyclohexane, 15 parts of hexafluoropropylene, 15 parts of tetrafluoroethylene, and 0.2 part of bis(trifluoromethylthio) mercury. The vessel is sealed and heated to 200°C., at which temperature the pressure is about 300 atmospheres. At this point the temperature spontaneously increases to about 250°C. and the pressure to about 350 atmospheres.

The tube is then cooled to 200°C., at which temperature the pressure is 150 atmospheres. The temperature of the tube is then raised in steps of about 25°C., the temperature being held for about 20 minutes at each step, to a peak temperature of about 325°C. Upon cooling and discharging the tube and evaporating off the liquid component there is obtained 8 parts of dark grey solid polymer, having a specific melt viscosity of 2.34×10^5 poises, an MIT flex life of 9,096 cycles, and a specific IR ratio of 5.15. A strip of film molded from a sample of the product is somewhat elastic.

Example 2: A stainless steel tube of about 85 parts water capacity is flushed with nitrogen, evacuated, and charged with about 40 parts of perfluorodimethylcyclohexane, 30 parts of hexafluoropropylene, 5 parts of tetrafluoroethylene, and 0.2 part of bis(perfluoromethylthio) mercury. The vessel is sealed and heated to 180°C., at which point the temperature rises spontaneously to 205°C. and the pressure to 325 atmospheres. The tube is then maintained at 190° to 205°C. for two hours, during which time the pressure decreases to 150 atmospheres. Upon cooling and discharging the tube there is obtained 1.2 parts of solid polymer, having a specific IR ratio of 2.43.

COPOLYMERS — GENERAL, CURABLE

Fluorinated Alpha-Olefins

M.I. Bro; U.S. Patent 2,943,080; June 28, 1960; assigned to E.I. du Pont de Nemours and Company describes copolymers of tetrafluoroethylene and fluoroolefins containing from 0.1 to 5% by weight of the fluoroolefin in the polymer chain. It is in this range that the most significant decrease in melt viscosity is obtained. The melt viscosity of the copolymers of this process may be further decreased by introducing a larger proportion of the long chain fluorinated olefin into the copolymer. However, the further decrease in melt viscosity of the copolymer is significantly smaller as compared to the decrease obtained in the preferred range and not necessary to impart melt fabricability to the copolymer. Furthermore, increasing the proportion of the fluorinated olefin in the copolymer beyond the preferred concentration will cause a significant lowering of the high temperature properties of the copolymer and is for that reason not beneficial.

The fluorinated olefins employed in this process are terminally unsaturated perfluorinated olefins and terminally unsaturated fluorinated olefins containing one hydrogen in the end position to the double bond which have at least 4 and not more than 12 carbon atoms. It has been found that the presence of one hydrogen in the end position of the olefin does not seriously affect the inertness of the monomer during polymerization nor the inertness of the resulting copolymer. Olefins containing more than one hydrogen or containing other substituents act as chain-transfer agents in the polymerization and cause the formation of undesirable low molecular weight copolymers if present in large quantities.

Perfluorinated olefins employed in this process can be prepared by various methods, such as shown in U.S. Patent 2,668,864. The hydrogen-containing fluoroolefins are prepared by methods illustrated in U.S. Patents 2,559,628, 2,559,629 and 2,668,864. The fluoroolefins are prepared by polymerizing tetrafluoroethylene in the presence of methanol, which causes the formation of fluorine-containing alcohols. The fluorine-containing alcohols are oxidized to the acid, and the sodium salt of the acid is pyrolytically decarboxylated to result in the hydrogen-containing fluoroolefin.

The polymerization of tetrafluoroethylene with the fluorinated olefins may be carried out by various procedures. Thus, the polymerization may be carried out in an aqueous medium employing a peroxidic catalyst. The temperature in such a system may be varied from 25° to 200°C. and the pressure of the gaseous tetrafluoroethylene may be varied from atmospheric pressure to pressures exceeding 500 atmospheres. A preferred method employed for the copolymerization of tetrafluoroethylene with the fluorinated olefin comprises the polymerization of the monomers in an inert perfluorinated liquid diluent employing preferably a fluorinated peroxide as the catalyst. Perfluorinated liquid diluents suitable are perfluorinated straight chain, cyclic and branched hydrocarbons.

The ratio of the tetrafluoroethylene to the fluorinated olefin comonomer in the polymer is controlled primarily by the ratio of the two monomers in the feed and the polymerization temperature. The low reactivity of the fluoroolefin towards polymerization requires the presence of relatively large quantities of the comonomer in the feed. The reactivity of the comonomer varies with each individual comonomer employed, and generally decreases as the number of carbon atoms in the fluorinated olefin is increased. Because of the low reactivity of fluorinated olefins containing more than 12 carbons, they are less desirable as comonomers, since the reaction time required to obtain high molecular weight polymers is greatly increased. This process is illustrated by the following examples in which the fluorinated olefin content of the copolymer was determined by infrared spectrometry and substantiated by pyrolysis of the copolymer and analysis of the pyrolysis gases.

Example 1: Into a 330 ml. stainless steel pressure vessel was charged 25 ml. of perfluorocyclohexane, 20 grams of perfluoroheptene-1, 0.0037 gram of perfluorobutyryl peroxide in 0.25 ml. of perfluorodimethylcyclohexane and 20 grams of tetrafluoroethylene. The reaction vessel was agitated for two hours at autogenous pressure and maintained at a temperature of 35° to 60°C. On cooling, removal of excess monomer and filtering there was obtained 6.6 grams of solid polymeric material. The polymer was found to contain 0.8% of perfluoroheptene-1 with a melting point of 312° to 315°C., and a melt viscosity of 3.2×10^5 poises at 380°C. The copolymer could be melt extruded and molded into tough films.

Example 2: Into a 330 ml. stainless steel pressure vessel was charged 20 grams of perfluoroheptene-1, 5 grams of tetrafluoroethylene and 0.017 grams of oxygen. The reaction mixture was heated under autogenous pressure to 135°C. and held at that temperature for 5.5 hours. The reaction mixture was then heated to 150°C. and held at that temperature for 1.5 hours. On cooling and removal of excess monomer there was obtained 4.5 grams of copolymer. Tough films were prepared from the copolymer by compression molding at 380°C. Analysis of the copolymer indicated the presence of 1.4% perfluoroheptene-1 in the copolymer. The melt viscosity of the copolymer was measured to be 3.5×10^6 poises at 380°C.

Example 3: Into a 330 ml. stainless steel pressure vessel was charged 23 ml. of perfluorodimethylcyclohexane, 40 grams of perfluorononene-1, 15 grams of tetrafluoroethylene and 0.075 gram of perfluorobutyryl peroxide in 0.5 milliliter of perfluorodimethylcyclohexane. The reaction vessel was agitated at 60°C. for a period of two hours. On cooling and removal of excess monomer and solvent there was obtained 23.6 grams of a polymeric material. The polymer was found to contain 5.5% of perfluorononene-1. The melt viscosity of the copolymer was 1×10^6 poises at 380°C. The copolymer could be molded into tough films.

Example 4: Into a 330 ml. stainless steel pressure vessel was charged 23 ml. of perfluorodimethylcyclohexane, 40 grams of perfluoropentene-1, 15 grams of tetrafluoroethylene and 0.075 gram of perfluorobutyryl peroxide in 0.5 milliliter of perfluorodimethylcyclohexane. The reaction vessel was agitated at autogenous pressure for 1.25 hours at 35°C. and 1.25 hours at 60°C. On cooling and removing excess monomer and solvent there was obtained 2.7 grams of a solid polymer material. The polymer was found to contain 1.7% of perfluoropentene-1. The polymer was melt extruded and found to have a melt viscosity of 1.8×10^5 poises at 380°C. and a melting point of 308° to 316°C. The polymer could be molded into tough films which could be creased 1,758 times before a break occurred.

1,1-Dihydroperfluoroalkene-1 Compounds

A process described by B.F. Landrum and C. Sandberg; U.S. Patent 3,043,815; July 10, 1962; assigned to Minnesota Mining and Manufacturing Company involves copolymerizing tetrafluoroethylene with a 1,1-dihydroperfluoroalkene-1 having between 4 and 8 carbon atoms, preferably between 4 and 6 carbon atoms, to produce a copolymer containing a minor amount of the 1,1-dihydroperfluoroalkene-1. These copolymers of tetrafluoroethylene and 1,1-dihydroperfluoroalkene-1 retain the desirable properties of the tetrafluoroethylene homopolymer, are relatively high molecular weight thermoplastic resins and additionally are capable of being worked, as in molding and extrusion operations, by conventional techniques. Among the 1,1-dihydroperfluoroalkene-1 compounds are 1,1-dihydroperfluorobutene-1; 1,1-dihydro-perfluoropentene-1 and 1,1-dihydroperfluorohexene-1.

The copolymers contain monomer units corresponding to between 65 and 99 mol percent of tetrafluoroethylene and corresponding amounts of 1,1-dihydroperfluoroalkene-1. Those copolymers containing between 80 and 97 mol percent of tetrafluoroethylene display outstanding molding and extruding characteristics and have a T_m of between 250° and 300°C., in addition to the resistance to chemical and thermal attack usually associated with completely fluorinated polymers, such as tetrafluoroethylene homopolymer. The heat stability of these copolymers is excellent, even though the copolymer "backbone" or carbon chain is only partially fluorinated, i.e., contains hydrogen substituents.

Thus, the copolymers of this process retain their physical properties at temperatures as high as 500°F. and higher. Tensile strengths of the copolymers of this process range from 2,600 to 3,500 psi, and percent elongations fall in the 225 to 700% range, usually between 400 to 600%. Extrusion into filaments can be effected using conventional techniques, such as forcing the copolymer through a die at temperatures of 300° to 325°C. under a pressure of 200 psi. The following example illustrates this process.

Example: The following recipe was charged to a 300 ml. polymerization vessel:

$CF_2{=}CF_2$	38	grams (0.38 mol)
$CF_3CF_2CF{=}CH_2$	3.28	grams (0.02 mol)
$C_7F_{15}COONH_4$	1.50	grams
H_2O	150	grams
$(NH_4)_2S_2O_8$	0.75	gram
$(NH_4)_2B_4O_7$	0.75	gram

The vessel was sealed and reaction allowed to proceed for 5.4 hours at 50°C. under autogenous pressure, during which time the pressure in the vessel dropped from 590 to 180 psig. When no further drop in pressure occurred, the reaction was considered essentially complete. Upon opening the polymerization vessel, 183 grams of latex containing 17 weight percent solids was recovered and filtered to remove impurities and any small particles of precoagulated latex. The filtered latex was diluted with two volumes of water and coagulated by slow freezing. After the coagulate was recovered, it was washed three times with hot water, three times with hot methanol, and dried under vacuum (60 hours at 75°C.).

Approximately 30 grams of dry powdery copolymer was obtained which, after being pressed into clear sheets at temperatures between 540° and 600°F., was found to have a ZST at 300°C. of 745 seconds, a tensile strength of 2,785 psi, a percent elongation of 460%, and a T_m of 265° to 270°C. Upon heat aging the pressed sheets at 500°F. for 72 hours, the tensile strength was determined to be 2,065 psi and the percent elongation to be 220%.

A weight loss of only 0.25% was noted after aging the powdery copolymer product at 500°F. for 168 hours, and extended heat aging at 500°F., for 288 hours resulted in a weight loss of only 0.75%. The copolymeric product is readily moldable using conventional techniques. The monomer $CF_3CF_2CF{=}CH_2$ was prepared by the route as shown on the following page. This method is described in greater detail in JACS 77, 3149 (1955), by E.T. McBee, D.H. Campbell and C.W. Roberts.

Polytetrafluoroethylene

$$CF_3CF_2CF_2CH_2OH + HSO_3\text{—}\bigcirc\text{—}CH_3 \longrightarrow CF_3CF_2CF_2CH_2OSO_2\text{—}\bigcirc\text{—}CH_3 \xrightarrow{NaI} CF_3CF_2CF_2CH_2I$$

$$CF_3CF_2CF_2CH_2I \xrightarrow[\text{Acetic Acid}]{Zn} CF_3CF_2CF{=}CH_2$$

1,2,3,3,3-Pentafluoropropylene

D. Sianesi, G.C. Bernardi and G. Diotallevi; U.S. Patent 3,350,373; October 31, 1967; assigned to Montecatini Edison SpA, Italy describe a highly fluorinated thermoplastic copolymer consisting of polymerized monomeric units derived from tetrafluoroethylene and 1,2,3,3,3-pentafluoropropylene. This process for preparing the above copolymer comprises polymerizing the monomers at a temperature ranging from about –30° to 200°C. in a liquid medium and in the presence of a free radical polymerization initiator. The product 1,2,3,3,3-pentafluoropropylene can be obtained by means of a process described in the literature, see R.N. Haszeldine, B.R. Steele, Journal of Chemical Society, 1592, 1953. The following example illustrates this process.

Example: A stainless steel 2,500 cm.^{3} autoclave was provided with an anchor stirrer. After a vacuum had been produced in the autoclave, 1,000 cm.^{3} of deaerated water containing 5.20 grams of perfluorooctanoic acid, 1.05 grams of $NaHCO_3$ and 0.064 gram of $Na_2S_2O_5$ were introduced. From a cylinder, 540 grams of a mixture of tetrafluoroethylene and 1-hydroperfluoropropylene having a 1-hydroperfluoropropylene content of 26% by mols were introduced. The autoclave was heated to 60°C. and, by means of a compressor, first 500 cm.^{3} of water and then further 100 cm.^{3} of water containing 0.240 gram $(NH_4)_2S_2O_8$ were added.

During the course of the copolymerization, more water was added so as to keep a constant pressure of 48 atmospheres. The total amount of water added was 2,116 cm.^{3}. About 75 minutes after the introduction of the catalyst, the autoclave was cooled. The residual gases were collected quantitatively and analyzed. They contained 49% by mols of tetrafluoroethylene. The clear, aqueous solution discharged from the autoclave was admixed with HCl, thus bringing about coagulation of the copolymer, which was then washed and dried. About 315 grams of white copolymer was obtained which was calculated to contain 13% by weight of copolymerized 1,2,3,3,3-pentafluoropropylene.

The crystalline melting point of the product, determined by means of a polarizing microscope with heated plate, ranged from 285° to 289°C. A ratio of 1.14 between the optical densities at 10.20 and 4.20 microns appeared in the infrared absorption spectrum. The copolymer may be molded into colorless, transparent, flexible plates by operating at temperatures above 300°C. under a pressure of approximately 100 kg./cm.^{2}. The dynamometric properties were determined, as described in the case of polytetrafluoroethylene in ASTM test methods (1961) D-1457-56T, on specimens consisting of copolymer plates 0.5 mm. thick, obtained by molding the powder at 320°C. with rapid cooling. At room temperature and at a rate of stretching of 50 mm./min., they showed: tensile strength, 345 kg./cm.^{2} and elongation at break, 320%.

A copolymer specimen, maintained at a temperature of 350°C., subjected to a pressure of 14 kg./cm.^{2}, flowed through a cylindrical orifice 2.1 mm. in diameter and 8 mm. long at a constant rate of 8.5 grams per hour. A copolymer specimen was kept at a temperature of 300°C. for 5 hours in a light current of air. After treatment, the copolymer, which lost approximately 0.7% by weight, had an ultimate tensile strength of 290 kg./cm.^{2} and an elongation at break of 330%.

Another copolymer specimen was kept at 350°C. for 5 hours under a light nitrogen stream. After such treatment, the copolymer, which lost 0.8% by weight, had an ultimate tensile strength of 320 kg./cm.^{2} and an elongation at break of 325%. Under the same extrusion conditions described above, the treated copolymer flowed at a constant rate of approximately 20 grams/hour. On plates 0.5 mm. thick, obtained by molding the copolymer at 320°C., thermal resistance tests were carried by heating the plates for a given period in an air-circulating oven over different periods of time. The results are reported below.

	150°C., 8 hrs.	200°C., 8 hrs.	230°C.		250°C.	
			6 hrs.	16 hrs.	8 hrs.	16 hrs.
Ultimate tensile strength (kg./cm.^{2})	315	325	300	300	300	327
Elongation at break (%)	340	327	380	380	370	377

Hydroxylated TFE-Isobutylene Polymers

J.N. Coker; U.S. Patent 3,475,391; October 28, 1969; assigned to E.I. du Pont de Nemours and Company describes a process whereby tetrafluoroethylene/isobutylene polymers containing hydroxyl groups and approximately equimolar amounts of each monomer are obtained by contacting isobutylene and 0.5 to 2.5 mols of tetrafluoroethylene per mol of isobutylene, alone or together with small amounts of other monomers, e.g., acrylic acid, at 50° to 120°C. and 20 to 50 atmospheres, in agitated aqueous media containing an emulsifier, a persulfate initiator, a sulfite or bisulfite reducing agent, a copper accelerator, a phosphate regulator, and preferably a chain transfer agent. Carboxylated species are stabilized with cationic metal bases. Tetrafluoroethylene/isobutylene polymers and their preparation in aqueous tert-butanol with organic peroxide initiators, at space-time yields of up to about 150 grams per liter of reaction medium per hour are described in U.S. Patents 2,468,664 and 3,380,974.

The process is described by means of the following illustrative and comparative examples in which, except as otherwise stated, all parts and percentages are by weight; all water employed as a reaction medium is distilled demineralized deoxygenated water; all space-time yields are expressed in grams/hour/liter of water or water/alcohol reaction medium initially charged; all percentages of combined acrylic acid in the products are based on titration of hot perchloroethylene solutions to a permanent pink phenolphthalein end point with alcoholic NaOH; all mol percentages of combined tetrafluoroethylene in the products are based on analyses of the product for fluorine and for other elements or groups characteristic of any monomers used other than isobutylene and tetrafluoroethylene; all melt indexes and visual homogeneities are determined by the procedures hereinbefore mentioned; all curing rates are determined by measuring an original melt index (MI_o) and melt index (MI) after holding the sample at temperature in the indexer for time t equal to one hour, and expressed in reciprocal hours, as the value A in the equation $MI_o = MI^{-At}$; all glass transition and crystalline melting points are determined by differential thermal analysis; and all absorptions in the infrared are determined on molded films of the polymer.

Example 1: This comparative example summarizes results obtained with tert-butanol reaction media and organic peroxide initiator in a series of experiments. Into a nitrogen-filled water-steam jacketed stirred stainless steel autoclave having a water capacity of ca. 7,500 parts were charged 2,000 to 2,500 parts by volume (1,560 to 1,950 parts by weight) tert-butanol, 2,000 to 2,500 parts water, 2,100 to 2,500 total parts of tetrafluoroethylene/isobutylene mixture in mol ratio of from ca. 0.7:1 to 2:1, 4 to 5 parts of benzoyl peroxide and 0 to 3 parts ammonium perfluorocaprylate.

The autoclave was closed and heated with stirring to 80° to 85°C. at which point pressure reached 50 to 100 atmospheres. Onset of reaction occurred after 15 minutes. Water, alone or together with a total of up to about 72 parts of acrylic acid, was then pumped in as necessary to maintain pressure at a predetermined level in the range of 50 to 100 atmospheres, and heating and stirring continued for 2 to 8 hours, after which the reactor was cooled, the pressure released, and the thick gelatinous reaction mixture dipped out of the autoclave and manually scrubbed from the autoclave walls.

The product mixture was diluted with water, and the polymer filtered off and washed thoroughly with water and methanol. Space-time yields of 10 to 150 and usually about 50 grams/liter/hour were obtained. The polymers contained 57 to 66 weight percent combined tetrafluoroethylene and 0 to 3 weight percent of combined acrylic acid. The polymers had glass transition temperatures in the range of 15° to 55°C. and melting points in the range of 110° to 195°C. The carboxyl-containing polymers, manifested stable melt indexes in the range of 2 to 250. The carboxyl-free polymers had similar stable melt indexes and showed no absorption peak at 2.75 microns wavelength.

Example 2: This illustrative example shows the effect of using an aqueous copper-accelerated, phosphate-regulated persulfate-bisulfite redox system together with an emulsifier for isobutylene at a tetrafluoroethylene/isobutylene mol ratio of about 1:1, to produce a dipolymer. Into a nitrogen-filled, water-steam jacketed, horizontally stirred stainless steel autoclave having a working capacity of ca. 35,000 parts were charged, as a reaction medium, having a pH of 6 to 8, as follows:

	Parts
Water	19,000
Na_2HPO_4	292
$CF_3(CF_2)_6COONH_4$	192
$(NH_4)_2S_2O_8$	200
$NaHSO_3$	26.5
$CuSO_4$	0.0333

The charge was then stirred and heated to 50°C., pressured with (1) 550 to 750 parts isobutylene and (2) an approximately equimolar amount of TFE, and further stirred and heated to 80°C., at which point the pressure reached 27 to 32 atmospheres and, after 15 to 25 minutes, commenced to decrease, indicating the onset of reaction. The pressure was thereafter maintained at 27 to 32 atmospheres during 3 to 5 hours with continued heating and stirring by periodic injection of tetrafluoroethylene and isobutylene in approximately equimolar amounts, to complete the reaction, after which the stirring rate was decreased, the reactor contents cooled, and the pressure bled down to atmospheric. In a series of four runs, there were obtained stable fluid aqueous dispersions containing ca. 10 to 30 weight percent dispersed polymer solids, corresponding to space-time yields of 50 to 100 grams/liter/hour. The dispersion was readily drained so as to leave a clean reactor. The dispersed polymer was coagulated by heating with sodium chloride. The coagulated solids were filtered off, washed with water and methanol, and dried overnight in an air oven at 50°C. They contained ca. 59.5 weight percent combined tetrafluoroethylene, were undiscolored and visually homogeneous and showed crystalline melting points of 132° to 148°C., glass transition temperatures of 20° to 25°C., stable melt indexes in the range of ca. 50 to 360, and strong absorption at 2.75 microns wavelength.

Example 3: This example shows the effect of having tetrafluoroethylene and isobutylene present in mol ratio of about 2:1 at the onset of polymerization in producing a dipolymer. The procedure of Example 2 was repeated except that the system was initially pressured with a 2:1 mol ratio tetrafluoroethylene/isobutylene mixture, and after the onset of reaction, an 0.8:1 mol ratio of tetrafluoroethylene/isobutylene was used to maintain pressure and the reaction continued for 1 to 2 hours.

In a series of four runs there were obtained stable fluid aqueous dispersions containing 33 to 34 weight percent dispersed polymer solids, corresponding to space-time yields of 330 to 495 grams/liter/hour. The dispersions were readily drained leaving a clean reactor. The polymers were undiscolored, visually homogeneous and showed ca. 61 weight percent combined tetrafluoroethylene, crystalline melting points of 181° to 186°C., glass transition temperatures of 37° to 40°C., stable melt indexes in the range of 0.8 to 3.0, and strong absorption at 2.75 microns wavelength.

Curable Olefin Copolymers

W.R. Brasen and C.S. Cleaver; U.S. Patent 3,467,635; September 16, 1969; assigned to E.I. du Pont de Nemours and Company describe copolymers of tetrafluoroethylene units, olefin units and optional cure-site units, the molar ratio of the tetrafluoroethylene units to the olefin units being about 1:0.6 to 1.2. The olefin units are selected from (1) 50 to 100 mol percent propylene, butene-1 or mixtures together with from 0 to 50 mol percent ethylene or isobutylene, and (2) ethylene and isobutylene in about 1:1 molar proportion. The following examples illustrate this process.

Example 1: To a 400 ml. silver-lined pressure vessel is charged under a gaseous nitrogen blanket 200 ml. deaerated distilled water; 1.1 gram ammonium persulfate; 3.0 gram sodium phosphate dibasic heptahydrate, 0.25 gram sodium bisulfite; 0.15 gram ammonium perfluoro-n-octanoate; and 0.5 gram of tert-butyl acrylate. The vessel is closed, cooled to Dry Ice/acetone bath temperature and evacuated to less than 1 mm. mercury pressure. There is introduced 15.6 gram of propylene and 38.3 gram of tetrafluoroethylene.

Under agitation the vessel is heated to 60°C. and the pressure is gradually built up to 2,000 psig by injecting additional water. The polymerization is conducted at 60°C. under 2,000 psig pressure for 4 hours. The vessel is cooled to room temperature, and the latex discharged. The white polymer product is isolated by freeze-coagulating the latex, washing the solid polymer with water several times, and drying it overnight in a vacuum oven at 90°C. The polymer is soluble in 1,1,2-trichloro-1,2,2-trifluoroethane which is commercially available as Freon 113. Infrared spectral analysis shows a strong band at 5.8 microns, indicating the presence of tertiary butyl acrylate in the polymer.

A 1.00 gram sample of the polymer product is heated for 5 minutes at 160°C. to eliminate isobutylene and liberate carboxyl groups. The resulting acidic copolymer is dissolved in 80 ml. of Freon 113 and titrated with 0.1807 N alcoholic KOH to a phenolphthalein end-point. 2.0 ml. of the KOH solution is required, which corresponds to the presence of 1.57% by weight of units derived from acrylic acid in the heated copolymer of 2.79% by weight of units derived from tert-butyl acrylate in the original polymer product. The fluorine content of the polymer product is 49.9% by weight. From these two analyses, the composition of the polymer product is calculated as follows: 45.7 mol percent units derived from tetrafluoroethylene, 52.7 mol percent from propylene and 1.7 mol percent from tert-butyl acrylate.

Example 2: To a 400 ml. silver-lined pressure vessel is charged under a gaseous nitrogen blanket, 200 ml. deaerated distilled water; 1.1 gram ammonium persulfate; 3.0 gram sodium phosphate dibasic heptahydrate; 0.25 gram sodium bisulfite; and 0.15 gram ammonium perfluoro-n-octanoate. The vessel is closed, cooled to Dry Ice/acetone bath temperature, and evacuated to less than 1 mm. mercury pressure. There is introduced 49 grams of tetrafluoroethylene and 12 grams of propylene.

Under agitation, the vessel is heated to 60°C. and under pressure of 2,125 psig is polymerized for 4 hours. The vessel is cooled to room temperature and the product is isolated as described in Example 1. The yield of tetrafluoroethylene-propylene dipolymer is 43.3 grams. The fluorine content of the dipolymer is 57.3% by weight which corresponds to a dipolymer having the composition 56.0 mol percent units derived from tetrafluoroethylene and 44 mol percent from propylene. A sample of the dipolymer is compounded on a two-roll rubber mill according to the following weight proportions:

Polymer	100
Magnesium oxide	10
Metaphenylene-bismaleimide	3
Precipitated silica filler	15
Dibenzoyl peroxide	4

The compound stock is molded under pressure for 2 hours at 110°C. When tested at room temperature, the vulcanized compound has a tensile strength of 950 psi, an elongation at break of 310%, and a permanent set of 34%.

Example 3: This example illustrates the heat stability of a copolymer of this process. The procedure of Example 2 is repeated except that 50 grams of tetrafluoroethylene and 11 grams of propylene are used and the pressure is 1,800 to 2,500 psig. The dipolymer, isolated as described above, has an inherent viscosity of 1.0, measured in Freon 113 at a concentration of 1 gram per 100 ml. at 30°C. Its fluorine content is 53.3% by weight which corresponds to a dipolymer having a composition of 50 mol percent units derived from tetrafluoroethylene and 50 mol percent from propylene. This dipolymer, when heated in an air oven at 288°C. for 72 hours loses only 11% by weight.

Example 4: The procedure of Example 2 is followed except with the exceptions that (1) the monomers charged are tetra-fluoroethylene, ethylene and propylene, in amounts of 43.1, 2.8 and 10 grams, respectively (equivalent to a mol ratio of tetrafluoroethylene/propylene/ethylene of 1/0.58/0.23), and (2) the polymerization pressure is built-up by water injection to 1,800 to 2,000 psig. The yield of isolated terpolymer is 22 grams. It is insoluble in aliphatic and aromatic hydrocarbons, aliphatic ketones, methylene chloride, triethyl amine, and ethylene diamine.

A sample of the polymer is compounded and cured as described in Example 2. The vulcanized elastomer has a tensile strength of 1,850 psi, and elongation at break of 240%, and a permanent set of 10%. The inherent viscosity of the ter-polymer is 1.12 as measured above. Its fluorine content is 53.4% by weight, corresponding to a terpolymer having 70.2% by weight of units derived from tetrafluoroethylene and, taking the average molecular weight of the olefin components as 35, this indicates the terpolymer to contain about 45 mol percent of units derived from tetrafluoroethylene. The pro-cedure above is repeated except that the mol ratio of tetrafluoroethylene/propylene/ethylene charged to the shaker tube is 1/0.33/0.42. A hard, powdery terpolymer product is obtained.

Example 5: A 400 ml. stainless steel-lined shaker tube is charged with 0.3 gram of benzoyl peroxide, 1.0 gram of acrylic acid, 100 ml. of tert-butyl alcohol, and 100 ml. of distilled deoxygenated water. The tube is cooled in a Dry Ice/acetone bath, evacuated, and charged with 60 grams of tetrafluoroethylene and 30 grams of 1-butene. Water is injected into the shaker tube, and the tube is heated with shaking so that at 80°C. the internal pressure is 2,100 psi. Heating at 80°C. is continued for 12 hours and pressure is maintained at 1,900 to 2,100 psi by additional water injection as needed.

The tube and contents are cooled, volatile materials are vented off, and the tube is opened. The residual, solid, air-dried polymer is soluble in hot and cold tetrahydrofuran. The product contains 43.7% fluorine by weight and has a neutral equivalent of 1,710 as determined by titrating a 1% solution in hot tetrachloroethylene with 0.01 M methanolic sodium hydroxide using a phenolphthalein indicator. These analyses correspond to a tetrachloroethylene/1-butene/acrylic acid content (by mol percent) of about 43.6/51.9/4.5.

Methyl Vinyl Ether and 2-Chloroethyl Vinyl Ether

D.B. Pattison; U.S. Patent 3,306,879; February 28, 1967; assigned to E.I. du Pont de Nemours and Company describes a curable elastomeric copolymer comprising (a) about 45 to 55 mol percent of:

$$-CF_2-CF_2-$$

units, (b) about 54.8 to 35 mol percent of:

$$\begin{array}{c} -CH-CH_2- \\ | \\ OR \end{array}$$

units and (c) about 0.2 to 10 mol percent of:

$$\begin{array}{c} -CH-CH_2- \\ | \\ O-CH_2-CH_2-X \end{array}$$

units where R is a saturated aliphatic hydrocarbon radical containing 1 to 18 carbon atoms or a radical $R'OCH_2CH_2-$, where R' is a saturated aliphatic hydrocarbon radical containing 1 to 4 carbon atoms and X is a halogen radical of an atomic number of 17 to 53 (chlorine, bromine and iodine), the hydroxyl group or a radical of the structure:

$$\begin{array}{c} O \\ \| \\ -NH-C-Y \end{array}$$

where Y is hydrogen or a saturated aliphatic hydrocarbon radical containing 1 to 8 carbon atoms. The presence of the tetrafluoroethylene units ($-CF_2CF_2-$) is an essential feature of this process. Copolymers in which part or all of the tetrafluoroethylene units are replaced by units derived from other halogen-containing olefins, such as chlorotrifluoro-ethylene and hexafluoropropene, have poorer low-temperature properties and tend to be plastics at room temperature. It has been found that in copolymerizing tetrafluoroethylene with the comonomers providing the (b) and (c) units, the tetrafluoroethylene units are usually incorporated in an amount corresponding to approximately one mol of tetrafluoroethylene units to a total of one mol of the units (b) plus (c). The units (c) corresponding to the following formula:

$$-CH-CH_2-$$
$$|$$
$$OCH_2CH_2-X$$

are essential to provide cure sites. If less than 0.2 mol percent of these units is present in the copolymer, the copolymer cannot be satisfactorily cured. On the other hand, if more than 10 mol percent of these units is present, the cost of the copolymer is increased, the thermal stability is usually inferior, and the elongation at break of the cured vulcanized elastomers tends to be undesirably low. The preferred proportions of (c) units is in the range of about 0.5 to 4 mol percent.

Examples of suitable monomers giving the (b) units of the above definition are methyl vinyl ether, ethyl vinyl ether, propyl vinyl ether and isopropyl vinyl ether. The preferred monomer is methyl vinyl ether because the copolymers prepared with this monomer in general show the most desirable properties. When the polymer contains, as the (b) component, both methyl vinyl ether units and units of an alkyl vinyl ether in which the alkyl group contains at least three carbon atoms, the copolymer shows improved low-temperature properties with some sacrifice in high-temperature stability. The preferred molar proportion of these higher alkyl vinyl ether units in the polymer is about 5 to 20 mol percent, based on the total molar amount of the (b) component present in the copolymer.

Examples of monomers which give (c) units of the above definition are 2-chloroethyl vinyl ether, 2-bromoethyl vinyl ether, 2-iodoethyl vinyl ether, 2-formamidoethyl vinyl ether, 2-acetamidoethyl vinyl ether, 2-octanoylaminoethyl vinyl ether, and 2-hydroxyethyl vinyl ether. The preferred comonomers are the chloroethyl, bromoethyl and iodoethyl vinyl ethers, the chloroethyl vinyl ether being most preferred. The copolymers of this process are prepared by copolymerizing a mixture of monomers using known techniques.

The copolymers may be used in the uncured state, or they may be compounded, fabricated, and cured in the same way as known fluoroelastomers. The presence in the copolymer of the units defined as (c) above make them even more readily curable than the known fluoroelastomers. Particularly the copolymers containing the pendant chloroethoxy, the bromoethoxy, or the iodoethoxy groups may be vulcanized by a wide variety of agents developed for other halogen-containing elastomers, such as vinylidene fluoride-hexafluoropropene copolymers and the polymers of chloroprene (2-chloro-1,3-butadiene).

The curing rate in most recipes increases in the following order: chloro < bromo < iodo so that a shorter time in the press is required for the copolymers containing iodoethoxy groups, and the postcure to develop the optimum properties can be reduced or even eliminated for some applications. Examples of suitable curing agents are as follows: organic peroxide, such as benzoyl peroxide, and bis (α,α-dimethylbenzyl) peroxide; aliphatic polyamines, such as hexamethylenediamine, and tetraethylenepentamine; and derivatives of aliphatic polyamines such as ethylenediamine carbamate, hexamethylenediamine carbamate, and N,N'-bis(arylalkylidene) alkylenediamines. Typical details of various methods of curing the known fluoroelastomers are given in the following references: U.S. Patents 2,951,832, 2,958,672, 2,965,553, 3,008,916 and 3,011,995. The following publications may also be consulted: Report No. 58-3, Viton A and Viton A-HV, A.L. Moran and T.D. Eubank, Elastomer Chemicals Department, E.I. du Pont de Nemours and Co., May 1958; and Report No. 59-4, Viton B, A.L. Moran, Elastomer Chemicals Department, E.I. du Pont de Nemours and Co., October 1959.

The copolymers of this process have many uses. The cured copolymers are characterized by excellent oil resistance, good solvent resistance, good thermal stability, excellent resistance to ozone, excellent flame resistance, and good electrical properties.

The method used in the following examples, unless otherwise indicated, is as given below. Part of the reaction mixture, as indicated in the examples is placed in a 400-ml. stainless steel shaker tube under nitrogen. The tube is cooled in a mixture of Dry Ice and acetone (about -78°C.), and is purged four times with nitrogen and vacuum, alternately, to remove air, the remainder of the polymerization system is then added and the tube is closed. The shaker tube is heated to the desired temperature with rapid agitation and is maintained at this temperature for the times indicated.

The initial pressure at the operating temperature is usually 150 to 1,000 psig, and the pressure falls rapidly as the polymerization proceeds. The resultant latex is cooled in a mixture of Dry Ice and acetone and frozen solid and then warmed to room temperature, the supernatant fluid removed by decantation, and the polymer is washed repeatedly with water in a Waring blendor. The polymer is dried on a rubber mill at 100°C. Inherent viscosities are measured at 30°C. using a solution of 0.1 gram of polymer in 100 milliliters of a solution consisting of 86 parts by weight of tetrahydrofuran and 14 parts by weight N,N-dimethylformamide. Unless otherwise stated, the compounding recipe used in curing the polymers is as shown in the following table.

Polytetrafluoroethylene

	Parts by Weight
Polymer	100
Medium thermal carbon black	20
Magnesium oxide	15
Hexamethylenediamine carbamate	1.5

The material is compounded by blending on a rubber mill. Test pieces are cured in a mold in a press for one hour at 180°C. and are then postcured in an oven as indicated in the various examples. The properties of the polymer are determined as follows: The stress-strain data (modulus, tensile strength, elongation and permanent set) are obtained at 25°C. using dumbbells 2.5 inches long and 1/16 inch wide at the narrowest section which are cut from a slab 0.04 to 0.05 inch thick. An Instron tensile testing machine, Model TT-B (Instron Engineering Corp.), is used at a cross-head speed of 10 inches per minute. The method followed is ASTM D 412-51 T. Yerzley resilience, hardness, and compression set are measured using pellets 1/2 inch thick and 3/4 inch in diameter. The following methods are used:

Yerzley resilience	ASTM D 945-59.
Compression set	ASTM D 395-55, Method B, 70 hr. at 121°C.
Hardness, Shore A	ASTM D 676-59T.

In the tables shown below the following abbreviations are used:

M_{100} – Modulus at 100% elongation, psi.
T_B – Tensile strength at the break, psi.
E_B – Elongation at the break, %.

Example 1:   The initial charge added to the shaker tube is a mixture consisting of the following:

Potassium carbonate	4	grams
2,2'-azobis(2-methylpropionitrile)	0.1	gram
Tertiary butyl alcohol	156	grams
2-chloroethyl vinyl ether	1.06	grams

After cooling and removing air, the following are weighed in:

Methyl vinyl ether	28 grams
Tetrafluoroethylene	48 grams

The shaker tube is heated to 75°C. and is maintained at this temperature for 3 hours with rapid agitation. The maximum pressure attained is 160 psig. 70 grams of product is obtained, which is a strong, tough elastomer. The inherent viscosity is 1.01. Analyses show the following:

	% by Weight
F	44.9
C	36.5
H	4.0
Cl	0.65

The composition of the polymer is approximately 50 mol percent tetrafluoroethylene units, 49 mol percent methyl vinyl ether units, and 1 mol percent 2-chloroethyl vinyl ether units. The polymer is compounded using the following recipe:

	Parts by Weight
Polymer	100
Medium thermal carbon black	30
Magnesium oxide	15
Hexamethylenediamine carbamate	1.5

A slab is cured one hour at 170°C. in a press. The table which follows shows the vulcanizate properties of the polymer prepared in this example (Polymer A) compared with those of a copolymer prepared in the same way as Polymer A except that it contained equimolar amounts of units of tetrafluoroethylene and methyl vinyl ether and no 2-chloroethyl vinyl ether (Polymer B).

Polytetrafluoroethylene

	Polymer A	Polymer B
No Post Cure:		
M_{100}	460	290
T_B	1,120	320
E_B	450	>1,000
Permanent Set, %	45	400
After Post Cure for 2 Days at 204°C.:		
M_{100}	1,760	480
T_B	2,590	1,020
E_B	220	430
Permanent Set, %	14	30

Example 2: The initial charge added to the shaker tube is a mixture consisting of the following:

Potassium carbonate	4	grams
2,2'-azobis(2-methylpropionitrile)	0.1	gram
Tertiary butyl alcohol	200	grams
2-chloroethyl vinyl ether	6.4	grams

After cooling and removing air, the following are weighed in:

Methyl vinyl ether	26 grams
Tetrafluoroethylene	50 grams

The shaker tube is heated to 75°C. and is maintained at this temperature for 3 hours. The yield is 68 grams of a white, tough elastomer having an inherent viscosity of 1.02. Analyses show the following:

	% by Weight
F	47.2
C	37.5
H	4.5
Cl	2.7

The polymer composition is approximately 50 mol percent tetrafluoroethylene units, 44 mol percent methyl vinyl ether units, and 6 mol percent 2-chloroethyl vinyl ether units. The polymer is compounded using the following recipe:

	Parts by Weight
Polymer	100
Medium thermal carbon black	20
Magnesium oxide	15
Ethylenediamine carbamate	1.25

The compound is cured as described in Example 1. The properties are shown in the table below.

Postcure	None	1 Day at 204°C.	7 Days at 204°C.	1 Day at 177°C.	14 Days at 177°C.
M_{100}	450	1,200	1,700	810	630
T_B	1,920	2,750	2,760	2,280	2,520
E_B	350	180	150	200	210
Permanent Set, %	17	6	3	6	9
Hardness, Shore A	–	73	–	69	–
Yerzley Resilience, 25°C., %	–	21	–	13	–
Compression Set, % (70 hrs. at 121°C.)	–	8	–	25	–

Dehydrofluorination of Tetrafluoromethyl Vinyl Ether–TFE Copolymer

T.J. Kealy; U.S. Patent 3,299,019; January 17, 1967; assigned to E.I. du Pont de Nemours and Company describes curable, partially dehydrofluorinated copolymers of trifluoromethyl vinyl ether and tetrafluoroethylene. The curable polymeric composition exhibits infrared absorption bands at 3.25 microns and at 3.4 microns, the absorption band at 3.25 microns having an intensity of about 10 to 40% of the intensity of the absorption band at 3.4 microns, the polymeric composition being a partially dehydrofluorinated copolymer which before dehydrofluorination consists essentially of about 50 to 75 mol percent of trifluoromethyl vinyl ether units and about 25 to 50 mol percent of tetrafluoroethylene units. The absorption band at 3.25 microns is indicative of the presence of:

$$H—C≡C—$$

units in the polymer structure. The absorption band at 3.4 microns is characteristic of:

units in the structure. The polymers to be used as starting materials in preparing the compositions of this process are co-polymers containing about 50 to 75 mol percent of trifluoromethyl vinyl ether units:

$$(CF_3O—CH—CH_2—)$$

and about 25 to 50 mol percent of tetrafluoroethylene units ($—CF_2—CF_2—$). In this composition range copolymers of optimum elastomeric properties are obtained. The most preferred range is from about 60 to 70 mol percent of trifluoromethyl vinyl ether units and 30 to 40 mol percent of tetrafluoroethylene units. In general, the proportion of the units of the trifluoromethyl vinyl ether and the tetrafluoroethylene in the polymer is about the same as in the mixture of co-monomers used as starting materials.

The initial copolymer before dehydrofluorination is prepared from tetrafluoroethylene and trifluoromethyl vinyl ether. Trifluoromethyl vinyl ether is prepared by dehydrohalogenation of 2-chloroethyl trifluoromethyl ether or 2-bromoethyl trifluoromethyl ether by means of an alkali metal hydroxide. The dehydrohalogenation is conveniently carried out by contacting the 2-chloro (or 2-bromo-) ethyl trifluoromethyl ether with at least an equimolar quantity of an alkali metal hydroxide, e.g., potassium hydroxide. While the use of an inert reaction medium is not essential in this dehydrohalogenation process, it is preferred that one be employed. Suitable solvents are absolute ethyl alcohol or denatured alcohol as they dissolve the alkali metal hydroxide, and the dehydrohalogenation is conveniently carried out at the reflux temperature of the mixture.

The dehydrohalogenation takes place over a wide range of temperature, but temperatures of 50° to 80°C. are very satisfactory. Either 2-chloroethyl or 2-bromoethyl trifluoromethyl ether may be employed in this process. When 2-bromo-ethyl trifluoromethyl ether is employed, the trifluoromethyl vinyl ether is conveniently isolated from the reaction mixture by fractional distillation. When 2-chloroethyl trifluoromethyl is used, it is more difficult to separate the desired trifluoromethyl vinyl ether from the by-product vinyl chloride because of the closeness of their boiling points. Preparative gas chromatography may be used if pure trifluoromethyl vinyl ether is desired.

Copolymerization of trifluoromethyl vinyl ether with tetrafluoroethylene is accomplished by bulk, solution, or emulsion polymerization methods in the presence of initiators yielding free radicals. Examples of suitable polymerization initiators are nitrogen fluorides (U.S. Patent 2,963,468), azo compounds (U.S. Patent 2,471,959) and organic or inorganic peroxy compounds, such as benzoyl peroxide and the salts of persulfuric acid. Examples of suitable solvents for the polymerization are octafluoro-1,4-dithiane and the cyclic dimer of hexafluoropropene. Aqueous emulsion polymerization may be carried out using known techniques. The following examples illustrate this process.

Example 1:   A Hastelloy-C pressure vessel of 400 ml. capacity is charged with 0.7 part of ammonium persulfate, 2 parts of disodium hydrogen phosphate heptahydrate, 0.15 part of sodium sulfite, 0.3 part of ammonium perfluorooctanoate and 200 parts of deoxygenated water under a nitrogen atmosphere. The vessel is closed under nitrogen, cooled to about −80°C. and evacuated. 36 parts of trifluoromethyl vinyl ether and 15 parts of tetrafluoroethylene are distilled into the vessel, and the vessel is closed and maintained at 60°C. for 2 hours.

The vessel is cooled to room temperature and vented. The copolymer is precipitated by addition of an aqueous sodium chloride solution. The polymer is separated from the supernatant fluid, washed with water and with methanol, and dried in air. A 100% yield of polymer is obtained, which is a copolymer containing 68 mol percent of trifluoromethyl vinyl ether units and 32 mol percent of tetrafluoroethylene units. It has an inherent viscosity of 0.90, determined using a solution of 0.1 gram of polymer in 100 ml. of acetone at 30°C. The polymer is only 91% soluble in acetone.

To a solution of 45.4 parts of copolymer in 622 parts of tetrahydrofuran is added 45 parts of piperazine. The mixture is then refluxed for 6.5 hours. The reaction mixture, containing some precipitated solid, is then poured into water to precipitate the polymer. The copolymer is isolated by filtration, washed with fresh water, and dried at about 25°C. in a nitrogen atmosphere to obtain 41 parts of pale yellow copolymer having an inherent viscosity (0.1 gram in 100 ml. of acetone at 30°C.) of 1.01. The infrared spectrum of this copolymer contains a band at 3.25 microns and one at 3.40 microns. The band at 3.25 microns is not present in the spectrum of the copolymer before treatment with piperazine. The intensity of this band is 22.5% of the intensity of the band at 3.4 microns. Analysis for fluoride ion in the filtrate and washings from the copolymer treatment indicate that 0.1 mol of HF is removed from the polymer by amine treatment. The tetrafluoroethylene content of the polymer sample before treatment is 0.135 mol. The copolymer is compounded using the following recipe:

	Parts by Weight
Copolymer	100
Medium abrasion furnace black	30
2,5-bis(tert-butylperoxy)-2,5-dimethylhexane	1.5
N,N'-m-phenylenebismaleimide	0.75

The compounded polymer is cured in a mold under pressure at 165°C. for one hour. The following table shows the tensile properties obtained at 25°C. using an Instron tensile testing machine at a cross-head speed of 20 inches per minute. The method followed is ASTM D 412-51 T.

Tensile strength at the break, psi	1,720
Elongation at the break, %	260
Modulus at 100 elongation, psi	480
Modulus at 200% elongation, psi	1,295

Similar results are obtained when any one of the following is used instead of piperazine for treating the copolymer: morpholine, diethylamine, triethylamine, n-butylamine or ammonia.

Example 2: A platinum tube is charged with 1.5 parts of trifluoromethyl vinyl ether, 1.5 parts of tetrafluoroethylene, 2 parts by volume of gaseous dinitrogen difluoride (N_2F_2) and 2 parts by volume of the saturated dimer of hexafluoropropene (as solvent). The tube is sealed and heated at 75°C. for 4 hours under an external pressure of 100 atmospheres. There is obtained 2.18 parts of a copolymer which analyzes for 62.04% fluorine. This indicates that it is a copolymer containing about 53 mol percent of trifluoromethyl vinyl ether units and 47 mol percent of tetrafluoroethylene units. When the polymer is treated with piperazine as described in Example 1 there is obtained a copolymer which exhibits a band in the infrared spectrum at 3.25 microns which has an intensity 32% of the intensity of the band at 3.4 microns. This copolymer is compounded and cured as described in Example 1 and gives similar results to those of Example 1.

MOLDING COMPOUNDS

Pulverization for Sheeting Material

P.E. Thomas and C.C. Wallace, Jr.; U.S. Patent 2,936,301; May 10, 1960; assigned to E.I. du Pont de Nemours and Company describe improved polytetrafluoroethylene granular powder having, among other desirable features, special suitability for molding into thin nonporous sheeting. It is known to improve the levelability of polytetrafluoroethylene granular powder by comminuting it to finely divided form (British Patent 638,328). It has also been known to improve the strength of unsintered objects made of polytetrafluoroethylene granular powder by preliminarily shearing the powder (U.S. Patent 2,578,523). Previously however, if sufficient shear stresses have been applied to the powder to improve the strength of preforms molded from it, it has either been impossible to level the product with sufficient uniformity to produce a preform of adequate uniform density in thin sections, or else it has been impossible to cause the sheared powder particles to coalesce to soundly fused articles during the sintering operation.

It has been found that the above limitations can be overcome by polytetrafluoroethylene granular powder having an air-permeability subsieve size of less than 5 microns, a wet-sieve size of less than 50 microns with less than 6 weight percent retained on a 230 mesh screen, and a shape factor of 5 to 12. The powders have a low apparent (bulk) density, in the range of 100 to 300, preferably 150 to 200 grams per liter. As seen under the microscope, they consist of small discrete relatively nonporous particles, a substantial number of which are in fibrous form. In addition to the above characteristics, and the ease with which they may be loaded, leveled, preformed, handled and sintered, they manifest high anisotropic expansion when preformed and sintered, having an anisotropic expansion factor of 1.16 to 1.28, which is apparently a consequence of the high proportion of microfibrous particles they contain.

The term "polytetrafluoroethylene granular powder" is used here in the conventional sense to refer to polytetrafluoroethylene resin in the form of rough irregular particles of supercolloidal size, having a total surface area of from 1 to 4 square meters per gram as measured by nitrogen adsorption. This value corresponds to a theoretical average particle diameter of 0.67 to 2.67 microns on the assumption that all particles are spherical. Such powders may be obtained by contacting tetrafluoroethylene in the absence of organic additives with an agitated aqueous solution of an inorganic peroxide catalyst, as described, for example, in U.S. Patent 2,393,967, and are ordinarily employed in molding operations. They are to be distinguished from the "fine powders" obtainable by the coagulation of aqueous dispersions of colloidal polytetrafluoroethylene, which have a much higher total surface area, and are not suited for general molding applications.

The granular powders may be conventionally prepared by a process which comprises pulverizing ordinary raw finely divided polytetrafluoroethylene granular powder by means of an enclosed bladed rotor rotating at peripheral speeds of about 10,000 feet per minute, in a vortex of air or other gaseous medium maintained at temperatures in the range of 19° to 327°C., preferably above 25°C. and below 250°C., where, as a result of collision, abrasion, and other disruptive forces present, the relatively large, heavy discrete, porous particles are disrupted into smaller, lighter, relatively nonporous particles comprising a high proportion of elongated, or fibrous shapes, and then introducing the resultant fluidal stream of the ground particles into a classifier at a temperature not exceeding about 90°C., where the desired relatively light small nonporous particles are separated.

The proportion of elongated particles depends upon the temperature and upon the total shear stresses induced and their rate of application, and may be controlled by regulating inlet air temperature, the peripheral speed of the rotor blades, the number, design and clearance of the rotor blades, the rate of throughput, and other factors involved in the art of fluid grinding in a gaseous medium. Apparatus especially suitable for grinding and classifying to obtain the powders of this process is commercially available under the name Hurricane Mill as a product of the Microcylomat Company.

In one preferred preparative process, a commercial grade of unsintered polytetrafluoroethylene granular molding powder having an apparent (bulk) density of about 500 grams per liter and a wet-sieve size of 300 to 1,000 microns is pulverized and classified at ambient grinding temperatures in the range of 25° to 250°C. and classifier temperatures maintained below 90°C., in a V-18 Hurricane Mill. The combined apparatus comprises a vertically motor driven rotor shaft positioned at the axis of a cylindrical housing 18 inches in diameter, the lower two-thirds of which houses five superposed grinding stages partially separated by horizontal discs, each stage containing flat vertically disposed blades mounted radially on a disc extending from the rotor shaft and adapted to clear the confining walls of the housing at least 1/8 and preferably about 1/2 inch, and the upper third of which houses a classifier comprising superposed horizontal centrally apertured discs 1/4 inch thick and 1/4 inch apart mounted on a support extending from the rotor shaft.

The stacked discs are surmounted in turn by (a) coarse particle entraining dispersing and recycling fan means comprising vertically disposed vanes mounted on the top classifier disc at an angle to its radius, (b) a plenum chamber partially separated from said fan means (a) by a stationary disc extending inwardly from the housing, and communicating with an air intake in the housing wall, and (c) fine particle entraining dispersing and discharging fan means similar to but of greater capacity than fan means (a), the vertically disposed blades being mounted on and dependent from a disc extending from the rotor shaft.

In operation, the powder is screw fed to the grinding section at about 50 pounds per hour, with the rotor shaft revolving at about 3,600 rpm, air entering the plenum chamber at about 1,200 cubic feet per minute, and air recycling through the recycle lines and pulverizer at about 250 cubic feet per minute. The powder fed is entrained by the recycle air stream and carried upwardly through the grinding stages to an exit communicating with the space adjacent the external periphery of the classifier plates, where the finer particles are entrained and drawn inwardly between the classifier plates, thence upwardly through the central aperture surrounding the rotor shaft, and finally to discharge through fan means (c) to a cyclone separator, and the coarser particles are entrained and drawn upwardly and outwardly through the recycle fan means (a) to return lines, then back to the lowest grinding stage for further grinding.

Adjustments are made to recycle air rate as necessary to maintain adequate pneumatic conveying back to and through the grinding stages, and powder feed rates are adjusted to maintain power input to the driving motor at 38 to 40 kilowatts. Discharge air pressure is held below 18 inches of water, classifier air inlet pressure below -2.5 inches of water, and recycle air pressure above -2 inches of water, by means of suitable damper and louver curtain controls. Discharge temperature is held below 90°C., by forced cooling of inlet air if necessary. The product obtained is a fluffy powder having an apparent (bulk) density of 100 to 300, preferably 150 to 200 grams per liter; an air permeability subsieve size of less than 5, preferably 2.8 to 4 microns, a wet-sieve size of less than 50, preferably 20 to 40 microns, with less than 6 weight percent of coarse particles retained on the 230 mesh sieve, and a shape factor of 5 to 12, preferably 8 to 10.

The anisotropic expansion factor is 1.16 to 1.28; 1.19 to 1.26 for the preferred powders. Upon microscopic examination of 500 diameters magnification or more the product appears to consist of small irregular relatively nonporous particles, a substantial proportion of which are in the form of short fibers having a length to diameter ratio of 5 or more. Distribution of particle sizes is fairly broad and uniform. The powder is readily fabricated by molding and sintering into flawless nonporous sheets in the thicknesses as low as 25 mils in sizes as large as 48 x 48 inches or more.

In a preferred procedure for making large molded sheets with the composition, the powder (after being gently sifted through a 6 to 12 mesh screen) is charged into a suitable deep mold cavity defined by a chase closely fitted over a bottom plate

Polytetrafluoroethylene

supported on a press base. In loading, care is taken to avoid piling in depths markedly greater than the height of the mold cavity. Uniform distribution of the powder is facilitated by its low bulb density, which affords greater working depth of bed. Leveling is accomplished by gentle front-to-back and side-to-side movements of a straight edge supported on the walls of the chase. The leveled charge is covered first with a 2 to 3 mil thick layer of film or foil, e.g., of polyethylene terephthalate, polyethylene or aluminum, then with an elastic pressure distribution sheet, e.g., natural gum rubber, 1/8 inch thick, of Shore hardness A-32, and Shore elasticity 88, cut to fit inside the mold cavity with about 1/16 inch clearance on all sides, and finally with a top plate of metal.

Pressure is then built up gradually during about 2 minutes and held during about 1 minute. Because of the greater coalescibility of powders, a preform of density as low as 1.81 can be sintered to the final density of at least 2.14, which is the minimum for nonporous sheeting. Such preform densities can be achieved at preforming pressures of as low as 500 pounds per square inch, with these powders. Compensation for minor loading and leveling errors can be made by using higher preforming pressures, up to 4,000 pounds per square inch, but pressures above 4,000 should not be used since they tend to give rise to the development of unhealable shear faults in the preform. Pressures of about 1,000 pounds per square inch are preferred.

In contrast, conventional powders must be compressed to a density of at least 2.00 in order to be capable of sintering to densities as high as 2.14 and therefore, ordinarily require preforming pressures of at least 2,000 pounds per square inch. The effect of preforming pressure and preform density on final density is illustrated in the following table for conventional granular molding powder and the new fluid-ground powder made from this process. The comparisons are based on 4 by 5 by 1/16 inch sheets, sintered 60 minutes at 370°C. and cooled to room temperature at 4°C. per minute.

Preforming Pressure, psi		Preform Density, g./cm.^{3}		Sintered Density, g./cm.^{3}	
Conventional	This Process	Conventional	This Process	Conventional	This Process
600	500	1.60	1.81	1.88	2.17
1,000	1,000	1.77	1.92	2.00	2.17
2,000	2,000	2.00	2.04	2.14	2.17
4,000	4,000	2.15	2.15	2.15	2.17

Thus, with given press equipment, the powder makes it possible to produce sheeting two to four times as large as was possible with conventional powders. In addition, preforms made of these powders at 1,000 pounds per square inch are roughly three times as strong as preforms made of conventional powders at 2,000 pounds per square inch, and hence are much more readily handled. For example, a 48 x 48 x 1/16 inch preform can be bowed 18 inches without developing cracks, whereas a similar preform of conventional powder cracks so as to yield a flawed sintered article when bowed 2 to 3 inches.

Choice of sintering and cooling conditions depends upon thickness, and upon the properties it is desired to emphasize in the final product. For high strength and toughness, preferred conditions for 25 to 60 mil sheeting involve sintering the preforms on flat plates at 380° to 390°C. for 60 minutes and then transferring them to a press where they are "coined" and cooled between cold plates under a pressure of about 1,000 pounds per square inch. For high resistance to permeation, preferred conditions for 25 to 60 mil sheeting involve sintering at 400°C. for 90 minutes, cooling rapidly to 340°C. further cooling from 340° to below 300°C. at less than 2°C. per minute, and further cooling to room temperature at any convenient rate while maintaining a uniform environmental temperature to avoid uneven contraction. Typical properties of 48 x 48 x 1/32 to 1/16 inch sheeting made via these coining and slow cooling procedures are shown in the following table.

	Coined	Slow Cooled
Density	2.14	2.18
Tensile strength	4,500	3,900
Flex life	3×10^6	10^5
Dielectric strength	610*	610*
CO_2 permeability	1×10^{-13}	3×10^{-16}

*No value below 400.

The powders are not only sinterable into nonporous sheeting in shorter times than conventional powders, but also may be sintered into higher quality sheeting than could previously be produced, as indicated in the following table, which compares the properties of sheeting obtained from the powder of this process and conventional powder under identical fabrication conditions. In the table, results are shown for 4 x 5 x 1/16 inch sheets, preformed at 2,000 psi, sintered at 375°C. for 90 minutes, and cooled at 2°C. per minute. Test methods and units are those previously described.

Polytetrafluoroethylene

	Conventional	This Process
Density	2.17	2.18
Tensile strength	2,700	4,400
Dielectric strength	425	625
N_2 permeability	1×10^{-13}	3×10^{-14}
Flex life	4×10^4	5×10^5

This process thus provides a polytetrafluoroethylene granular powder which can be readily loaded, leveled and compressed at low pressure to yield preforms of markedly improved strength and handleability which, in turn, can be readily sintered into large thin nonporous sheeting.

Water Cut Powder Techniques

P.E. Thomas; U.S. Patent 3,010,950; November 28, 1961; assigned to E.I. du Pont de Nemours and Company describes a polytetrafluoroethylene fine powder having special suitability for sheet calendering and paste molding operations. The polytetrafluoroethylene fine powder of high crystallinity contains a major proportion of submicroscopic bola-shaped particles, the powder having an infrared amorphous index of less than 0.15 and a negative 83/17 methanol-water wettability.

On examination by means of the electron microscope at 10,000 diameters magnification, the powders of this process appear mainly as a complex network of roughly spherical particles, generally ranging from 0.1 to 0.3 micron in diameter, at least half of which are connected to at least one other such particle by means of at least one elongated microfibrous portion, i.e., an elongated shape, generally of a diameter ranging from 0.01 to 0.05 micron and of a length ranging from 0.2 to 6 microns. The term "bola-shaped" is used as descriptive of these interconnected submicroscopic particles.

The term "infrared amorphous index," as used here, refers to a parameter which may be determined by measuring the absorbance of film, pressed from the powder, at a wavelength of 4.25 microns, and at a wavelength of 12.8 microns, and comparing the relative absorbances at these peaks. In a preferred test method, a 0.06 gram sample of powder is pressed at 5,000 pounds force in a chip mold having an area of about 2 square centimeters. The resulting thin disc is scanned at wavelengths in the ranges of 4 to 5 and 12 to 14 microns. It has been found from x-ray correlations that the peak absorption bands in the 4 to 5 micron range are fairly constant in polytetrafluoroethylene over the usual range of crystalline content, while the peak absorption bands in the 12 to 14 micron range are characteristic of amorphous material. The infrared amorphous index is calculated as the ratio of net absorbances at the peaks occurring at 12.8 and 4.25 microns, i.e., $A_{12.8}/A_{4.25}$, and is then, indirectly, a relative measure of crystalline content.

The term "83/17 methanol-water wettability" refers to a characteristic which apparently depends on the proportion of deformed particles contained in the powder. The presence or absence of the characteristic may be readily determined by dropping a 2 gram sample of powder on the surface of 10 ml. of a solution consisting of 83.0 weight percent absolute methanol and 17.0 weight percent distilled water in a 15 ml. graduated cylinder, and shaking the mixture gently for about 30 seconds. If the test is negative, the powder particles do not darken appreciably, and ordinarily remain floating on the surface of the liquid, indicating that they have not been wetted. However, if the test is positive, the particles turn grayish and sink to the bottom of the cylinder, indicating that they have been wet by the liquid. In carrying out the test, the proportions of methanol and water are quite critical, since the test is not reliable if the methanol-water percentages vary as much as 1%.

The powders may be prepared from suitable ordinary fine powders by a "water-cutting" procedure which in general, comprises slurrying such powders in a liquid medium, preferably water, and subjecting the slurry to the action of a high-speed bladed cutter until a major proportion of the polymer particles are deformed into the bola-shaped particles previously described. Pigments or fillers may be incorporated in the dispersion, or the slurry, or the final powders as desired. The following example illustrates this process.

Example: (A) An aqueous colloidal dispersion of polytetrafluoroethylene prepared by the procedure described in Example 1, runs D to K, of U.S. Patent 2,750,350, was used as a starting material. The dispersion was diluted, coagulated and dried by the process described in Example 2 of U.S. Patent 2,593,583 to obtain a fine powder. The powder obtained was typical of commercial fine powders, having a positive 83/17 methanol-water wettability, an infrared amorphous index of about 0.09, and consisting almost entirely of generally spherical particles averaging 0.1 to 0.3 micron in diameter, with only a few elongated or fibrous particle portions, i.e., about one elongated or fibrous particle portion to every 4 to 15 nonelongated particles.

(B) In a test of direct calenderability, the powder prepared by the procedure of paragraph (A) was hand-fed from above the nip of horizontally coplanar, X-alloy-surface, 6-inch diameter rolls, set about 12 mils apart, to yield a preformed product about 12 inches wide and 25 mils thick, at a rate of about one foot per minute. At a distance of about 2 feet from the bottom of the rolls, the product was picked up by an endless wire belt and thereby conveyed to and through an enclosed horizontal sintering oven about 10 feet long, heated by means of 30 end-to-end pairs of 2-kilowatt Chromalox infrared heaters, spaced along the oven length above and below the path of travel of the calendered product. The heaters were controlled by 10 powerstats adjusted so that the radiation intensity increased gradually to a maximum about 3/4 of the way through the oven, and decreased gradually thereafter. Peak temperature within the oven was 380° to 400°C. The product obtained contained visible cracks and flaws.

(C) In a modified calendering procedure, a compression plate of spring steel about one foot wide and 20 mils thick was covered first with a 20 mil thick felt of polyethylene terephthalate, and then with a 3 mil thick film of polyethylene terephthalate. Edge strips of synthetic rubber 1/8 inch thick and 1/14 inch wide were laid down to define a cavity which was filled with the powder of paragraph (A). The powder was then covered with upper sheets of film, felt and spring steel to form a sandwich, which was passed between a pair of vertically coplanar 6-inch diameter rolls set about 115 mils apart. On emergence of the sandwich the protective layers were carefully peeled away from the preformed sheet, which was transferred as peeled onto the conveyor belt and led to and through the sintering oven as described in paragraph (B). A 20-foot long nonporous sheet 12 inches wide and 35 mils thick was obtained via this semicontinuous process, which however, was considerably more laborious and time-consuming than was the direct calendering process previously described.

(D) In contrast to the foregoing, the water-cut powder of this process was found to be calenderable directly into nonporous sheeting by the process of paragraph (B). In preparing the water-cut powder, one part of a powder prepared as in paragraph (A) above was slurried with 8 parts of water in a cylindrical stainless steel vessel of 24 parts water capacity, and then water-cut for 10 minutes with a sharp-bladed propeller stirrer, mounted on a shaft driver from above the vessel. The propeller blades were of the Waring Blendor type, pitched upward at an angle of about 45° to the horizontal, positioned and sized to clear the retaining walls and bottom of the container vessel by 1 to 2 inches. The peripheral velocity of the blades was about 5,300 feet per minute.

After the treatment, the water-cut powder was filtered off onto cheese cloth, spread onto trays, and dried 12 hours in a circulating air oven. The resulting powder, which is typical of the preferred powders of this process, manifested a negative 83/17 methanol-water wettability, an infrared amorphous index of 0.10, and consisted mainly of the submicroscopic bola-shaped particles previously described. In separate tests, it was found that the powder when lubricated was no longer extrudable in the conventional sense, i.e., it did not yield a flawless extrudate when lubricated, preformed into a billet, and then forced through an orifice having a cross-sectional area less than about 1/30 of the cross-sectional area of the billet.

(E) In a number of direct calendering runs via the procedure of paragraph (D), nonporous films and sheetings one foot wide ranging in thickness from 25 to 125 mils were continuously prepared. In general, the final thickness of the sintered sheets in mils was about 12 plus 1.2 times the gap in mils between the calendering rolls. The densities of the unsintered calendered preforms ranged from about 1.60 at 125 mils to about 1.90 at 20 mils, as compared with the 1.54 minimum calendered density which appears necessary for the preparation of adequately dense sintered articles from fine powder. They were markedly less fragile, judging by qualitative bending tests, than calendered products obtained with conventional fine powder. The density of the sintered final products ranged from 2.18 to 2.22. They were free of holes as judged by high voltage spark test, with a short time dielectric strength of about 660 volts per mil, and a CO_2 permeability rate of 2.5×10^{-14} to 4.5×10^{-14} in the units previously set forth.

Permissible rate of production seemed limited only by the length of the sintering oven. In general, lowest permeability values were obtained with gently increased, extended heating, and gently decreased, extended cooling times, the extended gradual heating and cooling being particularly provided when the product was being taken through the 300° to 327°C. and 327° to 300°C. ranges. Maximum toughness, on the other hand, was achieved with gradual heating and quench cooling. Although it was also found that superior quality tape and film thinner than 20 mils could be calendered from the powders preferably using smaller diameter rolls heated to temperatures ranging from 100° to 300°C. for the lower thicknesses, the powders were considerably more difficult to feed to the rolls than (and in this were further distinguished from) ordinary fine powders. Accordingly, for products thinner than about 25 mils, the powders were preferable to the ordinary materials only in special circumstances, where higher than usual quality justified the extra handling expense.

W.P. Weisenberger; U.S. Patent 3,115,486; December 24, 1963; assigned to E.I. du Pont de Nemours and Company describes a method for producing a free-flowing polytetrafluoroethylene molding powder especially suitable for machine feeding. In this process, a polytetrafluoroethylene molding powder is prepared by polymerizing tetrafluoroethylene in an aqueous system using a free-radical initiator, water cutting the polytetrafluoroethylene at a temperature of 0° to 30°C., until the average weight particle size is from 200 to 700 microns, water washing the polymer at a temperature of 40° to 70°C., dewatering the polytetrafluoroethylene by passing the polymer-water slurry over a vibrating screen and drying the polymer powder by passing heated air through a bed of the dewatered polymer or by suspending the polymer powder in a stream of heated air. The following example illustrates this process.

Example: An aqueous polytetrafluoroethylene slurry, such as produced by the process described in U.S. Patent 2,394,243, was passed through a vibrating screen to remove adhesion polymer, and sufficient water was added until a ratio of polymer to water of 1:15 was obtained. The resulting slurry was passed through a cutting system comprising a commercially available water cutter (Fitzmill model K-14 cutter) equipped with a 1/16 inch screen. The cutter consists of a series of thick sharpened blades which rotate on a horizontal shaft at 1,450 revolutions per minute. Polymer is pumped as a slurry into the top of the mill and passes through the cutting zone and out of the cutter when it has been reduced enough in size to pass through a screen at the lower end of the mill. The slurry was dewatered by passing over a vibrating 100 mesh screen, thereby separating impurities contained in the water from the polymer.

The polymer was then again mixed with fresh water until a slurry concentration of water to polymer of 8:1 was obtained. The slurry was heated to a temperature of 50°C. and the polymer washed for a period of 8 hours while being agitated with a paddle agitator using a power of 10 hp./1,000 gal. The resulting slurry was dewatered and oversized agglomerates were removed. The polymer was then dried by suspending the polymer in a stream of hot air at a temperature of 120° to 180°C. moving at a linear velocity of 2,000 to 2,500 ft./min. Using substantially the same conditions, but washing the polymer at 20° to 25°C., a second batch of polymer was prepared.

The resulting polymer molding powders were tested in a Denison automatic preforming press, commercially available, by molding rings having a two inch outer diameter, a wall thickness of 1/16 inch and a height of 0.15 inch. In the operation of the Denison automatic preforming press, 15 lbs. of the molding powder are placed into the hopper of the press which feeds the polymer powder into a shuttle. The shuttle, containing enough polymer for a large number of molding shots, moves over the die comprising an annular cavity having an outer diameter of 2 inches, a wall thickness of 0.062 inch, and a depth of about 0.5 inch, which is then filled by the polymer in the shuttle.

The shuttle retracts underneath the hopper to be replenished by additional polymer powder in the hopper while an annular ram descends into the cavity preforming the polymer into the ring using a pressure of about 15,000 psi and a dwell time of 10 seconds. The ram then retracts and the preformed piece is ejected from the die to the level of the shuttle and as the shuttle moves forward, the preformed ring is moved aside by the forward moving shuttle, while the bottom of the mold retracts and the shuttle moves over the die. The shuttle then refills the die and the cycle is repeated. A total of 300 rings were made with each batch of polymer.

Every fifth ring was sintered at a temperature of 375°C. for one hour and cooled at a rate of 3°C. per minute after the sintering step. Ten rings were selected at equal intervals during the run and the tensile strength and ultimate elongation of these rings were determined. The height of each of the sintered rings was measured by measuring the height around the ring at three places and calculating the average height. From these measurements, the overall height variation was determined. The following results were obtained.

Polymer Powder	Particle Size, microns	Tensile Strength, psi[1]		Ultimate Elongation, psi[1]		Overall Height Variation, mils
		Average	Range	Average	Range	
Washed at 20° to 25°C.	490 – 700	1,320	782 – 1,440	4.5	2 – 7	16
Washed at 50°C.	415 – 625	2,220	2,060 – 2,530	140	75 – 180	7.5

[1]As measured on an Instron using a rate of 2 inches per minute.

It is apparent from the above data that the elevated temperature washing step results in a molding powder having substantially improved mechanical properties when fabricated in automatic feeding and preforming machines and also in a molding powder of substantially improved powder flow, as witnessed by the decrease in the height variation. The improvement in the mechanical properties and reproducibility of the dimension is the result of the improved powder flow of the molding powder produced by this process.

High Density Powder for Machine Feeding

A.L. Mathews, Jr. and R. Roberts; U.S. Patent 3,087,921; April 30, 1963; assigned to E.I. du Pont de Nemours and Company describe a process which comprises heating unsintered polytetrafluoroethylene powder to a temperature of 50° to 300°C. and preferably to a temperature of 60° to 200°C., compacting the polymer to a density of at least 2.15 grams per cc, and preferably to a density of 2.15 to 2.23 grams/cc and cooling and comminuting the compacted polymer. The density of the polymer obtained in compaction will depend on the temperature of the polymer, the pressure applied and the time for which the pressure is applied, increasing with increase of either of these variables.

At a temperature of 60°C. the minimum pressure required to achieve the critical density is 2,000 psi. The time for which the pressure is employed should be sufficient to relax the compressed polymer particles. Such time will decrease with increasing temperatures but should be at least one minute and preferably five minutes. The increase in density obtained by a further increase in pressure or time is small compared to the increase obtainable with an increase in temperature. At higher temperatures, such as 150° to 250°C., the critical density may be obtained with lower pressures and in shorter times. In the range of 150° to 250°C. preferred densities of above 2,200 are obtained with pressures of 1,000 to 3,000 psi. The preform obtained on the compacting step is extremely brittle and is readily shattered by impact and can consequently be comminuted into a free-flowing, high apparent density molding powder. The formation of the brittle preform and its ability to be comminuted into a free-flowing powder only occurs at the elevated temperatures. Molding powder prepared from comminution of polymer compacted at or below room temperature does not have the free-flowing characteristics of compacted polymer powder prepared at elevated temperatures.

This process is applicable to both dispersion and granular polytetrafluoroethylene, as a matter of fact slightly wet polymer as obtained directly on dewatering of the polymerization mixture can even be compacted to give rise to the high density preform which on comminution becomes a free-flowing molding powder. The disintegration or cutting of the polymer preform may be carried out in various ways. The polymer can be disintegrated by water cutting in which the polymer pieces are cut in an aqueous medium by a rapidly rotating blade. Other methods involve putting the polymer through commercially available dry disintegrators or cutters.

The compacted polymer obtained by this process is generally cut into a size which has an average diameter of smaller than 1,000 microns and preferably from 200 to 500 microns. Particles having these diameters are preferred since they are more readily compressed to preforms and on sintering give rise to structures which are substantially free of voids. The polymer particles although of irregular shape have relatively smooth surfaces. Polymer compressed at room temperature and disintegrated in a similar way does not have the same shape, but has rough irregular surface and is frequently sheared and fibrillated.

The polymer powders thus obtained have excellent flow characteristics and are extremely suitable for automatic feeding devices employed in ram extrusion machines and large molding presses. Preforms molded at room temperature from this improved molding powder are of sufficient strength to be handled and do not adhere to the mold and can therefore be readily removed and free-sintered. The molded articles obtained from such powders upon preforming and sintering have electrical and mechanical properties equivalent to those obtained with unprocessed granular molding powder. This process is illustrated by the following example.

Example: A sheet mold comprising a male and female part having a cavity of 4" x 5" x 1" was charged with a 40 gram sample of granular or dispersion polytetrafluoroethylene powder heated to the temperature indicated as preforming temperature in the table below. The powder was leveled with a straight edge and a silicone rubber caul followed by a heated steel plunger was placed on the level polymer. The mold lay-up was placed in a press having platens heated to the preforming temperature. The pressure on the mold was raised to 2,000 psi over a period of one minute and held at that pressure for 5 minutes.

The resulting preform was removed from the mold and cooled to room temperature. The density of the preform was determined by water displacement measurements. The preform was broken by hand and the resulting pieces were disintegrated into a fine powder by placing 40 grams of polymer in a Waring Blendor with 100 ml. of water and running the Blendor at high speed for a period of 2 minutes. The resulting polymer powder was filtered and dried.

Polymer type	Preforming temp., °C.	Preform density, g./cc.	Flow index, g./sec.	Leveling quality	Wet sieve, d_{50} in. microns	Apparent density g./l.	Appearance of powder on disintegration	Dielectric strength, volts/mil	Tensile strength, lbs./in.²	Elongation, percent	Porosity, percent
Molding powder properties								**Properties of sintered articles**			
Commercial powders:											
Granular	--------	--------	3-7	Poor	460-700	472-522	Fibrillated	287	2,167	173	1.6
Dispersion	--------	--------	(¹)	Extremely poor	350-650	400-600	do	Cannot be molded			
Densified:											
Granular	25	2.13	8	Poor	400-500	600-700	do	No improvement over commercial powder			
Dispersion	25	1.93	(¹)	Extremely poor	(¹)	(¹)	Difficult to disintegrate, very fibrillated.	Poorer than starting material			
Granular	60	2.16	20	Good	250-450	750-850	Distinct particles	304	2,150	180	1.7
Do	170	2.22	20	do	250-450	850-1,000	do	400	2,755	243	0.4
Dispersion	180	2.18	17	do	200-400	800-1,000	do	(250-300)	1,650	30	1.1

¹ Not measurable.   ² Very low (ca. 200–300).

NOTE.—Test methods: (1) Flow index; (2) dielectric strength, ASTM-D 149–44; (3) apparent density, (4) wet screen, (5) tensile properties, ASTM-D 1457–56T.

The polymer powder was evaluated with respect to its powder flow characteristic as shown in the table and compared to commercial polymer not subjected to the above described procedure and also compared to polymer powder preformed at room temperature and then disintegrated. The flow index of the polymer powder was determined by filling a polytetrafluoroethylene pipe 9 inches high and 2 inches in diameter and having a 6 mesh screen attached across the base of the pipe with the polymer powder and subjecting the pipe to a vibration having a frequency of 675 cycles per second and an amplitude of 0.3 inch.

The amount of powder flowing through the screen was continuously weighed and recorded. From the resulting curve the flow index was calculated as grams per second. The leveling quality of the polymer powder, which becomes extremely important when the powder is automatically fed, was determined qualitatively by the difficulty encountered in filling and leveling the mold employed to prepare the sintered article. The wet sieve and apparent density were determined by ASTM-D-1457-56T and are a measure of the particle size of the polymer powder and the bulk density of the polymer powder. The appearance of polymer powder when subjected to the disintegration is noted.

Polytetrafluoroethylene

The polymer powder was then molded into sheets by charging 43 grams of the respective treated or untreated, granular or dispersion polymer powders into the above described mold and preforming the mold under the same conditions as employed in the first preforming step. The resulting preform was sintered in an oven for one hour at 375°±7°C. The sintered sheet was cooled at the rate of 3°C. per minute. The dielectric strength, measured by the method disclosed in ASTM-D-149-44, tensile strength and elongation, as measured by ASTM-D-1457-56T employing a testing rate of 2 in. per min. and porosity of the sintered sheet, calculated from the following equation:

$$\left[\frac{\text{inherent density } - \text{ bulk density}}{\text{inherent density}} \right] \times 100$$

were determined and are listed in the table. As can be seen from the results listed in the table, the polymer powders which have been preformed to a density above 2.15 have markedly better powder-flow properties. It is also seen from the table that dispersion polymer powder which cannot ordinarily be compression molded, could be compression molded when preformed to a density above 2.15.

Dispersion polymer powder-flow characteristics could not be determined because of the fibrillated structure of the polymer and because of the adhesion of the polymer powder particles to each other and also could not be molded into sheets as indicated in the table because the preform cracked during the compression molding step and furthermore adhered to the mold and could therefore not be removed from the mold without being broken up. As can be seen from the table, the properties of sintered articles prepared from the improved flow powders are equal to and better than the properties obtained on sheets prepared with untreated powder. In the case of dispersion polymer powder the improved flow methods provide the only way of compression molding the polymer.

Polymerization in the Presence of Colloidal PTFE

K.L. Uhland; U.S. Patent 3,088,941; May 7, 1963; assigned to E.I. du Pont de Nemours and Company describes the preparation of polytetrafluoroethylene extrusion powder. This process comprises introducing, at a temperature of below 100°C., tetrafluoroethylene into an aqueous medium containing a water-soluble, free radical forming initiator, a dispersing agent and from 0.1 to 1% by weight of the aqueous medium of colloidal polytetrafluoroethylene dispersed in water, subjecting the resulting aqueous dispersion formed by polymerization to agitation until coagulation occurs, and recovering a polytetrafluoroethylene powder.

It was discovered that the presence of colloidal polytetrafluoroethylene in the aqueous medium prior to initiation will cause the formation of an aqueous polytetrafluoroethylene dispersion which on coagulation gives rise to an improved extrusion powder. The presence of the colloidal polytetrafluoroethylene is believed to cause the formation of colloidally dispersed polymer of larger particle size. The polytetrafluoroethylene polymer obtained on coagulation of the dispersion is improved with respect to its extrusion characteristics in that it is capable of being extruded at higher rates than the unmodified polymer, and in that the extrusion is subject to less pressure variation, thus giving rise to a more uniform caliper product. The extruded product obtained on sintering, furthermore, shows less flaws and, thus, allows for more uniform fabrication.

The addition of the colloidal polytetrafluoroethylene prior to polymerization is critical with respect to the upper limit employed. If the quantity of the added polytetrafluoroethylene exceeds 1 to 2% of the aqueous medium, coagulation of the entire dispersion occurs during the polymerization. Polymer coagulated during polymerization is unsuitable for extrusion and molding. It is believed that the presence of coagulated polymer during the polymerization acts as an active surface on which further polymer growth occurs. This further polymer growth adversely changes the structure and the nature of the resulting polymer particle. Furthermore, the presence of an amount of coagulated polymer in excess of the critical maximum quantity during the polymerization tends to cause the coating of the reactor walls and the fouling of reactor lines and valves, and for that reason must be avoided.

The best results are obtained when the quantity of the colloidal polytetrafluoroethylene added corresponds to 0.2 to 1.0% by weight of the aqueous medium. The only requirement of the added polytetrafluoroethylene is that it is in colloidal form. The polymerization of tetrafluoroethylene according to this process is otherwise carried out in accordance with known general procedures.

The colloidal polytetrafluoroethylene employed in this process comprises a high molecular weight, solid polymer of tetrafluoroethylene having a crystalline melting point at 327°±2°C. and substantially no melt flow at temperatures above its melting point. The colloidal polytetrafluoroethylene particle is spheroidal in shape and has an average diameter of 0.1 to 0.5 micron. The colloidal polytetrafluoroethylene is produced by such processes as described in U.S. Patent 2,559,752. The process is illustrated by the following example.

Example: Into a 10-gallon autoclave was charged water, 35% polytetrafluoroethylene dispersion, ammonium perfluoro-caprylate, disuccinic acid peroxide, paraffin wax, iron in the quantities tabulated below. The polytetrafluoroethylene particles in the dispersion were high molecular weight polytetrafluoroethylene having a crystalline melting point of 327°C.

Polytetrafluoroethylene

and were prepared by the process described in U.S. Patent 2,750,350. The reaction mixture was heated to 95°C. and pressured with tetrafluoroethylene until a pressure of 400 psig was obtained. The reaction mixture was agitated and pressure and temperature maintained until the quantity of monomer indicated in the table had been polymerized.

The resulting product was examined as to the solids dispersed and coagulum. A 900 g. sample of the polymer was lubricated by mixing it with 19% by weight of the total composition of 2 VM & P naphtha, a commercial hydrocarbon lubricant. The mixture was placed in a closed jar and rolled for a period of 30 minutes at a rate of 30 rpm. The lubricated polymer was employed in the coating of wire in a Jennings wire extruder, Model TF-1. An E-22 U.S. military specification wire [(M-12-W-16878A/5) (Navy)] was coated with a 10 mil layer of the polytetrafluoroethylene at the rate of 25 feet per minute according to the method disclosed in the article of G.R. Snelling and R.D. de Jong in Wire and Wire Products of June 1957. The wire yield is the percentage of coated wire in flaw-free pieces of 50 feet or more of the total length of the wire coated.

Polymerization Conditions and results	Run A	Run B	Run C	Run D	Run E	Run F
Water, Gallons	5.17	5.17	5.17	5.17	5.17	4.8
Polytetrafluoroethylene dispersion added on ml.	0	220	220	440	500	880
Percent Colloidal Polytetrafluoroethylene Charged on the Basis of Aqueous charge	0	.5	.5	1.0	1.15	3.0
Ammonium Perfluorocaprylate, lb.	0.067	0.067	0.067	0.134	0.067	0.067
Disuccinic Acid Peroxide, Gm.	14.2	14.2	14.2	14.2	14.2	14.2
Paraffin Wax in lbs.	1.68	3.36	1.68	1.68	1.68	1.68
Iron, gm.	0.04	0.04	0.04	0.04	0.04	0.04
Monomer, lb.	20	20	20	20	20	20
Percent Solids Obtained in resulting dispersion	33	35	36	35	27	
Coagulum					(1)	(2)
Wire Yield	0	30-80	28-75	35-80	100	

[1] Slightly coagulated.
[2] Completely coagulated

The results of this series of polymerizations clearly shows the criticality of the amount of colloidal polytetrafluoroethylene added to the polymerization. If the quantity is significantly increased above 1% of the aqueous medium, coagulation of the polymer in the subsequent polymerization occurs.

Controlled Nuclei Concentration

R.F. Anderson, W.L. Edens and H.A. Larsen; U.S. Patent 3,245,972; April 12, 1966; assigned to E.I. du Pont de Nemours and Company describe a process which comprises polymerizing tetrafluoroethylene in contact with a stirred aqueous medium containing a free radical initiator and solid nuclei in the form of agglomerated nonwaterwet particles having a total surface area per gram of greater than 3 square meters, characterized in that the medium contains at least 5×10^{-10} nuclei per milliliter, polymerization is continued until the particles have a reduced total surface area per gram of less than 9 square meters, and polymerization is discontinued before the total surface area of the particles per gram falls below 3 square meters.

The particles so obtained thus have a total surface area per gram of at least 3 and less than 9 square meters. As prepared they are irregular and stringy. However, they may be water-cut and washed by simple procedures to yield powders having excellent moldability and powder flow. In certain preferred aspects of this process there are obtained molding powders having a total surface area of at least 3 and less than 9 square meters per gram, a moldability index of less than 50 and an apparent (bulk) density, indicative of powder flow, of at least 400 grams/liter.

It has been found in the course of research leading to this process that when tetrafluoroethylene is contacted with an aqueous solution containing a free radical initiator, the tetrafluoroethylene dissolves in the water and polymerizes to form tiny solid waterwet nuclei, and nuclei formation continues for only a limited time, after which any further polymerization results in an increase in the size of the existing nuclei, rather than formation of new nuclei. When the waterwet nuclei have sufficiently increased in size, they are coagulated by the agitation applied to the aqueous medium, to agglomerated nonwaterwet particles, after which further polymerization takes place largely on the surface of the agglomerated nonwaterwet particles at an increased rate, apparently via direct contact of gaseous monomer with the nonwaterwet agglomerates, rather than by contact of dissolved monomer with waterwet particles.

The more rapid polymerization rate thus makes it more economical to produce powders by techniques involving direct contact of the monomer with the nonwaterwet aggregates, than to produce powders by dispersion techniques in which the difficultly soluble monomer must be dissolved in the medium before polymerization. There are however practical limits to this advantage, because the polymerization is highly exothermic, and as the reaction is accelerated it becomes increasingly difficult to remove the heat generated. For this reason it has previously been deemed desirable to carry out the preparation of granular powders in a violently agitated medium.

The violent agitation however has the added inherent function of consolidating the solid nuclei after their formation so that further polymerization fills the interstices of the agglomerates, rapidly converting them into large, solid particles of low total surface area per gram. By contrast, in the processes described here, the number of nuclei is made originally high and continues high during the further course of polymerization. In addition, in preferred aspects, the stirring is

42

carried out at a controlled power input and at a high ratio of power number to discharge coefficient so as to minimize shear, thus making possible the preparation of high yield of polymer of high total surface area while maintaining high rates of polymerization.

The total number of nuclei in the reaction medium can be controlled in various ways. For the purposes of this process, concentration of nuclei in a reaction medium is taken as the value calculated from the total surface area of the solid product on the assumption that the nuclei are solid spherical particles of uniform size having a density of 2.28. On this basis, the total volume of solid is calculated from the total weight of solid and the assumed density; the surface area of the individual nuclei is calculated from the total volume and surface area of the solid; the number of individual nuclei is calculated from the total surface area and the surface area of the individual nucleus, and this number is compared with the volume of the aqueous medium in ml. to determine the number of nuclei per ml.

The number of nuclei in the aqueous medium can be controlled by seeding the medium with solid nuclei having a total surface area per gram of greater than 3 square meters and preferably greater than 9 square meters, before commencing polymerization in accordance with this process. The solid nuclei may be composed of any material which is substantially insoluble in the reaction medium. Examples of materials which may be used as nuclei include glass, silica, carbon, and insoluble metal silicates and oxides such as titania, alumina, zirconia, and the like, as well as polymers and copolymers of highly fluorinated olefins such as tetrafluoroethylene, vinylidene fluoride, hexafluoropropylene, chlorotrifluoroethylene, perfluoropropyl perfluorovinyl ether, etc. For most ultimate uses however, polytetrafluoroethylene nuclei are preferred in order to take maximum advantage of the uniquely valuable combination of properties possessed by the unmodified polymer.

In each of the examples a solution containing 50 parts per million ammonium persulfate initiator and 5,600 parts per million sodium tetraborate buffer in distilled water is employed as the aqueous medium. 500 parts of this medium are charged to a stirred autoclave having a capacity of 700 parts. Stirring is then commenced and the charge is heated to the desired reaction temperature. The free space above the charge is then evacuated and purged with tetrafluoroethylene, after which the charge is pressured with tetrafluoroethylene and reacted at the indicated temperature, pressure and stirring rate until the indicated content of solids is produced.

At the conclusion of each run the pressure is released, and the aqueous slurry containing the product in the form of rough stringy particles is discharged. The solids are filtered from the slurry, and a 475 gram sample dispersed in 3,100 ml. water in water-cut for 25 seconds with a bladed stirrer rotating at 13,000 rpm. The water-cut slurry is then stirred for 2 hours in a 4-liter baffled steel beaker with just sufficient agitation to keep the nonwaterwet particles submerged. The solids are then filtered off, dried 16 hours at 120°C., and characterized by the tests indicated. In the polymerizations, the different agitators employed were a vertically disposed flat paddle, having a ratio of power number to discharge coefficient of 3.4; a stirrer with flat blades pitched at an angle of 15 degrees to the horizontal, having a said ratio of 1.45; a gas turbine having a said ratio of 1.65; a marine propeller having a said ratio of 1.60, and a bladed propeller having horizontal shear tips, having a said ratio of 1.3.

The tensile strengths and elongations recited are values obtained by the procedures described in ASTM D-1457-56T. In each of the illustrative examples the reaction proceeded initially at a slow rate, and then markedly increased in rate as the production of the nonwaterwet agglomerates occurred. In each of the illustrative examples, the total number of nuclei present is above 5×10^{10} per ml. The overall reaction time in each of Examples 3 to 5 is in the range of 20 to 40 minutes, and in Examples 1 and 2 is in the range of 5 to 10 minutes.

Example 1: The reaction temperature is 80°C., the pressure is 27 atmospheres of tetrafluoroethylene gas, the power applied in stirring is 0.001 kg.-m./sec./ml., the ratio of power number to discharge coefficient in stirring is 1.60, and the reaction is continued to produce 12% solids based on the weight of the aqueous medium. After cutting, washing and drying, the product has a total surface area per gram of 7.1 square meters, a standard specific gravity of 2.191, a moldability index of 1, an apparent density of 420 grams/liter, a tensile strength of 2,820 lbs./sq. in., and an elongation at break of 327%.

Example 2: The procedure of Example 1 is repeated except that the reaction is continued to produce 18% solids, based on the weight of the aqueous medium. After cutting, washing and drying, the product has a total surface area per gram of 4.2 square meters, a standard specific gravity of 2.173, a moldability index of 3, an apparent density of 500 grams per liter, a tensile strength of 4,130 lbs./sq. in. and an elongation at break of 340%. In a comparative example in which the total number of nuclei is less than 5×10^{10}/ml., and in which the reaction temperature is 65°C., the reaction pressure is 13 atmospheres, the power applied in stirring is 0.001 kg.-m./sec./ml., the ratio of power number to discharge coefficient in stirring is 1.3, and the reaction is continued to produce 30 weight percent solids based on the weight of the aqueous medium, the product after cutting, washing and drying has a total surface area per gram of 1.43 square meters, a standard specific gravity of 2,150, a moldability index of 50, an apparent density of 550 grams/liter, a tensile strength of 2,300 lbs./sq. in., and an elongation at break of 180%. Further treating the product by the procedure of U.S. Patent 2,936,301 lowers the moldability index, but at the same time also lowers the apparent density of the powder to below 330 grams/liter.

Example 3:   The reaction temperature is 65°C., the pressure is 20 atmospheres, the power applied in stirring is 0.001 kg.-m./sec./ml., the ratio of power number to discharge coefficient in stirring is 3.4, 200 parts ammonium 3,6-dioxa-2,5-bis-(trifluoromethyl)-undecafluorononanoate per million weight parts of water is included in the aqueous medium before commencing the polymerization, and the reaction is continued to produce 24% solids based on the weight of the aqueous medium. The product after cutting, washing and drying has a total surface area per gram of 5.13 square meters, a standard specific gravity of 2.169, a moldability index of 20, an apparent density of 590 grams/liter, a tensile strength of 3,290 lbs./sq. in. and an elongation at break of 288%.

Example 4:   The procedure of Example 3 is repeated except that the dispersing agent is 200 parts ammonium salt of ammonium omega-hydrohexadecafluorononanoate per million weight parts of water. The product has a total surface area per gram of 4.81 square meters, a standard specific gravity of 2.171, a moldability index of 18, an apparent density of 587 grams/liter, a tensile strength of 3,660 lbs./sq. in., and an elongation at break of 315%.

Example 5:   The procedure of Example 4 is repeated except that the dispersing agent is 200 parts of ammonium perfluoro-octanoate per million weight parts of water. The product has a total surface area per gram of 4.01 square meters, a standard specific gravity of 2.173, a moldability index of 22, an apparent density of 685 grams/liter, a tensile strength of 3,250 lbs./sq. in., and an elongation at break of 282%.

Free-Flowing Powder

M.B. Black III, E.E. Faust, W.S. Barnhart and R. Netsch; U.S. Patent 3,265,679; August 9, 1966; assigned to Pennsalt Chemicals Corporation describe methods for producing free-flowing granular polytetrafluoroethylene powders suitable for the production of molded articles. It has been found that this can be accomplished by a procedure which involves wetting fine granular polytetrafluoroethylene powder with a relatively low boiling liquid inert with respect to polytetrafluoroethylene and having a surface tension below about 45 dynes per centimeter and preferably between 15 and about 38 dynes per centimeter (at 20°C.); mechanically forming the wet powder into agglomerates (called glomerules) having a dry sieve size ranging from about 300 to 3,000 microns and drying the glomerules to produce a free-flowing powder having a markedly increased bulk density.

It thus becomes possible by means of the procedure outlined above to convert the original fine powder into loose agglomerates or glomerules of controlled size which have sufficient firmness to withstand all normal handling operations such as shipping, pouring from container to mold, etc., without disintegration, but which are readily deformable under preforming pressures such that the tensile strength of the preformed and sintered resin is not substantially reduced relative to the tensile strength obtainable from the original fine powder before treatment. The procedure thus not only produces a polytetrafluoroethylene granular molding powder which is vastly improved in flow characteristics and increased in bulk density over the original fine powder, but which also retains the ability of the original fine powder to produce high quality molded products.

An especially suitable class of liquids from the standpoint of effectiveness, cost, and ease of handling are halogenated hydrocarbons boiling between 30° and 150°C. which display surface tensions between 15 and 38 dynes per centimeter. In addition to being nonflammable and relatively nontoxic, these compounds are essentially water immiscible which facilitates their recovery when drying is carried out under reduced pressure by the use of a water driven aspirator.

Any suitable method may be employed to apply the liquid to the polytetrafluoroethylene powder which will uniformly wet the surfaces of the particles. Thus, for example, the wetting may be accomplished while the powder is stationery such as by soaking the powder with an equal volume or more of liquid and then draining off the excess on a filter or the like, or by spraying the liquid on the powder while spread in a thin enough layer so as to thoroughly wet all the particles. It is generally preferred, however, to apply the liquid with simultaneous agitation of the powder. A particularly preferred procedure is to tumble the powder, and while tumbling, apply the liquid in the form of a fine mist or fog. Using this latter procedure, the application of the liquid and the formation of the glomerules occurs at least to some extent simultaneously as will be explained in more detail below. Preferably the wetting is carried out at normal temperatures.

A preferred method of mechanically forming the wet powder into glomerules involves a simultaneous wetting and mechanical treatment step. Thus, for example, the powder may be placed in a slowly rotating vessel which gently rolls and tumbles the powder and while thus being mechanically agitated, the liquid is sprayed on the powder in the form of a fine mist or fog. As the powder becomes wet the glomerules begin to form. Depending upon the rate of liquid injection and other factors, glomerule formation may be completed during liquid injection or the mechanical agitation may continue for a period of time after all the liquid has been added in order to complete the glomerule formation.

The mechanical forming operation is preferably and most conveniently carried out at normal (i.e., ambient) temperatures but if desired, temperatures somewhat above or below ambient may be employed while, of course, avoiding temperatures at which the wetting liquid may be prematurely volatilized. In general, glomerule formation temperatures appreciably greater than ambient are disadvantageous, not only because the wetting liquid becomes more volatile, but because of the greater tendency of the glomerules to become unduly compacted.

Polytetrafluoroethylene

Following glomerule formation, the agglomerated powder is dried to remove all traces of liquid. The drying should be conducted at temperatures below 275°C. and preferably below 200°C. A moderate vacuum, e.g., 50 to 250 mm. Hg is advantageous in permitting rapid, complete removal of the liquid at moderate temperatures. The rate of drying should be such as to avoid overrapid evaporation of the liquid which would tend to cause disintegration of the glomerules. The drying may be conducted under static conditions or while the glomerules are subjected to gentle agitation such as slow tumbling in a rotating vessel to continually or intermittently expose new surfaces and thus speed the drying operation.

The intensity and duration of the agitation (if any) during drying must however be carefully controlled to avoid undue compaction and hardening of the glomerules which in turn leads to loss of tensile strength and other desired properties in the molded resin. In general the higher the drying temperature, the greater the tendency for compaction with agitation, and accordingly the less the amount of agitation permissible.

As is apparent from the foregoing, the ultimate hardness of the glomerules produced is affected both by the duration and intensity of the mechanical treatment which forms the wet powder into glomerules and the duration and intensity of the agitation if any during the drying operation. The forces exerted during these operations should be gentle enough to avoid undue compaction of the glomerules such that they remain readily deformable under usual preforming pressures, while at the same time the forces should be sufficient to produce a glomerule having sufficient hardness and coherence that it will not disintegrate under normal handling such as during packing, shipment, pouring of the powder from one container to the other or into the molds or the like.

The deformability of the glomerules is best determined by comparing the tensile strength (by Standard ASTM procedure D 1457-62T) of resin molded from the agglomerated powder to the tensile strength of resin molded from the original untreated powder. The agglomerated material by this test should provide molded resin which retains at least 80% and preferably at least about 90% of the tensile strength of the molded resin prepared from the untreated material.

If desired, after drying, the agglomerated product may be size fractionated, e.g., by dry sieving to remove fines and oversize glomerules. Preferably the size distribution is such that not more than 20% of the agglomerated product passes a 60 mesh (U.S. Standard) sieve while essentially 100% passes a 6 mesh sieve (by dry sieving). When using preferred procedures however such size fractionation is not necessary. Thus when using preferred liquid:powder ratios and preferred agglomeration techniques such as simultaneous tumbling and spraying, the product obtained from the agglomeration operation without any size fractionation, consists of glomerules of the desired dry sieve size (d_{50}) (i.e., in the range of from 300 to 3,000 and preferably 400 to 1,500 microns) and having a size distribution within the above preferred limits.

The free-flowing granular polytetrafluoroethylene molding powders that may be obtained in accordance with the procedures described consist of relatively large free-flowing glomerules having a dry sieve size of from 300 to 3,000 microns, and preferably from 400 to 1,500 microns; and are characterized by an uncompacted flow rating of at least 25 and preferably at least 36; a compacted flow rating of at least 9 and preferably at least 16; a bulk (apparent) density of from 400 to 850 and preferably from 500 to 750 grams per liter; and a glomerule hardness sufficient to withstand normal handling without disintegration but sufficiently deformable under preforming pressures to provide molded resin retaining at least 80% and preferably at least 90% of the tensile strength obtainable from the original untreated finely pulverized powder. The following example illustrates this process.

Example: Seven pounds of finely pulverized granular polytetrafluoroethylene molding powder having the properties of a Type IV powder as set out in ASTM Designation D 1457-62T is charged to an eight quart capacity liquid-solids blender. The finely pulverized powder has an air sedimentation size (d_{50}) of 28 microns with less than 2% greater than 60 microns (by air sedimentation measurement), and a low bulk density of 254 grams per liter. It has poor flow properties as indicated by an uncompacted flow rating of 4 and a compacted flow rating of 1. Because of its poor flow characteristics, such a powder is difficult or impossible to handle in many types of automatic molding equipment where a measured amount of powder must feed accurately and reproducibly from hoppers into mold openings without bridging or sticking. The low bulk density is a further disadvantage because of the increased size of molding equipment required.

The blender shell, containing the 7 pound charge of the above fluffy, poorly flowing powder, is rotated at a constant speed of 25 rpm thus subjecting the powder to a relatively gentle rolling and tumbling action. With the shell rotating at this speed trichloroethylene is fed into the shell through a pair of nozzles each consisting of a pair of spaced discs 3 1/2 inches in diameter and spaced apart 0.02 inch. The trichloroethylene issues from nozzles in a fine mist or fog induced by the rotation of the shaft turning the nozzles at a speed of 2,300 rpm.

Each of the two nozzles carry four fingers approximately 1/4 inch wide and 2 inches long which create rapid air currents in the vicinity of the nozzles thus maintaining circulation of the powder throughout the container and insuring uniform wetting of the powder with the trichloroethylene. In this manner, a total of 1,590 milliliters of trichloroethylene is introduced into the container over a period of 4 minutes, thus providing a liquid:powder ratio of 50 milliliters of trichloroethylene per hundred grams of powder. After 4 minutes, the high speed shaft carrying the nozzles is stopped and feed of trichloroethylene is discontinued. The container as a whole is rotated at 25 rpm for one additional minute, giving a total of 5 min. of tumbling time, during the first 4 min. of which trichloroethylene is sprayed into the tumbling powder.

Polytetrafluoroethylene

Following this treatment, the powder now consisting of roughly spherical agglomerates or glomerules is discharged from the container and dried in a forced draft convection oven at 175°C. for 4 hours. The dried product has a dry sieve size (d_{50}) of 700 microns and consists 100% of glomerules passing a 6 mesh sieve with 3% passing a 60 mesh sieve (by dry sieving). The product thus produced has more than double the bulk density of the original material (660 grams per liter) and has excellent flow properties, viz., an uncompacted flow rating of 36 and a compacted flow rating of 16 versus flow ratings of 4 and 1 for the original material.

The improvement in the flow properties of the molding powder produced by the procedures of this process in contrast to the flow properties of the starting material are further demonstrated by the following flow test. Two glass funnels having a 4 inch maximum diameter, a 30° mouth taper, and a straight stem 3 inches long; one having an internal stem diameter of 7 millimeters and the other a 14 millimeter internal stem diameter, are charged with 65 milliliters of resin powder with the bottom of the funnel stem blocked off. The funnel stem is unblocked and flow (if any) of the resin through the funnel stem is observed. In this test, the light, poorly flowing starting material does not flow out of either funnel even with repeated tapping of the funnel. The agglomerated product obtained according to the above procedure on the other hand readily flows out of both funnels without any tapping or shaking.

The excellent flow properties and bulk density of the molding powders is further illustrated by the following comparative tests in an automatic molding machine where various standard, commercially available granular polytetrafluoroethylene molding powders are compared to the agglomerated molding powders of this process. In these tests the powder is contained in a hopper-fed charging box open at the bottom which passes over the mold cavity whereupon the powder flows into the cavity by gravity. The mold cavity in each case is an annulus with a 2 inch outside diameter and a 1.5 inch inner diameter with a fill height of 2 inches.

Flow into the mold cavity is assisted by an eccentric revolving motion of the charging box. After filling the mold cavity, a preform is prepared at a preforming pressure of 4,380 psi. Three standard types of commercially available molding powders of decreasing particle size, as well as the agglomerated powder produced in accordance with the above procedure, are preformed (10 preforms made in each run) after which the mean weight of the preforms, and the percent standard deviations from the mean weight is determined. The results of these tests are tabulated below.

Molding Powder	Particle Size d_{50}	Number of Passes to Fill Mold	Mean Weight of Preforms	% Standard Deviation from Mean Weight
ASTM Type I*	600 microns (wet sieve)	1	20.036	4.0
ASTM Type II*	350 microns (wet sieve)	1	23.662	1.5
ASTM Type IV*	25 microns (air sedimentation)	2	10.686	0.8
Product of Example	700 microns (dry sieve)	1	25.703	0.4

*See ASTM Designation D 1457-62T.

As is apparent from data above the agglomerated product produced a preform having the smallest percent standard deviation thus indicating the most uniform and reproducible filling of the mold cavity. The finely pulverized ASTM Type IV powder required two passes of the charging box to fill the relatively shallow 2 inch mold cavity thus exhibiting the poorest flow of all the powders tested. Also because of its low bulk density the fine Type IV powder produced a preform 2.5 times smaller than that produced by the agglomerated product. The Type IV powder thus would require a mold 2.5 times as deep to produce the same size preform.

In addition to excellent flow properties and bulk density as demonstrated above, the agglomerated powder produced by the above procedure suffers no substantial loss in its ability to produce high quality molded resin. Thus, the agglomerated powder retains the ability to produce essentially void-free moldings at moderate preform pressures of 2,000 psi as well as the ability to produce moldings having high tensile ratings characteristic of finely pulverized powders. Thus, the microvoid rating of moldings produced by the agglomerated powder prepared as described above at a preform pressure of 2,000 psi is excellent and not detectably different from the microvoid rating of the finely pulverized starting material. The tensile strength of the molded resin produced from the agglomerated powder is 4,100 psi retaining 91% of the tensile strength of moldings produced by the original finely pulverized powder, viz., 4,500 psi.

While the individual glomerules are readily deformable under normal preforming pressures to produce void-free strong moldings, the glomerules are at the same time sufficiently strong to withstanding all the normal vibrations, shocks and attrition involved in shipping and handling. For example, the agglomerated powder prepared according to the above procedures shows no significant packing, disintegration or loss of its excellent flow characteristics when carried for approximately 5 days in a commercial delivery truck traveling a distance of over 800 miles with frequent starts, stops, loadings and unloadings.

A summary of the properties of the agglomerated product compared to three standard commercially available polytetrafluoroethylene molding powders is shown in the following table. As is apparent, the agglomerated powder has flow properties and bulk density as good as, or better than, the relatively coarse ASTM Type I and Type II powders, and vastly better than the highly pulverized, light, poorly flowing Type IV powder. At the same time the product provides a low porosity, high strength molded resin of the same order of excellence as that provided by the finely pulverized poorly flowing Type IV powder.

Granular polytetra-fluoroethylene molding powder	Particle size microns (d_{50})			Bulk density, gms./ liter	Flow Properties				Microvoid rating of molded resin at 2,000 p.s.i. preform pressure	Tensile strength of molded resin
	Wet sieve	Air Sedi-menta-tion	Dry sieve		Uncom-pacted flow rating	Com-pacted flow rating	Funnel flow test	Automatic mold filling percent standard deviation		
ASTM Type I..........	600	--------	--------	500	36	16	Will not flow through 7 mm. stem with tapping.	4	Poor............	2,500
ASTM Type II........	350	175	--------	536	36	4	.......do............	1.5	.......do...........	3,000
ASTM Type IV........	--------	28	--------	250	4	1	Will not flow through 14 mm. stem even with tapping.	[1] 0.8	Excellent........	4,500
Agglomerated product	--------	--------	700	660	36	16	Flows through 7 mm. stem without tap-ping.	0.4	.......do...........	4,100

[1] 2 passes over mold cavity required.

Internally Stressed Powders

In a process described by J. Hoashi, A. Matsuura, S. Koizumi and N. Horiuchi; U.S. Patent 3,413,276; November 26, 1968; assigned to Thiokol Chemical Corporation polytetrafluoroethylene fine powders, which have an average particle diameter in the range from 0.05 to 0.5 micron are produced by the coagulation of aqueous colloidal dispersions of the polymer. These powders may be internally stressed without substantial fibrillation of the polymer particles by kneading the polytetrafluoroethylene fine powder in the presence of an organic solvent capable of wetting polytetrafluoroethylene.

When the degree of stress (S_i) of the polytetrafluoroethylene fine powders exceeds 40, as measured by a polarizing micro-scope equipped with a tungsten light source, a polarizer, and an analyzer, using a photometer to determine the transmit-tance, there is a remarkable increase in the dielectric properties as well as improved molding properties of the "internally stressed" powder compared to the identical unstressed polytetrafluoroethylene fine powder. The degree of stress of poly-tetrafluoroethylene fine powders is defined by the formula $S_i = [I_1/I_2] \times 100$, where S_i represents the degree of stress, I_1 represents the transmittance of light of a given intensity from a tungsten lamp through a standard compression molded plate of the polytetrafluoroethylene fine powder, and I_2 represents the transmittance of parallel polarized light from the same light source through the standard compression molded polytetrafluoroethylene plate.

Selection of a suitable organic solvent for wetting polytetrafluoroethylene, in the presence of which organic solvent the unstressed polytetrafluoroethylene is kneaded until its degree of stress exceeds 40, may be made from any compound which possesses a strong affinity to polytetrafluoroethylene and is capable of penetrating (i.e., wetting) the powder particles. These solvents include ketones, such as acetone and methyl ethyl ketone; alcohols, such as methanol and ethanol; car-boxylic acids, such as acetic acid and propionic acid; esters, such as ethyl acetate and butyl acetate; ethers, such as diethyl ether and ethylpropyl ether; hydrocarbons, such as naphtha and decane; halogenated hydrocarbons, such as tri-chlorotrifluoroethane; and any other solvent or solvent system capable of wetting polytetrafluoroethylene powder. To avoid fibrillation of the polymer particles during kneading, the fine powder should be wetted by the organic solvent and kneaded while maintaining the temperature in the range from about −20° to 50°C., since higher temperatures have been observed to result in fibrillation of the powder and to promote the formation of an undesirable "wave" in unsintered tape of the fine powder.

Polytetrafluoroethylene fine powders having a degree of stress in excess of 40 and produced in accordance with this process yield wider tapes under ram extrusion than unstressed (or untreated) powder when both products are extruded through identical dies and then calendered under identical conditions. Moreover, unsintered tapes of these internally stressed polytetrafluoroethylene fine powders possess dielectric strengths of more than 6000 volts/0.1 mm. thickness, which is generally appreciably higher than unsintered tapes produced from conventional (or unstressed) polytetrafluoro-ethylene fine powders. The following examples are illustrative of this process.

Example 1:   1.5 liters of deionized pure water and 6.3% by weight of paraffin wax containing no unsaturated compounds were charged to a 4-liter stainless steel autoclave equipped with a stirrer, to which mixture was then added 0.15% by weight of ammonium perfluorooctoate. Oxygen was removed from the system under vacuum, and tetrafluoroethylene then was charged to the autoclave until the internal pressure was 20 atmospheres. The temperature of the reaction mixture was increased to 70°C., and a catalyst system consisting of 0.1% by weight of ammonium persulfate and 2 ppm of iron powder was charged into the reaction mixture, while running the agitator at a speed of 120 rpm.

The internal pressure of the autoclave was maintained constant at 20 atmospheres by adding make-up tetrafluoroethylene to the autoclave as the pressure began to drop due to polymerization of the monomeric tetrafluoroethylene. After the concentration of the polymer reached 40% by weight of the aqueous colloidal dispersion in the autoclave, the emulsion

Polytetrafluoroethylene

polymerization reaction was terminated by venting the unreacted tetrafluoroethylene. The aqueous colloidal dispersion was then removed from the autoclave before its temperature dropped below the melting point of the paraffin wax, and allowed to stand until a wax layer solidified upon cooling. After removal of the wax layer, the resultant colloidal dispersion was coagulated while stirring, and the coagulum (polytetrafluoroethylene particles) dried at 120°C. Under electron microscope magnifications of 4,000X, the powder was found to consist of particles having a diameter in the range from 0.2 to 0.3 micron.

Ten parts by weight of this polytetrafluoroethylene fine powder were mixed with 5 parts by weight of ethanol, and stirred for one hour in a kneader at a speed of 20 rpm, adding ethanol (if necessary) to maintain the ethanol concentration. The mixture was then added to 50 parts by weight of water, lightly stirred, and the sediment washed with water and then filtered. The product, which was designated as "PTFE Powder A" was then dried at 150°C. for 10 hours.

Example 2: 100 parts by weight of polytetrafluoroethylene fine powder, produced by the emulsion polymerization technique described in Example 1, were mixed with 50 parts by weight of methyl ethyl ketone and stirred for 40 minutes in a kneader operated at a speed of 50 rpm. The mixture was then added to 500 parts by weight of water, lightly stirred, washed with water and filtered. The resultant powder, which was designated as "PTFE Powder B" was then dried at 120°C. for 20 hours.

To test the effect of kneading unstressed polytetrafluoroethylene fine powder in an organic solvent containing water, the example was reproduced under identical conditions, except that the methyl ethyl ketone employed contained 20 parts by weight of water. The resultant powder, which was designated as "PTFE Powder C," was dried at 120°C. for 20 hours.

Example 3: Conventional (unstressed) polytetrafluoroethylene fine powder and each of the polytetrafluoroethylene fine powders produced in Examples 1 and 2 (PTFE Powders A, B and C) were separately calendered into tapes using a ram extruder having a die 15 mm. in length and 5 mm. internal diameter, the cylinder of the ram extruder having a 30 mm. internal diameter. 100 parts by weight of the sample powder and 26 parts by weight of liquid paraffin were thoroughly mixed, charged into a cylinder, preformed under a pressure of 100 kilograms per square centimeter for 3 minutes, heated at 50°C. for 10 minutes, and then ram extruded at a speed of 17 millimeters per minute.

The extruded product was passed between two rolls of 6 inches kept at 50°C., rotating at a speed of 5 rpm. The distance between the rolls was adjusted to 3, 1.5, 1.0, 0.5, 0.3, 0.2 and finally to 0.1 millimeter, and the product successively passed through the rolls each time the distance was narrowed. The final width of the tape of each sample, all of which possessed a thickness of 0.1 millimeter, was measured and recorded as its "tape width." The dielectric strength, in volts per 0.1 millimeter thickness, was determined for each sample, using the test designated for "tetrafluoroethylene resin tapes" in JIS-K-6887. The degree of stress of each sample was computed for each sample, using the technique previously described. The results of these tests are summarized below.

Properties of Tapes Prepared from Internally Stressed and
from Unstressed Polytetrafluoroethylene Fine Powders

Product	Tape Width (mm.)	Dielectric Strength (volts/0.1 mm.)	Degree of Stress
Conventional PTFE Powder	42	4,000	35
PTFE Powder A	55	7,000	60
PTFE Powder B	52	9,000	72
PTFE Powder C	48	6,500	–

As shown, kneading the polytetrafluoroethylene fine powder in the presence of an organic solvent capable of wetting polytetrafluoroethylene increased the degree of stress of the polymer particles, which resulted in a decrease of the ability of the product to slide in the direction of roll rotation upon calendering. Moreover, the anisotropism of the resultant tapes was lower with respect to both their length and width, while the width of the tape and its dielectric strength were increased.

Interestingly, when polytetrafluoroethylene fine powders having a degree of stress in excess of 40 and produced in accordance with this process are compression molded, less cracks appear than with conventional (unkneaded and unstressed) polytetrafluoroethylene fine powders, probably because the attraction for micelles is sufficiently strong because of the high internal energy level to resist the normal stresses of sintering. Electron microscope magnifications of 4,000X of conventional polytetrafluoroethylene fine powders and the stressed fine powders reveal little difference which indicates that this technique for stressing the powder avoids agglomeration of the polymer particles, yet the differences in the degree of stress of each product may be clearly demonstrated by examination in a polarizing microscope.

Polytetrafluoroethylene

Polymerization in the Presence of Trifluorotrichloroethane

Y. Kometani, S. Koizumi, S. Fumoto, S. Tanigawa and T. Nakajima; U.S. Patent 3,462,401; August 19, 1969; assigned to Daikin Kogyo Co., Ltd., Japan describe a method of preparing nearly spherical granular polytetrafluoroethylene molding powder. This process comprises polymerizing tetrafluoroethylene at a temperature of 0° to 100°C. at a pressure of 1 to 40 atmospheres in the presence of water and an organic liquid boiling below 150°C., such organic liquid being one in which chain transfer does not readily take place and which disperses in the water in the form of drops. A water-soluble catalyst is employed, the reaction system being stirred until the organic liquid is completely dispersed in the water and the polymerization being conducted until the resulting polymer ranges between 0.1 and 5 grams per ml., based on the volume of the organic liquid.

When the polymerization reaction is carried out according to this process, the adhesion of polymer to such as the polymerization vessel wall or agitator does not occur and moreover the polymer obtained can be used as a molding powder by merely removing the organic liquid, washing and drying without the need for any such post-treatments as pulverizing. The organic liquid to be used in this method must be one in which chain transfer takes place with relative difficulty when polymerizing tetrafluoroethylene, and another important condition is that it must be almost insoluble in water.

For obtaining nearly spherical granular polytetrafluoroethylene powder, the organic liquid must be drops in the aqueous medium in the dispersed form. This is important, for if an organic liquid which is soluble in an aqueous medium is used, it would not become a dispersion in the aqueous medium and hence spherical granular particles could not be obtained. Further, it is preferred that the organic liquid has a boiling point of below 150°C., and more preferably below 100°C. The reason for this is that although the organic liquid must be a liquid under the polymerization conditions, the elimination from the polymer particles after completion of the polymerization becomes difficult in the case of those boiling above 150°C.

Further, the use of an active organic liquid is unsuitable, because of the formation of low molecular weight polymers by the setting up of so-called chain transfer during the polymerization reaction by means of the reaction of the polymer chain with said organic liquid. As examples of the organic liquids which satisfy these conditions, included are the fluorine-containing halogenated saturated hydrocarbons such as tetrachlorodifluoroethane ($CCl_2F—CCl_2F$), trichlorotrifluoroethane ($CCl_2F—CClF_2$), dichlorotetrafluoroethane ($CClF_2—CClF_2$), trichlorofluoromethane (CCl_3F), monochlorohexafluoropropane ($HCF_2CF_2CF_2Cl$) and monochlorooctafluorobutane ($HCF_2CF_2CF_2CF_2Cl$). The following examples illustrate this process.

Example 1: One million parts of deoxygenated water and 300,000 parts of trichlorotrifluoroethane were charged in a polymerization vessel equipped with an anchor-type agitator. After the air was thoroughly eliminated, tetrafluoroethylene was supplied at 3°C. until the pressure in the vessel had built up to 6 atmospheres during which time the agitation was continued at a speed of 400 rpm. This was followed by the addition of 10 parts of ammonium persulfate, 5 parts of sodium bisulfite and 5 parts of ferrous sulfate.

As a drop in the pressure occurred with the start of the polymerization reaction, tetrafluoroethylene was continuously supplied in accordance with its rate of consumption and the pressure was maintained at 6 atmospheres. After having consumed 300,000 parts of tetrafluoroethylene in about 2 hours, the unreacted tetrafluoroethylene was liberated and the reaction was terminated. The reaction vessel was then opened and 300,000 parts of polytetrafluoroethylene containing trichlorotrifluoroethane were taken out.

No adhesion of polymer to the walls of the polymerization vessel was observed. After elimination of the trichlorotrifluoroethane, the polymer powder obtained by washing and drying the reaction product had an average particle size of 1,000μ and practically all the particles were granules of nearly spherical shape having a softness such that they could be flattened between the fingers. The other properties of the powder were as follows: Angle of repose, 30.2°, specific surface area of nitrogen adsorption, 3.0 square maters per gram; surface smoothness, 60 seconds; tensile strength, 2.6 kilograms per square millimeter; elongation, 300%; average dielectric strength, 9200 volts/0.1 mm.; anisotropic expansion factor, 1.13; molecular weight, 8,200,000.

Example 2: The polymerization reaction was carried out under identical conditions as in Example 1, except for the following changes. Instead of 300,000 parts of trichlorotrifluoroethane, 300,000 parts of monochlorohexafluoropropane were used; instead of 10 parts of ammonium persulfate, 5 parts of sodium bisulfite and 5 parts of ferrous sulfate, 8 parts of ammonium persulfate and 4 parts of sodium bisulfite were used; a temperature of 25°C. was used instead of 3°C., and besides the same agitator as in Example 1 but operating at 600 rpm, another agitator equipped with propeller-type vanes and having an agitation speed of 800 rpm was installed in the bottom of the vessel.

No polymer adhesion to the walls of the polymerization vessel was noted. The polymer powder washed and dried after elimination of the monochlorohexafluoropropane had an average particle size of 700μ, and practically all the particles were granules of nearly spherical shape. The other properties of the resulting powder were as follows: Angle of repose, 35°; specific surface area of nitrogen adsorption, 3.0 square meters per gram; tensile strength, 2.5 kilograms per square millimeter; elongation, 280%; average dielectric strength, 8500 volts/0.1 mm.; molecular weight, 7 million.

Polytetrafluoroethylene

In related work, J. Hoashi; U.S. Patent 3,345,317; October 3, 1967; assigned to Thiokol Chemical Corporation has found that trifluorotrichloroethane, when incorporated in an aqueous emulsion reaction medium containing the fluoroolefin, an emulsifying agent for the fluoroolefin, and a water-soluble catalyst system, is particularly effective in stabilizing the emulsion during the polymerization reaction. The trifluorotrichloroethane, which is preferably the isomer 1,1,2-trifluoro-1,2,2-trichloroethane, should be refined to remove any olefinic impurities prior to use in this process, since even micro-quantities of an olefinic impurity are sufficient to copolymerize with the tetrafluoroethylene and impair the thermal stability of the resultant polymer. The following example is illustrative of this process.

Example: 1.5 liters of deionized water containing 0.9% by weight of $H(CF_2)_6COONH_4$ and 10% by weight of 1,1,2-trifluoro-1,2,2-trichloroethane were charged to a four-liter autoclave fitted with an agitator, and the system purged with nitrogen to remove all oxygen. The autoclave was then evacuated and tetrafluoroethylene introduced under pressure until the internal pressure in the autoclave reached 6 atmospheres. After raising the temperature of the aqueous reaction medium to 30°C., a water-soluble catalyst system of 0.01% by weight of ammonium persulfate, 0.0005% by weight of sodium acid sulfate and 2 ppm of ferrous sulfate was added to the reaction medium, and the mixture agitated at a speed of 120 rpm, under which conditions the emulsion polymerization of tetrafluoroethylene occurred. During the polymerization, make-up tetrafluoroethylene was added to the autoclave to maintain a constant internal pressure of 6 atmospheres.

The polymerization reaction was stopped after about 30 hours, at which time the solids concentration of the polymer was about 30% by weight of the weight of the resultant aqueous dispersion. Any unreacted tetrafluoroethylene in the autoclave was recovered for further use. After the specific gravity of the emulsion was adjusted to 1.08, the polytetrafluoroethylene was coagulated by agitation, using a turbine-type agitator, the total time for coagulation being only 10 min. The resultant powder was recovered from the liquid aqueous emulsion and dried at 120°C.

Excellent results were also obtained when the example was repeated, using small amounts of polymerizable comonomeric fluoroolefins, such as hexafluoropropene and trifluorochloroethylene, in conjunction with tetrafluoroethylene. Moreover, when the reaction was repeated again using a reaction temperature of 40°C., the resultant emulsion contained 35% by weight of polymer solids 30 hours after the onset of polymerization. Furthermore, when $F(CF_2)_7COONH_4$ was used as the emulsifying agent instead of $H(CF_2)_8COONH_4$, all other factors being equal, an emulsion containing 35% by weight of polymer solids in about 30 hours of polymerization resulted.

Test were conducted to compare the physical properties of polytetrafluoroethylene powder produced by conventional emulsion polymerization and by this process. Each polymer was subjected to an extrusion test which consisted of thoroughly mixing 100 parts by weight of the polytetrafluoroethylene and 25 parts by weight of white oil, following which the mixture was charged to a cylinder having a 30 mm. diameter with a 5 mm. die diameter. The cylinder was immersed in a water bath at 60°C. for 10 minutes, and the mixture was extruded at the rate of 17 mm. per minute, using a Universal Testing Machine RS-10 (manufactured by Shimazu Seisakusho Co.) and the extrusion pressure measured. In addition, the two polytetrafluoroethylene powders were sintered at 360°C., and any discoloration noted. The results of these tests are set forth below.

Comparison of Two Polytetrafluoroethylene Powders Produced by Emulsion Polymerization

Polytetrafluoroethylene	Extrusion (kg./cm.^{2})	Properties	
		Coloration After Sintering	Appearance of Extrusion
This Process	24	None	No cracks
Conventional Process	45	Light brown	Some cracks

Copolymerization with Hexafluoropropene

A process described by Y. Kometani, M. Tatemoto and S. Fumoto; U.S. Patent 3,331,822; July 18, 1967; assigned to Thiokol Chemical Corporation involves an improvement in a process for the production of polytetrafluoroethylene molding powder by the aqueous suspension polymerization of tetrafluoroethylene, in which monomeric tetrafluoroethylene is polymerized by bringing the monomer into contact with an aqueous medium containing a free radical catalyst system, which comprises increasing the bulk density and enhancing the flow characteristics of the polymer by copolymerizing the monomeric tetrafluoroethylene with from 0.1 to 10% by weight of a perfluoroolefin containing from 3 to 4 carbon atoms through the addition of the perfluoroolefin to the aqueous polymerization medium, thereby producing a molding powder substantially composed of spheroidal particles having a high bulk density and characterized by improved resistance to coalescence. The following examples are illustrative of this process.

Example 1: 1.5 liters of demineralized water were added to a 3-liter autoclave equipped with a screw-type agitator, and the autoclave deaerated under vacuum. After complete deaeration of the water and evacuation of the autoclave, a gaseous mixture consisting of tetrafluoroethylene and 1% by weight of hexafluoropropene was injected into the auto-clave at a pressure of 5 kg./cm.^{2} gauge, while maintaining the temperature of the aqueous polymerization medium at 60°C. The agitator was then started and 0.01 gram of ammonium persulfate added to the reaction medium. After the pressure in the autoclave started to decrease, tetrafluoroethylene (with 1% by weight of hexafluoropropene) was con-tinuously injected into the autoclave to maintain a steady state reaction pressure.

When the reaction had proceeded for 3 hours, during which period 400 grams of tetrafluoroethylene had been consumed, the pressure was reduced and the polymerization reaction halted. The polymer was a milky white powder which was ob-served, under microscopic examination, to be composed substantially of spheroidal particles having an average particle size of 200 microns. The powder rapidly settled to the bottom of the aqueous polymerization medium and was easily re-covered. After washing with distilled water and drying, the powder was found to have a bulk density of 0.6 gram per cubic centimeter and superior flow characteristics. This powder can be used to produce a uniformly smooth film of 2 mm. thickness on a thin substrate, and possessed excellent mechanical properties.

Example 2: Using identical reaction conditions to those described in Example 1, tetrafluoroethylene was copolymerized with 2.5% by weight of hexafluoropropene over a period of 4 hours, yielding 400 grams of a powder which was similar in appearance to the product obtained in Example 1. Upon examination, the powder was found to be composed substan-tially of spheroidal particles having a bulky density of 0.7 grams per cubic centimeter.

Difluoroketene Units

E.N. Squire and W.F. Gresham; U.S. Patent 3,525,724; August 25, 1970; assigned to E.I. du Pont de Nemours and Company describe fluorocarbon polymers comprising at least 0.1 mol percent difluoroketene units, $-CF_2-C(O)-$, which are useful as molding resins. The process for preparing the polymers involves reacting fluorocarbon homo- or copolymers comprising alkyl trifluorovinyl ether units, $-CF_2-CF(OR)-$, with a Lewis acid selected from the class consisting of sulfur trioxide, aluminum trichloride, and titanium tetrachloride, whereby at least part of the ether units are converted to the ketene units. The reaction can be conducted in the presence of a liquid medium, such as a fluorocarbon, in an inert atmosphere at a temperature in the range of 20° to 200°C. for about 1 to 75 hours. Tetrafluoroethylene is referred to as TFE; methyl trifluorovinyl ether, as MTVE; perfluoro-2-methylene-4-methyl-1,3-dioxolane, as PMD; and difluoro-ketene units, $-CF_2C(O)-$, as DFK.

The starting polymer in this process, i.e., a fluorocarbon polymer or copolymer comprising alkyl trifluorovinyl ether units, can be synthesized as follows. First, the alkyl trifluorovinyl ether monomer, $CF_2=CF(OR)$, can be prepared according to U.S. Patent 2,917,548. Then that ether monomer can be allowed to polymerize with itself, or can be copolymerized with TFE according to U.S. Patent 3,159,609. Moreover, comonomers other than TFE and an alkyl trifluorovinyl ether can additionally be employed in the synthesis of the starting polymer.

The alkyl trifluorovinyl ether starting polymer should be supplied to the reaction system in a state of subdivision which will allow an appreciable rate of conversion. Hence, suspensions of the starting polymer as prepared may be employed, and are preferred in this process. Should the starting polymer have been recovered and dried prior to reaction, it is pre-ferred that its state of subdivision be reduced, for example, in a blender or in a reaction flask equipped with a chopper blade stirrer.

In the following examples, quantities of monomer units in the polymers are expressed in terms of mol percent unless otherwise noted. The monomer content of the ether polymers before reaction with a Lewis acid and of the ketene poly-mers was usually determined by infrared analysis. The results of the examples which follow are given in the table below. These results were obtained by the following method, unless otherwise indicated. Films of the polymer about 1-mil thick (about 1 to 4 mils thick for polymers comprising PMD monomer) were pressed as described in each example and infrared spectra were obtained on either a Perkin-Elmer Infracord Spectrometer or a Perkin-Elmer Model 21 Infrared Spectrometer. The mol percentage of MTVE in these polymers is proportional to the infrared absorbance at about 10.4 microns, that of TFE at about 4.25 microns, that of DFK at about 5.6 microns, and that of PMD at about 9.95 microns.

Example 1: The preparation of a TFE/DFK copolymer employing sulfur trioxide as the Lewis acid is described herein. (A) A TFE/MTVE copolymer was prepared by placing in a 330 ml. shaker tube 100 ml. of CCl_2FCClF_2 and 0.04 gram perfluoropropionyl peroxide. Then the tube was cooled to -80°C. and evacuated. There was added to the tube 20 grams of MTVE and 40 grams of TFE. The tube was sealed and shaken for 64 hours at room temperature. The tube was opened and vented, and a small sample of the suspension was dried in a vacuum oven at 110°C. and pressed into a film at 330°C. and 600 psi. Infrared analytical data on the film are found in the table below.

(B) To the remainder of the suspension obtained in (A) there was added 20 ml. of sulfur trioxide and the shaker tube was resealed. The mixture was heated for 5 hours at 180°C. under autogenous pressure and similar agitation. The tube was opened and the volatile materials removed in vacuum at room temperature. The solid polymer was recovered and stirred with 100 ml. of concentrated nitric acid for one hour, then filtered and washed with three 50 to 100 ml. portions of boiling water. The product was dried in vacuum at 110°C. for 16 hours, then a sample was pressed at 175°C. and about 300 pounds per square inch into a transparent film. Infrared analytical data are found in the table.

Example 2: The preparation of a TFE/MTVE/DFK terpolymer, employing a sulfur trioxide as the Lewis acid is described. (A) A TFE/MTVE copolymer was prepared by placing in a 330 ml. shaker tube 150 ml. of CCl_2FCClF_2, 0.02 gram of perfluoropropionyl peroxide, and 21.4 grams of MTVE. The shaker tube was flushed with dry nitrogen, then 60 grams of TFE was pressured into the shaker tube. The tube was closed, shaken at 60°C. for one hour. The suspension was removed from the tube, and the polymer recovered and dried at 110°C. in a vacuum oven at 16 hours to yield a light, white powder. A portion thereof was pressed into a film at 300°C. and 300 psi. Infrared analytical data are found in the table.

(B) To a similar shaker tube there was added 20 grams of the copolymer obtained in (A), 200 ml. of CCl_2FCClF_2, and 5 ml. of sulfur trioxide, under anhydrous conditions. The tube was closed and heated for 2 hours at 140°C. with agitation. Then volatile materials were removed in vacuum at room temperature. The solid polymer was then cleaned with nitric acid and water as in Example 2 (B) and dried in vacuum at 110°C. for 16 hours to yield 19.9 grams of polymer. A sample of the polymer was pressed into a transparent colorless film at 175°C. and about 300 psi. Infrared analytical data are found in the table.

Example 3: The preparation of a TFE/PMD/DFK terpolymer, employing sulfur trioxide as the Lewis acis, is described. (A) A TFE/MTVE/PMD terpolymer was synthesized as follows. In a 330 ml. stainless-steel tube there was placed 150 ml. of CCl_2FCClF_2 and 0.5 gram perfluoropropionyl peroxide. The tube was then cooled to -80°C. and evacuated. Then there was added to the tube 3.2 grams of MTVE, 5.0 ml. of PMD, and 60 grams of TFE while the tube was held at -80°C. The tube was then sealed, warmed to 60°C., held at that temperature for 30 minutes while being shaken, then cooled to room temperature. The solid polymer was recovered by filtration, then dried in vacuum at 110°C. for 16 hours. A sample was pressed into a film at 340°C. and 300 psi. Infrared analytical data are found in the table. The particle size of the polymeric product was then reduced to a finer state of subdivision by placing 20 grams thereof and 150 ml. of CCl_2FCClF_2 in a Waring Blendor and shearing the mixture for 30 minutes at the highest speed at which the blender could be run without causing the polymer to be ejected from the blender, thus producing a suspension.

(B) The suspension produced in (A) was transferred to a 330 ml. shaker tube and 20 ml. of sulfur trioxide was introduced also. The tube was flushed with nitrogen, sealed, then shaken for 2 hours at 180°C. Then the tube was allowed to cool to 55°C. and the volatile materials were removed in vacuum at that temperature. The polymer was washed with a 100 ml. portion of concentrated nitric acid at 40° to 80°C. and three 50 to 100 ml. portions of distilled water at 40° to 80°C. The polymer was then dried in vacuum at 110°C. for 16 hours. A sample of the polymer was then pressed at 340°C. and about 600 psi into a transparent film. Infrared analytical data are found in the table below.

Composition of Polymers as Determined by Infrared Analysis

Example	Mol Percentage of Monomer Units in Polymer			
	TFE	MTVE	DKF	PMD
1 (A)	63.6	36.4	-	-
1 (B)	63.6	0	36.4	-
2 (A)	69	31	-	-
2 (B)	71	25	4	-
3 (A)	92.5	3.5	-	4.0
3 (B)	92.5	0	3.5	4.0

MISCELLANEOUS

Gas Phase Polymerization in the Presence of Finely Divided Catalytic Support

A process described by R. Fuhrmann and D. Jerolamon; U.S. Patent 3,304,293; February 14, 1967; assigned to Allied Chemical Corporation involves the polymerization of tetrafluoroethylene in the presence of a finely divided catalytic support which can act as a filler. Several processes have been discovered in which tetrafluoroethylene can be polymerized to a high molecular weight polymer. One of these is the method described in U.S. Patent 2,847,391. In this patent a technique is described whereby homogeneous filled compositions of polytetrafluoroethylene can be obtained by utilization of a finely divided catalyst such as silica gel, silica clays, alumina impregnated with boric or phosphoric acid, granular magnesia, and hydrated orthosilicate at atmospheric pressure and at temperatures between -20° and 150°C. However, the production rates fall off very sharply as the reaction is continued to high levels of polymer production such as 3:1 by weight of polymer:catalyst and higher. It has been found that upon limited mixture with salts of oxy acids of hexavalent chromium, such as magnesium chromate, commercial silica gels having surface area of at least 100 m.^{2}/g.

and average pore diameter of at least 40 A. catalyze the polymerization of tetrafluoroethylene at good rates over prolonged periods. The sum of weights of aluminum and chromium, mixed with the silica gel, is about 1 to 10% by weight based on the sum of the weights of the silica and the admixed alumina and chromium-containing salts.

The process is illustrated by the following examples. The tetrafluoroethylene monomer used in the examples was kept in a cylinder surrounded by an aluminum sleeve whose diameter was about 2 inches wider than the cylinder. The entire system was kept in a box packed with Dry Ice, the temperature in the space between the aluminum sleeve and the cylinder being maintained at −30° to −40°C. The tetrafluoroethylene cylinder was connected to a glass Y joint, one arm of which admitted dry nitrogen. The second arm of the Y led to a flowmeter, and then to one neck of a 500 ml. 3-necked flask. The gases passed out via another neck of the flask, fitted with a straightbore air condenser, to a tube packed with glass wool, through a second flowmeter, and from there through a mineral oil bubbler to the hood chimney. The center neck of the reactor carried a polytetrafluoroethylene stuffing box provided with a glass stirring rod with polytetrafluoroethylene stirring blade.

The polymerization catalyst used in the table below was prepared by adding to the gel an aqueous solution of magnesium chromate in amount to form a slurry and at concentration to provide the desired chromium content. The slurry was agitated at about 100°C. several hours, until the gel appeared dry; then the solids were crushed. The crushed material was activated by packing about 20 grams of it between glass wool plugs in a Pyrex tube, and inserting the tube into a tubular oven which was heated to about 400°C. A thermocouple encased in a glass sleeve was used to measure the temperature of the catalyst during the heating.

The activation process was carried out first in an atmosphere of air and then in nitrogen. Both gases were predried by passing them successively over anhydrous calcium sulfate and phosphorus pentoxide. A bubble counter was affixed to one end of the Pyrex activation tube. After the desired flow rate of one bubble per second was set the catalyst was heated in air for 15 to 16 hours and then in nitrogen for 1 to 2 hours. Following the activation the Pyrex tube was permitted to cool under nitrogen flow, then stoppered and placed in a "dry box," i.e., a large container swept by pure dry, oxygen-free nitrogen to exclude air. A dry, stoppered reaction flask, complete with stirrer and stuffing box was also placed in the dry box and tared. It was then flushed with pure nitrogen and loaded with the activation catalyst and again weighed to give the weight of the catalyst by difference.

When removed from the dry box the reaction flask was rapidly connected to the source of tetrafluoroethylene. A good flow of pure nitrogen was started and the exit neck of the flask was then connected to the condenser. The flask was then immersed in a thermostat kept at the desired temperature and the nitrogen flow was decreased with a concomitant increase in tetrafluoroethylene flow. Once a flow of tetrafluoroethylene was established, the nitrogen was turned off and the starting time of the reaction was noted. To stop the reaction nitrogen was turned on, the tetrafluoroethylene was turned off, and a nitrogen purge of 5 minutes was then applied to the reaction flask. The reaction flask was then stoppered and weighed.

The results are presented in the following table where temperatures are in degrees centigrade and weight is in grams. The catalyst designated Houdry S-65" is a commercial silica-alumina cracking catalyst (about 90:10 by weight silica:alumina; average pore diameter of about 40 to 50 A., surface area about 375 square meters per gram). That designated Syloid G-72 is a commercial silica gel containing 99.2% silica (ignited basis); and having average pore diameter of about 120 A. and surface area about 370 square meters per gram.

Tetrafluoroethylene Polymerization Using Silica and Silica Alumina Catalysts

Ex.	Catalyst	Activation				Polymerization Temp.	Polymerization Time, Hrs.	Weight Catalyst	Weight Gain	Rate, g./g./hr.	Percent Polytetrafluoroethylene in Product
		Air		Nitrogen							
		Hrs.	Temp.	Hrs.	Temp.						
1	Houdry S-65	16	444	1.5	387	28±2	5.0	20.6	36.8	0.36	64.1
2	Houdry S-65	16	449	1.5	390	49±1	5.0	20.6	28.0	0.27	57.6
3	Houdry S-65 plus 0.1% MgCrO₄	15	510	1.5	400	26–34	5.0	19.8	42.0	0.42	68.1
4	Houdry S-65 plus 0.1% MgCrO₄	15	500	1.5	400	25 / 49	0.25 / 5.75	19.9	50.3	0.42	71.6
5a	Syloid G-72 plus 15% MgCrO₄	16	500	2.0	400	48	5.0	11.9	14.6	0.24	55.0
5b	Syloid G-72	16	500	2.0	400	25	5.0	11.9	4.0	0.07	25.0

NOTE — The very low rate in 5b above compared to the other runs shows the effectiveness of the alumina and/or chromium-containing salt additives in the catalysts.

Gas Phase Reaction Using Carbonyl Initiator

A process described by E.T. Clocker; U.S. Patent 3,475,306; October 28, 1969; assigned to Ashland Oil & Refining Company comprises forming a gaseous mixture of a polymerizable halogenated olefin and a carbonyl compound having the general formula, as shown on the following page.

Polytetrafluoroethylene

R_1 is a monovalent hydrocarbon radical, monovalent halogenated hydrocarbon radical, halogen, or hydrogen; R_2 is an R_1 radical or a hydroxyl group when R_1 is a monovalent hydrocarbon or halogenated hydrocarbon radical. The compound is then exposed to actinic radiation or to temperatures in the range of 200° to 400°C. in the absence of actinic radiation and a polymer of the halogenated olefin is recovered. Specific examples of suitable initiators include acetic acid, acetaldehyde, formaldehyde, acetyl chloride, acetyl fluoride, acetyl bromide, dimethyl ketone, methylethyl ketone, trifluoroacetic acid and chlorodifluoroacetic acid. The halogeanted olefins which can be polymerized by this process include all common halogenated olefins found to be polymerizable by free radical initiation.

Examples 1 – 4: Into a nitrogen-purged evacuated 250 ml. Vicor 7910 glass reaction chamber was charged a mixture of the gases indicated in the table below until atmospheric pressure was established in the reaction system. The reaction chamber was connected to a calibrated reservoir containing a floating plunger which allowed measurement of gas volume removed from the reservoir. This reservoir was also filled to atmospheric pressure with the gas mixture shown in the table.

The reaction chamber was then exposed to actinic radiation from a Hanovia lamp and the rate of gas depletion corresponding to the rate of tetrafluoroethylene polymerization was measured. The polymerization was permitted to go to completion. In each instance the tetrafluoroethylene was polymerized to a solid polymer weighing about one gram.

Example	Reaction Mixture	Relative Polymerization Rate
1	95% TFE;* 0% CF_3COCl	40 ml./5 min.
2	90% TFE; 10% CF_3COCl	50 ml./5 min.
3	80% TFE; 20% CF_3COCl	55 ml./5 min.
4	90% TFE; 10% CH_3COCl	80 ml./5 min.

*Tetrafluoroethylene

Example 5: A square glass reactor, the top side of which comprised a removable quartz plate, was equipped with gas charging and evacuating means, hot plate, and thermocouple. A metal panel was placed on the hot plate. The reactor was sealed, purged with nitrogen and evacuated. The metal panel was then heated to 250°C. and exposed to light of a Hanovia lamp through the quartz plate. The reactor was then charged to substantially atmospheric pressure with a 10% trifluoroacetyl chloride, 90% tetrafluoroethylene mixture. Additional quantities of the mixture were charged to maintain the pressure as polymerization proceeded. A powdery coating formed on the metal panel. The resulting coating was heated to 327°C. to form a coherent continuous coating on the metal panel.

Fibrous Powders for Filter Paper

Y. Kometani, S. Koizumi, K. Kubota and T. Nakazima; U.S. Patent 3,513,144; May 19, 1970; assigned to Daikin Kogyo Co., Ltd., Japan describe fibrous polytetrafluoroethylene powders which are suited for the production of a strong paper predominantly of polytetrafluoroethylene whose thickness is less than 200 g./m.^{2} and also thick but pliable thick sheets and paper boards and filter fiber. The polytetrafluoroethylene fibrous powders are characterized by having an average fiber length of 100 to 5,000 microns, an average shape factor of not less than 10 and an anisotropic expansion factor of 1.3 to 7.0. The following examples illustrate this process.

Example 1: (A) Commercial grade polytetrafluoroethylene fine powder having a specific surface area of 9 m.^{2}/g. (B) Polytetrafluoroethylene having a specific surface area of 11 m.^{2}/g. polymerized in the vapor phase tetrafluoroethylene by means of gamma rays from cobalt 60. (C) Polytetrafluoroethylene having a specific surface area of 3.5 m.^{2}/g. obtained by the suspension polymerization in the gaseous state at 60°C. of a gaseous tetrafluoroethylene whose proportion by weight of hexafluoropropylene at the start of the polymerization was 1%, using ammonium persulfate as catalyst. (D) Polymer having a specific surface area of 2.5 m.^{2}/g. obtained by suspension polymerization in water at 3°C. of liquid tetrafluoroethylene, using ammonium persulfate, acid sodium bisulfite and ferrous sulfate. The foregoing four types of polytetrafluoroethylene were ground using the following two types of pulverizers.

(1) Micron Mill — A pulverizer of the type which grinds chiefly by the action of a shearing force resulting from the rotation of multibladed rotor (a product of Hosokawa Iron Works, Ltd.).

(2) Ultramizer — A pulverizer of the type which grinds chiefly by the action of impact force resulting from the pounding and crushing action of hammers fitted in such a fashion to the circumference of rotating disks that they are capable of freely moving within the plane of the rotating disks. The properties of the ground products obtained are given in the table as shown on the following page.

Polytetrafluoroethylene

Specimen	Pulverizer	Grinding conditions				Properties of ground product			
		Peripheral speed of blades or hammer, m./sec.	Temperature, °C.	Classifier	Form	Average fiber length, μ	Average shape factor	Anisotropic expansion factor	Paper formability
Experiment:									
1......A	Micron mill	45	100	Used	Fibrous	950	38	5.2	Good.
2......A	...do	35	30	...do	...do	850	30	1.70	Do.
3......A	...do	35	30	Not used	...do	2,600	38	1.75	Do.
4......C	...do	25	95	Used	...do	900	36	5.1	Do.
5......C	...do	25	80	...do	...do	800	30	4.5	Do.
6......B	...do	35	40	...do	...do	360	25	1.72	Do.
7......B	...do	35	85	Not used	...do	1,900	35	5.3	Do.
8......D	...do	35	30	Used [1]	Nonfibrous	—	[3] 5	1.22	Unsatisfactory.
9......A	Ultramizer	95	30	...do [2]	...do	—	[3] 3	1.20	Do.
10......B	...do	95	30	...do	...do	—	[3] 3	1.20	Do.

[1] Basket type classifier.
[2] Centrifugal type classifier.
[3] Values obtained by measurement by method described in U.S. Pat. 2,936,301, since these were nonfibrous.

When the thermal properties of the fibrous powder obtained in Experiment 1 was measured at the rate of a rise in temperature of 10°C. per minute using a differential thermal analyzer Model DT-10 produced by Shimadzu Seisakusho Ltd., shoulders of heat absorption were observed 25°C. after the heat absorption peak of 348°C. ascribable to the melting of the crystals.

Example 2: Papers were molded using the various powders obtained from Example 1 and fibers cut from one of the commercial grades of polytetrafluoroethylene spun fibers to about 1 mm. length by fiber cutting procedures. Three grams of the fibrous powders or cutting fibers were added to about 300 cc of carbon tetrachloride and stirred well. The dispersions obtained were placed in 60-mesh metallic sieves 144 mm. in diameter. The sieves were immersed in advance in Petri dish containing carbon tetrachloride. Thus, by shaking the sieves, the dispersion is caused to spread out uniformly in the sieves. Then the sieves are removed from the Petri dish containing the carbon tetrachloride and dried, following which the powders or felts are heated along with the sieves to 345°C. in an air oven for about 30 minutes.

Papers 144 mm. in diameter and of a uniform thickness of about 0.2 mm. were obtained by this operation from the fibrous powders obtained in Experiments 1 to 7 of Example 1. On the other hand, papers could not be formed from the powders obtained in Experiments 8, 9 and 10 and the fibers obtained by fiber cutting. Namely, in the case of the nonfibrous powders, the self-bonding between the particles was poor and further because they were not fibrous they could not be formed in the fashion of a paper but became a weak foraminous film. In the case of the fibers obtained by fiber cutting, the shrinkage of the fibers was great and hence paper-like products did not result, there occurring unevenness of spotty nature over the surface.

By a similar method, the fibrous powders obtained in Experiments 1 through 7 of Example 1 could be formed into thick sheets 1 mm. in thickness. Not only were these papers and sheets pliable and could be folded, but their strength was also several times that of the ordinary paper. A particular feature is the point that there was no difference in their strength whether in water or air. This paper was effectively used particularly as filter paper in the filtration of strong acids and alkalies.

Polymerization in the Presence of Irradiated PTFE

P.E. Muehlberg, G.C. Jeffrey and L.W. Harriman; U.S. Patent 3,170,858; February 23, 1965; assigned to The Dow Chemical Company describe a process for preparing tetrafluoroethylene polymers suitable for making fluid-permeable filters. U.S. Patent 2,819,209 describes a process for making fluid-permeable filters by sintering a layer of particles of polytetrafluoroethylene or polychlorotrifluoroethylene while confining the layer under a pressure of from 30 to 50 psi at a temperature above the softening point and below the melting and decomposition temperatures of the particles.

It has been discovered that fluid-permeable filters can readily be prepared from tetrafluoroethylene polymers which have been prepared by polymerizing monomeric tetrafluoroethylene in contact with particles of a finely divided solid tetrafluoroethylene polymer that has previously been subjected to the action of high energy ionizing radiation in a reduced atmosphere not exceeding 100 millimeters, and preferably at about 100 microns or less, of Hg, absolute pressure, with an effective dose to initiate polymerization of the monomeric tetrafluoroethylene, continuing the polymerization until an amount of the monomer corresponding to at least 50% by weight of the starting polymer is polymerized, then irradiating the resulting polymer with high energy ionizing radiations in a reduced atmosphere not exceeding 100 millimeters, preferably not greater than 100 microns of Hg absolute pressure, with an effective dose to initiate polymerization of monomeric tetrafluoroethylene, and thereafter continuing the polymerization of monomer until the amount of total polymer corresponds to at least about an equal weight of the initial irradiated polymer starting material. The cycle of operations is preferably repeated for a plurality of times such that the weight of the final product is several times the weight of the initial polymer starting material. The following examples illustrate this process.

Example 1: A charge of 10 grams of powdered polytetrafluoroethylene was placed in a 4 liter capacity glass vessel. The vessel was evacuated to about 5 microns of Hg absolute pressure, then was filled with gaseous tetrafluoroethylene monomer and was reevacuated to 5 microns of Hg absolute pressure. The vessel was placed adjacent to a 500 curie cobalt 60 source and the polymer therein subjected to the action of gamma rays in a field of an intensity of 0.05 megarad for a period of one hour. The radiation source was removed. The reaction vessel was filled with gaseous monomeric tetrafluoroethylene to 10 pounds per square inch gauge pressure. Flow of monomeric tetrafluoroethylene into the vessel was maintained at 10 pounds per square inch gauge pressure.

After the rate of polymerization had decreased to a value corresponding to a flow rate of about 2 ml./min. of monomer, the reaction vessel was reevacuated to about 5 microns of Hg absolute pressure, the polymer therein was reirradiated to activate the same, the vessel was then filled with gaseous tetrafluoroethylene to 10 pounds per square inch gauge pressure, and the cycle of polymerization of the monomer repeated. In the experiment the polymer in the reaction vessel was subjected to six such cycles of evacuation, irradiation and monomer polymerization. It was observed that the volume of polymer in the reaction vessel increased during each polymerization period.

At the end of the sixth polymerization period, there was recovered 1,685 grams of product as a friable particulate material. It was pulverized and found to have a "sandy" or "granular" feel, distinctly different from the "fibrous" or "greasy" feel of the polytetrafluoroethylene starting material. The product was polytetrafluoroethylene by analysis. A portion of the product was formed into a sheet about 1/16 inch thick by pressing with 3,000 lbs. per square inch gauge pressure, and was sintered by heating the pressed sheet in an oven at a temperature of 380°C. for a period of 40 minutes, then was cooled to room temperature. The sintered sheet was porous and permeable to the flow of fluids, e.g., air. A summary of the experiment is reported in the table below. The table identifies the cycle, gives the amount of polymer in the reaction vessel at the start of the cycle, and the irradiation dose in megarads to which the polymer was subjected before polymerization of monomer was continued.

Cycle No.	Polymer in Reaction Vessel, grams	Irradiation Dose, megarads
0	10	0.055
1	19.5	0.047
2	68	0.0155
3	234	0.0062
4	736	0.0029
5	1,230	0.001
6	1,685	-

Example 2: The experiment of Example 1 was repeated for five cycles of evacuating, irradiating, and polymerizing of monomer, for a period of 24 hours, then on the sixth cycle the polymerization was allowed to continue for a period of 8 days. A summary of the experiment is reported below.

Cycle No.	Polymer in Reaction Vessel, grams	Irradiation Dose, megarads
0	10	0.055
1	33.7	0.045
2	152	0.0173
3	480	0.0033
4	757	0.0041
5	962	0.0041
6	3,212	-

The polymer product was obtained as a friable cake. It was ground to a particulate form in a Waring Blendor. The granular polymer had a sand-like grainy feel or hand. Sheets prepared by pressing and sintering the pressed sheet at 380°C. were found to be permeable to fluids. The polytetrafluoroethylene prepared in the foregoing examples was found to be suitable for making porous sheets by pressing, then sintering the pressed sheet by heating at a temperature of 380°C. for a period of 40 minutes. The porosity or pore size of the sheet is controlled in part by the pressure employed to press the powder or granular material into a sheet.

Example 3: In each of a series of experiments, a portion of the combined batches of polytetrafluoroethylene prepared in Examples 1 and 2 above was pressed into a sheet 1/16 inch thick under a pressure as stated in the following table, then was sintered by heating the sheet at a temperature of 380°C. for a period of 40 minutes. The sintered sheet was tested to determine its tensile strength, its yield strength and the percent elongation. Other pressed and sintered test sheets were tested to determine the rate of flow of air and of water therethrough under a stated pressure differential, and to determine the diameter of the pores in the sintered sheet.

The table below identifies the experiments and gives the pressure in pounds per square inch gauge employed to press the granular polytetrafluoroethylene into a sheet, after which it was sintered by heating to a temperature of 380°C. for a period of 40 minutes. The sintered sheet was tested to determine its tensile strength, its yield strength and the percent elongation. Other pressed and sintered test sheets were tested to determine the rate of flow of air and of water therethrough under a stated pressure differential, and to determine the diameter of the pores in the sintered sheet. The table identifies the experiments and gives the pressure in pounds per square inch gauge employed to press the granular polytetrafluoroethylene into a sheet, after which it was sintered by heating to a temperature of 380°C. for a period of 40 minutes. The table also gives the mechanical properties determined for the sintered sheet, the rate of flow of air at room temperature through the sheet at several pressures, the rate of flow of water at room temperature through the sheet at several pressures, and the size of the pores in the sintered sheet.

Run No.	Pre-Form pressure, p.s.i.	Tensile Strength, lbs./sq. in.	Yield Strength, lbs./sq. in.	Elongation, Percent	Flow of Air cu. ft./min./sq. ft. at ΔP, p.s.i.			Flow of Water, gal./min./sq. ft. at ΔP, p.s.i.			Pore Size, Microns
					ΔP=2	ΔP=4	ΔP=8	ΔP=2	ΔP=4	ΔP=8	
1	30	724	474	53	78	136	220	1.8	7.4	28.5	15
2	50	1,304	688	109	52	100	184	0.6	2.3	17.0	19.5
3	100	1,750	943	167	36.4	64.5	153	1.2	4.2	12.1	19.2
4	1,000	1,818	838	100	8.5	9.1	24.2	0.2	0.4	0.85	7.25
5	5,000	2,187	1,456	114	0.96	2.56	3.6	0	0.005	0.17	5.35

Living Polymers — Addition of Difluorodiazirine

R.A. Mitsch and P.H. Ogden; U.S. Patent 3,493,629; February 3, 1970; assigned to Minnesota Mining and Manufacturing Company describe a living polymer comprising recurring units of the formula $-(CF_2)-$ forming a polymer chain, one end of the chain having an inactive terminal group and consisting of a fragment of a boron trifluoride catalyst attached to a $-CF_2-$ group, and the other end of the chain having an active terminal group consisting of the remainder of the boron trifluoride catalyst attached to a $-CF_2-$ group. The activity of the active terminal group is characterized by the polymer's capability to resume its growth at the active terminal group upon the addition of difluorodiazirine. It has been discovered that when exposed to the effect of ultraviolet light or heated to the point of pyrolysis, a mixture of difluorodiazirine,

and a Lewis acid, e.g., boron trifluoride, reacts to provide "living" polydifluoromethylene in substantial yield. The polymeric material provided by this process is represented by the formula, $X(CF_2)_nCF_3$, where X is a viable terminal group derived from a Lewis acid enabling the polymeric material to resume its growth at the viable terminal group upon the addition of difluorodiazirine under the influence of photolysis or pyrolysis, and n is a number representing the average number of carbon atoms in the polymer chains.

Although the insolubility of the polymer obtained precludes accurate determination of molecular weight values, it has been observed that molecular weights are directly proportional to monomer concentration and inversely proportional to Lewis acid concentration and reaction temperature. The reaction conditions employed also have an effect on molecular weights. Thus the value of n in the formula, $X(CF_2)_nCF_3$, can be varied within broad limits by altering the ratio of Lewis acid to difluorodiazirine and by employing various different reaction conditions. Some indication of this variation is found by reference to the following table where the Lewis acid is boron trifluoride, in that when the products are liquids, n is a relatively small number; where they are higher melting solids, n may be 50 or more.

Mols BF$_3$	Mols CF$_2$N$_2$	Reaction Conditions	Polymer Properties
1	10	Photolysis at 25°C. in trichlorotrifluoroethane.	Melted with decomposition at 250°C.
1	100	Photolysis at 25°C. in trichlorotrifluoroethane.	Melted without decomposition at 230°C.
1	1,000	Photolysis at 25°C. in trichlorotrifluoroethane.	Melted without decomposition at 300°C.
1	10	Photolysis at 25°C. in methylene chloride.	Mixture of yellow oil and white solid.
1	10	Photolysis at 25°C. in vapor phase.	Mixture of yellow solid and yellow oil.

The catalytic activity of the particular Lewis acid employed should depend on its ability to behave as an electron acceptor species for the charged linear intermediate,

$$:\bar{C}F_2N{=}\overset{+}{N}$$

However, because difluorodiazirine acts as a fluorinating agent under the conditions employed, it is preferred that a Lewis acid be selected which is not susceptible to fluorination. Several Lewis acids will undergo replacement in the presence of difluorodiazirine (e.g., BCl$_3$ to BF$_3$) and although the resulting fluorinated Lewis acid will catalyze the desired polymerization reaction, a corresponding amount of diazirine will have been expended uselessly in the replacement side-reaction. Thus the preferred Lewis acids are those which are not susceptible to fluorination such as boron trifluoride. This process can be carried out conveniently in the vapor phase or in a solvent. Suitable solvents are those which are inert to difluorodiazirine and the Lewis acid employed, with which they must also be compatible. They should not absorb ultraviolet light of wavelength between about 3000 and 3600 A. Halocarbons generally have been found to be the most suitable, particularly fluorocarbon solvents such as trichlorotrifluoroethane; however, non-fluorinated solvents such as carbon tetrachloride, chloroform, methylene chloride, etc., also have been found to be satisfactory.

The polymerization reaction is readily effected by mixing a catalytic amount of a suitable Lewis acid with difluorodiazirine, either in the presence of a compatible solvent, or in the vapor phase, and exposing the mixture to ultraviolet light or pyrolytic heat. If photolysis is to be employed the useful wavelengths of ultraviolet light may vary between about 3000 and 3600 A. Such actinic light can be conveniently obtained from most commercially available ultraviolet light sources. The most favorable results are obtained when a powerful source such as a General Electric 1000 watt BH-6 lamp is utilized. When employing pyrolysis the temperature range within which polymerization will proceed with acceptable results is between about 110° and 170°C. The most favorable results are obtained at about 120°C.

In addition to their utility based on catalytic activity whereby the polymers of this process can be employed to make block copolymers, or to catalyze polymerization of, e.g., diazoalkanes, the polymers can, when of relatively high molecular weight, be formed into desired configurations, as by sintering powdered polymer. Polymers of low molecular weight can be used as inert oils or greases, e.g., as lubricants in severe environments. The following examples illustrate this process.

Example 1: Difluorodiazirine (0.35 gram, 4.5 millimols) and boron trifluoride, BF_3 (0.03 gram, 0.45 millimol) were condensed into a Pyrex tube containing 10 ml. of trichlorotrifluoroethane and then photolyzed with ultraviolet light from a BH-6 lamp for 4 hours at room temperature. After removal of the volatile material, a yellow solid remained [0.14 gram, i.e., 60% of theoretical yield for $(-CF_2-)_n$]. The infrared spectrum and x-ray powder photograph of this material were identical with those of polytetrafluoroethylene. The material melted with decomposition at 250°C. Analysis—C, 23.3; F, 73.0; $(CF_2)_n$ requires C, 24.02; F, 75.98.

Example 2: Difluorodiazirine (0.35 gram, 4.5 millimols) was condensed into a Pyrex tube containing 20 ml. of trichlorotrifluoroethane and a trace of the polymer described in Example 1, and irradiated with ultraviolet light from a BH-6 lamp for 4 days at room temperature. After removal of all volatile material a brown solid remained [0.134 gram, 64% theoretical yield of $(-CF_2-)_n$], the infrared spectrum and x-ray powder photograph of which were identical with those of polytetrafluoroethylene. The material melted sharply at 300°C. and solidified at 285°C. Analysis—C, 23.98; F, 73.50; $(CF_2)_n$ requires C, 24.02; F, 75.98.

Condensation Products with Sulfur Pentafluoride

H.L. Roberts; U.S. Patent 3,063,972; November 13, 1962; assigned to Imperial Chemical Industries Ltd., England describes a process for polymerizing tetrafluoroethylene and for making telomers of tetrafluoroethylene and sulfur chloride pentafluoride which comprises subjecting a mixture of tetrafluoroethylene and sulfur chloride pentafluoride to the action of ultraviolet light.

The molecular weight of the polymers and telomers of tetrafluoroethylene obtained in this process is largely governed by the proportion of sulfur chloride pentafluoride in the reaction mixture. When this proportion is small, for example, in the range of 0.1 to 0.001% molar and less, the polymers are solids with properties substantially indistinguishable from polytetrafluoroethylene made by known methods. Up to about 3% molar solids are obtained having softening points above 300°C. but with greater proportions lower melting point solids are formed while with proportions greater than about 30% molar mainly liquid products are obtained.

In one way of carrying out this process a silica reaction vessel is connected to stainless steel filling equipment by means of which the vessel may be pressurized with a mixture of tetrafluoroethylene and sulfur chloride pentafluoride up to a maximum of about 10 atmospheres. The silica vessel is irradiated with a low-pressure mercury vapor lamp and polymerization sets in rapidly, the rate being roughly proportional to the pressure. The following example illustrates this process.

Example: In an apparatus as described above a mixture of tetrafluoroethylene and sulfur chloride pentafluoride containing approximately 0.001% molar of the latter was introduced into the silica vessel, which had a volume of about 300 ml., and allowed to react at a maximum pressure of 7 atmospheres and at room temperature for six hours. A white solid was obtained having a melt viscosity at 440°C. of more than 10^7 poises and a sulfur content of less than 0.1% by weight and showing by infrared analysis the characteristic bands attributable to polytetrafluoroethylene and no bands at a wave number of 890 cm.$^{-1}$. At 380°C. the weight loss of the solid was 0.0028% per hour.

Polytetrafluoroethylene

Color Stabilization by Addition of Copper Fluoborate

A process described by S.C. Dollman; U.S. Patent 3,328,343; June 27, 1967; assigned to Allied Chemical Corporation involves the improvement of the color of sintered polytetrafluoroethylene by incorporating a small quantity of copper fluoborate, $Cu(BF_4)_2$, in the polymer. The copper fluoborate may be incorporated in the polymer in a variety of ways. The preferred method involves adding the copper fluoborate to an emulsion of the polymer prior to coagulation. The copper fluoborate-containing polymer, upon sintering, exhibits a substantially white color in contrast to a gray to black color if no copper fluoborate is added. Since copper fluoborate is employed in low concentration, the fluoborate incorporated in the polymer does not affect the polymer properties, including stability, reactivity, etc.

Decolorization of Color-Coded Wire

In a process described by A.J. Perreault; U.S. Patent 3,440,235; April 22, 1969; assigned to Haveg Industries, Inc. polytetrafluoroethylene which is color-coded has the color removed by the use of nitric acid. This process can be employed for example to remove any of the following coding pigments from polytetrafluoroethylene employed as insulation on wire, e.g., copper or silver wire.

Pigment	Name	General Chemical Composition
Yellow	Cadmium yellow	Cadmium sulfide.
Red	Selenium red	Cadmium and selenium sulfide.
Green	Chrome green	Chromium and cobalt compounds.
Blue	Cobalt blue	Chromium, cobalt and aluminum compounds.
Orange	Mercadmium sulfide	Cadmium and mercuric sulfides.
Purple	Composite of cobalt blue and selenium red.	Cadmium and selenium sulfides, chromium, cobalt and aluminum compounds.
Violet	Mineral violet	Manganese, cobalt and aluminum compounds.

Nitration of any of these pigments converts them to water-soluble salts which can be removed by washing with water. The following example illustrates this process.

Example: 6,000 feet of copper wire insulated with Teflon which was color-coded yellow with cadmium yellow was placed in a stainless steel tank provided with valves to add nitric acid and water and to drain spent acid and wash water. 95% nitric acid was pumped into the tank at 67°F. to cover the coils of Teflon coated wire. The wire was allowed to dwell in the tank 5 minutes to thoroughly wet the coils and at least partially nitrate the inorganic pigment. The acid was partially drained to a spent acid carboy so that one inch of acid was allowed to remain in the tank. The water line valve was opened carefully and a small quantity of water allowed to flow on the acid. This liberated oxides of nitrogen such as N_2O_3, NO, N_2O and NO_2.

The temperature of the bath was about 80°F. after this preliminary dilution. The N_2O_4 fumes liberated completed the nitration of the inorganic pigment to water-soluble salts. This was observed by the disappearance of color. Water was introduced into the tank to cover the coils. The water was then drained out. The coils were removed from the tank and washed acid free in fresh water using Congo red paper as the indicator. The code-free Teflon insulated wire was dried and repooled ready for use either for recoding or code-free insulated wire.

VINYLIDENE FLUORIDE ELASTOMERIC PRODUCTS

VINYLIDENE FLUORIDE HOMOPOLYMERS

Polymerization in the Presence of an Alkylene Oxide

M. Hauptschein; U.S. Patent 3,012,021; December 5, 1961; assigned to Pennsalt Chemicals Corporation has found that if vinylidene fluoride is subjected to polymerizing conditions in the presence of an alkylene oxide, up to 100% of the vinylidene fluoride will be polymerized. In addition, by conducting the reaction in the presence of an alkylene oxide, the polymer product is obtained as a free flowing powder, well suited for immediate processing.

The alkylene oxides especially suitable for use with the process are those having not more than about 10 carbon atoms in the molecule, such for example as ethylene oxide, propylene oxide, isobutylene oxide, and 1,2-epoxyoctane. As to the proportions of alkylene oxide, it is found that even a trace, e.g., on the order of 1/100 of a mol per mol of vinylidene fluoride suffices to give a very substantial increase in conversion. The use of proportions up to about 10 mols of alkylene oxide per mol of vinylidene fluoride improves the physical characteristics of the product, giving a free flowing mass of discrete particles of uniform size which is suitable for use in various applications without grinding or shredding.

In the following examples, which illustrate the process, the vinylidene fluoride, a free radical forming initiator, and the alkylene oxide were put into a stainless steel bomb of about 250 ml. capacity, and heated to 135°C. ±5°C. The pressure ranged from about 100 to about 1,500 psig. Further experimental details are presented in the following table.

Example	Mols $CF_2=CH_2$	$C_2H_4O/CF_2=CH_2$ Ratio	DTBP[1], Wt. % $CF_2=CH_2$	Reaction Time (Hrs.)	% Conversion (based on $CF_2=CH_2$)	Analysis, % by wt. C	H
1	0.52	0	9.1	140	53	38.5	3.5
2	0.59	0	7.9	60	54	-	-
3	0.52	0.04	9.1	24	84	-	-
4	0.59	0.14	7.9	19	80	38.2	3.3
5	0.59	0.46	1.3	60	71	-	-
6	0.62	0.50	7.9	18	90	-	-
7	0.72	1.4	6.8	18	83	-	-
8	0.48	1.7	10	96	100	39.6	3.6
9	0.22	4.8	14	60	87	39.6	3.5
10	0.23	5.1	6.7	19	80	38.2	3.2

[1]DTBP = di-tert-butyl peroxide.

Theoretically, $-(CF_2CH_2)_n-$ requires C, 37.51%; H, 3.15% by weight. The slightly higher percentages of carbon and hydrogen found in the examples are thought due to the incorporation of end groups from the DTBP catalyst. In all the examples, the polymer obtained was washed for several hours with boiling water. No decrease in weight was noted, indicating the absence of an ethylene oxide homopolymer. The infrared spectra of the polymers obtained in Examples 9 and 10 were examined and found to be identical with the spectra of Example 1, where no ethylene oxide was used. Thus, only homopolymerization of vinylidene fluoride occurred.

Polymerization in the Presence of an Olefinic Compound

A process described by J.D. Calfee and L.E. Erbaugh; U.S. Patent 3,022,278; February 20, 1962; assigned to Monsanto Chemical Company comprises polymerizing a monomeric charge consisting of vinylidene fluoride together with a small amount e.g., less than about 1.5 weight percent, of an olefinic compound other than vinylidene fluoride.

By contrast with the hard and brittle vinylidene fluoride homopolymers prepared in conventional processes, the polymers obtained by this method are readily processable, e.g., on hot mill rolls at temperatures of 170°C. Molded specimens of the polymer require no quenching step, but are directly obtained as tough and extensible materials exhibiting a marked yield point prior to break in tensile tests, and having high elongation values, whereas the conventional vinylidene fluoride homopolymers have no such properties when molded into standard thickness specimens in the absence of other modifying steps taken to change the physical properties of the polymer. The following example illustrates the process.

Example: In this and the following experiments there were used stainless steel autoclave reactors fitted with a thermowell and adapted to be held in a heater-shaker assembly. In most of the runs there was used an autoclave of 1.5 inches inside diameter and 8 inches inside length with a capacity of 250 ml.; a similar, smaller autoclave was used in other runs. The procedure used was as follows. Prior to each run the autoclave is thoroughly cleaned, sealed and evacuated. If solid catalyst is used, it is placed in the autoclave before sealing. After evacuation is complete, any liquid components of the polymerization charge, such as dissolved catalyst or olefinic compound, where this is a liquid, are drawn into the autoclave.

Gaseous olefinic compounds or other gaseous additives are next introduced into the autoclave, after which the autoclave is refrigerated and the vinylidene fluoride is added in the form of a gas. The charged autoclave is then placed in the heater-shaker assembly and connected to a high-pressure water pump through a manifold. The autoclave is heated and pressured simultaneously, to minimize the danger of overpressuring. The volume of water introduced into the autoclave is adjusted in accordance with the desired pressure, there generally being used about 100 c. of water at 5,000 pounds per square inch (psi) pressure in the larger autoclave described above, and lesser or greater amounts of water correspondingly at other pressures. When the desired temperature and pressures are reached, automatic controls hold them constant. At the close of the reaction, the heater is turned off and the autoclave allowed to cool for about two hours before opening; thus unreacted monomer can be vented safely.

The added olefinic compound, in the amount shown in the tables below, was introduced into the autoclave with or after the catalyst and prior to the introduction of vinylidene fluoride and pressuring of the bomb. All of the runs were made with 70 grams vinylidene fluoride charged, except that 100 grams was used in the 0.5% propylene run, and at a pressure of 5,000 pounds per square inch gauge, except the last four runs in the table which were at 25,000 psi. For comparison, runs made in the absence of olefinic compounds are noted in the first table.

Olefinic compound [1]	Weight Percent	Catalyst [2]	Mg.	Temp., °C.	Time, hrs.	Conversion, Percent	n_{sp}	d	Tensile, p.s.i.		Elongation, Percent		Remarks
									Yield	Break	Yield	Break	
None		PPS	50	70	12	63	0.338	1.75		5,305		18	Homopolymer.
Et	0.2	PPS	50	70	12		.192			5,519		12	Not enough olefin.
None		PPS	100	80		42	.207	1.76		4,907		5	Homopolymer.
Et	0.4	PPS	100	80	12	36	.084	1.75	6,502	5,146	18	191	Processable.
Et	0.6	PPS	100	80	12	67	.091	1.75	6,239	5,855	20	312	Do.
Et	0.7	PPS	100	80	12	8	.097	1.75	6,486	6,634	31	193	Do.
Et	2.1	PPS	50	70	12	None							Too much olefin.
S	1.4	PPS	100	80	12	53	.101	1.76	6,813	5,277	19	118	Processable.
S	2.8	PPS	100	80	12	Trace							Too much olefin.
VTS	1.1	PPS	100	80	12	38	.112	1.77	6,715	5,735	18	162	Processable.
VTS	2.3	PPS	100	80	12	Trace							Too much olefin.
Pr	0.5	PPS	100	80	13	43	.122	1.75	5,795	4,521	19	155	Processable.
Pr	0.9	PPS	100	80	14	7	.062						Too much olefin for good yield.
None		3% H₂O₂	6,500	115	12	34	Insol.			5,491		17	Homopolymer.
Et	0.6	3% H₂O₂	6,500	115	13	Trace							Too much olefin for this catalyst.

[1] Et is ethylene; S is styrene; VTS is vinyl trimethyl silane; and Pr is propylene.
[2] PPS is potassium persulfate; buffer, sodium acid pyrophosphate, Na₂H₂P₂O₇, was used in amount of 50% of the catalyst weight in each of these runs with potassium persulfate as catalyst in the suspension polymerization.

Olefinic compound [1]	Weight Percent	Catalyst [2]	Mg.	Temp., °C.	Time, hrs.	Conversion, Percent	n_{sp}	d	Tensile, p.s.i.		Elongation, Percent		Remarks
									Yield	Break	Yield	Break	
Et	1.0	PPS	200	80	13	29	0.086	1.73	5,300	3,948	12	285	Processable.
Et	3.0	PPS	400	80	16	7	0.007						Too much olefin.
Et	3.0	PPS	800	80	15	9	0.029						Do.
Et	3.0	PPS	800	80	12	13	0.033						Do.
Et	3.0	PPS	400	80	16	3	Insol.	1.69		3,534		10	Do.
Et	3.0	PPS	200	80	13	3	Insol.	1.65		5,416		12	Do.
Et	3.0	PPS	100	80	14	2	0.071						Do.

[1] Et is ethylene.
[2] PPS is potassium persulfate; buffer, sodium acid pyrophosphate, Na₂H₂P₂O₇, was used in amount of 50% of the catalyst weight in each of these runs with potassium persulfate as catalyst in the suspension polymerization; exceptions are last three runs, each of which used 400 mg. sodium acid pyrophosphate.
NOTE.—Last four runs at 25,000 p.s.i.; all others at 5,000 p.s.i.

The difference in the polymer produced in the presence and the absence of the added olefinic compounds under otherwise identical conditions is immediately evident from the tables. On inclusion of olefinic compounds in the polymerization charge, the specific viscosity is at least halved; however, the yield of polymer in the process need not be sacrificed to obtain a polymer of the desired characteristics. As shown in the tables, the polymers of the process, in contrast to the hard, inextensible, unprocessable vinylidene fluoride homopolymers prepared in the absence of an olefinic compound, are readily processable on hot mill rolls at 170°C., and molded specimens of the polymers are found to exhibit yield points and have high ultimate elongations at break. Consistent results are obtained with a variety of olefinic compounds, i.e., ethylene, propylene, styrene, vinyltrimethylsilane, indicating that the process is independent of the particular olefinic compound utilized.

It will be noted from the data that in the polymerizing system used, namely, water-soluble catalysts in aqueous systems, there is evidently a marked sensitivity to the concentration of olefinic compound, and that while a minimum of olefinic compound must be present to alter the resultant polymer properties as desired, excess olefinic compound inhibits formation of the polymer. The quantity of olefinic compound operating to inhibit the conversion to polymer naturally varies somewhat with the particular olefinic compound in a given catalyst system, and also varies with the catalyst system.

Perchlorobenzenecarboxylic Acids as Dispersing Aids

H. Iserson; U.S. Patent 3,031,437; April 24, 1962; assigned to Pennsalt Chemicals Corporation describes the preparation of polyvinylidene fluoride in latex form in an aqueous system in the presence of a perchlorobenzenecarboxylic acid or one of its salts as a dispersing agent and a water soluble peroxy compound as a polymerization initiator.

Despite the absence of a long-chained hydrophobic group, the perchlorobenzenecarboxylic acids are highly efficient as dispersing agents for the preparation of polyvinylidene fluoride in latex form as may be seen from the examples below. Additionally, they are readily available, low-cost, materials for this purpose. Whereas the fluorine-containing dispersing agents of the prior art must be synthesized by special procedures, the perchlorobenzenecarboxylic acids are available as commercial products at a fraction of the cost.

The perchlorobenzenecarboxylic acids useful in the process include particularly perchlorobenzoic acid, perchloro-1,2-benzenedicarboxylic acid, perchloro-1,3-benzene dicarboxylic acid, and the perchloro-1,4-benzenedicarboxylic acid and the ammonium, sodium, potassium, and lithium salts of these acids. The sodium salt of perchloro-1,2-benzenedicarboxylic acid is preferably used. The perchlorobenzenedicarboxylic acids are also commonly known as perchlorophthalic acids, i.e., perchloro-o-phthalic, perchloroisophthalic acid, and perchloroterephthalic acid. The following example illustrates the process.

Example 1: One part of perchloro-o-phthalic anhydride is added to 200 parts of water. The mixture is made slightly alkaline with 6 normal sodium hydroxide and charged into a stainless steel autoclave. 0.2 part of ammonium persulfate is added. The reactor is cooled in ice, evacuated and heated to 86°C. It is then pressured to 400 psig with vinylidene fluoride and agitated at about 86°C. for 35 minutes. During this period the pressure is held constant by addition of vinylidene fluoride. The autoclave is then vented and the reaction mixture removed. A latex of polyvinylidene fluoride in water is obtained. The latex is highly stable at ambient temperatures. Subjecting the latex to freezing conditions results in coagulation of only a small amount of the dispersed polymer. Melting the frozen latex and boiling it thereafter causes very little additional precipitation. Addition of methanol and boiling of the latex-methanol mixture are also ineffective as means of coagulation. The polymer, however, can be coagulated by addition of 5% aqueous sodium chloride to the latex followed by boiling of the mixture for several minutes. The coagulated polymer is then filtered, washed with hot water and dried. The dry weight is about 8 parts.

The dried polyvinylidene fluoride has a plasticity index number of about 1,600. The plasticity number is an empirical index found useful for comparison of vinylidene fluoride polymers in the absence of a method for measuring molecular weight. The plasticity number is the area in square millimeters of a plaque made by pressing 0.5 g. of polyvinylidene fluoride at 2,500 psig and 225°C. in a Carver press for one minute. The relationship between the area of the plaque and the apparent molecular weight is that as the molecular weight increases, the area of the plaque decreases. In other words, polymer having a low molecular weight will flow to form a larger-sized area than will polymer having a high molecular weight. Polymer having a low plasticity number is therefore indicated to be high in apparent molecular weight. Polymer having a plasticity number in the range 1,500 to 3,000 has been experimentally determined to be suited for fabrication purposes. Polymer having a plasticity number of 1,800 to 2,800 has optimum physical properties and is especially preferred.

Example 2: Using the same equipment and procedure as in Example 1, vinylidene fluoride is polymerized in the presence of 200 parts of deionized deoxygenated water containing 1 part ammonium persulfate and 1 part of perchloroterephthalic acid and made slightly alkaline with ammonium hydroxide. The polyvinylidene fluoride polymer is recovered in the form of a stable latex.

Di-Tertiary Butyl Peroxide Catalyst

A process described by M. Hauptschein; U.S. Patent 3,193,539; July 6, 1965; assigned to Pennsalt Chemicals Corp. involves the polymerization of vinylidene fluoride in the presence of di-tertiary butyl peroxide as a polymerization catalyst. The catalytic polymerization of vinylidene fluoride to high molecular weight polymers has been known for some time, being described, for example, in U.S. Patent 2,435,537. A wide variety of catalysts have been suggested for the polymerization of this monomer. Both inorganic peroxy compounds such as potassium persulfate, preferably in combination with reducing agents such as sodium bisulfite, and organic peroxides such as dibenzoyl peroxide and acetyl peroxide, have been suggested.

Of the two types of catalysts, i.e., inorganic peroxy compounds as exemplified by potassium persulfate, and organic peroxy compounds as exemplified by dibenzoyl peroxide, it has been found that the organic type of catalyst provides a higher quality polymer than the inorganic peroxy type, the organic peroxy catalysts having been found to provide polymers of considerably higher thermal stability and processability.

While the organic peroxy catalysts provide in general better polymers, catalysts of this type previously tried, such as dibenzoyl peroxide or acetyl peroxide, are known to require extremely high pressures, for example, pressures of the order of 10,000 lbs./in.^{2} to produce appreciable yields of polymer. Even at these high pressures, the yields of the polymer are often relatively poor, for example, from 10 to 20%. Because of the expensive equipment required, the cost of carrying out the polymerization at such high pressures is exorbitantly high. An organic catalyst has been found, namely di-tertiary-butyl peroxide of the formula

$$CH_3-\underset{\underset{CH_3}{|}}{\overset{\overset{CH_3}{|}}{C}}-O-O-\underset{\underset{CH_3}{|}}{\overset{\overset{CH_3}{|}}{C}}-CH_3$$

which is capable of providing excellent yields of high quality vinylidene fluoride polymer at low, commercially practicable pressures, for example, of the order of 300 to 1,000 lbs./in.^{2}. It appears that di-tertiary-butyl peroxide is unique among organic catalysts in providing high conversions under mild pressure since numerous attempts to obtain similar results with other organic catalysts, even catalysts as similar as tertiary-butyl hydroperoxide have been unsuccessful. The following example illustrates the preparation of polyvinylidene fluoride of high molecular weights using di-tertiary-butyl peroxide catalyst at moderate pressures.

Example: A 300 milliliter stainless steel autoclave is charged with 100 milliliters of deionized and deoxygenated water and 0.8 gram of di-tertiary-butyl peroxide. The autoclave is evacuated, cooled in liquid nitrogen and then charged with 35 grams of vinylidene fluoride ($CF_2=CH_2$) by gaseous transfer in vacuo. The autoclave is then sealed and placed in an electrical heating jacket mounted on a horizontal shaking apparatus and held at 122° to 124°C. for 18 1/2 hours. A maximum pressure of about 800 lbs./in.^{2} gauge is reached after the first 2 hours of heating and the pressure then decreases as the polymerization proceeds.

After the reaction period, the autoclave is cooled, vented and opened. The contents consist of precipitated polyvinylidene fluoride suspended in a liquid phase having a pH of 2.5. The polymer is vacuum filtered and washed on the filter funnel with methanol, then washed ten times with distilled water, and given a final wash with methanol. It is then dried in a vacuum oven at 102°C. The washed, dried polymer weighs 29 grams, representing an 83% conversion of the monomer to polymer. The polymer, which melts at 160° to 165°C. has a high molecular weight as indicated by its plasticity number of 3,020. The polymer has excellent thermal stability. It does not decompose when heated to 320°C. and shows substantially no discoloration when exposed in a circulating air oven for 100 hours at 200°C. It has good low temperature properties as shown in a test in which the polymer is flexed 180° over a 1/8" mandrel at -70°C. without breaking.

Disuccinic Acid Peroxide Catalyst

H. Iserson; U.S. Patent 3,245,971; April 12, 1966; assigned to Pennsalt Chemicals Corporation describes a polymerization system in which the combination of four essential ingredients is required namely, (a) the use of an aqueous polymerization medium, (b) the use of a water soluble peroxide of a dibasic acid as the polymerization catalyst, (c) a small amount of a fluorinated surfactant, and (d) a small amount of iron powder.

The water-soluble dibasic acid peroxides useful in the process include peroxides of saturated aliphatic, dibasic acids having four to five carbon atoms, and thus include, disuccinic acid peroxide, monosuccinic acid peroxide, diglutaric acid peroxide, an monoglutaric acid peroxide. Particularly preferred is disuccinic acid peroxide. The catalyst is used in concentrations of from about 0.01 to 5%, and preferably from 0.05 to 2% by weight based on the total weight of monomer employed in a given run. The examples on the following page illustrate the process.

Examples 1 to 7: These examples demonstrate the necessity for using both the fluorinated surfactant and the iron powder. Seven runs are conducted under substantially identical conditions. In two runs both the iron powder and the fluorinated surfactant are employed (Examples 1 and 2); in another two runs both of these ingredients are omitted (Examples 3 and 4); in another two runs the iron powder is omitted (Examples 5 and 6); and in the final run the fluorinated surfactant is omitted (Example 7).

In each of these examples a 300 milliliter stainless steel autoclave is charged with 100 milliliters of deionized and deoxygenated water and 0.4 gram of disuccinic acid peroxide. The autoclave is evacuated, cooled in liquid nitrogen and then charged with 35 grams of vinylidene fluoride (CF_2=CH_2) by gaseous transfer in vacuo. In Examples 1 and 2 the autoclave is also charged with 0.5 gram of ammonium perfluorooctanoate, $C_7F_{15}COONH_4$ and 0.0007 gram of metallic iron powder prepared by the reduction of a pure iron oxide, equivalent to 7 parts of iron powder per million parts by weight of water. In Examples 3 and 4 both the fluorinated surfactant and the iron powder is omitted. In Examples 5 and 6 only the surfactant is employed, while in Example 7 only the iron powder is used.

After loading, the autoclave in each case is then sealed and placed in an electrical heating jacket mounted on a horizontal shaking apparatus and held at a temperature of about 80°C. for a period of 17 to 21 hours. A maximum pressure of about 750 lbs./in.^{2} gauge is reached and the pressure then decreases as the polymerization proceeds. After the reaction, the autoclave is cooled, vented and opened. Where polymer was obtained, it is obtained in the form of a colloidal dispersion which is frozen and then thawed to precipitate the polymer, which is then filtered by suction.

The polymer is washed with water and methanol and then dried in a vacuum oven at about 70°C. The polymer is obtained as a fine white free flowing powder. The polymer obtained in each run is then tested to determine molecular weight and thermal stability. Since it has not been possible to obtain a true solution of the polymer, absolute molecular weight determinations are not possible. A reliable indication, however, of relative molecular weight is obtainable by measurement of the "plasticity number" of the polymer. Vinylidene fluoride homopolymers having plasticity numbers over about 3,500 have only limited practical utility and preferably the plasticity number should be 2,500 or less.

To determine the thermal stability of the polymer samples of compression molded films of each polymer are placed in a forced draft oven for 4 hours at a temperature of 200°C. and the appearance of the film noted after this treatment. A rating of excellent indicates no change in clarity or color. A rating of fair indicates yellowing of the film. Results of the seven examples are tabulated below. Note that in Examples 1 and 2 where both the fluorinated surfactant and the iron powder is employed, a good conversion to high molecular weight polymer (indicated by low plasticity number) was obtained which displayed excellent heat stability. In Examples 3 and 4 where both the fluorinated surfactant and the iron powder is omitted, and in Example 5 where the iron is omitted, no polymer of any kind is obtained. In Example 6 where the iron powder is omitted and in Example 7 where the fluorinated surfactant is omitted, only a very small conversion to polymer is obtained which displays only a fair thermal stability.

| | | | | | | | | Polymer Product | | |
Example	CH_2=CF_2, grams	H_2O, ml.	DSAP* catalyst, grams	$C_7F_{15}COONH_4$, grams	Fe powder, ppm.**	Temp., °C.	Time, hours	Weight, grams	Plasticity number	Heat stability, 4 hrs. at 200°C.
1	35	100	0.4	0.5	7	82	18.5	22.3	1,257	Excellent
2	35	100	0.4	0.5	7	82	18	19.8	1,257	Excellent
3	34.3	100	0.4	-	-	80	17	0	-	-
4	35	100	0.4	-	-	81	17	0	-	-
5	35	100	0.4	0.5	-	80	21.5	0	-	-
6	35	100	0.4	0.5	-	80	17.5	1.2	1,700	Fair
7	35	100	0.4	-	7	80	17.5	0.4	2,100	Fair

*Disuccinic acid peroxide.
**Parts per million parts by weight of H_2O.

Diisopropyl Peroxydicarbonate Catalyst

A process described by G.H. McCain, J.R. Semancik and J.J. Dietrich; U.S. Patent 3,475,396; October 28, 1969; assigned to Diamond Shamrock Corporation comprises polymerizing vinylidene fluoride in an aqueous medium containing small quantities of a fluorinated surfactant and in the presence of a catalytically effective amount of diisopropyl peroxydicarbonate, the polymerization being conducted at pressures less than 1,000 psig, at a temperature of 65° to 85°C. and for a time period of from 0.5 to 6 hours. From this process there is obtained from 90 to 100% yields by weight of high molecular weight, thermally stable polymer which is useful for the production of coatings, films, fibers, moldings and other finished plastic articles as fabricated from commercial vinylidene fluoride polymers. A particular advantage of the process is the fact that it provides high conversion of polymer in a much shorter reaction time than is possible when employing the previously available low pressure processes.

The amount of diisopropyl peroxydicarbonate employed in the polymerization recipe under the stated reaction conditions is somewhat critical and cannot be varied over wide limits if optimum yields of high quality polymer are obtained. The catalyst concentration employed should range from 0.125 to about 1.0% by weight of total monomer, with the preferred concentration ranging from 0.25 to about 0.5% by weight of monomer. The following examples illustrate the process.

Example 1:   A one-gallon stainless steel autoclave prechilled to 15°C. is charged with 740 g. of deoxygenated deionized water, 2.59 g. of diisopropyl peroxydicarbonate and 13.0 g. of sodium perfluorooctanoate (one-half the total quantity of surfactant). The autoclave is evacuated, cooled in a Dry Ice-acetone bath and then charged with 518.4 g. of vinylidene fluoride by transfer bomb under vacuum. The autoclave and the agitated contents are then brought rapidly up to a temperature of about 75°C. (in about 15 minutes) by means of an internal heating coil heated by a circulating oil bath. The remaining 13.0 g. of sodium perfluorooctanoate surfactant along with water is introduced at a steady rate by means of a water pump connected to the autoclave. During the reaction (a total of 5 hours) a pressure of 600 psig is maintained on the reaction mixture by periodically pumping in deoxygenated deionized water as needed.

After the reaction, the water pump is shut down, the autoclave is cooled, vented and opened. The polymer latex is removed and the polymer product is isolated by freezing the latex. The polymer product is separated by filtration, washed thoroughly with distilled water and finally dried at 50°C. under vacuum. The dried polymer product weighs 482 g. (93% monomer conversion). The flow number of this product is 1,520. When molded at 325° and 450°F. it yields colorless films which are strong and tough.

Example 2:   Using the procedure as described in Example 1, a vinylidene fluoride polymer is prepared in 5 hours at a temperature of 75°C. and a pressure of 600 psig. In this experiment, the proportions of all ingredients used is the same as in Example 1 with the exception of the diisopropyl peroxydicarbonate. In this example, 1.295 g. of the catalyst is used (0.25%, by weight of total monomer). The polymer product is isolated, washed and dried as in Example 1. The total polymer obtained is 492 g. (95% monomer conversion). The polymer product has a flow number of 1,260 and produces a colorless molding at 450°F.

Example 3:   The general procedure of Example 1 is repeated with the same quantities of deoxygenated, deionized water, vinylidene fluoride monomer, diisopropyl peroxydicarbonate and sodium perfluorooctanoate. Additionally, 0.06 g. of ethylene oxide is added to the water initially as chain transfer agent (equivalent to 0.01% ethylene oxide, by weight of total monomer). The reaction is conducted at 75°C. and a pressure of 600 psig. After the reaction mixture is raised to the reaction temperature, the reaction is continued for an additional 45 minutes, making a total reaction time of 1 hour. The yield of polymer product is 87%, by weight of total monomer. This polymer has a flow number of 2,120 and produces completely colorless films and moldings when molded at 325° and 450°F.

Example 4:   The experiment as described in Example 3 is repeated with the same proportions of ingredients and reaction conditions except that the total reaction time is 4 hours. The polymer which is recovered in 90% yield has a flow number of 1,810. It produces colorless films and moldings when molded at 325° and 450°F.

Example 5:   To compare the physical properties of vinylidene fluoride polymers of this process with those exhibited by commercially available vinylidene fluoride polymeric materials, each of the polymer products of Examples 1 to 4 was extruded at 470°F., pelletized and fabricated into test specimens by injection molding. Samples of two types of commercial polymer were similarly molded. Testing in accordance with the indicated ASTM procedures gave the following values.

Polymer Product	Tensile Strength,[1] p.s.i.	Tensile Modulus,[1] p.s.i.×10³	Elongation, Percent	Flex Strength at 5% Def.,[2] p.s.i.	Flex Modulus,[2] p.s.i.×10³	Hardness Shore D [3]
Example 1	6,638	116	68			79
Example 2	6,973	122	86			78
Example 3	6,310	139	101	8,576	210	77
Example 4	6,348	131	70	8,557	206	77
Kynar* Type 1875 Pellets	6,184	95	86	7,600	193	74
Kynar* Type 1875 Powder	6,284	117	87			74
Kynar* Type L-1900 Pellets	7,122	116	78			77

[1] ASTM D-638-61T.
[2] ASTM D-790-63.
[3] ASTM D-1706-61.
*Kynar—Reg. Trade Mark—Pennsalt Chemicals Corp.

As the above values indicate, the polymer products of this process compare favorably with commercial vinylidene fluoride polymer products.

Stabilization with Barium and Strontium Compounds

H. Iserson; U.S. Patent 3,154,519; October 27, 1964; assigned to Pennsalt Chemicals Corporation has found that vinylidene fluoride polymer containing a minor amount of a perfluoroaliphatic or perfluorocycloaliphatic sulfonic acid salt can be stabilized for use at high temperatures by the inclusion of small amounts of water soluble salts of barium and strontium. These strontium and barium salts are effective heat stabilizers for vinylidene fluoride polymer containing a residual amount of a perfluoroaliphatic or perfluorocycloaliphatic sulfonic acid salt and which have a plasticity of 3,000 or lower. Plasticity is an indication of the molecular weight of the polymer.

The conventional heat stabilizers for halogen-containing plastics such as polyvinyl chloride proved to be of little or no value when incorporated in vinylidene fluoride polymers and tested at temperatures as high as 270°C. Inorganic acid derivatives, metal oxides, long-chain organic acid metal salts, amine-type and phenol-type antioxidants and dialkyldithiocarbamic metal salts proved ineffective as stabilizers for vinylidene fluoride polymer at high temperatures. Incorporation of as much as 3 parts of these additives per 100 of polymer failed to maintain the polymer clear and free of color even at temperatures as low as 200°C. for periods of 2 to 8 hours. The following example illustrates the process.

Example: Polyvinylidene fluoride was prepared in latex form by means of a polymerization recipe which included potassium perfluorooctylsulfonate as a surfactant. The potassium perfluorooctylsulfonate analyzed as follows on an anhydrous basis: carbon, 17.38; sulfur, 6.07; potassium, 7.03. The polymer latex was coagulated with sodium chloride solution, filtered by centrifugation, washed thoroughly on a centrifuge with distilled water and dried at 120° to 125°C. The polymer had a plasticity number of about 2,000. The polymer was rewetted with methanol and water, and sucked dry on a Buchner funnel. A 50 gram sample of the wet cake was stirred 3 minutes in a Waring blender with 150 ml. of each of the solutions shown in the table below. The stirred mixture was poured on a Buchner funnel, and the residual polymer in the blender was washed onto the funnel with 100 ml. wash solution. The polymer was filtered under vacuum, dried at 120°C. overnight in an oven and tested for heat stability first at 250°C., then at 270°C. A light color after the heat treatment indicates satisfactory stabilization while a tan, dark brown or black color indicates excessive polymer degradation.

Wash Solution	Polymer Appearance after Heat Stability Test at—	
	250°C. for 2 hours	270°C. for 1 hour
Cold water	Black	–
Methanol	Dark-brown to black	–
5% aq. NH$_4$OH	Dark-brown to black	–
1% aq. Ba(OH)$_2$·8H$_2$O	No discoloration	Light colored
Sat. MgCO$_3$	Black	–
1% Thermolite in methanol	Black	–
1% aq. tetraethylurea	Black	–
Hot water	Dark-brown to black	–
1% aq. BaCl$_2$	–	Light colored
1% aq. urea	–	Black

ELASTOMERIC COPOLYMERS — HEXAFLUOROPROPENE

General Process

In a process described by D.R. Rexford; U.S. Patent 3,051,677; August 28, 1962; assigned to E.I. du Pont de Nemours and Company copolymers are prepared by a method which results in elastomeric compositions containing from 70 to 30% by weight of vinylidene fluoride units and from 30 to 70% by weight of hexafluoropropene units. When the copolymer contains less than 30% of hexafluoropropene units it tends to become nonelastic. When the polymerization of the two monomers is carried out by the batch process, an excess of the hexafluoropropene is employed over that present in the resulting polymer, so that ordinarily in the batch process from 60 to 15 parts by weight of vinylidene fluoride will be copolymerized with from 40 to 85 parts by weight of hexafluoropropene to give products containing from 30 to 70% copolymerized hexafluoropropene.

Since hexafluoropropene does not homopolymerize under the conditions described, the use of amounts of hexafluoropropene in excess of 85% by weight in a batch process does not produce copolymers containing more than 70% by weight of copolymerized hexafluoropropene. In the continuous process, the copolymerization will be carried out employing substantially the same amounts of the monomers desired in the final copolymer. The polymerization is preferably carried out at temperatures of about 80° to 120°C., where the reaction is rapid. At temperatures below 80°C. a very low rate of conversion results. The process may be carried out at temperatures above 120°C., but at such temperatures the process is less economical. The polymerization is preferably carried out in a stainless steel pressure reactor or other type of equipment which will not be reacted upon during the process, such as an enamel-lined pressure vessel, etc. In the batch process the vessel is flushed free from oxygen with a gas such as nitrogen, then it is charged with deoxygenated water and the polymerization initiator. After closing and evacuating the vessel, it is then charged with the gaseous reactants and heated to a reaction temperature of from 80° to 120°C. under agitation. When the reaction is completed, the mass is cooled to room temperature and any unreacted gas removed. The partially coagulated copolymer is then completely coagulated with acids or salts in the customary manner, and is discharged from the reactor.

As illustrated in Figure 2.1, hexafluoropropene is metered from cylinder 1 and vinylidene fluoride is metered from cylinder 2 into line 3, and the mixture passes into compressor 4 where it is pumped into reaction vessel 5. Simultaneously, a solution of the polymerization initiator and dispersant in tank 6 is pumped by pump 7 into reaction vessel 5. The reaction vessel 5 is preferably filled completely with liquid. Cylinder 8 contains nitrogen or other inert gas and this gas applies pressure to a pressure control valve 9. When the internal pressure in reaction vessel 5 exceeds the applied pressure on valve 9, valve 9 functions to permit the copolymer emulsion to escape from the reaction vessel into receiver 10 from which the latex is removed and coagulated for further processing.

FIGURE 2.1: VINYLIDENE FLUORIDE-HEXAFLUOROPROPENE COPOLYMER PROCESS

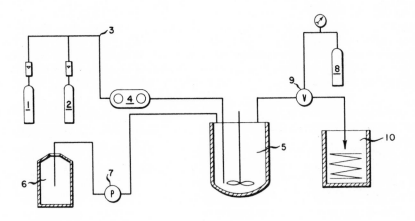

Source: D.R. Rexford; U.S. Patent 3,051,677; August 28, 1962

The elastomeric copolymers obtained from the reaction are elastomeric in nature, but can be further treated to produce elastomers of exceptionally good physical and chemical properties. This subsequent treatment is a curing process which probably causes cross-links to be established throughout the copolymer. Curing and compounding of polymers is well known in the art and can conventionally be carried out by the usual methods for this type of polymer.

The elastomeric copolymer of this process can be used in the manufacture of films, foils, tapes, fibers and articles of any desired shape, and can be used as coatings for wires, fabrics, ceramics, etc., and for the impregnation of felt which may be made from various fibers since the products can be extruded and molded under pressure. Ordinarily these copolymers are preferably extruded at temperatures not substantially higher than 190°F., although this temperature will vary depending upon the particular constitution of the copolymer. The following examples illustrate the process.

Example 1:   A conditioned stainless steel pressure vessel is swept with nitrogen and charged with 125 parts of deoxygenated distilled water containing 0.16 part of ammonium persulfate, 0.03 part of sodium bisulfite and 0.33 part of disodium phosphate heptahydrate. The closed vessel is cooled to −80°C. and purged of oxygen by three alternate cycles of producing a vacuum in the vessel and then pressuring with oxygen-free nitrogen. The nitrogen is then removed, and, while the system is under reduced pressure, 35 parts each of gaseous hexafluoropropene and vinylidene fluoride is bled into the pressure vessel. The system is agitated and the temperature inside the reaction chamber raised to 100°C. over a 15 minute period. The autogenous pressure is observed to increase to about 700 psig, which drops to 300 psig after two hours. After an additional heating period of 12 hours to ensure that the reaction is completed, the reaction mass is allowed to cool to room temperature and the pressure chamber vented to the atmosphere. The partially coagulated latex product is removed and coagulation completed by the addition of a small amount of dilute hydrochloric acid. The coagulated crumb is washed thoroughly with water and rolled on a hot rubber mill at about 140°C. to obtain 63 parts (90% conversion) of an off-white elastomer in rolled sheet form. Analysis of this elastomer for carbon, hydrogen and fluorine by combustion analysis indicates that the product copolymer contains about 45% hexafluoropropene and about 55% vinylidene fluoride by weight.

Example 2: Example 1 is repeated, but the pressure vessel charged with 30 parts of hexafluoropropene and 40 parts of vinylidene fluoride. The elastomer obtained is similar to, but somewhat harder than, the product of Example 1. Combustion analysis indicates that this copolymer contains about 30% hexafluoropropene and about 70% vinylidene fluoride by weight.

Example 3: When 0.01 part of the ammonium salt of omega-hydroperfluoroheptanoic acid is added to the reaction kettle charge of Example 1, the product obtained after reaction is entirely dispersed throughout the aqueous system. The product is coagulated by the addition of sodium chloride, filtered, washed and dried on a rubber mill. A product substantially identical to that of Example 1 is obtained.

Example 4: The elastomer of Example 1 is compounded at 25°C. on a rubber mill to contain the following ingredients:

	Parts
Copolymer	100
Benzoyl peroxide	3
Zinc oxide	5
Dibasic lead phosphite	5

The compounded stock is cured by pressing in a mold for one hour at 120° to 150°C. and baked for an additional 16 hours at 100° to 150°C. The resulting vinylidene fluoride-hexafluoropropene elastomer thus obtained is extremely tough and has the following physical properties:

Properties at 170°F.	Initial	After 72 hrs. at 400°F.
Tensile at break, psi	1,710	1,490
Elongation at break, percent	625	615
Modulus—300% elongation, psi	500	420
Hardness (Shore A)	58	-
Permanent Set (ASTMD-412-51T(h), percent	35	31

Example 5: The copolymer of Example 1 is compounded at 25°C. on a rubber mill to contain:

	Parts
Copolymer	100
Benzoyl peroxide	3
Zinc oxide	5
Precipitated silica*	20

  *Available as "Hi-Sil" 202 from Columbia-Southern Chemical Corp.

The stock is cured at 120°C. for one hour and baked at from 120° to 150°C. for 16 hours. The vinylidene fluoride-hexafluoropropene elastomer thus obtained shows good tensile strength, as exemplified in the following table:

Temperature of Water	25°C.	70°C.	100°C.
Tensile at break, psi	2,350	1,850	1,050
Elongation at break, percent	390	490	500
Modulus—300% psi	1,450	800	475

Low Molecular Weight Copolymers

G.A. Gallagher; U.S. Patent 3,069,401; December 18, 1962; assigned to E.I. du Pont de Nemours and Company describes low molecular weight copolymers which are obtained from hexafluoropropene, vinylidene fluoride and aliphatic chain transfer agents. The copolymers of this process may be prepared by the general procedure of heating a mixture of hexafluoropropene, vinylidene fluoride and the chain transfer agent under pressure, either in the presence or absence of a solvent and in the presence of a nonmetallic, organic peroxide. By using the chain transfer agent it is possible to prepare a copolymer of controlled molecular weight. If desired, the copolymer may be prepared by heating the hexafluoropropene, vinylidene fluoride and chain transfer agent in an aqueous emulsion system using a redox catalyst system such as ammonium persulfate and sodium bisulfite. In preparing these copolymers the temperatures used may range from about 100° to about 250°C. under autogenous pressure which is generated. The time for carrying out the procedure is variable in that it may range from about one-half hour, or less, in a continuous process to about 24 hours in a bomb.

The copolymers of this process are characterized as being viscous oils or semisolid greases at room temperature, i.e., about 20° to 25°C., which melt to clear oils when heated to not more than 120°C. Further, these copolymers are capable of being cured to form solids. These copolymers may be further characterized as having inherent viscosities in the range of from about 0.025 to 0.25 at 30°C. when dissolved in an anhydrous solvent consisting of 86.1% by weight of tetrahydrofuran and 13.9% by weight of dimethylformamide at a concentration of 0.1% by weight.

The ratio of vinylidene fluoride and hexafluoropropene units in the copolymer is a matter of importance. The ratio, by weight, of these components should range from about 70:30 to 45:55 which is equivalent to a range of molar ratios of vinylidene fluoride to hexafluoropropene units of approximately 85:15 to 65:35. Within this range, the copolymers of this process in the upper part of the molecular weight range are curable to elastic solids and in the lower part are curable to solids which are less elastic and more plastic. As the proportion of vinylidene fluoride becomes appreciably higher than the 70:30 by weight upper limit, the copolymer, on curing, tends to lose its elastic or resilient property and becomes brittle. At the other end of the range, it becomes increasingly troublesome to introduce more hexafluoropropene units into the copolymer. Considerable excesses of hexafluoropropene must be used and recovered. The 45:55 by weight limit may be considered a practical lower limit of vinylidene fluoride content rather than an absolute lower limit.

The chain transfer agents which are useful in the process are saturated aliphatic compounds containing not more than about 8 carbon atoms. These chain transfer agents may be straight or branch chain aliphatic alcohols, ketones or carboxylic acid esters.

The copolymers of this process represent copolymers of vinylidene fluoride and hexafluoropropene of controlled, low molecular weight and which are capable of being cured to highly useful solids. They are useful as plasticizers for fluorocarbon elastomers. They have the advantage of being cured by the elastomer curing agents, such as hexamethylene diamine carbamate, so that they become integral parts of the cured elastomer and cannot migrate. They can be compounded with filling agents and curing agents and used as plastic caulking compositions where the solid fluoroelastomers cannot be used because of the difficulty in positioning the elastomer. These copolymers may be compounded with curing agents and used as pourable, curable compositions for forming complex castings for which the conventional solid fluoroelastomer compositions are not suitable.

The following examples illustrate the process. The inherent viscosity was determined in an anhydrous solvent consisting of 86.1% tetrahydrofuran and 13.9% dimethylformamide. For convenience this solvent will be referred to as the THF/DMF solvent.

Example 1: A mixture of 4.5 ml. of ethyl acetate, 1.0 ml. of di-t-butylperoxide, 90 g. of hexafluoropropene and 90 g. of vinylidene fluoride was heated to 130°C. in a 400 ml. stainless steel bomb for sixteen hours. The resulting copolymer was dissolved in acetone and the solution was drum dried on a drum at 5 lbs. steam pressure yielding 142 g. of a tacky semisolid having an inherent viscosity in THF/DMF at 30°C. of 0.109 which corresponds to a molecular weight of about 8,400. The vinylidene fluoride content of the copolymer is 56.4% by weight, the hexafluoropropene content is 42.5% by weight and the ethyl acetate content is 1.1% by weight.

Example 2: A mixture of 335 ml. of ethyl acetate, 8.5 ml. of di-t-butylperoxide, 315 g. of hexafluoropropene and 315 g. of vinylidene fluoride was heated in a 1,350 ml. stainless steel bomb to 130°C. The temperature rose rapidly to 250°C. at which point the reaction was complete. The resulting solution was drum dried at 5 psig steam pressure to yield 450 g. of a tacky grease with an inherent viscosity in THF/DMF of 0.059 which corresponds to a molecular weight of about 2,800. The vinylidene fluoride content of the copolymer is 66.9% by weight, the hexafluoropropene content is 30.2% by weight and the ethyl acetate content is 3.2% by weight.

Example 3: (A) Into an autoclave of 1,430 ml. free volume agitated at 545 rpm was pumped at the rate of 1.8 liters per hour a catalyst solution consisting of 980 ml. of ethyl acetate and 20 ml. of di-t-butylperoxide. Vinylidene fluoride was added at the rate of 1.5 lb. per hour and hexafluoropropene at the rate of 1 lb. per hour. The pressure was 900 lbs. per sq. in. The polymer solution was removed continuously through a pressure let-down valve.

The temperature of the autoclave was maintained at 125° to 130°C. for the first hour and the product was collected. During the second hour the temperature was held in the range of 130° to 137°C. and the product was collected separately. The third hour, the temperature was held at 137° to 139°C. and the product collected. The three polymer solutions thus obtained were drum dried at a steam pressure of 20 lbs. per sq. in. gauge to yield tacky greases.

The data for the three runs are as follows:

Hour	Temp., °C.	Wt. Dried Product	Brookfield Viscosity, 90° C.
1	125-130	312	15,000
2	130-137	468	4,000
3	137-139	569	1,000

It is obvious that the molecular weight of the grease decreases with increasing temperature. (B) A composition of material produced at the several temperatures in (A) above had an inherent viscosity in THF/CMF of 0.089 which corresponds to a molecular weight of about 4,700. The vinylidene fluoride content is 63.9% by weight, the hexafluoropropene content is 34.3% by weight and the ethyl acetate content is 1.9% by weight.

The grease was heated in an oven at 204°C. for 24 hours to remove any volatile material and then 20 parts was milled into 100 parts of a copolymer of 60 parts vinylidene fluoride and 40 parts of hexafluoropropene. The Mooney viscosity of the mixture at 100°C. was 27 after 10 minutes compared to 60 for the unplasticized vinylidene fluoride-hexafluoropropene copolymer. The vulcanizate properties of the cured plasticized elastomer were the same as those of the cured control containing no plasticizer as shown in the following table. (The control and plasticized elastomer were both cured, after the addition of curing agents, in a press for 1 hour at 150°C., followed by oven cures of 1 hour at 100°C., 1 hour at 140°C. and 24 hours at 203°C.)

	Control	Plasticized Copolymer
Vinylidene fluoride/hexafluoropropene copolymer	100	100
Composite of Part B of this Example	-	20
Magnesium Oxide	15	15
Medium Thermal Black	18	18
Hexamethylenediamine carbamate	1.0	1.5
Modulus at 200% elongation, lbs./sq. in.	1,210	1,340
Tensile strength at break, lbs./sq. in.	2,100	2,150
Elongation at break, percent	290	290

Polymerization in the Presence of Silica

E.S. Lo; U.S. Patent 3,023,187; February 27, 1962; assigned to Minnesota Mining and Manufacturing Company describes the polymerization of fluoropropene and vinylidene fluoride in the presence of a siliceous material. The siliceous material employed as an ingredient of the polymerization system is an adsorptive silica which may be in uncombined or combined form. The process has particularly profound effects on the properties and especially on the physical and mechanical properties of copolymers of hexafluoropropene and vinylidene fluoride.

When hexafluoropropene and vinylidene fluoride are copolymerized in the presence of the adsorptive silicas, the tensile strength, tear strength, etc. of the resultant polymer products are significantly greater than such properties of the polymer prepared in the absence of the silica. Further the polymers also possess improved properties such as greater tensile strength than the polymers produced in the absence of the adsorptive silica but to which the silica has been added after the copolymer has been prepared. The improvement in the physical and mechanical properties realized by the process are attained without any deleterious effect on the outstanding chemical and other properties of the copolymer of hexafluoropropene and vinylidene fluoride. The following examples illustrate the process.

Example 1: (A) A stainless steel autoclave was charged with the following ingredients of an aqueous emulsion catalyst solution: 7,000 ml. of deionized water, 35 grams of potassium persulfate, 35 grams of perfluorooctanoic acid and 140 grams of disodium hydrogen phosphate heptahydrate ($NaHPO_4 \cdot 7H_2O$). The autoclave was then evacuated and connected to a steel cylinder containing a mixture of hexafluoropropene and vinylidene fluoride in an amount equivalent to 30 mol percent of the hexafluoropropene and 70 mol percent of the vinylidene fluoride.

The steel cylinder containing the monomer mixture was equipped with a pressure gauge and a needle valve located between the autoclave and the steel cylinder. The contents of the autoclave were then heated to 50°C. The needle valve between the steel cylinder and the polymerization autoclave was then opened and the aforesaid monomer mixture of hexafluoropropene and vinylidene fluoride was fed into the bomb at a rate sufficient to maintain the polymerization pressure at 200 pounds per square inch gauge. The polymerization reaction was allowed to run for 10 hours during which time the monomer mixture of hexafluoropropene and vinylidene fluoride was fed into the autoclave at a rate sufficient to maintain the pressure within the autoclave at a constant pressure of 200 psig, while the reaction temperature was maintained constant at about 50°C. At the end of the 10 hour period the needle valve was closed and the autoclave vented to atmospheric pressure. During the ten hour polymerization run it was found that approximately 3,929 grams of monomer mixture had reacted. The reaction mixture in the autoclave was a latex having a transparent blue appearance which was coagulated by freezing at about -70°C. in Dry-Ice chest. The solid rubbery copolymer product obtained thereby was broken into small pieces and washed about four times with cool water, followed by washing four times with hot water. The polymer was then dried to constant weight in an air oven, overnight at 50°C. yielding a rubbery product (3,719 grams) in 95% yield based on the total monomers charged to the polymerization zone. The copolymer contains approximately 30 and 70 mol percent of hexafluoropropene and vinylidene fluoride, respectively.

(B) A 100 gram aliquot of the hexafluoropropene/vinylidene fluoride elastomeric copolymer prepared in accordance with the procedure of part (A) of this example was placed on a two-roll mill and worked until a continuous band was formed. The rolls were heated to a temperature of about 60°C. to hasten the formation of the band. After the copolymer was banded, there were added thereto 1.5 grams of benzoyl peroxide, 10 grams of zinc oxide and 10 grams of Diphos ($2PbO \cdot PbHPO_3$) with cutting and turning of the copolymer as it was banded on the rolls. During the blending operation the temperature was maintained at about 100° to about 150°F. After the addition of all the ingredients the batch was thoroughly mixed on the rolls and sheeted out for the molding operation which in this case involved the preparation of standard ASTM test sheets. These test sheets were prepared by taking a sheet of stock approximately 10% thicker than that desired and placing it in a mold.

The mold was placed in a suitable press having platens heated to 300°F. and was maintained at this temperature for a period of 30 minutes. The stock was then placed in an oven and heated to a temperature of 300°F. for a period of 16 hours. The vulcanized sample was a snappy rubber having a tensile strength of 1,345 pounds per square inch, a tear strength of 123 pounds per square inch and a percent set at break of 10.

Example 2: A 100 gram sample of the hexafluoropropene/vinylidene fluoride raw copolymer produced in accordance with part (A) of Example 1 above was mixed with 10 grams of synthetically refined free silica (SiO_2) on a rubber mill. The sample was then vulcanized as described in part (B) of Example 1 above by mixing the raw copolymer containing the silica with 1.5 grams of benzoyl peroxide, 3 grams of zinc oxide and 3 grams of Diphos. The resultant mixture was then placed in a press the platens of which were maintained at a temperature of 300°F. for one-half hour. The pressed cured stock was next placed in an oven and heated at a temperature of 300°F. for a period of 16 hours. The vulcanized sample was a snappy rubber having a tensile strength of 1,700 pounds per square inch.

Example 3: (A) After flushing a 300 ml. Aminco polymerization bomb with nitrogen the following ingredients were charged to the bomb freezing the contents of the bomb after the addition of each ingredient: 73 ml. of deionized water containing 3 grams of dissolved disodium hydrogen phosphate heptahydrate and 0.75 gram of perfluorooctanoic acid; 60 ml. of deionized water containing 0.75 gram of dissolved potassium persulfate; and 17 ml. of an aqueous solution containing 30% by weight of dispersed synthetically refined free silica (SiO_2). The bomb was then connected to a gas transfer system and evacuated at liquid nitrogen temperature, and was then charged with 25 grams of hexafluoropropene and 25 grams of vinylidene fluoride corresponding to a total monomer charge containing 30 mol percent of hexafluoropropene and 70 mol percent of vinylidene fluoride.

The polymerization bomb was then closed and placed in a mechanical shaker. The polymerization reaction was carried out with constant shaking of the bomb in a constant temperature bath maintained at a temperature of 50°C. for a period of 22 hours under autogenous pressure. At the end of 22 hours the polymerization bomb was vented to atmospheric pressure and the unreacted monomers were removed. The polymer latex was coagulated by freezing in a Dry-Ice chest. The coagulated product was collected, washed thoroughly with hot water to remove residual salts and dried in vacuo at 35°C. A tough white elastomeric product was obtained in about an 88% conversion and contains approximately 68 mol percent of combined vinylidene fluoride and 32 mol percent of combined hexafluoropropene. This raw copolymer product was found to have a tensile strength of 1,700 pounds per square inch and a tear strength of 210 pounds per square inch.

(B) The copolymer produced in accordance with the procedure of part (A) of this example was vulcanized according to the procedure of part (B) of Example 1 above by mixing 100 parts of the raw copolymer with 10 parts of zinc oxide, 10 parts of Diphos and 1.5 parts of benzoyl peroxide. The resultant mixture was then placed in a press for 1/2 hour at 300°F. The pressed stock was then placed in an oven and heated at a temperature of 300°F. for a period of 16 hours. The vulcanized sample was a tough and snappy rubber having a tensile strength of 2,200 pounds per square inch and a tear strength of 240 pounds per square inch.

Comparison of the tensile strengths of the products of Example 3 (A) with Example 1 (B) shows that the tensile strength (i.e., 1,700 psi) of the raw unvulcanized copolymer of hexafluoropropene and vinylidene fluoride which is prepared in the presence of silica is greater than the tensile strength (1,345 psi) of the vulcanized copolymer prepared in the absence of silica by about 26%. Comparison of the results of Example 3 (B) with Example 1 (B) shows that the tensile strength (2,200 psi) of the vulcanized copolymer of hexafluoropropene and vinylidene fluoride prepared in the presence of silica is greater by about 64% than the tensile strength (1,345 psi) of the vulcanized copolymer prepared in the absence of silica.

Further comparison of the results of Example 3 (B) with the results of Example 2 shows that the tensile strength (2,200 psi) of the vulcanized copolymer prepared in the presence of silica is greater by about 30% than the tensile strength (1,700 psi) of the copolymer which has been prepared in the absence of silica but which has been vulcanized in the presence of mixed silica.

It is to be noted that the vulcanization recipe employed in Examples 1 (B), 2 and 3 (B) were the same and that the weight percent of silica employed in Examples 2 and 3 (A) were the same, that is, in Example 2 the amount of the silica mixed with the raw copolymer was 10% by weight and in Example 3 (A) about 10% by weight of the silica based on total monomers charged was present during the polymerization reaction.

Partially Crystalline Copolymers

A process described by E.S. Lo; U.S. Patent 3,178,399; April 13, 1965; assigned to Minnesota Mining and Manufacturing Company involves the preparation of copolymers which contain hexafluoropropene (or omega-hydroperfluoropropene) and vinylidene fluoride in varying comonomer ratios from about one to about 15 mol percent, preferably from about one to about 13 mol percent.

In carrying out the polymerization reaction it has been found that the copolymers having above about 15 mol percent of hexafluoropropene are essentially completely amorphous elastomers and are particularly outstanding for their low torsional modulus and retention of their rubbery properties over a wide range of temperatures, i.e., between about –30°F. and about 600°F. without embrittlement, degradation or hardening. Those copolymers containing between about 6 and about 15, preferably between about 6 and about 14, mol percent of combined hexafluoropropene are partially crystalline thermoplastic resins with outstanding high tensile and reversible elongation properties. Those copolymers having between about 1 and 3 mol percent of hexafluoropropene are normally resinous thermoplastic materials at room temperature and also retain their flexibility over a wide range of temperature without embrittlement. Those copolymers containing between about 4 and about 6 mol percent of combined hexafluoropropene are valuable resinous thermoplastic materials having slightly rubbery characteristics.

The copolymerization reaction between hexafluoropropene and vinylidene fluoride is carried out in the presence of a polymerization promoter which may be a free radical-forming or an ionic-type promoter. The free radical-forming promoters or initiators comprise the organic and inorganic peroxy and azo compounds. The ionic initiators comprise inorganic halides of the Friedel-Crafts catalyst type, and mineral acids. The initiator is generally employed in an amount between about 0.001 and about five parts by weight per 100 parts of total monomers employed, and preferably are employed in an amount of between about 0.01 and about 1.0 part by weight.

The hexafluoropropene-vinylidene fluoride copolymers of the process are suitable and useful as durable, flexible coatings for application to metal or fabric surfaces. The copolymers are dissolved in a suitable solvent and applied to the surfaces by spraying, brushing, or other such conventional coating techniques. Particularly useful solvents for this purpose comprise the relatively low molecular weight and volatile aliphatic carboxylic acid esters such as methyl acetate, ethyl acetate and butyl acetate. The following examples illustrate the process. The Gehman stiffness values given in the following examples were determined according to ASTM Designation D-1053-49T.

Example 1: After flushing a 300 ml. Aminco polymerization bomb with nitrogen, the bomb was charged with the following aqueous emulsion polymerization catalyst system, freezing the contents of the bomb after the addition of each ingredient:

(1) 15 ml. of water containing 0.3 gram of dissolved sodium metabisulfite;
(2) 90 ml. of water containing 0.75 gram of dissolved potassium perfluorooctanoate, the pH having been adjusted to 12 by the addition of a 5% aqueous potassium hydroxide solution; and
(3) 45 ml. of water containing 0.75 gram of dissolved potassium persulfate.

In a separate experiment, the final pH of this polymerization catalyst system was found to be about 7. The bomb was then connected to a gas transfer system and evacuated at liquid nitrogen temperature. The bomb was then charged with 12.4 grams of hexafluoropropene and 47.6 grams of vinylidene fluoride to make up a total monomer charge containing 10 mol percent of hexafluoropropene and 90 mol percent of vinylidene fluoride. The polymerization bomb was then closed and placed in a mechanical shaker. The polymerization reaction was conducted for a period of 24 hours at a constant temperature 50°C. under autogenous pressure with continuous shaking of the bomb. At the end of the 24 hours the bomb was vented to atmospheric pressure to remove the unreacted monomers. The polymer latex was coagulated by freezing it at liquid nitrogen temperature. The coagulated product was collected, washed thoroughly with hot water to remove residual salts and dried in vacuo at a temperature of 35°C. A white resinous thermoplastic was obtained which was very slightly rubbery, and was obtained in a 58% conversion. This product was analyzed and was found to contain 61.3% fluorine corresponding to 5 mol percent of combined hexafluoropropene and 95 mol percent of combined vinylidene fluoride. Upon analysis for carbon content, the product was shown to contain 35.48% carbon corresponding to 6 mol percent of combined hexafluoropropene and 94 mol percent of combined vinylidene fluoride. The weight percent of combined hexafluoropropene and vinylidene fluoride, based upon the carbon analysis, was found to be 13 and 87 weight percent, respectively. X-ray analysis of this product shows it to be highly crystalline.

This very slightly rubbery hexafluoropropene-vinylidene fluoride copolymer when milled in a rubber mill at 25°C. yielded a nontransparent plastic sheet. When pressed between chrome-plated ferrotype plates in an electrically-heated Carver press at 325°F. for a period of 5 minutes at 10,000 pounds per square inch, a clear, flexible plastic sheet was obtained.

The resistance of the resinous thermoplastic copolymer of this example to fuels and strong corrosive chemicals was excellent as evidenced by the fact that the copolymer exhibited a volume swell of only 1% after exposure at 25°C. to a fuel containing 30% by volume of toluene and 70% by volume of isooctane for 4 and 7 days, and only a 4% volume swell after exposure at 25°C. to red fuming nitric acid for 4 and 7 days.

The results of further preparation of copolymers of hexafluoropropene and vinylidene fluoride are set forth in Table 1. The procedure followed included charging a feed cylinder with a monomer mixture of the proportions indicated. After flushing with nitrogen an autoclave was charged with 2,700 grams water, 13.5 grams $Cl(CF_2CFCl)_3CF_2COOH$, 13.5 grams $K_2S_2O_8$ and 28.8 grams Na_2HPO_4. The autoclave was pressured four times with nitrogen, vented, then pumped to a 20 inch mercury vacuum. After heating to 64°C. with agitation the autoclave was pressured with the given monomer mixture to 125 psig and the feed regulator was set to continue demand feeding at this pressure throughout the reaction period. After the feed cylinder pressure dropped markedly, indicating that the liquid phase was exhausted, the polymerization was stopped and the polymeric latex was recovered. Polymer samples were obtained by freeze coagulation, followed by washing with hot water, dissolution in acetone and reprecipitation in water. Polymer properties are given in Table 2.

TABLE 1: COPOLYMER PREPARATION

Run #	Mol % CF_2CH_2 in charge	CF_2CH_2 charged, g.	CF_3CFCF_2 charged, g.	Reaction conditions hr. at 64°C.*	Reaction pressure	Yield, g.	η in acetone	%C	Wt. % CF_2CH_2	Mol % CF_2CH_2
1	87	556	195	4.2	125	612	0.93	34.1	74.8	87.4
2	89	570	165	3.2	125	625	0.96	34.3	76.3	88.3
3	91	582	135	2.9	125	602	0.90	34.9	80.7	90.7
4	93	595	105	4.3	125	500	0.88	35.8	87.4	94.2
5	83	36	184**	2.6	125	360	0.86	34.2	70.8	87.3

*Water—100, $Cl(CF_2CFCl)_3CF_2COOH$—0.5, $K_2S_2O_8$—0.5, Na_2HPO_4—1.06, Monomers 25-50.
**$HCF_2CF_2CFCF_2$ was the comonomer.

TABLE 2: COPOLYMER PROPERTIES

Run #	Mol % charged	CF_2CH_2, found*	Comonomer	η in acetone	Unoriented film Tensile, psi	Modulus,** psi	Elongation,*** psi
1	87	87.4	CF_3CFCF_2	0.93	4,140	4,060	710
2	89	88.3	CF_3CFCF_2	0.96	4,160	5,080	530
3	91	90.7	CF_3CFCF_2	0.90	5,350	9,790	550
4	93	94.2	CF_3CFCF_2	0.88	6,150	24,100	530
5	83	87.3	$HCF_2CF_2CFCF_2$	0.86	1,410	5,260	510

*Based on percent carbon analysis.
**Modulus calculated at extension rate of 0.2 in./min.
***Elongation based on extension rate of 2 in./min.

Measurement of tensile strength and reversible elongation were made on 40 mil thick specimens pressed at 350° to 400°F. and quickly cooled to room temperature. Elongation was measured on the sample by pulling at an extension rate of 2 inches per minute until the sample broke. The length of the broken sample at break and after retraction was observed and the reversible elongation of the sample was calculated from the data obtained. The copolymers prepared by co-polymerization of omega-hydroperfluoropropene or hexafluoropropene and vinylidene fluoride under the polymerization conditions discussed possess unique and highly desirable physical, mechanical, and chemical properties. They have high tensile strength and good reversible elongation and are remarkably resistant to embrittlement, degradation and discoloration at temperatures between about –30° and about 600°F. and higher.

They are resistant to swell by aromatic and aliphatic oils and fuels, and ester-type hydraulic fuels such as relatively high molecular weight alkyl esters of organic dicarboxylic acids having typical oily characteristics. In addition they are resistant to swell and chemical breakdown when exposed to strong powerful corrosive agents such as red fuming nitric acid and ozone. In addition, they do not swell when exposed to carbon bisulfide, carbon tetrachloride, 2,5-dichloro-benzotrifluoride, diethyl ether and are stable to attack by sunlight, fungi and microorganisms in general.

ELASTOMERIC POLYMERS — CURING AGENTS

Dicumyl Peroxide

E.L. Yuan; U.S. Patent 3,025,183; March 13, 1962; assigned to E.I. du Pont de Nemours and Company describes the use of dicumyl peroxide as a curing agent. The process involves mixing or blending dicumyl peroxide with a copolymer of about 70 to 30% by weight of vinylidene fluoride and about 30 to 70% by weight of hexafluoropropene, shaping the mixture and heating the shaped mixture to cure the coating. The heating step is preferably carried out at from about 300° to about 400°F. for a length of time sufficient to cure the copolymer to the point of insolubility in methyl ethyl ketone at room temperature.

The data in the table below shows the advantages of curing the copolymer with dicumyl peroxide in comparison with benzoyl peroxide. For example, the dicumyl peroxide cured compositions exhibit superior retention of elastomeric properties on heat aging. Further, the dicumyl peroxide cured compositions lose less weight on heat aging than the same composition cured with benzoyl peroxide.

Example	1	2	3	4
Composition—Parts By Weight:				
Copolymer of vinylidene fluoride and hexafluoropropene*	100	100	100	100
Zinc Oxide	10	10	–	–
Dibasic Lead Phosphite	10	10	–	–
Dicumyl Peroxide	–	3	–	3
Benzoyl Peroxide	3	–	3	–
Initial Tensile Properties:				
Tensile (psi)	1,670	1,420	1,150	1,050
Modulus 300% (psi)	820	380	330	270
Elongation (percent)	450	670	530	770
Hardness, Shore A	62	64	57	60
Heat Aging, 15 days at 400°F.:				
Tensile (psi)	900	1,500	810	1,180
Modulus 300% (psi)	610	410	240	270
Elongation (percent)	400	720	570	700
Hardness, Shore A	60	63	57	66
Weight Loss (percent)	7.8	5.5	3.5	1.5
Heat Aging, 30 days at 400°F.:				
Tensile (psi)	920	1,450	680	1,150
Modulus 300% (psi)	700	440	240	280
Elongation (percent)	370	670	590	750
Hardness, Shore A	61	61	56	64
Weight Loss (percent)	8.9	6.5	3.6	2.0
Heat Aging, 60 days at 400°F.:				
Tensile (psi)	1,090	1,460	520	970
Modulus 300% (psi)	610	430	200	160
Elongation (percent)	410	680	550	750
Hardness, Shore A	59	61	57	64
Weight Loss (percent)	15.3	9.6	5.8	4.3

*60 parts of vinylidene fluoride and 40 parts hexafluoropropene were mixed on a two roll rubber mill, then press cured at 275°F. for 1 hr. and oven cured at 320°F. for 16 hrs.

Dicumyl Peroxide and Diallyl Terephthalate

J.F. Smith; U.S. Patent 3,011,995; December 5, 1961; assigned to E.I. du Pont de Nemours and Company describes a process for preparing a cured fluorocarbon elastomer which comprises incorporating, per 100 parts by weight of the fluorocarbon elastomer, (a) from about 5 to 20 parts by weight of magnesium oxide, (b) from about 2 to 10 parts by weight of dicumyl peroxide and (c) from about 1 to 3 parts by weight of diallyl terephthalate or tetraallyl terephthalamide; followed by heating the compounded stock at temperatures between about 100° and 200°C. so as to effect a cure. The use of diallyl terephthalate or tetraallyl terephthalamide in conjunction with the other curing agents results in highly improved cured products. The following example illustrates the process.

Example: Vinylidene Fluoride (VF$_2$)/Hexafluoropropene (HFP) Copolymer — Copolymer A is a 60/40 weight percent copolymer of VF$_2$ and HFP. It has an inherent viscosity (0.1 g. copolymer in 100 cc of an 87/13 weight percent tetrahydrofuran (THF)/dimethyl formamide (DMF) mixture at 30°C.) of 0.95±0.05, a Mooney viscosity (ML 10 at 100°C.) of 75±6 and a number average molecular weight of about 100,000. The general procedure for preparation of a copolymer of this type is given in "Industrial and Engineering Chemistry" 49, 1687 (1957).

Curing procedure — Five stocks (1A-1E) were compounded on a cold rubber roll mill according to the recipes given below. The dicumyl peroxide was added last. The stocks thereby obtained were heated in 3 x 6 x 1/4" molds in a press at 150°C. for one hour. Subsequently they were removed from the molds and heated in a circulating air oven at 200°C. for 18 hours to complete the cure. The stress-strain properties of the various vulcanizates were then measured in accordance with ASTM Test Method 412-51T. The data obtained, which are displayed in the table below, show that the use of diallyl terephthalate or tetraallyl terephthalamide produces vulcanizates with superior physical properties.

Component			Stocks		
	1A	1B	1C	1D	1E
Copolymer A	100	100	100	100	100
Medium Thermal Black	–	–	18	18	18
"Hi Sil 202"*	15	15	–	–	–
Magnesium Oxide	15	15	15	15	15
Diallyl Terephthalate	2	–	–	2	–
Tetraallyl Terephthalamide	–	–	–	–	2
Dicumyl peroxide (100%)	2	2	2	2	2
Property (at 25°C.):					
M100 (psi)	350	400	Sponged	250	280
M200 (psi)	900	700	Sponged	600	620
T$_B$ (psi)	2,500	1,000	Sponged	1,850	2,080
E$_B$ (percent)	480	480	Sponged	500	510

*Precipitated hydrated silica of very fine particle size (has a particle size of about 0.022 micron and a surface area of 150 square meters per gram and contains 10.7% of water of hydration, corresponding to 0.073 gram of water per 100 square meters of surface area). Commercially available from Columbia-Southern Chemical Corporation, Pittsburgh, Pennsylvania.

Benzoyl Peroxide and N,N'-Methylenebisacrylamide

L.P. Dosmann and G.L. Barnes; U.S. Patent 2,944,995; July 12, 1960; assigned to United States Rubber Company have found that the cure of a hexafluoropropylene-vinylidene fluoride elastomeric copolymer with peroxidic compounds is accelerated by N,N'-methylenebisacrylamide. By using this accelerator, it is possible to cure the hexafluoropropylene-vinylidene fluoride copolymer rapidly, under the ordinary conditions usually used for vulcanizing rubber, and the resulting product is found to be free from voids or blows, and has excellent chemical and mechanical properties. The following example illustrates the process.

Example: The fluorocarbon elastomer employed was a commercial material known as "Viton A," by the Du Pont Company. The molar ratio of vinylidene fluoride to hexafluoropropylene was 4 to 1. It was in the form of a gum having the following properties:

Specific gravity	1.85
Fluorine content	65%
Color	White, translucent

(continued)

(continued)

Solubility	Ketones
Storage stability	Excellent — no change in Mooney viscosity after 30 days at 100°F.
Mooney viscosity ML-4/212°F.	35–55
Williams plasticity	110
Recovery	15

The gum is mixed on a rubber mill with other ingredients, to provide the stocks shown in the table below. The elastomer is suitably milled at a mill roll temperature of 120° to 150°F. Magnesium oxide and N, N'-methylenebisacrylamide are added and blended in by milling. The peroxide is then added and blended by further milling. The compounds can be sheeted from the mill (face roll temperature 180°F.) or calender. The stocks are then cured in an oven in an air-ammonia atmosphere at a temperature of 270°F. for 2.8 hours. The physical properties of the resulting vulcanizates were then determined, with the results shown below.

	Parts			
Ingredients	Stock 1	Stock 2	Stock 3	Stock 4
Viton A	100	100	100	100
Magnesium Oxide	10	10	10	10
Benzoyl Peroxide	2.5	2.5	2.5	2.5
N, N'-Methylenebisacrylamide	–	5	10	15

Physical Properties (2.8 hour cure)

Tensile (psi)	664	1,013	1,262	1,930
Elongation (percent)	1,075	570	388	106
Modulus 100% Elongation	160	373	820	1,898
Modulus 200% Elongation	190	551	1,049	–
Modulus 300% Elongation	215	719	1,200	–

It will be apparent that the cure proceeded much more rapidly in the presence of N, N'-methylenebisacrylamide, and the vulcanizates made with this accelerator had superior physical properties.

In related work, L.P. Dosman; U.S. Patent 2,944,927; July 12, 1960; assigned to United States Rubber Company describes the use of N, N'-methylenebisacrylamide for curing chlorofluorocarbon resins.

Hexamethylenediamine Carbamate and N, N'-Disalicylidenepropylenediamine

In a process described by A.L. Moran; U.S. Patent 2,951,832; September 6, 1960; assigned to E.I. du Pont de Nemours and Company vinylidene fluoride-hexafluoropropylene copolymer elastomers having improved scorch-resistance and other improved processing characteristics with good tensile strength, compression set and high temperature properties when cured, may be produced by homogeneously incorporating into the elastomer prior to curing from 1.0 to 6.0%, based on the weight of the elastomer, of an N, N'-diarylidenealiphaticdiamine, such as N, N'-dibenzylidene (or substituted dibenzylidene) alkanediamine, or by homogeneously incorporating into the elastomer materials which will form the compound in situ. These compounds have the general formula: Ar—CH=N—R—N=CH—Ar where Ar is an aromatic radical of the benzene series, and R is a saturated aliphatic or cycloaliphatic hydrocarbon radical of from 4 to 18 carbon atoms.

These compounds are produced in known manner by the reaction of an aromatic aldehyde with an aliphatic or cycloaliphatic diamine. They are also produced by reacting a diarylidene-1, 2-propylene-(or ethylene-) diamine with a C_4 to C_{18} aliphatic or cycloaliphatic diamine or diamine carbamate, which diamines or diamine carbamates have previously been used as curing agents for these elastomers (see e.g., U.S. Patent 2,181,121). When produced in situ in the elastomer an amount of from 0.5 to 3.0%, based on the weight of the elastomer, of the N, N'-diarylidene-1, 2-propylene-(or ethylene-)diamine is used with preferably at least a molar equivalent of the longer carbon chain aliphatic or cycloaliphatic diamines or diamine carbamates normally used in the curing of these elastomers. Where the aldehyde itself is added, about 2 mols should preferably be used per mol of diamine or diamine carbamate.

The preferred curing agent in the process is the combination of a disalicylideneethylenediamine or a disalicylidene-1, 2-propylenediamine with hexamethylenediamine carbamate. The curing agent of this process gives very satisfactory cures at the relatively low temperature of from 150° to 200°C. without imparting to the compounded copolymer elastomer any undesirable scorching properties. At higher temperatures, such as above 200°C., very fast cures are obtainable.

Temperatures as high as 320°C. for 15 hours have been found to be satisfactory without detrimental effect to the cured elastomer. The following example illustrates the process.

Example: A copolymer consisting of 40% by weight of hexafluoropropylene and 60% by weight of vinylidene fluoride was compounded according to the following formulations:

| | Parts by Weight | |
	A	B
Fluoroelastomer	100	100
Magnesium oxide ("Maglite" D)	15	15
Thermax (carbon black filler)	60	60
N, N'-disalicylidenepropylenediamine*	-	1.3
Hexamethylenediamine carbamate	1.25	1.25
Mooney Scorch Test at 250°F. (minutes to a 10 point rise)	7	26

*Added as a 65% solution in an aromatic solvent such as xylene.

The above compositions were cured in a press at 275°F. for 30 minutes and subjected to an aftercure in an air oven at 212°F. for 1 hour, 250°F. for 1 hour, 300°F. for 1 hour, 350°F. for 1 hour and finally at 400°F. for 24 hours.

	A	B
Stress Strain at 70°F.:		
Tensile at break (lbs./sq.in.)	2,400	2,875
100% Modulus (lbs./sq.in.)	1,225	1,625
Elongation at break (percent)	210	180
Shore A hardness	85	86
Stress Strain at 300°F.:		
Tensile at break (lbs./sq.in.)	584	1,023
Elongation at break (percent)	100	60
Compression Set: 70 hrs. at 250°F., Method B, ASTM	58.7	29.7

The compounded elastomers were subjected to a test for blistering as follows: 1 1/2" square sample pieces of the cured elastomers (75 mils thick) were submerged in "Turbo Oil No. 15" (a diester lubricant—MIL 7808) and the submerged samples placed in an autoclave which was heated to 400°F. and pressured with nitrogen to 800 lbs./sq. in. After 8 hours, the samples were removed and observed visually for blisters. It was observed that sample A was badly blistered, while no blisters were seen on sample B.

Similar results to those shown above were obtained, using from 0.65 part to 1.95 parts of disalicylidenepropylenediamine, based on 100 parts by weight of the elastomer. As illustrated in the above example, it is often desirable to carry out the cure by stepwise progression to the final curing temperature. This method allows the elimination of any gas formed during the cure or which may be incorporated during the mixing process, without causing porosity in the vulcanizate.

Dithiol and Aliphatic Tertiary Amine

J.F. Smith; U.S. Patent 3,008,916; November 14, 1961; assigned to E.I. du Pont de Nemours and Company describes a process for curing an elastomeric copolymer consisting of 30 to 70% by weight of vinylidene fluoride and 70 to 30% by weight of hexafluoropropene which process comprises compounding with each 100 parts of elastomer (a) 0.2 to 2 parts of an organic aliphatic or aromatic dithiol, (b) 0.2 to 1 part of an aliphatic or cycloaliphatic tertiary amine, and (c) 2.5 to 25 parts of magnesium oxide, followed by curing the compounded elastomer at about 150°C. for about 60 minutes and completing the curing process at about 200°C. for about 2 to 24 hours. The following examples illustrate the process.

Example 1: 100 parts of a copolymer of 59% by weight of vinylidene fluoride and 41% by weight of hexafluoropropene was compounded on a water-cooled rubber mill with 15 parts of magnesium oxide, 18 parts of MT (medium thermal) carbon black, 0.5 part of dimethyldodecylamine, and 1 part of dimercaptomethylsulfide ($HS-CH_2-S-CH_2-SH$). The compounded elastomer was then placed in a mold at 150°C. for 60 minutes. The molded article was then removed from the mold without difficulty and the cure completed in an oven at 200°C. for 24 hours. The cured polymer had exactly copied the contour of the mold giving clean, sharp impressions. The physical properties of the cured elastomer follows.

Tensile at break (lbs./in.^{2})	2,300
Elongation at break (percent)	210
Modulus—200 (lbs./in.^{2})	2,080

Example 2:   Following the procedure of Example 1, 100 parts of a copolymer of 39% by weight of vinylidene fluoride and 61% by weight of hexafluoropropene was compounded with 15 parts of MgO, 18 parts of MT black, 0.4 part of dimethyldodecylamine and 0.6 part of ethylenebismercaptoacetate. After curing in the mold at 150°C. for 60 minutes, the product was cured further in an oven at 204°C. for 18 hours. The physical properties of the final product follows:

Tensile at break (lbs./in.^{2})	2,320
Elongation at break (percent)	360
Modulus—200 (lbs./in.^{2})	850

100 parts of the same copolymer used above was compounded with 15 parts of magnesium oxide, 18 parts of MT black and 1 part of hexamethylenediamine carbamate, and cured simultaneously with the above sample. The following properties were obtained:

Tensile at break (lbs./in.^{2})	1,680
Elongation at break (percent)	360
Modulus—200 (lbs./in.^{2})	850

Comparison of the properties of the two samples demonstrates the superiority in tensile strength properties of the dithiol cured stock.

Example 3:   Following the procedure of Example 1, three samples of a copolymer containing 59% by weight of vinylidene fluoride and 41% by weight of hexafluoropropene were compounded as follows:

	A	B	C
Fluoroelastomer	100	100	100
Magnesium Oxide	15	15	15
MT Black	18	18	18
Hexamethylenediamine Carbamate	–	–	1
Dimethyldodecylamine	0.5	0.5	–
Ethylenebismercaptoacetate	0.8	0.8	–

The compounded elastomers were subjected to cures as follows:

	A	B	C
Mold Cure:			
Minutes	60	60	60
Temp.°C.	150	150	150
Oven Cure:			
Hours	None	2	24
Temp.°C.		204	204

On evaluation of stress-strain properties, the cured elastomers gave the following values:

	A	B	C
Tensile at break (lbs./in.^{2})	1,800	2,350	1,470
Elongation at break (percent)	410	300	270
Modulus—200 (lbs./in.^{2})	860	1,350	940

It can be seen that the combination of dithiol, tertiary amine and magnesium oxide give a rapid curing system. After only one hour in the mold and without an oven cure, very good properties are developed (Sample A). When cured further in an oven for only two hours, the properties were better than those obtained by a 24-hour cure with the control hexamethylenediamine carbamate curing agent (B vs. C).

Methylenediamine Carbamate and Tertiary Amines

A process described by J.F. Smith; U.S. Patent 3,080,336; March 5, 1963; assigned to E.I. du Pont de Nemours and Co.

involves curing fluorocarbon elastomers with hexamethylenediamine carbamate and certain tertiary amines. It is known that satisfactory vulcanizates can be obtained by curing fluorocarbon elastomers with a combination of hexamethylenediamine carbamate and magnesium oxide. It would be desirable, however, to improve this process so as to increase the scorch resistance and to obtain a better cure.

This process is based on the discovery that when a tertiary amine of certain type is added to a curing recipe for fluorocarbon elastomers containing hexamethylenediamine carbamate and magnesium oxide, the amount of hexamethylenediamine carbamate required to attain a particular state of cure can be reduced. At the same time, the Mooney scorch time of the stock is significantly increased. The use of the process thus permits attainment of a satisfactory state of cure with a high degree of processing safety.

The tertiary amines used in the process are selected from the group consisting of tertiary alkylamines, tertiary cycloalkylamines, N-alkyl piperidines, N, N'-dialkyl piperazines, and N, N, N', N'-tetraalkylalkylenediamines. To be useful in this process the tertiary amine should be sufficiently basic so that at least one of the amine groups has a pKb value at 25°C. or less than 4.50. The following example illustrates the process. The tests were conducted in accordance with the following methods:

Test—	ASTM test method
Mooney scorch	D 1077–55T
Modulus at 100% extension	D 412–51T
Compression set	D 395–55 (Method B)

Fluorocarbon elastomer A is a 60/40 weight percent copolymer of vinylidene fluoride and hexafluoropropene. It has an inherent viscosity (0.1 gram copolymer in 100 cc of an 87/13 weight percent tetrahydrofuran/dimethyl formamide mixture at 30°C.) of 0.95±0.05, a Mooney viscosity (ML 10 at 100°C.) of 75 ±6 and a number average molecular weight of about 100,000. The general procedure for preparation of polymers of this type is given in "Industrial and Engineering Chemistry," 49, 1687 (1957).

Example: Five stocks (1A–1E) were compounded on a rubber roll mill at 25°C. The table below gives the recipe and the Mooney scorch data for each stock. After these stocks had been heated in a press at 150°C. for one hour, they were removed and heated in a circulating air oven at 204°C. for 18 hours.

	Stocks				
	1A	1B	1C	1D	1E
Component:					
Fluorocarbon Elastomer A	100	100	100	100	100
Medium Thermal Black	18	18	18	18	18
Magnesium Oxide	15	15	15	15	15
Hexamethylenediamine Carbamate	1	1	0.5	0.25	0
Dimethyldodecylamine	0	0.1	0.25	0.5	1.0
Property:					
Mooney Scorch (min. to 20 pt. rise)	13	19	20	25	>45
Modulus at 100% Extension at 25°C. (psi)	430	650	400	330	sponged
Compression Set, 22 hrs., at 70°C. (percent)	31	16	24	27	sponged

The following typical observations taken from the above example illustrate the improvement provided by the process. When one part of hexamethylenediamine carbamate was compounded with a vinylidene fluoride/hexafluoropropene copolymer stock containing magnesium oxide and carbon black, the time taken for a 20 point rise in the Mooney scorch test was 13 minutes; the vulcanizate obtained had a modulus at 100% extension of 430 psi and a compression set of 31%. When one part of dimethyldodecylamine, a representative tertiary amine, was used in place of the hexamethylenediamine carbamate, an undercured sponged vulcanizate resulted under the same curing conditions.

When, however, 0.5 part of hexamethylenediamine carbamate was used as a curing system in combination with 0.25 part of dimethyldodecylamine, the Mooney scorch was greatly improved (20 minutes to a 20 point rise), the modulus at 100% extension of the vulcanizate was 400 psi, and the compression set was 24. The state of cure thus obtained was at least equivalent to that of the conventional stock containing twice as much hexamethylenediamine carbamate.

At the same time the processing safety was much enhanced. It is highly significant that the above-specified tertiary amines can be substituted for part of the hexamethylenediamine carbamate and that the combination display a synergistic action.

Amine Cure in the Presence of Ammonia

A process described by L. Dewey; U.S. Patent 2,941,987; June 21, 1960; assigned to United States Rubber Company involves the finding that if the amine cure of the fluoroelastomer is carried out in the presence of ammonia the vulcanization can be effected rapidly without detriment to the physical properties, and with positive enhancement of the tensile strength.

The vulcanization may be carried out in an oven or autoclave at elevated pressure in an atmosphere of air and ammonia. The curing conditions will vary somewhat, depending on the exact properties and degree of cure desired in the final article, the size of the article, etc. In general, useful cures are obtainable over much the same time and temperature ranges as may be employed in ordinary rubber vulcanization. The temperature may range from about 150° to 350°F. The higher the temperature, the less is the time required for the cure. The amount of ammonia present in the air-ammonia atmosphere may vary from 4 to 100% by volume. A preferred range is from 4 to 33% ammonia by volume. The following examples illustrate the process.

Example 1:   Preparation of a Batch Using Viton A — The rubbery copolymer of vinylidene fluoride and hexafluoropropylene employed in this and Examples 2 and 3 was a commercial material known as "Viton A," made by the Du Pont Company. The molar ratio of vinylidene fluoride to hexafluoropropylene was 4 to 1. It was in the form of a gum and had the following properties:

Specific gravity	1.85
Fluorine content	65%
Color	White, translucent
Solubility	Ketones
Storage stability	Excellent—no change in Mooney viscosity after 30 days at 100°F.
Mooney viscosity ML-4/212°F.	35–55
Williams plasticity	110
Recovery	15

This gum was mixed on a rubber mill with other ingredients to provide the stock A given below.

Stock A

Viton A	100	parts
Philblack A	20	parts
Magnesium oxide	5	parts
Hexamethylene diamine carbamate	1.25	parts
Total	126.25	parts

The fluoroelastomer is milled at a mill roll temperature of 120° to 150°F.

Example 2:   Curing According to Standard Procedure — A. 126.25 parts of Stock A were placed in a heater and subjected to an air pressure cure at a pressure of 40 psi and a temperature of 275°F. for one hour. After this preliminary curing a post-cure was given as follows: one hour at 212°F., one hour at 250°F., one hour at 300°F., one hour at 350°F., and twenty-four hours at 400°F. The post curing was carried out in a circulating air oven, the temperature being brought up to 400°F. in a step-wise fashion as shown above. The total time for cure was twenty-nine hours.

B.   Alternatively, the Stock A formulation of Example 1 is molded for thirty minutes at 275°F. in a press and then post-cured for twenty-eight hours in precisely the same manner as the post-cure above.

Example 3:   Ammonia-Air Cure Technique — A.   The Stock A formulation of Example 1 is placed in a dry heater that has been prewarmed to 160°F. The temperature of the heater for the succeeding 5 1/2 hour period is adjusted as follows:

30 minutes—temperature held at 160° to 180°F.
1 hour—temperature gradually raised to 230°F.
1 hour—temperature gradually raised to 262°F.
30 minutes—temperature gradually raised to 276°F.
1 hour and forty minutes—temperature held constant at 276°F.
40 minutes—heater permitted to cool down
10 minutes—pressure bled to zero

Total time 5 1/2 hours.

The pressure is gradually raised to 45 psi during which time ammonia is metered in until the atmosphere is 8% ammonia by volume. The time required to introduce the ammonia and raise the pressure is about 1 hour. The above conditions are maintained for about 4 hours, after which the oven is gradually cooled down and the pressure gradually reduced to atmospheric. Total time: 5 1/2 hours.

B. Alternatively, the Stock A formulation of Example 1 is molded for 30 minutes at 275°F. in a press and then subjected to the same 5 1/2 hour cure of Example 3A. The following table gives comparative properties for the twenty-nine hour cure (Example 2) and five and one-half hour cure (Example 3).

	Tensile (psi)	Elongation (Percent)	Hardness (Shore)
Twenty-nine hour cure:			
A. Air pressure cure at 275°F. and 40 psi			
followed by 29 hour post-cure	1,512	340	85
Aged 50 hours at 550°F.	1,232	300	85
Aged 100 hours at 550°F.	984	215	90
B. Molded 30 mins. at 275°F. followed by			
29 hour post-cure	1,794	230	78
Aged 50 hours at 550°F.	1,446	227	81
Aged 100 hours at 550°F.	1,298	160	85
Five and one-half hour cure:			
A. Directly after curing	2,195	360	83
Aged 50 hours at 550°F.	1,911	290	87
Aged 115 hours at 550°F.	1,618	80	97
B. Molding 30 mins. at 275°F. followed by			
5 1/2 hour post-cure	1,783	250	78
Aged 50 hours at 550°F.	1,538	175	82
Aged 100 hours at 550°F.	1,493	115	90

Considering the data given above, the comparative results are striking. It is to be noted that in every instance the tensile strength is higher when ammonia cure is used as opposed to the twenty-nine hour cure. Further, the hardness is approximately the same in both instances, and the elongation does not suffer appreciably. Most importantly, the time for the vulcanization has been decreased significantly. The cure process is also applicable to copolymers of vinylidene fluoride and monochlorotrifluoroethylene.

Sulfenamides

A process described by J.F. Smith; U.S. Patent 2,955,104; October 4, 1960; assigned to E.I. du Pont de Nemours and Company involves curing a fluoroelastomer prepared by copolymerizing 70 to 30% by weight of vinylidene fluoride with 30 to 70% by weight of hexafluoropropene. The process comprises the steps of (1) compounding the elastomer with 2 to 5% of its weight with a sulfenamide having the structure

where R is an aromatic hydrocarbon radical or a heterocyclic radical, (2) 5 to 20% magnesium oxide, and, (3) heating the compounded elastomer to a temperature between 150° and 260°C. for a time from 1 to 25 hours. The following example illustrates the process.

Example: A suspension of 33.4 grams of 2-mercaptobenzothiazole in 300 cc of carbon tetrachloride was refluxed and chlorine passed through the refluxing suspension until a clear solution was obtained. Then 16.8 parts of piperazine in 100 parts pyridine was added. After agitating for an hour the precipitated solid was filtered off, washed with methanol and crystallized from benzene. It was a white compound which melted at 197° to 198°C. and analyzed correctly for $C_{18}H_{16}N_4S_4$ which corresponds to the structure

which may be called N,N'-piperazine-bis-2-thiobenzothiazole. 150 parts of a copolymer of 60% by weight vinylidene fluoride and 40% by weight hexafluoropropene was compounded with 23 parts of magnesium oxide, 27 parts of medium thermal carbon black and 4 parts of N,N'-piperazine-bis-2-thiobenzothiazole. Mooney scorch tests on the compounded elastomer carried out at 120°C. showed no rise in viscosity even after 45 minutes. The compounded elastomer was pressed at 260°C. for 1 hour, removed from the mold and then oven cured at 200°C. for 24 hours. The final cured polymer that was obtained had the following physical properties:

Tensile strength at break	2,050 lb./in.^{2}
Elongation at break	300 percent
Modulus at 100% elongation	470 lb./in.^{2}

Cycloaliphatic Polyamines

In a process described by <u>J.S. Rugg; U.S. Patent 2,933,481; April 19, 1960; assigned to E.I. du Pont de Nemours and Company</u> a nonscorchy yet thermally curable stock of a vinylidene fluoride-hexafluoropropene copolymer containing from 30 to 70% by weight of vinylidene fluoride units and from 70 to 30% by weight of hexafluoropropene units can be produced by homogeneously incorporating from 0.5 to 2 parts, per 100 parts by weight of the copolymer, of a compound of the group consisting of diaminocyclohexanes and bis(4-aminocyclohexyl) alkanes as the curing agent. The diamino-cyclohexanes may be the 1,2-diamino-, 1,3-diamino- or the 1,4-diaminocyclohexanes, and the alkane group in the bis(4-aminocyclohexyl) alkane contains from 1 to 3 carbon atoms including the 1,1- and 2,2-propanes. The resulting stocks are characterized by being nonscorchy, that is, safe processing stocks, in that they may be milled, extruded, calendered or otherwise processed and shaped at temperatures up to about 250°F. without premature curing, and in being curable to elastic, rubbery products by heating at temperatures of from 300° to 500°F.

The quantity of the cycloaliphatic polyamine employed will depend primarily on the state of cure desired in the vulcanizate. Normally, quantities of from 0.5 to 2 parts, and preferably 0.75 to 1.5 parts, based on 100 parts of the copolymer will be employed. These curing agents are easily and readily dispersed in the stocks by the usual rubber milling procedures. The following examples illustrate the process.

Example 1:   Hexafluoropropene is passed at a rate of 2 pounds per hour and vinylidene fluoride at 3 pounds per hour into a one gallon stainless steel constant volume reactor which is equipped with an agitator, two inlet lines and an outlet line leading to a receiver. The above monomers are introduced simultaneously through one inlet line which extends to near the bottom of the reactor. Simultaneously the initiator-dispersant solution made up of:

> 135 parts of ammonium persulfate
> 27 parts of sodium bisulfite
> 284 parts of disodium hydrogen phosphate
> 30 parts of ammonium perfluorooctanoate, and
> 21,000 parts of deoxygenated water

is pumped at a rate of 1.06 gallons per hour into the reactor through the other inlet line near its top. The reaction mass is agitated continuously and maintained at 100°C. and 900 psi. The outlet line, which leads from the top of the reactor, is connected to a receiver for the product through a pressure release valve. A pressure of 900 psi is maintained in the reactor. When the pressure in the reactor exceeds the 900 psi, a steady stream of copolymer emulsion escapes from the reactor through the outlet line into the receiver (which is conveniently at atmospheric pressure). The latex obtained after the second hour of operation is coagulated by the addition of sodium chloride, and the water-insoluble copolymer is collected, washed with water and dried. Analysis shows the copolymer to be made up of 40% hexafluoropropene units and 60% vinylidene fluoride units by weight.

Example 2:   (A) A sample of the copolymer prepared as described above is compounded on a rubber mill at 25° to 50°F. with the following ingredients, which are added in the order listed below:

Formulation:	Parts
Raw polymer	100
Zinc oxide	10
Dibasic lead phosphite	10
MT carbon black	18
1,3-diaminocyclohexane	1

The safe-processing character of this compounded stock, that is, its resistance to premature cure during subsequent processing, is shown by Mooney scorch data obtained for this stock according to the procedure described in ASTM D 1077-49T, using the small rotor at 250°F. For example, the time required for the Mooney viscosity to rise 10 units from the low reading of 20 at 250°F. is 25 minutes, which correlates with the processability of this stock in extrusion tests.

(B) In a comparative test in which the above formulation is employed except that the diamino cyclohexane is replaced by 1.0 part of triethylene tetramine (a typical polyamine originally used in curing this copolymer), there is a 10 point rise from a low reading of 24 in Mooney viscosity in only 4 minutes at 250°F., and this stock is not processable, e.g., extrudable, without scorching.

(C) The above compounded formulation of this process, that is, containing 1,3-diaminocyclohexane, is valuable for the preparation of mechanical goods of varied uses mentioned above, as indicated by the properties given below of the vulcanizate derived therefrom. The stock is placed in a mold and is press cured for 30 minutes at 275°F. The mold is allowed to cool to about 150°F., the molded stock removed, and then placed in a 400°F. oven for 24 hours. The product shows the following properties:

Tensile strength, psi	2,375
Elongation at break, percent	380
Hardness, Shore A	75
Permanent set at break, percent	2
ASTM Compression set—method B	
(70 hrs. at 250°F.), percent	24

(D) Replacing the 1,3-diaminocyclohexane by 1.4 parts of bis(4-aminocyclohexyl)methane in the above formulation produces a stock which in the Mooney Scorch Test referred to above shows a 10 point rise in Mooney viscosity in 40 minutes at 250°F. (a ten-fold increase in scorch resistance over that observed for the stock containing triethylene tetramine). Subjected to the curing conditions described above, this stock is transformed into a rubbery product having a tensile strength of 2,500 psi.

(E) Further increase in process safety may be achieved using smaller quantities of the cycloaliphatic diamines. For example, use of 0.7 part of 1,3-diaminocyclohexane in the above formulation produces a stock which does not show a 10 point rise from a low reading of 21 in Mooney viscosity in 45 minutes at 250°F. in the Mooney Scorch Test, and which is also curable to a rubbery product (tensile strength 2,000 psi) under the conditions of cure given above.

Quaternary Ammonium Salts

W. Flavell and A. Nodar-Blanco; U.S. Patent 3,403,127; September 24, 1968; assigned to Yarsley Research Laboratories Limited, England describe the use of quaternary ammonium salts for curing fluoro-rubbers. Examples of the bifunctional trialkyl quaternary ammonium salts suitable for use in the process are

$$[(CH_3)_3-N-CH_2-C_6H_4-CH_2-N-(CH_3)_3]^{++}, 2Cl^-$$

(1,4-phenylene dimethylene)bis(trimethyl ammonium chloride)

$$[(CH_3-CH_2)_3-N-CH_2-C_6H_4-CH_2-N-(CH_2-CH_3)_3]^{++}, 2Cl^-$$

(1,4-phenylene dimethylene)bis(triethyl ammonium chloride)

$$\{[CH_3-(CH_2)_3]_3-N-CH_2-C_6H_4-CH_2-N-[(CH_2)_3-CH_3]_3\}^{++}, 2Cl^-$$

(1,4-phenylene dimethylene)bis(tributyl ammonium chloride)

Suitable monofunctional trialkyl quaternary ammonium salts are exemplified by the following:

$$[(CH_3)_3-N-CH_2-C_6H_5]^+, Cl^-$$

trimethyl benzyl ammonium chloride

$$[(CH_3-CH_2)_3-N-CH_2-C_6H_5]^+, Cl^-$$

triethyl benzyl ammonium chloride

$$\{[CH_3(CH_2)_{11}](CH_3)_2-N-CH_2-C_6H_5\}^+, Cl^-$$

dimethyl dodecyl benzyl ammonium chloride

It is an advantage of the method that the cure times necessary for the attainment of a particular predetermined combination of properties may be made considerably shorter than those required with the conventional curing agents, the final properties being similar to, or better than those of fluoro-rubbers cured in the conventional way. Thus, for example, "Viton A" may be cured with (1,4-phenylene dimethylene)bis(triethyl ammonium chloride) in a time as short as seven to eight hours (compared with the time of cure of twenty-four hours necessary with dicinnamylidene hexamethylene diamine. The example on the following page illustrates the process.

Example: 100 parts of a copolymer of vinylidene fluoride (4 mols) and hexafluoropropene (1 mol) (sold under the trademark "Viton A"), 20 parts of carbon black, 15 parts of magnesium oxide and varying proportions of a mixture of equal parts by weight of calcium oxide and (1,4-phenylene dimethylene)bis(triethyl ammonium chloride) were blended together on a 2-roll mill.

The components were compounded as shown below and vulcanized in a mold for 45 minutes at 160° to 170°C. followed by an oven cure comprising a gradual rise from 100° to 205°C. (25°C. per hour) followed by 16 hours at 205°C. The properties of the cured fluoroelastomers are shown in the table.

	(a)	(b)	(c)	(d)	(e)
Components, parts by weight:					
Vinylidene fluoride/hexafluoro-propene copolymer ("Viton A")	100	100	100	100	100
MT carbon black	20	20	20	20	20
Magnesium oxide	15	15	15	15	15
Dicinnamylidene hexamethylene diamine	3				
Calcium oxide (1 part) (1,4-phenylene dimethylene) bis (triethylammonium) chloride (1 part)		2	4	6	8
Physical properties after press cure and oven cure:					
Ultimate tensile strength (p.s.i.)	2,320	2,375	2,300	1,880	1,870
Modulus at 100% elongation (p.s.i.)	800	460	1,090	1,660	
Elongation at break, percent	190	330	180	130	110
Hardness	75	62	65	70	
Weight loss during cure, percent	1.9	1.1			
Shrinkage, percent:					
In press cure	2.5	2.2			
In oven cure	1.7	1.3			
Total shrinkage	4.2	3.5			
Resistance to compression set:					
24 hours under compression at 160° C	20.3		9		
24 hours under compression at 250° C	97		45		
Physical properties after ageing 24 hours at 250° C.:					
Ultimate tensile strength	2,300	2,500			
Modulus at 100% elongation		650			
Elongation at break, percent	150	280			
Resistance to compression set:					
24 hours under compression at 160° C	7.5		3.6		
72 hours under compression at 160° C	20		14		
500 hours under compression at 160° C	50		42		

The composition (a) in the above table represents the prior art and the data is given for comparison with the compositions (b), (c), (d) and (e) which are in accordance with this process.

Schiff's Bases and RTV Sealant

R. G. Spain; U.S. Patent 3,424,710; January 28, 1969; assigned to the U.S. Secretary of the Air Force describes compositions which contain at least (1) a vulcanizable hydrofluorocarbon homopolymer, such as chlorotrifluoromethylene and/or copolymer, such as vinylidene fluoride and hexafluoropropylene, (2) an azadiene such as a Schiff base, which in the presence of water, acts as vulcanizing agent, and (3) a lower alkyl ketone as a vulcanization inhibitor. Other components may be present in the compositions, including reinforcing agents such as carbon black and stabilizing agents such as magnesium oxide.

The method of vulcanizing these compositions consists of the steps of substantially removing the ketone and adding water. Normal atmospheric moisture may also serve as the source of the water. Demonstrative of the improved stability and mechanical properties of fluorocarbon polymers, prepared and vulcanized according to the process, are the results achieved from the following examples:

Example A:

Component	Parts by weight
Copolymer of vinylidene fluoride and hexafluoropropylene (Vitron A sold by Du Pont)	100.0
Carbon black	20.0
Magnesium oxide	15.0
Ethylene diamine-carbamate	0.8
Methylethyl ketone (as a solvent)	540.0

Example B:

Component	Parts by weight
Copolymer of vinylidene fluoride and hexafluoropropylene (Viton A sold by Du Pont)	100.0
Carbon black	20.0
Magnesium oxide	15.0
Hexamethylene diamine-carbamate	1.0
Methylethyl ketone (as a solvent)	540.0

Example C:

Component	Parts by weight
Copolymer of vinylidene fluoride and hexafluoropropylene (Viton A sold by Du Pont)	100.0
Carbon black	20.0
Magnesium oxide	15.0
Methylethyl ketone (as a solvent)	540.0
2,4,9,11-tetramethyl-5,8-diazo-4,8-dodecadiene	4.0

The gel time for the above formulations mixed at room temperatures without addition of water but exposed to normal atmospheric humidity was found to be 14 days as to Example A, 5 days as to Example B and greater than 60 days for Example C. It is thus demonstrated that the stability of the coating solution containing the azadiene is greater by far than those containing the diamine-carbamates.

Cast films 0.010 inch thick of the above examples after exposure to normal atmospheric humidity for 7 days at 75°F. were found to have a tensile strength of 1,000 pounds per square inch in the case of Example B and of 1,700 pounds per square inch in the case of Example C, Example A having failed to cure. It is also apparent that the tensile strengths developed by case films subsequently exposed to the atmosphere is much greater where the coating solution contains a diimine or azadiene instead of a diamine carbamate. The formulations of the process are therefore considerably more stable when not in contact with moisture and provide yet substantially greater mechanical properties upon exposure to moisture.

Tropolone and 1,10-Phenanthroline as Accelerators

A.L. Barney and W. Honsberg; U.S. Patent 3,502,628; March 24, 1970; assigned to E.I. du Pont de Nemours and Company describe the acceleration of the vulcanization of saturated, fluorinated polymers by the use of tropolone and 1,10-phenanthroline. The accelerators enable a more rapid cure with conventional amine-based vulcanization systems and also enable a satisfactory rate and state of cure with weakly basic bisnucleophiles otherwise incapable of adequately curing fluorinated polymers. The following example illustrates the process.

Example: Procedure for Compounding, Vulcanizing and Testing Fluorine-Containing Copolymers — On a cool, 2-roll rubber mill, 100 parts of a copolymer of vinylidene fluoride and hexafluoropropene containing 60 and 40 weight percent, respectively, are compounded with 20 parts of medium thermal carbon black and 15 parts of magnesium oxide. Except as noted, 135 parts of this composition are used for further compounding with the vulcanizing agents and accelerators described in the examples. The compositions are sheeted off the mill and specimens for physical testing are prepared.

An oscillating disc rheometer (ODR) is used to determine rapidity of cure after selected times that a test piece is held at curing conditions. The ODR measures the relative viscosity of an elastomer by oscillating (e.g., at 900 cpm) a grooved conical disc through 3 degrees of arc while pressed tightly between two test pieces. The amount of torque required to oscillate the disc is reported as the measure of viscosity. Vulcanizates are prepared by compression molding of appropriate samples in a press for 30 minutes at 163°C. followed by removing them from the mold and "post-curing" by heating in an air oven to 204°C. over a 4-hour period and then an additional 24 hours at this temperature.

The procedure described above is followed using 1.5 parts HMDAC and the following sequestering compounds as accelerators:

	A	B	C	
Parts of 1,10-phenanthroline	0*	5	-	
Parts of tropolone	0	-	5	(continued)

Vinylidene Fluoride Elastomeric Products

Oscillating disc rheometer at 160°C. (inch pounds of torque)

	A	B	C
After 2.5 minutes	3	64	9
After 5 minutes	10	94	29
After 10 minutes	38	104	67
After 20 minutes	55	112	83

*Outside the process—for comparison only.

Polyethylene Glycols and Polyhydroxy Aromatic Nucleophiles

A.L. Barney and W. Honsberg; U.S. Patent 3,524,836; August 18, 1970; assigned to E.I. du Pont de Nemours and Company describe the vulcanization of saturated, fluorinated elastomeric polymers with a polyhydroxy aromatic nucleophile in a weakly basic system in the presence of about 0.1 to 20 parts of an accelerator, per 100 parts of polymer, of the formula

$$R-O-\left[(CH_2)_m-\underset{\underset{R'}{|}}{CH}-O\right]_n R$$

where R is hydrogen or C_1 to C_4 alkyl, m is from 1 to 3, R' is hydrogen or methyl, and n is an integer sufficient to give a molecular weight of at least about 150.

This process is applicable to saturated interpolymers of vinylidene fluoride (VF_2) with other fluorine-containing ethylenically unsaturated monomers. Of particular interest are the vinylidene fluoride/hexafluoropropene copolymers containing from about 70 to about 30 weight percent vinylidene fluoride and about 30 to about 70 weight percent hexafluoropropene (see U.S. Patent 3,051,677).

Representative ethers are bis[2-(2-methoxyethoxy)ethyl]ether, where m is 1, n is 4, R' is hydrogen, and R is methyl; diethyleneglycol diethylether where m is 1, n is 2, R is ethyl, and R' is hydrogen; and polypropyleneglycol (MW400), where m is 1, R is hydrogen, R' is methyl, and n is about 7. Other ethers of this type are polytetramethylene ether glycol (MW 1000), where R and R' are hydrogen, m is 3 and n is sufficient to give a molecular weight of about 1000, monomethoxy polyethylene ether glycol of a molecular weight of about 350 and a formula $CH_3-O-(CH_2-CH_2-O)_n H$ (Carbowax 350), a polyethylene ether glycol of molecular weight of about 400 and a formula $HO-(CH_2-CH_2-O)_n H$ (Carbowax 400), and 1,2-bis(2-methoxyethoxy)ethane.

The accelerators of this process enable the rapid and satisfactory vulcanization of saturated, fluorinated polymers by using poly(nucleophiles) in a basic system. Representative poly(nucleophiles) are 2,2-bis(4-hydroxyphenyl)propane (bisphenol-A), 2,2-bis(4-hydroxyphenyl)perfluoropropane (bisphenol AF), resorcinol, 1,3,5-trihydroxybenzene, 1,7-dihydroxynaphthalene and their alkali or alkaline earth metal salts. The following examples illustrate the process.

Example 1: This example illustrates the curing of a copolymer of VF_2, TFE and HFP in the following respective weight percentages: 45, 30 and 25 (prepared according to U.S. Patent 2,968,649) with either the dipotassium salt of bisphenol AF or a mixture of the dipotassium salt with bisphenol AF.

On a cold rubber mill, the copolymer is compounded with carbon black, magnesium oxide (Maglite D), monomethoxy polyethylene ether glycol of MW about 350 (Carbowax 350), and the dipotassium salt of bisphenol AF, as shown in the table. A 3" x 6" x 0.75" slab and small pellets are cured in a press for 30 minutes at 165°C., and post-cured in an oven for 24 hours at 204°C. The elastomers are strong and well cured as shown by the good tensile properties, measured at 25°C., and good compression set, measured at 25°C.

Ingredients	Parts	
	1-A	1-B
Copolymer	100	100
Thermax MT carbon black	20	20
MgO	10	10
Accelerator	2	4
Dipotassium bisphenol AF	2	1
Bisphenol AF		0.83
Test data:		
Modulus, 100% elong	700	600
Tensile strength, p.s.i.	2,300	2,100
Percent elong. at break	205	210
Shore hardness	71	68
Compression set, 70 hr., 25° C	24	22
Compression set, 70 hr., 204° C	50	49

Example 2: This example illustrates the curing of a copolymer of 60% VF$_2$ and 40% HFP (prepared according to U.S. Patent 3,051,677) and the copolymer of Example 1 with various samples of a potassium salt of bisphenol AF, but using as accelerator either a polyethylene ether glycol of MW about 400 (Carbowax 400) or 1,2-bis(2-methoxyethoxy)-ethane.

As before, the copolymer is compounded as shown in the table and test specimens are cured by the method of Example 1. The potassium salt of bisphenol AF, 65% basic, is prepared by stirring until soluble 43 g. of potassium hydroxide pellets (85+percent purity, 0.65 mol), 300 ml. methanol, and 168 g. of bisphenol AF (0.50 mol). The solution is evaporated in a spinning flask, finishing at 40°C. under 5 mm. pressure. The solid product is ground, and dried again at 40°C. under 5 mm. pressure. The yield is 167 g. of a white powder, which was ground as a fine powder before use. The potassium salt of bisphenol AF, 35% basic, is prepared in a similar fashion from 13.2 g. of potassium hydroxide pellets (0.20 mol), 200 ml. methanol, and 96 g. of bisphenol AF (0.286 mol). The yield is 108 g. of a white powder.

Ingredients	Parts	
	2-A	2-B
VF$_2$/HFP copolymer	100	
Copolymer of Example 1		100
MT carbon black	30	30
MgO	4	4
Polyethylene ether glycol (M.W. 400)	2	
Potassium salt of bisphenol AF, 65% basic	1.5	2.5
Potassium salt of bisphenol AF, 35% basic	1	
1,2-bis(2-methoxyethoxy)ethane		2
Test Data (at 25° C.):		
Modulus, 100% elong.	520	400
Tensile strength, p.s.i.	1,350	1,950
Percent elongation at break	220	285
Shore hardness	70	70
Compression set, 70 hr., 25° C	17	25
Compression set, 70 hr., 121° C	10	26
Compression set, 70 hr., 204° C	41	49

Alkali Metal Salts of Polyphenols

P.O. Tawney and R.P. Conger; U.S. Patent 3,243,411; March 29, 1966; assigned to United States Rubber Company have found that fluorocarbon elastomers can be vulcanized in the presence of vulcanizing agents which are chemicals that ionize in water to yield an entity which has two or more negative charges with a basic strength greater than that of the acetoxy ion. Chemicals which function in this manner include the following, in order of decreasing preference:

(1) alkali metal salts of polyphenols
(2) alkali metal salts of polythiols
(3) alkali metal salts of mercaptophenols
(4) alkali metal salts of mercaptoalkanols
(5) alkali metal salts of polyalcohols

Example 1: This example illustrates the curing of Viton A-HV in solution using a bisphenoxide [namely the disodium salt of Bisphenol A (4,4'-isopropylidenebisphenol)]. To a solution of 12.5 g. of Viton A-HV in 125 ml. of 2-butanone at room temperature was added a slurry of the disodium salt of Bisphenol A in ethanol, prepared by reaction of 1.2 g. of sodium hydroxide in 20 ml. of ethanol with 3.42 g. of Bisphenol A. As soon as some of the solid had dissolved in the Viton A-HV cement, the solution became gelled. A portion was dried and was found to have a swelling index in 2-butanone of 17.8 with 10% being soluble. Therefore, it was crosslinked by the disodium salt of Bisphenol A.

Example 2: This example illustrates the curing of black-filled Viton A-HV by a bisphenoxide (the disodium salt of Bisphenol A). A black-filled Viton A-HV master batch was prepared using the recipe given below. To this was added varying amounts of the disodium salt of Bisphenol A (prepared from 6.84 g. of Bisphenol A and 2.40 g. of sodium hydroxide in ethanol) on a mill with the stock temperature being kept below 220°F. Samples were cured in the press for 60' at 325°F. Stock 2A, which is in contrast with those (2B, 2C, 2D) which illustrate this process, was processed and tested like the others except it contained no bisphenoxide curing agent.

Stock Code	2A	2B	2C	2D
Viton A-HV	100	100	100	100
Philblack A	20	20	20	20
Magnesium oxide	5	5	5	5
Disodium Bisphenol A		2.42	7.26	12.10
	Cured 60' at 325° F. in a press			
Scott Tensile at R.T.,[1] p.s.i.	(2)	414	2,260	2,080
Elongation at Break at R.T., percent		810	275	240
Modulus at 200% Elongation at R.T., p.s.i.		200	1,350	1,575
Scott Tensile at 300° F., p.s.i.		34	375	450
Elongation at Break at 300° F., per-cent		27	140	140
Durometer at R.T.		60	68	74

[1] Room temperature.
[2] Not cured.

This example shows that (1) without this curing agent (Example 2A) the Viton A-HV masterbatch is essentially uncured, (2) a cure is obtained with as little as 2.42 phr (parts per hundred of rubber), and (3) certain physical properties of the vulcanized sample can be regulated by varying the amount of the curing agent used — the upper limit of amount of chemical used being determined by the certain physical property (if it is in the range obtainable) desired in the end product. For example, the modulus at 200% elongation at RT of this masterbatch can be varied from 200 to 1,575 psi by using from 2.42 to 12.10 phr of the disodium salt of Bisphenol A.

Post-Curing Technique

J.K. Sieron; U.S. Patent 3,452,126; June 24, 1969; assigned to the U.S. Secretary of the Air Force describes a process for the preparation of low compression, high temperature resistant, benzoyl peroxide cured fluoroelastomer composition where the composition is heated at about 200°F. at about atmospheric pressure. The composition is then heated in the range of about 200° to 400°F. under reduced pressure to remove substantially all volatiles and further post-cured by heating in the range of 400° to 550°F. at atmospheric pressure. The following example illustrates the process.

Example: A vinylidene fluoride-perfluoropropylene copolymer (100 parts by weight), magnesium oxide (15 parts), MT carbon black (30 parts), and benzoyl peroxide (3 parts) were mixed. The composition was formed into button-shaped gaskets which were vulcanized by press curing for 60 minutes at 260°F. using known procedures. The gaskets were then post-cured in the following sequence of steps:

Step	Time, hrs.	Temp.° F.	Pressure, mm. Hg.	Set, percent	Tensile, p.s.i.	Elongation, percent	Hardness Shore A, pts.
1	24	200	760				
2	24	200	1				
3	24	250	1				
4	24	300	1				
5	24	350	1				
6	24	400	1	72.8			
7	8	500	760	56.7			
8	16	500	760	41.8			
9	96	500	760	26.3	2,285	385	76
10	168	550	760	Substantially unchanged			

As indicated, the gaskets were tested for compression set. In this test the gaskets are held at 400°F. for 72 hours under a 25% compression. The pressure is then released and the gaskets are allowed to expand freely. If the gasket button expands and returns to its original height, it has zero compression set. If it does not expand and retains its compressed height, it has 100% compression set. Test gaskets reached a compression set of 26.3% which, under the severe conditions of the test, is very good.

Control gaskets which were not heated in vacuum as indicated but which were post-cured at atmospheric pressure showed substantially 100% compression set and could not be heated to 500°F. Comparable diamine cured polymer buttons were hard and had substantially 100% compression set. The tensile of 2,285, the elongation of 385, and the Shore hardness of 76 in the tested gasket indicates that the latter still retains the general properties characteristic of elastomers before they have been subjected to adverse environments.

Modification with Dimethylformamide

In a process described by G. Wood; U.S. Patent 3,507,844; April 21, 1970; assigned to Minister of Aviation in Her Britannic Majesty's Government of the United Kingdom of Great Britain, England a fluorocarbon elastomer is dehydrofluorinated and rendered more readily curable by dissolving it at a normal or moderately elevated temperature in a substituted amide solvent, for example, dimethylformamide or dimethylacetamide.

The substituted amide acts as both solvent and reagent and it has been found that the process takes place more readily and can even be effected at ambient temperatures, if a salt is present which is soluble in the substituted amide. Salts which may be used include alkali (which includes ammonium) cyanides and thiocyanates and alkaline earth metal nitrates. Specific examples of these salts are sodium, potassium and ammonium cyanide, sodium, potassium and ammonium thiocyanate, and calcium nitrate. The following examples illustrate the process.

Example 1: 30 g. of Viton A were dissolved in 450 ml. of dimethylformamide and refluxed for one hour. The gum was precipitated by pouring as a thin stream into a larger volume of cold water. The gum was a distinct amber color and was more easily milled than the original Viton A.

Example 2: The dehydrofluorinated Viton A as prepared in Example 1 was incorporated in the following composition which was compounded on a rubber mill:

Dehydrofluorinated Viton A	100 g.	Medium thermal black (carbon black)	20 g.
Magnesium oxide	15 g.	Divinyl benzene in 55% w./w. ethyl vinyl benzene solution	1.5 g.

The composition was cured in the course of a treatment which involved 1/2 hour in a press at 150°C. followed by treatment in an oven for 1 hour at 150°C., 2 hours at 175°C. and then 24 hours at 200°C. As a result a high-quality vulcanizate was obtained which had a tensile strength of 2,000 lbs. per sq. in., an elongation to break of 170% and a modulus for a 1% strain of 492 lbs. per sq. in.

Example 3: A 20% w./w. solution of Viton A was refluxed in dimethyl formamide for fifteen minutes and the gum precipitated by pouring into water. The dehydrofluorinated gum was dissolved in acetone and reprecipitated and then dried in an oven at 60°C. for one week. 100 g. of the dehydrofluorinated Viton A were mixed with 15 g. of magnesium oxide and 2 g. of hexamethylenediamine. The composition vulcanized on standing at room temperature; the progress of vulcanization is shown by the test results given in the following table.

Time in hours at room temperature	Fraction soluble in acetone at 28°C., %	V_r in acetone at 28°C.
0	100	–
6	38	0.06
10	30	0.12
20	20	0.155
25	17	0.17
40	13	0.18
60	9	0.186
80	7	0.19

V_r is the volume fraction of elastomer in the swollen phase.

The compound is virtually fully vulcanized after 48 hours at room temperature. These figures compare very favorably with known fluorocarbon elastomer sealants in which full vulcanization takes as long as twelve days to complete.

Bisulfite Stabilization

M.F. Kasparik; U.S. Patent 3,325,447; June 13, 1967; assigned to E.I. du Pont de Nemours and Company describes the stabilization of fluoroelastomers with bisulfites. It has been found that as little as 0.025 part of the bisulfite will produce a significant improvement in color stability of the fluoroelastomer, both during milling and on subsequent exposure to high temperatures. More than 2 parts of the salt is not required. The preferred amounts range from about 0.1 to 0.5 part. All these "parts" are by weight based on 100 parts by weight of fluoroelastomer. The stabilized fluoroelastomers may be compounded and cured by any of the conventional techniques used with fluoroelastomers of this type. The tensile properties of the vulcanizates are not adversely affected by the presence in the elastomer of sodium bisulfite or ammonium bisulfite. The following examples are illustrative of the process. The fluoroelastomers used in these examples are made as follows:

Fluoroelastomer A: This is a copolymer containing, by weight, 60% of vinylidene fluoride units and 40% of hexafluoropropene units. It is prepared essentially as described in Example 11 of U.S. Patent 3,051,677 except that the catalyst solution does not contain disodium hydrogen phosphate and ammonium perfluorooctanoate. The polymer, isolated by centrifuging, is in the form of white crumbs.

Fluoroelastomer B: This fluoroelastomer is a copolymer containing, by weight, 45% of vinylidene fluoride units, 30% of hexafluoropropene units, and 25% of tetrafluoroethylene units and is prepared essentially as described in U.S. Patent 3,039,992, columns 5-6. It is isolated by centrifuging and is in the form of white crumbs.

Example 1: Varying amounts of sodium bisulfite are incorporated into Fluoroelastomer A by the following method. An aqueous solution (0.5 to 3%) of the sodium bisulfite is mixed with the elastomer crumbs and the mixture is dried in a vacuum oven at 70°C. The resulting composition is then milled on a standard rubber mill at 100° to 110°C. for 4 to 7 minutes. (The temperature, 100° to 110°C., is selected to facilitate the milling process.) After the milling, the sample color is compared visually with the color of the control which has been treated in the same way except that no NaHSO3 has been added to the copolymer. After milling, samples of the test and control compositions are stored in an air oven maintained at 100°C. The color of the samples is observed after 16 hrs. The results are as follows.

Parts of NaHSO₃ per 100 parts of fluoroelastomer	Color	
	After Milling	After storage at 100° C. for 16 hours
0	Light amber	Amber.
0.25	Light tan, much lighter than control.	Light Amber.
1.5	White	White.

Example 2:   In this example Fluoroelastomer B is used. Varying amounts of ammonium bisulfite are added as a water solution when the polymer is being milled. After milling at 100° to 110°C. until the polymer is dry, the control sample to which no ammonium bisulfite is added shows severe brown discoloration. Samples of fluoroelastomer containing 0.1, 0.2, 0.3, and 0.4 part, per 100 parts of copolymer, show only slight discoloration.

COPOLYMER MODIFICATION FOR VULCANIZATION

Carboxy Terminated Copolymers

E.F. Cluff; U.S. Patent 3,147,314; September 1, 1964; assigned to E.I. du Pont de Nemours and Company describe copolymers which have a molecular weight corresponding to an inherent viscosity of about 0.04 to about 0.25 at 30°C. in 1.0% by weight solution in anhydrous, reagent grade acetone, which have a carboxyl content of from 0.2 to 3.75% by weight, based on the weight of the copolymer and which consist essentially of a multiplicity of $-CH_2-CF_2-$, and units

$$-\underset{\underset{CF_3}{|}}{CF}-CF_2-$$

or a multiplicity of

$$-CH_2-CF_2-,\ -\underset{\underset{CF_3}{|}}{CF}-CF_2-\ \text{and}\ -CF_2-CF_2-$$

It is apparent from the above definition that the copolymers of this process have a specified carboxyl, i.e., $-COOH$ content and are made up of units derived from vinylidene fluoride along with units derived from hexafluoropropene, and, if desired, tetrafluoroethylene.   These copolymers may be characterized as being fluids or semisolids and may be cured to form highly useful solids.

In preparing the copolymers of this process the high molecular weight copolymer is dissolved in an inert solvent. Suitable solvents include tetrahydrofuran, acetone and methylethyl ketone. A nitrogen base is then added and the mixture heated to a temperature of from about 35° to 70°C. for a period of time of from about 6 to 24 hours, until the requisite number of olefinic carbon-to-carbon double bonds have been introduced into the copolymer. Heating in the presence of a nitrogen base effects dehydrohalogenation of the copolymer. Suitable nitrogen compounds are those having a K_B of at least 1×10^{-5}. Representative compounds include amines such as n-butylamine, diethylamine, triethylamine and piperidene. Ammonia, ammonium hydroxide or quaternary ammonium hydroxides are also included.

The length of time required to effect dehydrohalogenation will be dependent on several variables, particularly the nature of the solvent, the temperature of heating, the nature of basicity of the specific nitrogen compound involved and the amount of this compound. These variables may be adjusted as desired by the skilled chemist. The preferred amount of nitrogen base to be used ranges from about 0.1 to 0.4 mol per 100 grams of the high molecular weight copolymer starting material.

The high molecular weight copolymer, which now contains a plurality of olefinic carbon-to-carbon double bonds is now oxidized to cleave these bonds with the formation of carboxyl groups. The oxidation step is carried out by heating the dehydrohalogenated polymer at a temperature of from about 50° to 70°C. for a period of time of from about 3 to 24 hours in a solvent in the presence of an oxidizing agent. Suitable solvents include acetone or acetic acid. Suitable oxidizing agents include potassium permanganate or fuming nitric acid. The resulting carboxyl containing copolymer is then isolated. The following examples illustrate the process.

Example 1:   400 parts of a copolymer of vinylidene fluoride and hexafluoropropene, the moieties being in the weight ratio of 60:40, having an inherent viscosity of 0.9 at 30°C. in 0.1% concentration in an anhydrous solvent consisting of 86.1% of tetrahydrofuran and 13.9% of dimethyl formamide (called the THF/DMF solvent), was dissolved in 1,330 parts of tetrahydrofuran. 80 parts of triethylamine was then added and the solution was heated to reflux for about 20 hours. The resulting solution was poured into a large excess of water and the copolymer precipitated. The copolymer was collected from the water and placed on a corrugated rubber wash mill where it was washed with water to remove residual solvent, amine and amine hydrofluoride.

The copolymer was then dissolved in 1,584 parts of cold acetone and 126 parts of potassium permanganate was added gradually while agitating. The mixture gelled in about 10 minutes and was then allowed to stand overnight. The next morning the gel had broken. The mixture was then heated to reflux for about 3 hours when the color of permanganate had completely disappeared. 29 parts of water was added and hydrogen chloride was bubbled through the thick suspension until the viscosity appeared to decrease no further and the solid matter settled to the bottom. The mass was filtered, and the filtrate was evaporated under reduced pressure to recover the copolymer. The copolymer was then heated on a steam bath and finally at 90°C. in a vacuum oven to remove residual solvent.

398 parts of a sticky, grease-like copolymer was obtained. It had an inherent viscosity at 30°C. in 1.0% solution in anhydrous acetone of 0.096. It showed a strong infrared absorption at 5.63 microns which is characteristic of the $-CF_2COOH$ group. It had a neutralization equivalent of 7,440. The copolymer has a carboxyl content of 0.61% by weight.

100 parts of the copolymer was compounded on a rubber mill with 6 parts of magnesium oxide. The compounded mass, suitable for use as a caulking compound, was cured in a mold in a press at 150°C. for 45 minutes. The cured elastic product was immersed in the following solvents for 3 days at room temperature with the results shown:

Solvent	% Increase in Weight
Toluene	8.4
Hexane	0.3
ASTM Oil #3	0.35

Example 2:   267 parts of a copolymer of vinylidene fluoride and hexafluoropropene (weight ratio 60/40) having an inherent viscosity of 0.9 in the THF/DMF solvent at 0.1% concentration at 30°C. was dissolved in 792 parts of acetone in an enamel-lined autoclave. 10 parts of 28% ammonium hydroxide was added, the autoclave was closed and the contents were stirred for 13 hours at 55°C. The autoclave was then vented and nitrogen was bubbled through for 25 minutes to remove residual ammonia. To the reaction mass was added 21.1 parts of potassium permanganate and the mass was stirred at 55°C. for 3 hours and then cooled to room temperature. 10 parts of water was added to the resulting slurry and hydrogen chloride was passed into the suspension until the solid settled to the bottom. The mass was filtered, the filtrate was dried with magnesium sulfate, filtered again, and the solvent was evaporated. The last traces of solvent were removed by heating in a vacuum oven at 60°C. 265 parts of putty-like copolymer was obtained.

The copolymer had an inherent viscosity of 0.21 in anhydrous acetone at 1.0% concentration at 30°C. and a neutralization of 19,400. This is equal to a carboxyl content of 0.23% by weight. 20 parts of this copolymer and 100 parts of a high molecular weight copolymer, having an inherent viscosity of 0.9 in 0.1% solution in the THF/DMF solvent, of vinylidene fluoride and hexafluoropropene (weight ratio 60/40) were blended together on a rubber roll mill. The mixture had a Mooney viscosity of 36 (ML-10 at 100°C.) whereas the unplasticized high molecular weight copolymer had a Mooney viscosity of 60.

The plasticized copolymer (A) and the unplasticized copolymer (B) were compounded as shown below and cured in a press for 1 hour at 150°C. and then held in an oven for 1 hour at 140°C. and then 24 hours at 204°C. The properties of the cured fluoroelastomers are shown in the table.

Components	Parts by Weight	
	(A)	(B)
VF$_2$/HFP copolymer (B)	-	100
Plasticized copolymer (A)	100	-
Magnesium oxide	16	16
Medium thermal black	18	18
Ethylene diamine carbamate	1.5	1.5
Properties:		
Tensile strength at the break (lbs./sq.in.)	1,800	2,200
Modulus at 100% elongation, (lbs./sq.in.)	505	560
Elongation at break, percent	190	200

It is evident that the use of the carboxyl containing copolymer as a plasticizer for the fluoroelastomer reduces the viscosity of the stock to a satisfactory level for compounding and molding and at the same time detracts very little from the properties of the cured fluoroelastomer. Furthermore, since it is so similar in structure, it is cocured and will not bloom to the surface.

Carboxy and Ester Termination — Sealant Intermediates

D.E. Rice and C.L. Sandberg; U.S. Patent 3,438,953; April 15, 1969; assigned to Minnesota Mining and Manufacturing Company describe a process for the preparation of carboxyl or carboxyl ester terminated copolymers of vinylidene fluoride and perfluoropropene which are characterized by the formula as shown on the following page.

(1)

$$RO\overset{O}{\underset{\|}{C}}(R_f)\text{--}\left[\left(CF_2CH_2\right)_x\text{--}\left(\underset{\underset{CF}{|}}{\overset{\overset{CF_3}{|}}{C}}F_2CF\right)_y\right]_m(R_f)\overset{O}{\underset{\|}{C}}OR$$

R_f is a perfluoroalkylene group containing from 1 through 15 carbon atoms,

R is hydrogen; or an organic radical such as an α, α-dihydroalkyl radical containing not more than 20 carbon atoms and not more than 14 hydrogen atoms, the only other substituents in said alkyl radical being fluorine, or an aryl radical containing from 6 through 12 carbon atoms, and which may be substituted with fluorine,

m is a positive whole number of at least 5 and preferably less than 500 and more preferably less than 100,

x and y are positive numbers, y being 1, and the average ratio of x to y in a copolymer molecule is from about 1:1 to 10:1.

The copolymers of vinylidene fluoride and perfluoropropene are prepared by contacting a liquid mixture of these two monomers with bis-(ω-carboxyl ester perfluoroacyl) peroxides. In one example, vinylidene fluoride and perfluoropropene in the proportions required to obtain the ratio of x to y outlined above in Formula 1 are contacted with an aqueous solution of a peroxide selected from the group consisting of alkali metal and alkaline earth metal peroxides, and a perfluoro acyl halide. In this process, the peroxide is generated in situ. The reaction is illustrated by the following equations:

(2)

$$2R_f\begin{bmatrix}\overset{O}{\underset{\|}{C}}\text{--}Cl\\\underset{\|}{\underset{O}{C}}\text{--}OR\end{bmatrix}\xrightarrow[\text{Aqueous}]{Na_2O_2} RO\overset{O}{\underset{\|}{C}}(R_f)\overset{O}{\underset{\|}{C}}\text{--}O\text{--}O\text{--}\overset{O}{\underset{\|}{C}}\text{--}(R_f)\overset{O}{\underset{\|}{C}}OR \longrightarrow 2RO\overset{O}{\underset{\|}{C}}(R_f)\cdot + 2CO_2 + 2NaCl$$

(2B)

(3)

$$2RO\overset{O}{\underset{\|}{C}}(R_f)\cdot + xCF_2{=}CH_2 + yCF_2{=}\underset{\underset{CF}{}}{\overset{\overset{CF_3}{|}}{C}}F \longrightarrow RO\overset{O}{\underset{\|}{C}}(R_f)\text{--}\left[\left(CF_2CH_2\right)_x\left(\overset{\overset{CF_3}{|}}{C}F_2CF\right)_y\right]_m(R_f)\overset{O}{\underset{\|}{C}}OR$$

In Equations 2 and 3 above R_f, R, m, x and y are as defined above in reference to Formula 1. Preferably, in this process as described in Equations 2 and 3, R is an organic radical such as is described in connection with the definitions of Formula 1 above. Since Compound 2B above must be soluble in the fluorocarbon monomer liquid phase of the polymerization mixture, it is required that R be such a group as will render Compound 2B soluble in the polymerization liquid mixture. Preferably, therefore, R has the indicated definitions as defined above.

Alternatively, the bis-(ω-carboxyl ester perfluoroacyl) peroxides (the structure of Formula 2B above) can be first prepared in fluorocarbon solution, and then this solution can be used directly in the polymerization reaction described in Equation 3 above.

The polymerization reaction described in Equation 3 is carried out under liquid phase conditions preferably at autogenous pressures. A preferred operating temperature is in the range of 20° to 40°C. although those skilled in the art will appreciate that temperatures in the range from about 0° to 100°C. or even higher or lower can be employed dependent upon the particular processing conditions, apparatus etc., employed in any given circumstance. The following examples illustrate the process.

Example 1: Perfluoroglutaric anhydride (56.6 grams) is charged into a flask fitted with a reflux condenser, addition funnel, and stirrer, and then the so charged flask is cooled to -30°C. Anhydrous methanol (10.2 cubic centimeters) is then added dropwise at a rate such that the reaction temperature does not exceed -20°C. Distillation of the reaction mixture gives 57.5 grams $CH_3O_2C(CF_2)_3COOH$, boiling point 88° to 94°C. at 1 mm. Hg identified by its infrared spectrum and neutralization equivalent. This material is then added dropwise to PCl_5 (47.5 grams) and the mixture stirred for one hour at room temperature. Distillation yields 47.4 grams $CH_3O_2C(CF_2)_3COCl$, boiling point 42° to 44°C. at 15 mm. Hg n_D^{26} (refractive index) 1.3515, identified by its infrared and nuclear magnetic resonance spectra, and neutralization equivalent.

Example 2: A 500 cubic centimeter flask fitted with a condenser, stirrer, addition funnel and thermometer is charged with 138.5 grams (0.50 mol) of perfluoroglutaryl chloride and then the so-charged flask is cooled to -15°C. To this is added dropwise 16 grams (0.50 mol) of methyl alcohol while keeping the reaction mixture at -10° to -15°C. After the addition is complete the reaction mixture is stirred for another 30 minutes and then distilled.

A yield of 57.5 grams of

$$ClC(CF_2)_3COOCH_3$$

which has a boiling point of 64° to 68°C. at a pressure of 34 mm. Hg and having a n_D^{20} (refractive index) of 1.3519 is obtained and is further identified by its infrared spectrum and neutralization equivalent. In addition there is obtained 41.8 grams of unreacted perfluoroglutaryl chloride and 34.1 grams of dimethyl perfluoroglutarate.

Example 3: A one gallon stainless steel autoclave is cooled to -30°C. and charged with 213 grams $CF_2=CH_2$, 1,000 grams C_3F_6 and 228 grams of

$$CF_3CH_2OC(CF_2)_3CCl$$

After warming to about -5°C., a solution of 450 ml. water, 90 ml. of H_2O_2 and 28.9 grams NaOH is added. The reactor is warmed and stirred at 24° to 30°C. for three hours. The reactor pressure is observed to drop from a maximum of 220 psig down to 120 psig during the reaction. The unreacted gases are vented from the vessel and the water phase is discarded. The product is stirred for 16 hours in 1,000 ml. of acetonitrile, then the acetonitrile is decanted and the polymer filtered. It is next washed with water and then placed in 1,000 ml. water and the water is refluxed for 16 hours. The product is separated from the water by decanting then azeotropically dried using benzene, and the benzene is then removed by heating under vacuum.

A very viscous syrup is obtained, 510 grams total which has an inherent viscosity ($<\eta>$) of 0.063 in xylene hexafluoride, $\overline{M}_n$ (number average molecular weight) is 3,990 by vapor pressure osmometry. Analysis for percent carbon indicates that the product contains 65 mol percent copolymerized $CF_2=CH_2$. When 100 parts of this product is mixed with 1.8 parts of pentaerythritol and heated for 24 hours at 300°F., the resultant product is a tough rubber, insoluble in xylene hexafluoride.

Example 4: An example of a cured polymer formulated with filler materials is as follows:

	Parts
Copolymer of Example 3 above (65:35 mol ratio, molecular weight is about 4,000 ($-CF_2)_3COOH$ termination)	100
Zirconium silicate	20
Silicone oil-treated silica to control viscosity	5
Pentaerythritol	1.5

The above compounds are blended on a rubber mill and cured in open molds by the following heating cycle:

Hours:	°F.
24	150
24	200
24	300

The resulting cured rubber is tough, exhibits good adhesion to metals, and is stable at elevated temperatures in the presence of hydrocarbon fuels. The formulation is thus suitable for sealing the fuel tanks of high speed aircraft.

Vinylidene Fluoride-Perfluoropropene-Acrylic Acid Terpolymers

In a process described by C.L. Sandberg and J.M. Mullins; U.S. Patent 3,080,347; March 5, 1963; assigned to Minnesota Mining and Manufacturing Company perfluoropropene and vinylidene fluoride are copolymerized together to form a substantially saturated polymer which is of low molecular weight and varies in physical form from normally liquid to a waxy material having a melting point below about 60°C. The above comonomers may be polymerized in the presence of other monomers in small proportions, such as less than 10 mol percent, to modify the polymer in such a way as to vary the physical characteristics or provide additional sites for vulcanization and thereby produce ultimately a new and useful material. These low molecular weight polymers are placed upon surfaces or preformed under moderate pressures and temperatures and then cross-linked or vulcanized in situ to form a suitable solid adherent surface or solid article, which surface or article is substantially rigid and tough, such as a solid rubber or elastomeric material, and is not fluid at temperatures as high as 150°C. and higher.

The proportion of perfluoropropene monomer units in the ultimate copolymer will be between about 15 mol percent and about 50 mol percent depending upon the monomer mixture and physical properties desired. The proportion of vinylidene fluoride units will be between about 50 and about 85 mol percent. The perfluoropropene-vinylidene fluoride polymer of the process appears to have a tendency to be crystalline. As a result, the liquid range is relatively narrow and the grease and wax has a tendency to be stiff and hard.

The above can be overcome to some extent by inclusion in the polymer of side chains which tend to prevent crystallinity. This is accomplished by polymerizing vinylidene fluoride and perfluoropropene in the presence of an alpha-beta unsaturated carboxylic acid, such as acrylic acid, methacrylic acid and crotonic acid, to produce a terpolymer. The acid radicals in the polymer also serve as sites for vulcanization or cross-linking. Only a relatively small proportion of the carboxylic acid unit in the polymer is necessary, usually less than 10 mol percent and generally in the range of 0.1 to 2 mol percent, based on the total monomer unit content of the product. The total upper limit of the other monomers in the system will be decreased corresponding to the amount of third component included.

The saturated polymer is of low molecular weight of less than about 50,000 and is in the physical form of a normally liquid, grease or wax having a melting point below about 60°C. The preferred molecular weight range is from about 2,000 to about 10,000. The following examples illustrate the process. In the examples the viscosity $<\eta>$ is equal to

$$\frac{\ln \dfrac{n \text{ solution}}{n \text{ solvent}}}{C}$$

where C is concentration of polymer in grams per 100 ml. of solution, and n solution and n solvent are viscosities in consistent units.

Example 1:   Perfluoropropene, vinylidene fluoride and acrylic acid were polymerized to produce a liquid polymer as follows. The recipe for the polymerization included 100 parts of total monomers, 200 parts of water, 1 part of potassium persulfate, 4 parts of the ammonium salt of perfluorooctanoic acid, 1.6 parts of acrylic acid, and 0.5 part of dodecyl mercaptan. The monomers were in a 70:30 mol ratio of vinylidene fluoride to perfluoropropene in the above mixture except for the acrylic acid. Polymerization was effected in a polymerization tube at a temperature of 60°C. for a time equivalent to 100 hours. The tube was shaken during the polymerization. After polymerization, the tube was emptied and a liquid polymer was obtained having an average molecular weight of about 3,000 with a substantial conversion.

Example 2:   Vinylidene fluoride, perfluoropropene and acrylic acid were polymerized together at a temperature of about 80°C. for sixteen hours in a mol ratio correspondint to about 70:30:3, respectively. The polymerization recipe included the following:

> 6.4 grams of vinylidene fluoride
> 15 grams of perfluoropropene
> 0.422 gram of acrylic acid
> 42.8 grams of water
> 0.64 gram of potassium persulfate
> 0.85 gram of the ammonium salt of perfluorooctanoic
>   acid
> 0.012 gram of carbon tetrachloride

The polymer obtained was a low molecular weight terpolymer having a viscosity $<\eta>$ of 0.29 corresponding to an average molecular weight of about 10,000. The polymer was readily dissolved in acetone. The latex from the polymerization mixture was freeze coagulated, the coagulated polymer was dissolved in acetone and the solution of polymer was precipitated with water. The polymer could be made fluid at temperatures between 50° and 60°C.

Such a polymer is then vulcanized to produce a rigid elastomeric material by mixing with 10 parts of triethylene tetra-amine per 100 parts of polymer and heating at an elevated temperature about 130°C. for four hours. The resulting vulcanized or cured polymer has the following typical properties:

	Percent
Weight loss, 119 hours at 400°F.	1.1
Weight loss, 117 hours at 500°F.	5
Percent weight gain in 70/30 fuel, 72 hours at 180°F.	1.6

Vinylidene Fluoride Elastomeric Products

Methylol-Phenol Groups

In a process described by R.P. Conger; U.S. Patent 3,142,660; July 28, 1964; assigned to United States Rubber Company a fluorocarbon elastomer is reacted with an alkali metal salt of a phenolic material selected from the group consisting of 2,6-dimethylol-4-lower-alkyl phenols where alkyl has 1 to 10 carbon atoms, and 2,2'-methylene-bis(4-chloro-6-methylolphenol). It is believed that in the resulting products some of the fluorine groups are replaced by methylolphenol groups, so that the polymer chains are composed of recurring fluorocarbon groups and methylolphenoxy carbon groups.

The resulting product is characterized by the ability to be cured or cross-linked simply by heating, preferably in the presence of air. The cured products are especially valuable because they do not continue to evolve hydrogen fluoride as do fluorocarbon rubbers vulcanized by other methods. Further, films of modified fluorocarbon rubbers can be cured while exposed to air, which is not possible with other vulcanizing methods. The process is illustrated by the examples which follow.

Example 1: This example illustrates the modification of Viton A-HV fluorocarbon elastomer (copolymer of vinylidene fluoride and hexafluoropropene in mol ratio of 4:1) by potassium 4-tert-butyl-2,6-dimethylolphenoxide. To 200 grams of Viton A-HV, dissolved in 2,000 ml. of 2-butanone, was slowly added 400 ml. of an ethanol solution containing the dissolved reaction product of 0.120 mol of potassium hydroxide (6.73 grams) and 0.120 mol of 4-tert-butyl-2,6-dimethyl-olphenol (25.2 grams). This is a loading of 12.6 parts of methylolphenoxide ion per 100 grams of Viton A-HV.

The solutions were combined and refluxed at 74° to 76°C. for 1 hour, then cooled. A film of the cooled solution was cast on a glass plate and a white, rubbery polymer which resembled untreated Viton A-HV was obtained. The remainder of the solution was precipitated with water, then purified by successively redissolving in 2-butanone and reprecipitating by water 3 times.

By analysis it contained 57.5% F, 36.2% C, and 2.1% H. This corresponds to 11.0 grams of 4-tert-butyl-2,6-di-methylolphenol attached to every 100 grams of Viton A-HV elastomer (87.5% yield). Further evidence of the modification of Viton A-HV is shown by the fact that the dried film cast from the reaction mixture cross-linked tightly (i.e., became insoluble in hot 2-butanone) when heated 30 min. at 325°F. in an air oven.

Example 2: This example illustrates the modification of Viton A-HV fluorocarbon elastomer by the monopotassium salt of 2,2'-methylene-bis(4-chloro-6-methylolphenol). To 12.5 grams of Viton A-HV, dissolved in 250 ml. of 2-butanone, was added slowly 50 ml. of an ethanol solution containing the dissolved reaction product of 0.003 mol of potassium hydroxide (0.182 gram) and 0.003 mol (1.05 grams) of 2,2'-methylene-bis(4-chloro-6-methylolphenol). This is a loading of approximately 8.4 parts of the methylolphenoxide ion per 100 grams of Viton A-HV.

The combined solutions were refluxed 1 hour at 74° to 76°C., then cooled. A film was cast on a glass plate to obtain rubbery polymer resembling Viton A-HV. The polymer from the remainder of the solution was precipitated, purified, and dried as in Example 1. By analysis the purified elastomer contained 64.13% F, 33.8% C, and 2.1% H. This corresponds to 3.03 grams of the phenoxy moiety attached to every 100 grams of Viton A-HV (36% yield). Further evidence of the modification of Viton A-HV by this reagent is the fact that the dried film cast from the reaction solution cross-linked tightly (i.e., became insoluble in hot 2-butanone) after being heated 30 min. at 330°F. in an air oven.

Example 3: This example shows the air cure of a fluorocarbon elastomer without isolation of the modified soluble polymer. One hundred twenty-five grams of a masterbatch containing 100 parts of Viton A-HV elastomer, 20 parts of FEF black (Philblack A) and 5 parts of acid acceptor (magnesium oxide), were dissolved in 500 ml. of 2-butanone. Into this cement was slowly stirred 200 ml. of an ethanol solution containing 3.02 grams of potassium hydroxide and 10.75 grams of 4-tert-butyl-2,6-dimethylolphenol. This corresponds to a loading of 10 parts of the phenol moiety per hundred of elastomer. The solutions were combined then refluxed at 74°C. for 1 hour, followed by the spreading of a film, 0.020 inch thick after drying, on glass by successive passes with a small knife spreader. The dried film was given an air oven cure of 30 min. at 325°F. The cured physical properties are as follows:

Tensile	2,000 psi
Elongation at break	366 percent
Modulus (100%)	598 psi
Modulus (200%)	1,224 psi
Modulus (300%)	1,758 psi

Solubility—24 Hrs. at Room Temperature in 2-Butanone

Percent sol	5.0
Swelling index	2.9

Amine Modified RTV Sealant

W.R. Griffin; U.S. Patent 3,041,316; June 26, 1962; assigned to the U.S. Air Force describes a method of vulcanizing selected fluorine-containing copolymers by first providing chemically reactive sites on the copolymers at elevated temperatures followed by cross-linking through these sites at room temperatures. By providing chemically reactive sites on the polymer chain at elevated temperatures, room temperature vulcanization can be effected by reacting the modified polymer with polyfunctional cross-linking agents. The formation of the reactive site is accomplished by using various amines. The amines which can be employed in this process are primary amines such as ethylamine, propylamine and the like; secondary amines such as diethylamine, diisopropylamine, di-n-butylamine, diisobutylamine, di-n-octylamine, diallylamine; tertiary amines such as triethylamine, tripropylamine, triamylamine and tetramethylguanidine. It is theorized that these agents provide reactive sites in the halogenated polymer such as commercially available Viton A (a copolymer of hexafluoropropylenevinylidene fluoride). All of these amines provide double bonds on the polymer chain and in the case of the unsaturated primary and secondary amines, the double bonds of the amines are additional active sites.

Following the formation of the reactive sites, the modified fluorinated polymer is vulcanized by employing polyfunctional cross-linking agents active toward the modified polymer and capable of reaction at room temperature. Examples of cross-linking agents are dimercaptans such as hexamethylene dithiol and glycol dimercaptoacetate; heterocyclic amines such as piperazine; and alkane diamines such as N, N'-dimethyl hexamethylene diamine.

Example: A vulcanizate was prepared from the following composition divided into two parts for storage stability.

	Parts by weight
Part A:	
Viton A[1]	100.00
Medium thermal carbon black	30.00
Magnesium oxide	10.00
Diallylamine	1.00
Methyl ethyl ketone	141.00
Part B:	
Hexamethylene dithiol	1.50
Tri-n-amylamine	0.50

[1]Viton A is a hexafluoropropylene-vinylidene fluoride copolymer having a molecular weight of about 60,000 and a monomer ratio of 30% vinylidene fluoride and 70% hexafluoropropylene.

The hexafluoropropylene-vinylidene fluoride copolymer is a water clear, rubbery gum which easily forms a band on conventional rubber mixing equipment. The rubbery gum is banded on a mill with two parallel rolls adjusted to provide a rolling bank in the nip of the rolls. The remaining ingredients of Part A are weighed using the dry powdered carbon black to absorb the liquid diallylamine. The diallylamine is thereby in a less mobile state and incorporates into the rubber at a faster rate with less danger of loss. The carbon black-diallylamine mix is added to the revolving band and is immediately followed by the magnesium oxide. The mixing action is continued until all the ingredients are uniformly blended.

The blended rubbery mass, Part A, remains "warm to the touch" from a balance between the cold water cooling effect and the heat generated by the mixing action. The blended rubbery mass is next heated to drive the diallylamine-polymer reaction, under conditions which prevent volatilization of the diallylamine. Two hours at 300°F. are sufficient to drive this reaction to completion. The volatilization of the diallylamine is prevented by placing the blended rubbery mass in a standard ASTM cavity mold which has been liberally coated with a surfactant mold release agent. The mold is then placed in a hydraulic press at 300°F. under sufficient pressure to keep it tightly closed. Usually, hydraulic pressures of about 500 pounds per square inch of mold area are sufficient. After a reaction time of two hours the pressure is released and the blended rubbery mass is removed from the mold while hot. The mold release agent and other foreign materials are washed from the mass with cold water. The water is allowed to evaporate.

Part A is now refined, as before, by end passing on a tight cool mill in order to break up any agglomerates and to prepare the rubbery mass for rapid attack by the solvent. Generally, ten passes are sufficient and result in a smooth material similar to the original mixture before the heat treatment. The refined material in the form of a thin crumpled sheet is weighed and placed in a container with an equal weight of methyl ethyl ketone solvent.

As indicated in the above formulation, Part B is a physical mixture of hexamethylene dithiol, referred to as HMDT and tri-n-amylamine referred to as TAA. The HMDT was selected in preference to others because it is in a liquid state and has a low vapor pressure. The TAA was chosen as the catalyst because it is liquid at room temperature, strongly basic and found to have little effect in the vulcanizate after high temperature exposure. The HMDT and TAA are water-like in appearance and are completely miscible in the proportions used. The vulcanizate was prepared from a uniform dispersion of Parts A and B in which 1 part by weight of B was mixed into 141 parts by weight of A. In the presence of moisture a cross-linking reaction takes place which leads to a vulcanized elastomer in approximately seven days. The catalyzed solution may be ejected from a pressure gun used for applying sealants, worked into place by using spatulas and the like or flowed out onto a flat sheet of metal or other surface to evaporate the solvent and vulcanize. The rate of the cross-linking reaction depends upon the concentrations of TAA and moisture.

In the near absence of moisture TAA does not function as a catalyst; however, normally sufficient moisture is contained by the rubber. The increase of either of the two catalysts TAA or moisture, will result in an increase of rate of reaction. The extent of the cross-linking reaction is dependent upon the number of active sites and how many of these are used in cross-linking. Desirable conditions would provide just enough HMDT to react with all of the reactive sites.

The properties of the vulcanizate were determined during the cross-linking period. A portion of catalyzed solution was flowed out onto a clean steel plate. After it had reacted for 24 hours at room temperature (75° ±5°F.) it had sufficient strength to be stripped from the plate. A small part of this flowout was placed in methyl ethylketone. It swelled but did not dissolve, thus indicating that a substantial amount of cross-linking had occurred. The rate of vulcanization at room temperature was traced by measuring tensile strength, percent elongation, hardness and permanent set after break of the specimen. The test specimens were micro-dumbbells measuring 1/8" x 5/8" in the constricted portion, with 1/2" tabs. The tests except hardness were performed on a modified Twing-Albert paper tensile tester, used because of its sensitivity to small changes in tensile strength. The test data obtained in the above referred to tests are presented in Table 1.

TABLE 1: RATE OF CROSS-LINKING OF ROOM TEMPERATURE VULCANIZING VITON A

Time After Catalyzing, hrs.	Tensile Strength, p.s.i.	Elongation percent	Break Set, percent	Hardness, Shore A
24	73	1,100	240	20
48	235	1,500	260	45
120	810	850	80	55
144	1,180	790	70	55
168	1,470	730	50	57
192	1,590	700	40	58
21 Days	1,600	500	20	62

The test results outlined in Table 1 indicated a good state of vulcanization after seven days at room temperature. Therefore, test specimens, exhibiting this state of vulcanization were subjected to high temperature tests. Micro-dumbbells, similar to those used in the tests outlined in Table 1 were exposed to JP-4 jet engine fuel in a pressure bomb at 400°F. for 70 hours. The room temperature vulcanized elastomers displayed excellent resistance to the hot fluid and also showed little additional vulcanization occurring at this higher temperature. Similar test specimens were exposed to 500°F. air for 70 hours in a circulating air oven. The results of the high temperature tests are summarized in Table 2.

TABLE 2: AGING RESISTANCE OF ROOM TEMPERATURE VULCANIZING VITON A

Aging Conditions	Tensile Strength, p.s.i.	Elongation, Percent	Hardness, Shore A	Volume Change, Percent
Original [1]	1,500	700	57	....
70 Hours at 400° F., JP-4 Fuel	1,800	550	53	+7
70 Hours at 500° F., Air Oven	1,450	320	65	−13

[1] Vulcanized 168 hours at room temperature.

The above tests indicate that excellent physical properties are obtained by the room temperature vulcanization system and that a useful fluorinated elastomeric product is produced which fills a need on high speed aircraft.

Perfluoroalkyl Perfluorovinyl Ethers

A process described by J.R. Albin and G.A. Gallagher; U.S. Patent 3,136,745; June 9, 1964; assigned to E.I. du Pont de Nemours and Company involves elastomeric copolymers of perfluoroalkyl perfluorovinyl ethers and vinylidene fluoride, which show excellent low temperature properties as well as good thermal stability.

The perfluoroalkyl perfluorovinyl ethers that may be used in the preparation of the copolymers have the general structure $R-O-CF=CF_2$ in which R is a perfluoroalkyl group having one to three carbon atoms. This includes perfluoromethyl perfluorovinyl ether, perfluroethyl perfluorovinyl ether, and perfluoropropyl perfluorovinyl ether or mixtures of any of these. The preferred monomer is perfluoromethyl perfluorovinyl ether because the copolymers prepared from this compound in general show the most desirable properties. Usually, lengthening the perfluoroalkyl chain of the perfluorovinyl ether serves no useful purpose, and perfluoroalkyl groups having more than three carbon atoms are undesirable since the resulting copolymers are relatively inferior in thermal stability and are more costly. These perfluoroalkyl perfluorovinyl ethers may be prepared by the pyrolysis of 2-(perfluoroalkoxy)perfluoropropionic acid or related derivatives. This acid has the following structure

$$CF_3-CF-COOH$$
$$|$$
$$OR$$

where R has the same meaning as above. In a preferred method, the ethers are prepared by pyrolysis of the alkali metal salt of the 2-(perfluoroalkoxy)perfluoropropionic acid at a temperature in the range of 100° to 250°C. The dry salt by itself may be pyrolyzed, in which case a temperature of 170° to 250°C. is used. The pyrolysis may also be carried out in the presence of polar or nonpolar solvents. In the presence of polar solvents, such as 1,2-dimethoxyethane and benzonitrile, the decomposition is generally carried out at temperatures of 100° to 180°C.

The presence of the vinylidene fluoride unit is an essential feature of the copolymers, both to provide curability and to insure the desired combination of properties. Copolymers of the perfluoroalkyl perfluorovinyl ether with a hydrogen-containing olefin other than vinylidene fluoride, such as vinyl fluoride and trifluoroethylene, have poor low temperature flexibility, poor thermal stability, or both. The preferred copolymers are those which contain about 72 to 80 mol percent of vinylidene fluoride units and about 28 to 20 mol percent of perfluoroalkyl perfluorovinyl ether units.

The copolymers prepared in accordance with this process are elastomers having outstanding low temperature flexibility combined with good thermal stability. Their stiffening points, determined by the Clash-Berg test (ASTM D1043-51), are in the range -23° to 31°C. For comparison, the commercially available fluoroelastomers prepared from vinylidene fluoride and hexafluoropropene have stiffening points in the range of -18° to -12°C. At the same time, the copolymers of this process are at least equivalent in thermal stability to the known fluoroelastomers. The cured copolymers require at least 70 hours at 288°C. to lose 10% of their weight and most of them require 100 to 170 hours. For comparison, commercially available fluoroelastomers prepared from vinylidene fluoride and hexafluoropropene lose 10% of their weight in 50 to 120 hours. Representative examples illustrating the process follow. The preparation and evaluation of the polymers in the examples which follow are carried out as described below.

A. Preparation of Polymers: Conditions used for effecting the copolymerization are as follows. Into a 400-ml. Hastelloy C bomb are placed the desired amounts of ammonium persulfate, ammonium perfluorocaprylate, and deoxygenated distilled water while maintaining the whole operation under a blanket of nitrogen. After closing the bomb and freezing for 10 to 15 minutes in a "Dry Ice"-acetone bath (about -78°C.) it is evacuated to a pressure of less than 1 mm. Hg.

The desired monomers, preweighed into loading cylinders, are then added to the evacuated, cold bomb in order of their boiling points, starting with the highest boiling material. The loaded bomb is then placed in a shaker unit and heated to 60°C. while shaking in a reciprocal motion at 180 cycles per minute. Heating and shaking is continued for two hours after the last observable pressure drop and is then discontinued. After cooling to room temperature, any small amount of residual gas is recovered by attaching an evacuated cylinder, cooled to about -78°C. to the bomb. The contents of the bomb are then removed.

The bomb contents, usually in the form of an emulsion, are placed in a stainless steel beaker which is partially immersed in a "Dry Ice"-acetone bath until the contents are frozen solid. Upon warming the beaker and contents to room temperature, the copolymer is obtained as a coagulum. The aqueous material is removed by filtration to remove the bulk of initiator and emulsifier residues. The coagulum is then washed with 200-ml. portions of distilled water using an Osterizer blender until two successive washes are acid free. The wet coagulum is dried in a vacuum oven at 70°C. The inherent viscosity of the polymer is measured at 30°C. using a solution of 0.1 gram of copolymer in 100 ml. of a mixture consisting of 87 parts of tetrahydrofuran and 13 parts of N,N-dimethylformamide.

B. Polymer Evaluations: (1) Raw film — The films are made by compressing one to two grams of the dry raw copolymer between two aluminum sheets for one to two minutes at 150°C. and a pressure of 2,000 pounds per square inch.

(2) Vulcanized strips — Ten grams of the raw copolymer is worked on a 2" x 6" two-roll rubber mill, and the compounding ingredients are milled in using conventional rubber-compounding techniques. The recipe used is:

	Parts by weight
Copolymer	100
Magnesium oxide	15
Medium thermal carbon black	20
Hexamethylenediamine carbamate	As indicated

The compounded strips are placed in a standard 1" x 5" x 0.075" cavity mold and heated at 150°C. for 0.5 to 1 hour while maintaining the mold under a pressure of 1,000 to 4,000 pounds per square inch. The time of the press cure is indicated in the examples. The mold is then cooled under pressure by passing cold water through the press platens, and the molded slab is removed. The cure is completed by placing the slabs in an oven, heating from 25° to 204°C. over a period of 2 hours and then heating at 204°C. for 18 to 24 hours. Strips of one-quarter inch width are cut from the slabs for testing.

(3) The copolymers are tested as follows — Stress-strain properties are obtained by pulling a strip on an Instron tester at a rate of 10 inches per minute at 25°C. To determine resistance to thermal degradation, strips one-quarter inch in width are hung for 48 hours in a circulating air oven heated to 288°C. The strips are then cooled to room temperature and the stress-strain properties obtained as described above. Also the length of time required for samples of both the raw and the cured polymer to lose 10% of their original weight when submitted to a temperature of 288°C. is observed.

Low-temperature properties are obtained qualitatively by placing the raw film or a bent loop of the vulcanizate (made by joining the ends of a strip of dimension 0.075" x 5" x 0.25") in a freezer maintained at -25°C. for 24 hours and observing the degree of rigidity. Quantitative measurements are made by the Clash-Berg test (ASTM D1043-51), results of which are expressed in the temperature, °C., at which the modulus of the sample, on gradual cooling, reaches 6,000 psi (stiffening point).

Example 1: Using the general procedure described in paragraph A above, 350 ml. of water, 0.53 g. of ammonium persulfate, and 0.14 g. of ammonium perfluorocaprylate are charged to the bomb. The monomers charged are 6.4 g. of vinylidene fluoride and 14.2 g. of perfluoromethyl perfluorovinyl ether, a weight ratio of monomers of 31/69 which is a molar ratio of 54/46. The time between the start of the reaction and the last observed pressure drop is 0.75 hour. The maximum pressure attained is 250 psig and the pressure drop is 250 psig. There are no off gases. The product weighs 18.6 g. and is a rubbery polymer having an inherent viscosity of 0.76.

Analysis of the product shows that it contains 25.9% carbon, which indicates that the copolymer has a composition, by weight, of 27% of vinylidene fluoride units and 73% of perfluoromethyl perfluorovinyl ether units, which corresponds to a molar ratio of these units of 49/51. The raw copolymer is flexible at -25°C. and requires 200 hours at 288°C. to lose 10% of its weight.

The elastomer is cured by the method of paragraph B above using 2 parts of hexamethylenediamine carbamate and a press-cure of 60 minutes. The cured copolymer requires 113 hours at 288°C. to lose 10% of its weight. The stiffening point by the Clash-Berg test is -28°C.

Example 2: Using the general procedure described in paragraph A above, 360 ml. of water, 0.53 g. of ammonium persulfate, and 0.14 g. of ammonium perfluorocaprylate are charged to the bomb. The monomers charged are 8.9 g. of vinylidene fluoride and 9.5 g. of perfluoromethyl perfluorovinyl ether, a weight ratio of 48/52 which is a molar ratio of 71/29. The time between the start of the reaction and the last observed pressure drop is 1.5 hours. The maximum pressure attained is 320 psig and the pressure drop is 250 psig. There is 0.8 g. of off-gas, corresponding to 4.3% of the weight of monomers charged. The product weighs 13.7 g. and is a rubbery copolymer having an inherent viscosity of 1.00.

Analysis of the product shows that it contains 29.9% carbon, which indicates that the copolymer has a composition, by weight, of 52% of vinylidene fluoride units and 48% of perfluoromethyl perfluorovinyl ether units, which corresponds to a molar ratio of these units of 74/26. The raw copolymer is flexible at -25°C. and requires 205 hours to lose 10% of its weight.

The copolymer is cured by the method of paragraph B above using 1 part of hexamethylenediamine carbamate and a press-cure of 30 minutes. The cured copolymer requires 101 hours to lose 10% of its weight. The stiffening point by the Clash-Berg test is -29°C. The tensile properties of the cured copolymer are:

	Original	After heat aging
Tensile strength at the break, psi	910	480
Elongation at break, percent	640	570
Modulus at 200 — elongation, psi	380	210

1, 2, 3, 3, 3-Pentafluoropropylene

A process described by D. Sianesi, G.C. Bernardi and A. Reggio; U.S. Patent 3,331,823; July 18, 1967; assigned to Montecatini Edison SpA, Italy involves polymers which are prepared by copolymerizing 99 to 5 parts by weight of vinylidene fluoride with 1 to 95 parts by weight of a mixture of olefins having the general formula C_3F_5H containing at least 1% and up to 100% of 1, 2, 3, 3, 3-pentafluoropropylene, the remainder consisting substantially of 1, 1, 3, 3, 3-pentafluoropropylene. It has been observed that while polyvinylidene fluoride is a highly crystalline resinous product which, at room temperature, shows no elastic characteristics, a gradual appearance of the elastomeric characteristics is brought about by introducing into its macromolecules increasing amounts of copolymerized units of 1, 2, 3, 3, 3-pentafluoropropylene, which progressively destroy chain symmetry and hence crystallinity.

Quite similar modifications are obtained when monomeric units of 1, 2, 3, 3, 3-pentafluoropropylene and 1, 1, 3, 3, 3-pentafluoropropylene are simultaneously introduced into the macromolecules of polyvinylidene fluoride. 1, 2, 3, 3, 3-pentafluoropropylene ($CF_3-CF{=}CFH$) which, when copolymerized with vinylidene fluoride, provides the products of this process, is a known compound, obtainable by the processes indicated in the literature (see for instance R. N. Haszeldine, B. R. Steele, J. Chem. Soc. 1592 (1953). The following examples illustrate the process.

Example 1:   Into a stainless steel autoclave of 120 cc capacity were introduced the following components in the order named: 0.300 g. of $Na_2HPO_4 \cdot 12H_2O$, 0.076 g. of KH_2PO_4, 0.017 g. of $(NH_4)_2S_2O_8$, 0.004 g. of $Na_2S_2O_5$, 150 mg. of a mixture of products of the formula $H(CF_2-CF_2)_nCOONH_4$ where n ranges from 4 to 7, and 50 g. of air-free H_2O. While introducing these substances, the autoclave was maintained at the temperature of $-78°C$. in a bath of Dry Ice and acetone, and water was gradually added after introduction of each salt, in such a manner that the various salts were separated from each other by layers of frozen water.

The autoclave was then closed and a high vacuum produced therein. 1.98 g. of 1, 2, 3, 3, 3-pentafluoropropene and 18.4 g. of vinylidene fluoride were then distilled under vacuum into the autoclave. The closed autoclave was placed into a 70°C. oil bath and kept at this temperature for 16 hours under reciprocal agitation. After 16 hours reaction, the residual monomers were discharged and the polymerization was practically quantitative. The autoclave was opened and a milky copolymer was discharged. The copolymer was coagulated by the addition of 10 cc of concentrated HCl, and then it was thoroughly washed with distilled water and dried under a vacuum of 15 mm. Hg at 100°C. The dried copolymer weighed 20.0 g. When examined under a hot stage polarizing microscope, it showed a crystalline melting range of 124° to 135°C. The copolymer could be compression molded at temperatures, e.g., about 150°C. into transparent, flexible plates showing the characteristic behavior of thermoplastic products.

The intrinsic viscosity measured in dimethylformamide at 110°C. was found to be 0.91 (100 cc/g.). Stretching tests carried out on specimens of the type described in ASTM D 412 D, but 15 mm. long in the middle, at a stretching rate of 50 mm./minute, gave the following results:

Ultimate tensile strength	310 kg./cm.^{2}
Elongation at break	370 percent

Elementary analysis showed a carbon content of 36.48% by weight, which corresponds to a content of 5% by mols of units derived from 1, 2, 3, 3, 3-pentafluoropropene in the copolymer. The copolymer was soluble in organic solvents, such as dimethylformamide, at room temperature. By evaporation of these solutions, settling of the polymer in the form of a homogeneous, transparent and flexible film could be easily obtained.

Example 2:   Into the autoclave described in Example 1 and according to the same procedure, were introduced 0.034 g. of $(NH_4)_2S_2O_8$, 0.008 g. of $Na_2S_2O_5$, 0.300 g. of $Na_2HPO_4 \cdot 12H_2O$, 0.074 g. of KH_2PO_4, 0.150 g. of ammonium perfluorocaprilate, and 50 g. of air-free H_2O. 11.88 g. of 1, 2, 3, 3, 3-pentafluoropropene and 13.44 g. vinylidene fluoride were then introduced into the closed autoclave by vacuum distillation, so as to form a monomeric mixture consisting of 30% by mols of pentafluoropropylene and 70% by mols of vinylidene fluoride. The autoclave was then placed into a 70°C. oil bath and kept at this temperature for 16 hours under reciprocal agitation.

The polymerization was practically quantitative. At the end, a milky suspension of copolymer was discharged from the autoclave, from which suspension the copolymer was precipitated by the addition of 10 cc of HCl. This copolymer was thoroughly washed with distilled water, dried at 100°C. under a vacuum of 15 mm. Hg and calendered. It weighed 24.20 g. and had the typical aspect of unvulcanized rubber. The copolymer, containing 65.2% by weight of fluorine and 32.75% of carbon, therefore had an average composition of 29.4% by mols of units derived from 1, 2, 3, 3, 3-pentafluoropropene. It showed an intrinsic viscosity of 2.64 (100 cc/g.) in dimethylketone at 30°C. 100 parts of this copolymer were mixed at room temperature, on a conventional roll-mixer, with MgO (15 parts), MT (medium thermal) carbon black (20 parts) and dicinnaoxylidenehexaneethylenediamine (3 parts). The resulting mix was molded into sheets about 1 mm. thick, by pressing at 150°C. for 30 minutes, under a stress of about 50 kg./cm.^{2}. These sheets were then kept in an air oven at 230°C. for 4 hours after having reached said temperature within two and a half hours. Specimens as described in Example 1 were obtained from them. Stretching tests on said specimens, carried out at 25°C. with a stretching rate of 50 mm./minute, gave the following characteristics:

Ultimate tensile strength	174 kg./cm.^{2}	Modulus at 100%	29 kg./cm.^{2}
Elongation at break	307 percent	Modulus at 200%	77 kg./cm.^{2}
Residual set	10 percent (10 min. after breaking)		

Specimens of the same vulcanized product, obtained as described above, 1 mm. thick, were maintained immersed in various solvents or chemical agents at 25°C. for 192 hours. The specimens were then dried and weighed, and then vacuum dried in an oven and weighed again. The obtained data are reported below.

Solvent or agent	Change in weight (percent)	Change in weight after drying (percent)
Acetone	+159.6	−4.75
Ethyl acetate	+191.4	−4.35
n-Heptane	0.00	−0.09
Benzene	+14.2	+1.43
Sulphuric acid 98%	+3.2	+2.30
Nitric acid 65%	+12.8	+5.56
Aqueous NaOH 10%	+0.2	−0.03

TERPOLYMERS FOR COATINGS

Tetrafluoroethylene and Vinyl Benzoate

J.R. Chalmers and F.B. Stilmar; U.S. Patent 3,451,978; June 24, 1969; assigned to E.I. du Pont de Nemours and Company describe normally solid interpolymers of (a) vinyl fluoride or vinylidene fluoride, (b) about 0.05 to about 1.5 mols, per mol of (a) of tetrafluoroethylene, and (c) about 0.01 to about 0.7 mol, per mol of (a) of at least one monovinyl ester of an aromatic carboxylic acid derived from benzene or naphthalene. The interpolymers of this process are useful for molding, fiber and self-supporting film applications and are particularly useful as easily applied, weatherable, protective coatings for a variety of substrates.

Copolymers consisting only of vinylidene fluoride and tetrafluoroethylene within the composition limits of this process are not highly soluble in common organic solvents and yield poorly coalesced, cloudy, somewhat brittle solution-cast films unless baked. Copolymers consisting only of tetrafluoroethylene and a vinyl aromatic carboxylate, such as tetrafluoroethylene/vinyl benzoate copolymers, are brittle and considerably less soluble in organic solvents than the polymers of this process. Vinyl aromatic carboxylates inhibit the polymerization of vinylidene fluoride and therefore, two-component vinylidene fluoride/vinyl aromatic carboxylate copolymers are not readily prepared. Even when the amount of vinyl aromatic carboxylate is minimized, the copolymerization proceeds poorly.

For example, the copolymerization of vinylidene fluoride and vinyl benzoate in the ratio of 120/2 by weight gives very poor results. When larger amounts of vinyl benzoate relative to the vinylidene fluoride are employed, the copolymer yield is even less. However, when tetrafluoroethylene is added as a third monomer, copolymerization proceeds well. Similarly, the free radical initiated copolymerization of vinyl fluoride and vinyl aromatic carboxylates gives only low yields of copolymer. Addition of tetrafluoroethylene as a third monomer makes possible good yields of readily soluble terpolymer which form well-coalesced solution-cast films at room temperature. Mixtures of vinyl fluoride and tetrafluoroethylene within the composition limits of the process polymers copolymerize well but the resulting copolymers are not appreciably soluble at room temperature and cannot be solution-cast to form well-coalesced films at room temperature.

The incorporation of both vinyl fluoride and vinylidene fluoride along with tetrafluoroethylene and the vinyl alkane carboxylate in the interpolymer results in the best-all-around balance of properties for use as a coating composition. The presence of the vinyl aromatic carboxylate is necessary to give soluble polymers which are coalescible at room temperature. A three-component interpolymer, vinyl fluoride, vinylidene fluoride, and tetrafluoroethylene (in the ratio 70/70/35), although soluble in hot dimethylacetamide, gives a powdery, noncoalesced coating when cast from the dimethylacetamide solution.

The vinyl esters of aromatic carboxylic acids useful in the process have the structure $CH_2=CHO_2CAR$. AR in the structural formula represents a C_6 to C_{16} carbocyclic aromatic group derived from benzene or naphthalene. The benzene or naphthalene groups may be free of substituents other than hydrogen, such as in vinyl benzoate or vinyl naphtholate, or may carry one or more substituents attached to the benzene or naphthalene ring. These substituents, however, must not interfere with free radical polymerizations.

The preferred vinyl aromatic carboxylate is vinyl benzoate. Of the substituted vinyl benzoates, the vinyl chlorobenzoates are particularly useful. Vinyl esters of aromatic acids containing three or more fused benzene rings are not useful in the polymers of the process because they are either not readily copolymerizable to interpolymers or yield products which are too high melting and intractable to meet the requirements for protective coating compositions. Representative examples illustrating the process follow. Polymer compositions were based on elemental analysis for carbon, hydrogen, fluorine, chlorine, and phosphorus using established analytical methods. Acid content was determined by titration with base, and vinyl aromatic carboxylate content was determined by difference or was based on the infrared spectrum using a correlation between absorption at 14 microns (aromatic ring) and the concentration of vinyl aromatic carboxylates in polymers containing C^{14}-tagged vinyl aromatic carboxylate.

Accelerated weathering tests were conducted in an "Atlas Weather-Ometer," Model XW (Atlas Electrical Devices Co., Chicago, Ill.). The salt-fog machine is an Industrial Corrosion Test Cabinet, Type 411, 1ABC. Melt flow rate in the following examples was determined using ASTM Method D-1238-62T. The "sticking temperature" is a softening temperature which is measured by a determination of the lowest temperatures at which a polymer in contact with the heated brass block leaves a molten trail when moved across the block. Film or coating hardness was determined by the Pencil Method as illustrated in the following articles by W.T. Smith, Official Digest, 28, 232-7 (1956); H.A. Gardner and G.G. Sward, Paint Testing Manual, 12th ed., Gardner Laboratory, Inc.

Example 1:   Vinylidene fluoride/tetrafluoroethylene/vinyl benzoate/bis(2-chloroethyl) vinylphosphonate/itaconic acid pentapolymer — A 400 ml. pressure vessel was flushed with nitrogen and charged with 250 ml. of trimethyl phosphate, 2 ml. of vinyl benzoate, 0.5 g. itaconic acid, 0.5 ml. bis(2-chloroethyl) vinylphosphonate, and 0.65 ml. of a 75% solution of tertiary-butyl peroxypivalate in mineral spirits. The vessel was closed, cooled in Dry Ice-acetone, and evacuated. One hundred twenty grams of vinylidene fluoride and 30 g. of tetrafluoroethylene were added. The pressure vessel and its contents were shaken and heated to 60°C. at autogenous pressure (1,000 psi). When the temperature reached 60°C., the temperature was increased to 65°C. over a period of 30 minutes and then was maintained at 65°C. for one hour.

The pressure vessel and its contents were cooled to room temperature and, after venting any unreacted gases, the contents were discharged. The reaction mixture was a clear viscous solution. When the reaction mixture was poured into 400 ml. of chloroform, the polymer which precipitated was separated by filtration and then was washed twice in a blender with 300 ml. of chloroform. The polymer was then dried in a vacuum oven at 60° to 70°C. to yield 34 g. of solid polymer.

The polymer was shown to contain 0.2% phosphorus and 59.2% fluorine by elemental analysis. These analyses correspond to a 1.5% by weight content of bis(2-chloroethyl) vinylphosphonate-derived groups, i.e., 0.005 mol per mol of vinylidene fluoride. The presence of groupings within the polymer derived from the vinyl benzoate was indicated by absorption in the infrared at 5.8, 6.2, 6.3, 6.7, and 14.1 microns. The concentration of the vinyl benzoate in the polymer was 2.2% by weight or 0.0115 mol per mol of vinylidene fluoride.

Itaconic acid-derived groupings were shown to be present in a concentration of 0.2% by weight by a neutral equivalent determination. This amounted to a concentration in the polymer of 0.00115 mol of itaconic acid-derived groups per mol of vinylidene fluoride. Elemental analysis of the polymer showed that 13 parts by weight were from tetrafluoroethylene groups, i.e., 0.10 mol of tetrafluoroethylene per mol of vinylidene fluoride.

A solution containing 15% solids was readily prepared in dioxane and a 13% solids solution was obtained using a mixture of equal volumes of methyl ethyl ketone and cyclohexanone as the solvent. The inherent viscosity of the polymer as a 0.5% solution by weight in dimethylformamide at 30°C. was 0.68. The polymer exhibited some crystallinity and the crystalline melting point was 182°C. Coatings of the polymer on bright aluminum, galvanized and Bonderized steel exhibited excellent adhesion in both dynamic and static wet adhesion tests. Solutions (15 to 20% solids) of the polymer in methyl ethyl ketone/dioxane, in dioxane, in dioxane/cyclohexanone, and in dioxane/4-methyl-4-methoxypentan-2-one were prepared and brush applied to give clear coatings on various woods, including redwood and cedar. Solutions (10 to 12% solids) of the polymer in the same solvents were readily spray applied to wood. These coatings were air dried to give clear, continuous films which exhibited good resistance to weathering tests.

A tetrapolymer containing all the monomers above except vinyl benzoate was prepared using the above procedure, solvent and initiator. The amounts of the monomers charged were 120 g. vinylidene fluoride, 30 g. tetrafluoroethylene, 0.5 ml. bis(2-chloroethyl) vinylphosphonate, and 0.5 g. itaconic acid. This polymer was soluble in methyl ethyl ketone and the solvents listed above but gave only cloudy, poorly coalesced, weak films which were unsuitable as protective coatings.

Example 2:   Vinyl fluoride/vinylidene fluoride/tetrafluoroethylene/vinyl benzoate tetrapolymer — Following the same procedure described in Example 1, a 400-ml. pressure vessel was charged with 225 ml. of acetic acid, 15 ml. of vinyl benzoate, 0.4 ml. of tertiary-butyl perbenzoate, 70 g. of vinyl fluoride, 70 g. of vinylidene fluoride, and 35 g. of tetrafluoroethylene. With continuous agitation, the vessel and its contents were heated at 95° to 100°C. for 2 hours, 100° to 102°C. for 3 hours, 102° to 104°C. for 2 hours, and successively for 2-hour intervals at 106°C., 108°C., 110°C., and 115°C. The reaction mixture was then cooled to room temperature, vented, and the product was discharged. After precipitation in ethanol and thorough washing in a blender with ethanol, 100 g. of white solid polymer was obtained after drying in a vacuum oven. Elemental analysis revealed that the polymer contained 28% vinyl fluoride and 29% vinylidene fluoride. Based on one mol of the mixture of vinyl fluoride and vinylidene fluoride, the results showed that the polymer contained 0.265 mol of tetrafluoroethylene (28%) and 0.095 mol of vinyl benzoate. The polymer was soluble in methyl ethyl ketone and in dimethylacetamide. Films cast from either of these solvents were bright, clear, continuous, and well coalesced after drying at room temperature. The dried polymer exhibited a melt flow rate of 14.1 g. per 10 minutes at 190°C. with a load of 2,160 g. and an orifice size of 0.082 inch diameter and 0.319 inch length. A 1-mil coating of the polymer on aluminum had a pencil hardness of B. When the above procedure is repeated except that an equivalent amount of vinyl naphthoate is substituted for the vinyl benzoate, a polymer containing a similar balance of properties is obtained.

Examples 3-9:   A series of other polymers containing vinyl fluoride, vinylidene fluoride, tetrafluoroethylene, and a vinyl aromatic carboxylate were prepared in the same manner as above and are tabulated in the table below. All preparations were carried out under autogenous pressure. All of the polymers have inherent viscosities in the range of 0.1 to 3.0 as 0.5% by weight solutions in dimethylformamide at 30°C.

Vinyl Fluoride/Vinylidene Fluoride/Tetrafluoroethylene/Vinyl Aromatic Carboxylate Interpolymers

Ex. No.	Monomers	Grams Charged	Initiator	Medium	Temp. (°C.)	Time (hrs.)	Wt. percent	Mole/mole VF₂	Yield (g.)	Hardness [a]	Solubility [b]
3	Vinyl fluoride	70	t-Butyl peroxypivalate, 0.9 ml.	t-Butanol/acetic acid, 125/125 ml.	60-75	10	33.0	....	89	2B	G
	Vinylidene fluoride	70					30.0	....			
	Tetrafluoroethylene	35					30.0	0.250			
	Vinyl benzoate	6					7.0	0.059			
4	Vinyl fluoride	70	do.	do.	60-75	13	18.0	....	57	B	G
	Vinylidene fluoride	70					20.0	....			
	Tetrafluoroethylene	35					27.0	0.385			
	Vinyl benzoate	30					35.0	0.340			
5	Vinyl fluoride	60	t-Butyl peroxypivalate, 0.5 ml.	Acetic acid 290 ml.	50-75	14			62	B	G
	Vinylidene fluoride	60									
	Tetrafluoroethylene	30									
	Vinyl 2,4-dichlorobenzoate.	10									
6	Vinyl fluoride	40	t-Butyl peroxypivalate, 1.2 ml.	Acetic acid, 295 ml.	45-70	15	21.0	....	45	F	G
	Vinylidene fluoride	80					21.0	....			
	Tetrafluoroethylene	20					23.0	0.29			
	Vinyl benzoate	20					35.0	0.31			
7	Vinyl fluoride	50	t-Butyl perbenzoate, 0.6 ml.	Acetic acid, 210 ml.	100-120	11	14.0	....	118	F	G
	Vinylidene fluoride	100					39.0	....			
	Tetrafluoroethylene	35					21.0	0.23			
	Chlorotrifluoroethylene.	15					11.5	0.11			
	Vinyl benzoate	10					14.3	0.11			
8	Vinyl fluoride	60	t-Butyl peroxypivalate, 0.75 ml.	Acetic acid, 270 ml.	60-75	11	20.0	....	57	HB	G
	Vinylidene fluoride	60					34.5	....			
	Tetrafluoroethylene	30					18.0	0.185			
	Vinyl benzoate	10					15.0	0.10			
	Dibutyl maleate	10					12.6	0.06			
9	Vinyl fluoride	80	t-Butyl perbenzoate, 0.5 ml.	Acetic acid, 225 ml.	100-120	16	34.0	....	77.5	B	G
	Vinylidene fluoride	80					31.0	....			
	Tetrafluoroethylene	20					14.0	0.11			
	Vinyl benzoate	20					20.0	0.11			
	Itaconic acid	3					0.9	0.006			

[a] Pencil Hardness.

[b] Solubility in cyclohexanone and/or dimethylacetamide at room temperature. G = Good, forms clear solution at 10% solids.

Example 10:   The polymers were readily pigmented and showed good pigment wetting properties. A general procedure for preparing pigmented compositions was as follows. To 80 ml. of the selected solvent was added 25 g. of polymer. After solution was obtained, it was occasionally necessary to add additional solvent to obtain a viscosity suitable for the milling operation. Two pigment formulations were used. Formulation A Standard Green — 12.5 g. pigment grade titanium dioxide, 2.0 g. Monastral Green B (Color Index No. 10006), 0.5 g. Columbian Carbon Lampblack No. 11. Formulation B Standard White — 15 g. pigment grade titanium dioxide. Depending on whether a green or a white pigmented composition was desired, Formulation A or Formulation B was added to the polymer solution described above and the mixture was rolled for 7 days in a 16 oz. jar with 100 ceramic 0.5 inch diameter mill balls. The resulting dispersion was then strained through a coarse wire mesh to separate the balls and stirred in the usual fashion before use. It was applied by brush in a normal manner or by spray techniques. Each of the polymers gave a dried pigmented finish which exhibited medium to high gloss.

Tetrafluoroethylene and Vinyl Butyrate

F.B. Stilmar; U.S. Patent 3,449,305; June 10, 1969; assigned to E.I. du Pont de Nemours and Company describes an interpolymer comprising (A) vinylidene fluoride, (B) from about 0.1 to about 0.4 mol of tetrafluoroethylene per mol of vinylidene fluoride, and (C) from about 0.05 to about 0.3 mol, per mol of vinylidene fluoride, of a vinyl ester of a C_2 to C_{18} alkane carboxylic acid.

The interpolymer has an inherent viscosity of at least 0.30 as a 0.5% solution by weight in dimethylformamide at 30°C., and can also contain from about 0.05 mol to about 0.4 mol, per mol of vinylidene fluoride, of at least one polymerizable mono-unsaturated monomer chosen from (a) C_3 fluorocarbon olefins, (b) C_2 to C_3 chlorofluorocarbon olefins, (c) C_2 to C_8

hydrocarbon monoolefins containing the group $CH_2{=}C{<}$, and (d) alkyl vinyl ether where the alkyl group is 1 to 16 carbon atoms.

Among the useful copolymerizable vinyl esters of alkane monocarboxylic acids are vinyl esters of such straight chain acids as acetic, propionic, n-butyric, n-valeric, n-caproic, n-heptoic, pelargonic, lauric, and stearic acids. Vinyl esters of branched chain alkane monocarboxylic acids are also useful components in the polymers. The preferred vinyl ester is vinylbutyrate. Representative examples, illustrating the process follow.

The thermal properties of the polymers including the crystalline melting point T_m (a first order transition temperature) and the glass transition temperature T_g (a second order transition temperature) were determined by standard methods utilizing differential thermal analysis. The mechanical behavior of the interpolymers was measured by standard tensile stress-strain techniques using a testing machine of the constant-rate-of-crosshead-movement type. Mechanical properties determined were the yield point (T_y), tensile strength at failure (T_B), elongation at failure (E_B), and modulus of elasticity. Accelerated weathering tests were conducted in an "Atlas Weather-Ometer," Model XW (Atlas Electrical Devices Co.

A general batch procedure for the preparation of the interpolymers used in a majority of the following examples consisted of the charging under dry nitrogen to a nitrogen-purged, 420 ml. stainless steel-lined pressure vessel the liquid polymerization medium, the iniator and the liquid monomers. After sealing the pressure vessel and pressure testing with nitrogen, it was cooled in a mixture of Dry Ice and acetone, evacuated and charged by weight difference with vinylidene fluoride, tetrafluoroethylene and any other gaseous monomer desired in the polymer. The tetrafluoroethylene was freed of inhibitor prior to charging by passage through a column of silica gel. The polymerization mixture at autogenous pressure was heated with agitation to and maintained at the reaction temperature until the desired degree of polymerization was achieved as indicated by the decrease in pressure.

The polymerization mixture was then cooled to room temperature, the reaction vessel vented, and the product discharged. Polymer was isolated by addition to a rapidly agitated (e.g., a Waring blender) three-fold volume of methanol or ethanol. Following separation of the polymer from the alcohol by filtration, it was washed thoroughly with additional alcohol and was dried to constant weight in a vacuum oven at 60° to 80°C.

Example 1: Vinylidene fluoride/tetrafluoroethylene/vinyl butyrate terpolymer — According to the above batch procedure, a terpolymer was prepared from 120 g. vinylidene fluoride, 30 g. tetrafluoroethylene, and 21 g. of vinyl butyrate using 0.3 ml. of a 75% solution of t-butyl peroxypivalate in mineral spirits as initiator, and a solution of 180 ml. of acetic acid and 10 ml. of t-butanol as the polymerization medium. The initial reaction temperature of 50°C. was raised by 2° increments at 2-hour intervals to 65°C. giving a total reaction time of 14 hours. From the polymerization mixture was isolated 114 g. of dried terpolymer. A 0.5% by weight solution of the polymer in dimethylformamide had an inherent viscosity of 0.89 at 30°C. Elemental analysis of the polymer product gave the following results: Carbon, 36.3%; hydrogen, 3.4%; fluorine, 54.1%.

The infrared spectrum of the polymer showed an absorbance at 5.7 microns of 1.28 which corresponds to 15% by weight of the polymer chain units being derived from vinyl butyrate. From the vinyl butyrate content and the elemental analysis, the terpolymer was calculated to contain 63% by weight of chain units derived from vinylidene fluoride and 22% by weight from tetrafluoroethylene.

A film of the terpolymer cast onto glass from a 10% by weight solution in a 60/40 by volume mixture of methyl ethyl ketone and cyclohexanone was allowed to dry at room temperature. The tack-free dry film was removed from the glass to give an unsupported, colorless, tough, somewhat rubbery film. Another film sample of this terpolymer cast from cyclohexanone exhibited a tensile strength at yield of 330 psi, a tensile failure of 550 psi, and elongation at failure of 412%.

Clear, colorless, attractive 1-mil coatings of the terpolymer on redwood were obtained by spray application of a solution of 6 parts by weight of the terpolymer in 50 parts by volume of a 1/1 by volume mixture of methyl ethyl ketone/ cyclohexanone. After over one year exposure conditions, such as that provided by accelerated weathering in the "Atlas Weather-Ometer" the coating was unchanged in appearance except for a whitening of the wood beneath the coating. The panel was removed from the "Weather-Ometer" after 3,260 hours because some adhesive failure was noted at the wood-coating interface although the polymer coating itself remained unchanged in appearance. Brush-applied coatings 3 mils in thickness from the same polymer solution did not show incipient adhesive failure until after 4,340 hours in the "Weather-Ometer." Similar results were obtained from coatings on redwood laid down from the terpolymer solution in cyclohexanone and in butyrolactone.

A paint was prepared from the terpolymer using the following formulation: terpolymer, 25 g.; cyclohexanone, 80 ml.; titanium dioxide pigment, 12.5 g.; lampblack, 0.5 g.; "Monastral" Green B pigment (Color Index No. 10006), 2.0 g. The terpolymer was dissolved in the solvent, the pigment mixture was placed in a pebble mill, and the terpolymer solution was added. The slurry was milled for 7 days. The resulting paint was separated from the pebbles and was used to coat panels of a construction material comprising a filled polyamide having the surface prime-coated with an epoxy resin. The paint was applied by brush application and was allowed to dry at room temperature. On exposure of the coating in the "Weather-Ometer," the onset of chalking was noted, as evidenced by the first trace of white on black velvet rubbed firmly across the panel, at 3,000 hours. No change in the coating was noted after 6 months exposure out of doors. When an air-dried coating of the paint, which had been brush-applied to a pigmented polyvinyl fluoride film laminated onto aluminum, was exposed in the "Weather-Ometer," the onset of chalking was noted at 2,500 hours exposure. In comparison under the same "Weather-Ometer" exposure conditions, such commercial coatings as alkyd, epoxy, vinyl, and urethane paints showed definite failure by chalking in less than 500 hours.

Example 2:   Vinylidene fluoride/tetrafluoroethylene/vinyl acetate terpolymer — The batch procedure was repeated as in Example 1 using 180 ml. acetic acid, 10 ml. t-butanol, 120 g. vinylidene fluoride, 30 g. tetrafluoroethylene, 14 g. of vinyl acetate, and 0.3 ml. of a 75% solution of t-butyl peroxypivalate in mineral spirits. The polymerization mixture was heated at 50° to 65°C. for 14 hours. The isolated dry terpolymer weighed 111 g. A 0.5% by weight solution of the polymer in dimethylformamide had an inherent viscosity of 0.51 at 30°C. Elemental analysis of the polymer product showed: carbon, 36.3%, 36.0%; hydrogen, 2.7%, 2.5%; fluorine, 56%, 56.2%.

The infrared spectrum of the polymer, as determined in tetrahydrofuran solution as described hereinbefore, showed an absorbance at 5.7 microns of 1.30 which corresponds to 11.3% by weight of the polymer chain units being derived from vinyl acetate. This, together with the elemental analysis, shows that the terpolymer contains 67.9% units from vinylidene fluoride and 20.8% units from tetrafluoroethylene. The polymer was soluble in methyl ethyl ketone and cyclohexanone. A film cast from the latter solvent was soft, slightly cloudy, colorless, tough. Tensile measurements indicated a yield point of 508 psi and an elongation of about 700%.

Using the same formulation and procedure described in Example 1, a paint was prepared and one coat was applied by brush to composite laminated sheeting having a surface of polymethylmethacrylate. The coating was allowed to dry for about 2 weeks at room temperature. An X-cut was made through the coating, cellophane adhesive tape was pressed firmly over the X-cut and pulled away. No loss in coating adhesion occurred. The coated panel was then soaked in water at room temperature for 7 days and the adhesion at the X-cut remained excellent. A brush-applied coating of the paint over steel primed with a zinc-rich primer was allowed to air dry at room temperature for one week and then was placed in a carbon arc "Weather-Ometer" from which the Corex D glass filters had been removed. Chalking was first noted after 600 hours, whereas under the same "Weather-Ometer" conditions commercial coatings of epoxy, vinyl, and urethane paints showed the same levels of chalking after only 150 hours and an alkyd paint was at the same level of chalking at 250 hours.

Polar Monomers — Use as Surfacing Agents

E.N. Squire; U.S. Patent 3,194,796; July 13, 1965; assigned to E.I. du Pont de Nemours and Company describes the surfacing of various substrates with a fluorocarbon elastomer. The preferred fluorocarbon elastomers are vinylidene fluoride/hexafluoropropene copolymers, containing from 45 to 85 weight percent vinylidene fluoride, preferably 55 to 75 weight percent, and terpolymers of vinylidene fluoride, perfluoropropylene, and one of the group consisting of methacrylic acid, glycidyl methacrylate, ethylene dimethacrylate, and acrylonitrile.

The elastomers were used to surface molded polymethacrylate resin, polyethylene film, polyethylene terephthalate film, molded polymethyl methacrylate sheet, polymethacrylate cast acrylic resin, glass fiber-reinforced phenol formaldehyde panels, glass fiber-reinforced polyethylene terephthalate panels, alpha-methyl styrene-methyl methacrylate copolymer sheet, bis phenols and phosgene condensation product, polycarbonate resin, polyvinyl fluoride, polyvinylidene fluoride and copolymers of acrylonitrile and methyl methacrylate. Each of the above substrates surfaced with the fluorocarbon elastomers provides a highly scratch- and mar-resistant durable sheet or panel.

VINYL FLUORIDE

GENERAL PROCESS TECHNIQUES

Propylene Addition and Recycle of Waste Water

J.L. Hecht; U.S. Patent 3,265,678; August 9, 1966; assigned to E.I. du Pont de Nemours and Company describes a continuous process for the polymerization of vinyl fluoride. A particularly attractive process for the polymerization of vinyl fluoride involves charging a suitable reaction vessel equipped with agitating means with water and water-soluble reaction initiator, e.g., 2,2'-diguanyl-2,2'-azopropane dihydrochloride along with a stream of vinyl fluoride, the amount of water employed ranging from 0.1 to 20 and preferably from 3 to 20 times the weight of vinyl fluoride. The reactor is maintained at a superatmospheric pressure of 25 to 1,000 and preferably 150 to 1,000 atmospheres, and a temperature of 25° to 250°C., and preferably 50° to 250°C., the combination of conditions and concentration of the charge being arranged to provide polymerization of the vinyl fluoride.

A slurry of particulate vinyl fluoride polymer in water is removed from the reactor at a rate sufficient to maintain conditions within the reactor conducive to polymerization. After isolating the resulting vinyl fluoride polymer from the slurry by filtration, centrifuging or the like, the polyvinyl fluoride in the form of a powder or cake can be washed with water or an organic solvent and dried.

The vinyl fluoride monomer can be prepared by the hydrofluorination of acetylene according to the process described in U.S. Patent 2,118,901. The monomer can also be prepared by the dehydrofluorination of 1,1-difluoroethane using any processes described in U.S. Patents 2,480,560; 2,599,631 and 2,674,632.

The improvement in the polymerization of vinyl fluoride which is the basis of this process comprises introducing into the reactor to the extent of from 10 to about 10,000 parts by weight of a mono-olefin of at least 3 carbon atoms per million parts by weight of vinyl fluoride. Any mono-olefin having at least 3 carbon atoms can be employed in the process. Those mono-olefins having at least 3 but not more than 7 carbon atoms are preferred. The most preferred mono-olefins are those having 3 but not more than 5 carbon atoms, and particularly a mono-olefin selected from the group consisting of propylene and the butylenes, i.e., 1-butene, 2-butene and isobutylene.

The mono-olefin can be continuously introduced into the polymerization reactor either as a component of the vinyl fluoride feed stream or as a separate feed stream. While the mono-olefin can be employed to the extent of from 10 to about 10,000 parts by weight per million parts by weight of the vinyl fluoride fed into the reactor, in the case of the lower mono-olefins such as propylene and the butylenes, a level of from 250 to about 2,500 parts by weight per million parts by weight of vinyl fluoride has been found extremely effective in delaying the appearance of low molecular weight polymer in the effluent slurry. The resulting vinyl fluoride polymeric products will be useful in the form of films, foils, sheets, ribbons, bands, rods, tubing and molded objects. They will also be useful as coatings for fabrics, leather, cellulosic materials such as paper, etc. The process is illustrated by the following examples.

Example 1: A stream of filtered, deionized, substantially deoxygenated water containing 111 parts by weight of 2,2'-diguanyl-2,2'-azopropane dihydrochloride, as reaction initiator, per million parts by weight of water was continuously fed at a rate of 480.5 lbs./hr. into a stainless steel reactor operated at 97°C. and 4,000 psi. The pressure in the reactor was maintained at 4,000 psi by automatically controlling the rate at which the product was withdrawn. Simultaneously, vinyl fluoride at a rate of 47.3 lbs./hr. was fed into the reactor. Propylene at a concentration of 1,000 parts by weight per million parts by weight of vinyl fluoride had been added to the vinyl fluoride stream.

The contents of the reactor were agitated continuously to provide intimate mixture of the feed streams throughout the interior of the reactor. The product slurry composed of finely particulate vinyl fluoride polymer and water was removed at a rate to provide 22.8 lbs. of polymer/hour/cu. ft. of reactor volume. The polymer was separated from the slurry by filtration and recovered as a substantially dry powder. The melt-flow number of a polymer is the square of the average diameter in inches of a roughly circular film disc resulting from the pressing between two polished chromium plated steel plates of a 1 inch diameter wafer consisting of 1.00 ± 0.01 g. of the polymer in dried, particulate, compressed form for 5 minutes at $260° \pm 1°C$. under a total load of 12,250 lbs. The melt-flow number for the vinyl fluoride polymer of Example 1 ranged from 7.34 to 7.88 throughout the entire production run. The continuous polymerization was conducted for 38.8 hours before shutting down without signs of objectionably high melt-flow number.

Example 2: The process of Example 1 was repeated except that the vinyl fluoride feed rate was 48.5 lbs./hr. and the water feed rate was 481.6 lbs./hr. The propylene concentration remained the same while the reaction initiator concentration was 114 parts per million parts of water. The polymer production rate was 24.4 pounds of polymer/hour/cu. ft. of reactor volume. The melt-flow number throughout the entire run of 32.8 hours ranged from 7.24 to 8.08. The run was shut down without signs of objectionably high melt-flow number.

Example 3: The process of Example 1 was repeated except that the vinyl fluoride feed rate was 49.1 lbs./hr. and the water feed rate was 478.3 lbs./hr. The polymer production rate was 17.8 pounds of polymer/hour/cu. ft. of reactor volume. The reaction initiator concentration was 101 parts per million parts of water while the propylene concentration was gradually increased from 250 parts per million parts of vinyl fluoride at the start of the run until it reached 4,000 ppm parts of vinyl fluoride shortly before the run was terminated after 81 hours of continuous operation. The melt-flow number throughout the run ranged from 7.30 to 7.89. There was no sign of objectionably high melt-flow number at the termination of the run.

In control A, the process of Example 1 was repeated but without the addition of a mono-olefin. The vinyl fluoride feed rate was 46.8 lbs./hr. and the water feed rate was 479.2 lbs./hr. The polymer production rate was 22.2 pounds of polymer/hour/cu. ft. of reactor volume. The reaction initiator concentration in the feed water was 87.4 ppm. Polymerization was continued for 10.9 hours before the run had to be terminated because of the sudden appearance of polymer of excessively high melt-flow number, i.e., 8.07 to 11.53. The melt-flow number range of normal polymer, i.e., polymer produced during the "on control" portion of this run was 7.41 to 7.87.

In control B, the process of Example 1 was repeated but without the addition of a mono-olefin. The vinyl fluoride feed rate was 49.7 lbs./hr. and the water feed rate was 475.5 lbs/hr. The polymer production rate was 24.6 pounds of polymer/hr/cu. ft. of reactor volume. The reaction initiator concentration was 91 parts per million parts of water. Polymerization was continued for 24 hours before the run had to be terminated because of the sudden appearance of polymer of excessively high melt-flow number, i.e., 8.13 to 8.74. The melt-flow number range of normal polymer, i.e., polymer produced during the "on control" portion of this run was 7.11 to 7.84.

Using the same basic polymerization process noted above, V.E. James; U.S. Patent 3,129,207; April 14, 1964; assigned to E.I. du Pont de Nemours and Company describes the reuse of at least a portion of the waste water. Despite the fact that this water, contains substantially no catalyst, its use permits a reduction of from 10 to 25% or more in the amount of new catalyst necessary to obtain the same polyvinyl fluoride product. Additional savings are obtained if, in addition to recycling the waste water, the water is treated to increase its specific resistance to at least 50,000 ohm-centimeters, preferably to at least 200,000 ohm-centimeters, and to reduce its oxygen content to less than 50 parts per million. Although recycling of the waste water in any amount will reduce the amount of additional catalyst that is necessary under any particular set of conditions, the necessity to obtain polymerization of using some catalyst in the recycle water, either by addition of solid catalyst or catalyst dissolved in water, cannot be eliminated. The following examples illustrate the process.

Example 1: A stream of filtered, deionized, deoxygenated process water, water that had not previously been used in the polymerization, along with an equal amount of recycle water that was also filtered, deionized and deoxygenated were continuously fed into a stainless steel reactor, each at a rate of 240 lbs./hr. to provide a total feed water rate of 480 lbs./hr. Specifically, the water were passed through a full-flow in-line filter and a deionizer column, and then was deoxygenated by sparging with nitrogen before being fed to the reactor. The dihydrochloride of 2,2'-diguanyl-2,2'-azopropane, had been added to the process water to provide a concentration of 103 parts of the catalyst per million parts of the total feed water. The specific resistance of the water was 1,000,000 ohm-cm. Simultaneously, vinyl fluoride at a rate of 48 pounds/hour was fed to the reactor. The temperature of the reactor was maintained at 97°C. and the pressure in the reactor was maintained at 4,000 psi by automatically controlling the rate at which the product was withdrawn.

The contents of the reactor were agitated continuously to provide intimate mixture of the feed streams throughout the interior of the reactor. The product slurry composed of finely particulate polyvinyl fluoride and water was removed at a rate to provide 23 pounds of polymer/hr./cu. ft. of reactor volume. The polymer was separated from the water by filtration and recovered as a substantially dry powder. The water recovered from the filtration step was the recycle water, which was joined with additional process water to provide the feed water. The melt-flow number for the polyvinyl fluoride of Example 1 was 7.3.

Example 2: The process described in Example 1 was repeated except that the recycle water rate was reduced to 144 lbs./hr. and the process water rate was increased to 336 lbs./hr. so that the total feed water rate remained at 480 lbs./hr. To obtain the identical polyvinyl fluoride product (melt-flow number of 7.3) at the same production rate of 23 lbs./hr./ cu. ft. of reactor volume without changing any other conditions, it was necessary to use a catalyst concentration of 112 parts per million parts of total feed water.

In control A, the process of Example 1 was repeated but without using any recycle water. Instead, 480 lbs./hr. of process water was used as the feed water. To obtain the identical polyvinyl fluoride product (melt-flow number 7.3) at the same production rate of 23 lbs./hr./cu. ft. of reactor volume without changing any other conditions, it was necessary to use a catalyst concentration of 129 parts per million of feed water, an increase of 25.3% over the catalyst used in Example 1 and 15.2% over the catalyst used in Example 2.

In control B, the process of Example was repeated but without using any process water. Instead, 480 lbs./hr. of recycle water was used as the feed water. No catalyst was added to the recycle water and no polymer was produced.

Two-Stage Process with Particle Nucleation

J.L. Hecht and C.T. Hughes; U.S. Patent 3,464,963; September 2, 1969; assigned to E.I. du Pont de Nemours and Co. describe a two-stage rate controlling process which comprises nucleating polyvinyl fluoride particles in a first reaction zone for a sufficient time until the particle nucleation becomes nonrate limiting and then polymerizing the bulk of the vinyl fluoride in a second reaction zone.

Figure 3.1 shows a flow diagram for the two-stage vinyl fluoride polymerization process. In the process, filtered, de-ionized, substantially deoxygenated water containing a water-soluble reaction initiator is continuously fed through pump 10, maintained at a pressure of 300 to 500 psig higher than reactor 28, and line 11 to a point 12 where from about 25 to 50% of the stream is diverted through line 13, fed through heat exchanger 14 preferably maintained at a temperature substantially that of reactor 17 and then through line 15 to mixer 16 where it is joined with water and vinyl fluoride monomer prior to being introduced into nucleation reactor 17 maintained at a pressure of 1,500 to 9,000 psig and a temperature within the range of 60° to 90°C. The remainder of the initiator stream passes through line 18 to mixer 19 where it joins the vinyl fluoride monomer and water not sent to the nucleation reactor and the nucleated polyvinyl fluoride stream 20 from nucleation reactor 17.

Twenty-five to fifty percent of the vinyl fluoride monomer in storage vessel 21 is continuously fed under a pressure of 300 to 500 psig higher than reactor 28 supplied by pump 22 through line 23 to heat exchanger 24 maintained at a temperature substantially that of reactor 17; thence through line 25 to mixer 16 where initiator stream 15 and medium stream 25 are mixed prior to entrance to reactor 17. The remainder of the vinyl fluoride monomer is fed by pump 26 through line 27 directly to mixer 19 prior to entrance into bulk polymerizer 28 maintained at a pressure of 4,000 to 14,000 psig and a temperature of 90° to 140°C.

FIGURE 3.1: VINYL FLUORIDE POLYMERIZATION PROCESS

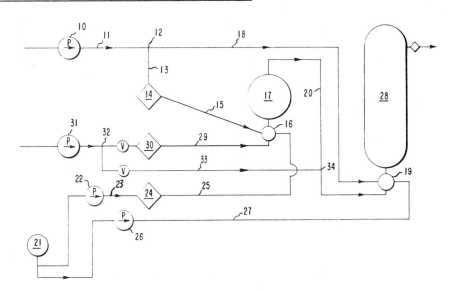

Source: J.L. Hecht and C.T. Hughes; U.S. Patent 3,464,963; September 2, 1969

Water is supplied to mixer 16 from line 29 after passing through heat exchanger 30 maintained at a temperature substantially that of prereactor 17. Pump 31 feeds the water at a pressure that of reactor 28 to point 32 where 25 to 50% of the water proceeds to nucleating reactor 17 through line 29 while the remaining water is fed through line 33 to a point 34 in the nucleated polyvinyl fluoride stream 20 from reactor 17 where it is combined with the nucleated polyvinyl fluoride stream. The combined stream joins monomer stream 27 and initiator stream 18 at mixer 19 and thus to polymerizer 28 where the bulk of the polymerization is completed.

It has been found that during the polymerization the polymer particles, once formed, act as the locus of the reaction. At the start of the reaction, there are no polymer particles. The vinyl fluoride dissolved in water reacts with the initiator to give a free radical which polymerizes in the aqueous phase until at some point a particle is formed. This particle then acts as the site for further reaction by virtue of the fact that the reaction proceeds much more rapidly on the particle. Initially, the rate of reaction is slow because there are no polymer particles. As particles are formed, i.e., nucleated, the more rapid reaction takes place on particles themselves and the rate increases until it levels off at the reaction rate for the conditions involved. At this point, there is sufficient particle surface and the reaction rate is limited by the chemical reaction itself.

In a continuous polymerization, polymer particles are being continuously removed; thus, at steady state conditions, new particles must be continuously nucleated. The conditions which favor nucleation are quite different from those which favor the polymerization. It has been found that the best method for continuously polymerizing vinyl fluoride is to nucleate in one reactor and to carry out the bulk polymerization in the second reactor.

To make the particle nucleation nonrate limiting for the subsequent polymerization, the conditions for the aqueous phase system with a water-soluble initiator and vinyl fluoride are as follows: broadly, the particle nucleation can take place within 15 seconds to five minutes, at 60° to 90°C. and 1,500 to 9,000 psi. Raising the temperature of the system decreases the rate of particle nucleation; therefore, lower temperatures produce more nucleation in a shorter time. The particle nucleation step should produce as much nucleation as practical, i.e., number of particles, to make nucleation nonrate limiting. On a practical basis, it is only necessary to produce less than about 1% of the total polymer ultimate basis. There is little polymerization at particle nucleation conditions. The particle nucleation step decreases the relative proportion of polymer formed by the aqueous phase polymerization thereby resulting in improved inherent weatherability of the final polyvinyl fluoride film.

The optimum bulk polymerization conditions should be within 90° to 140°C. at 4,000 to 14,000 psi for sufficient time to produce the optimum yield or about 0.5 to 15 minutes. Polymerization rates increase with increasing temperature. The rate controlling role of a nucleation reactor is illustrated in the table below. Four identical runs are made, except for the time spent at nucleating conditions, with a vinyl fluoride aqueous phase system containing 0.01 % 2,2'-diguanyl-2,2'-azopropane initiator.

Run No.	Nucleation time at 77° C. and 8,500 p.s.i. (min.)	Polymerization time at 97° C. and 8,000 p.s.i. (min.)	Polymer Wt. (gms.)
1	0	15	19
2	2	15	34
3	5	15	72
4	15	0	3

These runs show that when particle nuclei are formed at a low temperature prior to carrying out the bulk polymerization at the higher temperature, the reaction rate increases two to three fold. Thus, the continuous polymerization of vinyl fluoride should utilize two reactions for a higher molecular weight product: first, for particle nucleation, second, for bulk polymerization. The process is illustrated by the following examples.

Examples 1 and 2: In these examples, the equipment shown in Figure 3.1 is used. The process conditions are shown in the table below. In both examples and in both controls the initiator is 1,560 ppm 2,2'-diguanyl-2,2'-azopropane dihydrochloride in water of the rate of 30 to 34 lbs./hr.

	Control 1	Control 2	Example 1	Example 2
Aqueous phase rate (lb./hr.)	605	605	740	725
Monomer (VF) rate (lb./hr.)	65	65	85	100
Pre-reactor temp. (° C.)			79	80
Residence time in pre-reactor (min.)			2.0	1.6
Pre-reactor pressure (p.s.i.)			8,500	8,500
Percent solids exit pre-reactor			0.1	0.05
Polymer produced in pre-reactor (lb./hr.)			0.5	0.3
Reactor temp. (° C.)	97	97	97	97
Reactor pressure (p.s.i.)	8,000	8,000	8,000	8,000
Polymer produced (lb./hr.)	43	44	58	62
Inherent viscosity (η inh.)	1.14		1.47	

Films are prepared from the polymer obtained from Controls 1 and 2 and Examples 1 and 2 in accordance with the procedures of U.S. Patents 2,953,818 and 3,139,470. These films are tested for accelerated weatherability expressed as FS-O3 life. The accelerated weatherability of a particular polyvinyl fluoride is expressed as the number of hours that molecularly unoriented film prepared therefrom must be exposed to intense ultraviolet irradiation in a gaseous oxidative atmosphere at a temperature elevated above normal room temperatures before its elongation-at-break, expressed as a percent of the original length of the test specimen, is reduced to a value of less than 10%. The accelerated exposures are carried out in a test device containing 20 Westinghouse FS-20-T-12 fluorescent sun lamps sealed except for inlet and outlet bleeds for fresh air and ozone at 250 parts per million at a rate of 200 cu. ft./min.

Polymer	Weatherability (hrs.)
Control 2	440
Example 2	540

It is seen that an improvement of about 20% in accelerated weatherability is obtained.

Polymer	Weatherability (hrs.)
Control 1	460
Example 1	545

An improvement of about 20% in accelerated weatherability over Control 1 is observed; therefore higher inherent weatherability of polyvinyl fluoride is attained through the use of two-stage polymerization. Also observed is about a 40% increase in the space-time yield factor.

Activated Metallo-Organic Lead and Tin Catalysts

D. Sianesi and G. Caporiccio; U.S. Patent 3,513,116; May 19, 1970; assigned to Montecatini Edison SpA, Italy have found that certain specific agents, present in catalytic amount, are capable of promoting the catalytic activity of organic lead and tin compounds in initiating the polymerization and copolymerization of vinyl fluoride in the absence of the photochemical activation and at a temperature lower than 150°C.

Illustrative examples of these metallo-organic compounds can be: tetraethyl lead, tetraethyl tin, tetramethyl lead, diethyldipropyl lead, ethyltriphenyl lead, diethylmethylpropyl lead, chloromethyltriethyl lead, 2-hydroxyethyl-triphenyl tin and 3-diethylaminopropyl-triethyl lead.

The inorganic activating agents having oxidizing properties which can promote the catalytic activity of metallo-organic compounds of lead and tin in the polymerization and copolymerization of vinyl fluoride, can be of either neutral, basic or acidic nature. Illustrative examples of these compounds are: salts of alkali metals or alkali earth metals, ammonium salts, and in general salts of mono- or polyvalent cation with chromic, permanganic, periodic, chloric, perchloric, chlorous, hypochlorous, persulfuric acid; or oxidizing compounds of acid nature, e.g., nitric, perchloric acid, etc., or derivatives of metals, having multiple valence, present in a state of higher valence, such as, e.g., salts of tetravalent cerium, ferric salts, etc.

Polyvinyl fluoride, obtained according to the process, is a highly crystalline polymer. The temperature of complete crystalline melting is normally higher than 200°C., and usually between 200° and 235°C. It has a high molecular weight and its intrinsic viscosity normally appears to be higher than 0.3 and lower than 8 (100 cc/g.). These properties can suitably be regulated within desired limits for particular applications by regulating the normal operative factors of the polymerization, including temperature, monomer concentration, concentration of the catalytic agents.

Example 1: 150 cc of a 0.04 molar solution of $Pb(C_2H_5)_4$ in pure and oxygen-free tert-butyl alcohol are introduced by siphoning into a stainless steel autoclave, having capacity of 1,000 cc provided with mechanical agitator and thermostatic jacket. 250 g. of vinyl fluoride are then introduced into the autoclave. The mixture thus obtained is maintained at 40°C. while agitating and then, through a liquid feeding pump, 60 cc of a deaerated aqueous solution of 3% ammonium persulfate and 1% borax are introduced.

The reaction is controlled by keeping the temperature at 40°C. and the whole is reacted for 20 hours while continuously agitating. At the end of this time, the unreacted monomer is removed. The polymer formed is collected, washed with warm aqueous methanol acidified with HNO_3, washed with boiling pure methanol, dried to constant weight at 100°C. under a vacuum of 25 mm. Hg, and finally weighed. 162 g. of polyvinyl fluoride are obtained.

A sample of this polymer at the elementary analysis shows a C, H and F content perfectly corresponding to that calculated for C_2H_3F. The temperature of complete crystalline melting, determined on the polarizing microscope provided with a heating plate, appears to be between 205° and 215°C. Samples of the polymer, by heating to 160°C. wtih a fixed amount of dimethylsulfoxide, give homogeneous solutions. Upon cooling, polymer gels are obtained. They can be extruded at

200°C. and after removal of the residual solvent, flexible and resistant films and fibers which are stretchable at low temperature can be obtained. Molded laminae of the polymer after being annealed at 100°C. for 4 hours, result in a density of 1.39 g./cc at 25°C. as determined by flotation.

Examples 2 through 5: These examples are directed to the copolymerization of vinyl fluoride with other fluorinated olefins. Into a series of 50 cc stainless steel autoclaves cooled to –78°C. and kept in a nitrogen atmosphere, there are successively introduced at intervals of 5 minutes: 2 cc of 2% aqueous ammonium persulfate solution, 5 cc of tert-butyl alcohol, 0.2×10^{-3} g. mols of $Pb(C_2H_5)_4$. Then in the different autoclaves there are introduced, by vacuum distillation at –78°C., 0.2 g. mol of the gaseous mixtures indicated in the table below and obtained by mixing, so as to obtain a mixture containing 90% by mols of vinyl fluoride. The polymerization is carried out at 40°C. for 6 hours.

Ex. No.	Comonomer	Polymer obtained, residue from boiling CH_3OH)	Intrinsic viscosity 100 cc./g. DMF 110° C.	Temperature of complete crystalline melting, ° C.	Vinyl fluoride in the polymer
2	$H_2C=CF_2$	1.1	1.5	225–231	98.3
3	$F_2C=CF_2$	5.2	1.85	212–240	86.4
4	$F_3C—FC=CF_2$	0.8	0.93	160	90.6
5	$F_3C—CF=CFH$	0.95	0.98	150–170	90.7

Cyclic Azo Amidine Initiators

In a polymerization process described by J.E. Cook and O.L. Marrs; U.S. Patent 3,428,618; February 18, 1969; assigned to Phillips Petroleum Company a fluoroolefin monomer is contacted with a cyclic amidine free radical initiator. Specific compounds which can be employed as polymerization initiators in the process are:

> azobis-N,N'-methyleneisobutyramidine monoacetate
> azobis-N,N'-dimethyleneisobutyramidine
> azobis-N,N'-dimethyleneisobutyramidine dihydrochloride
> 2,2'-azobis-[N-p-tolyl-N,N'-tetramethylene-2-ethyl-2-
> (4-methylcyclohexyl)]acetamidine
> 2,2'-azobis-N-(2-phenethyl)-N,N'-(2-phenyltrimethylene)-2,2-
> diphenylacetamidine and
> 2,2'-azo-(N-ethyl-N-N'-o-phenylene-2,2-di-n-propylacetamidine)-
> (N-[2-ethylphenyl]-N,N'-o-naphthylene-2,2-dicyclohexylacetamidine)
> dihydrochloride.

The cyclic amidine initiators can be prepared, for example, by treatment of noncyclic amidines with diamines. The noncyclic amidines are in turn prepared by treatment of azobisnitrile compounds with alcohol and acid followed by treatment with ammonia. The following examples illustrate the process.

Example 1: α,α'-azodiisobutyronitrile was reacted with ethanol and hydrochloric acid to form an intermediate which was treated with dry ammonia to form azobisisobutyramidine dihydrochloride. Five grams of azobisisobutyramidine dihydrochloride was then reacted with 50 ml. of ethylenediamine. The azobisisobutyramidine dihydrochloride dissolved in the ethylenediamine, and after a few minutes stirring at 25°C. a white precipitate of the free base of azobis-N,N'-dimethyleneisobutyramidine was formed. The azobis-N,N'-dimethyleneisobutyramidine precipitate was filtered off and dried. An elemental analysis of the prepared azobis-N-N'-dimethyleneisobutyramidine showed that the material contained 51.4 weight percent carbon and 10.3 weight percent hydrogen. The infrared spectrum of this material was consistent with that for a cyclic amidine.

Approximately one-third of the dried precipitate was separated and removed to storage and the remaining two-thirds was dissolved in about 50 to 75 ml. of warm chloroform after which dry hydrogen chloride gas was bubbled through the solution for about one-half hour. The treatment with the dry HCl gas resulted in the formation of a precipitate identified as the dihydrochloride of azobis-N,N'-dimethyleneisobutyramidine. This precipitate was then filtered from the solution, washed with chloroform, and dried under vacuum.

Example 2: A series of polymerization runs was conducted in which vinyl fluoride was contacted with the free base and acid salt initiators prepared by Example 1. The particular initiator and the amount used in each of the several runs is reported in the following table. In each of these polymerization runs, the initiator was charged to a 1.1 liter reactor along with 700 ml. of water. As indicated in the table, the amount of initiator was varied for each of the runs to illustrate the operability of the different initiators. Vinyl fluoride was then pressured into the reactor at a temperature of about 30°C. until a pressure of 1,100 psi was obtained. The reactor was then heated to a temperature of about 97°C. and agitated by means of a rocker for the desired length of time. At the termination of the polymerization run, the reactor was opened after it had cooled.

In the runs using a cyclic amidine free base, 2,2'-azobis(N,N'-dimethylene)isobutyramidine, as the initiator, it was observed that the polymer had stratified into a layer on the surface of the water diluent. In the runs using the acyclic amidine, 2,2'-azobis(isobutyramidine)dihydrochloride, as the initiator, it was observed that the polymer was dispersed throughout the diluent. A slurry including water and polyvinyl fluoride was then removed from the reactor and the polymer separated by filtration. The polyvinyl fluoride was difficult to separate from the diluent when the prior art initiator was employed. When 2,2'-azobis(N,N'-dimethylene)isobutyramidine was used, the polymer was easily separated from the diluent. The polymer recovered was then washed with water and acetone and dried.

The melt-flow number, which is an indication of molecular weight, was determined for the polymers produced. The melt-flow number can be defined as the square of the average diameter in inches of a film formed from a 1 inch diameter disc of 1 g. of polymer by compressing the 1 inch disc between two platens for a period of 5 minutes at a temperature of 500°F. using a load pressure of 13,000 lbs. After cooling the platens the film was removed and its diameter measured. The square of the diameter of each film in inches is reported in the following table as the melt-flow number. It is apparent that the melt-flow numbers for each of the polymers produced by using the free base cyclic amidine initiators are much lower than the melt-flow numbers of the polymers produced with the initiators of the prior art.

A mixture of polymer and γ-butyrolactone for each of the polymers produced was compression molded in a 8"×8" press using 20,000 pounds ram pressure with a 4" ram at a temperature of 400°F. for two minutes to produce a polymer film. The percent elongation and tensile strength for each of the films thus produced was determined by pulling the polymer film on a commercial Instron machine using a 1/2 inch wide specimen approximately 4 mils thick, a jaw to jaw separation of 1 inch, and a crosshead speed of 1"/min. The tensile strength and percent elongation for each of the polymers produced by using the cyclic amidine initiators have a percent elongation which is much higher than the percent elongation of the polymers produced in accordance with the initiators of the prior art. This indicates that a much less brittle polymer was produced thus greatly expanding the utility of the product.

Initiator	Pressure, p.s.i.g.	Temperature (°C.)	Time (hrs.)	Amount of Initiator (grams)	Polymer Yield (grams)	Melt-Flow Number	Tensile Strength[1] (p.s.i.)	Elongation (percent)
Run No.:								
1 ... 2,2'-azobis(isobutyramidine) dihydrochloride.	5,000	97	4	0.2	80	11.2	6,360	17
2 ... do.	5,000	97	3	0.1	83	11.2	5,717	49
3 ... do.	5,000	97	4	0.05	27.6	8.7	6,420	80
4 ... do.	5,000	97	7	0.05	55.5	9.8	6,050	92
5 ... do.	5,000	97	7	0.025	43	9.9	---	---
6 ... do.	6,500	97	3	0.0062	32.5	7.6	5,927	202
7 ... 2,2'-azobis(N,N'-dimethylene-isobutyramidine)·2HCl.	5,000	97	4.5	0.2	58.3	10.7	5,893	18
8 ... do.	6,500	97	6	0.1	29	12.5	6,023	8
9 ... do.	6,500	97	6	0.05	26	10.9	5,855	27
10 ... do.	6,500	97	16	0.05	30	10.2	5,305	21
11 ... do.	6,500	97	6	0.025	22	8.7	5,304	21
12 ... do.	6,500	97	6	0.025	17	8.7	5,214	18
13 ... 2,2'-azobis(N,N'-dimethylene-isobutyramidine).	5,000	97	4.5	0.2	71.2	7.2	5,582	105
14 ... do.	5,000	97	4.5	0.2	44.9	7.2	5,920	109
15 ... do.	5,000	97	4.5	0.1	40	7.2	5,445	210
16 ... do.	6,500	97	6	0.05	45	6.4	5,437	251
17 ... do.	6,500	97	6	0.05	51	6.2	5,470	209
18 ... do.	6,500	97	6	0.05	35	7.2	6,000	116
19 ... do.	6,500	97	6	0.025	59	3.7	5,660	381

[1] Yield.

1,1,2-Trichlorotrifluoroethane as Accelerator

J.J. Dietrich; U.S. Patent 3,437,648; April 8, 1969; assigned to Diamond Shamrock Corporation describes a process for the polymerization of vinyl fluoride, using 1,1,2-trichlorotrifluoroethane as an accelerator.

Example: Into a 2 l. high pressure-stirred autoclave are charged lauroyl peroxide, an aqueous solution containing 4.95%, by weight, Methocel 65-HG-50 (a methoxylated propoxylated cellulose) and select polymerization aids. The autoclave is then brought under a vacuum at which time water and vinyl fluoride monomer are introduced into the autoclave. The reaction mixture is heated to 65°C. and the pressure is brought up to 5,000 psig and maintained at that pressure by the continuous addition of water. The reaction mixture is continuously agitated with a mechanical stirrer. The concentration of the reactants, expressed as parts by weight, and processing conditions are given in the following table.

Run No.	1	2	3	4	5	6	7
Vinyl fluoride monomer	90.80	90.80	90.80	90.80	90.80	90.80	90.80
Methocel solution[1]	5.80	5.80	5.80	5.80	5.80	5.80	5.80
Lauroyl peroxide	0.38	0.38	0.38	0.38	0.38	0.38	0.38
Isopropanol	4.05	4.05					
Cyclohexane			5.60				
Toluene				6.10			
Benzene					5.20		
Carbon tetrachloride						10.30	
1,1,2-trichlorotrifluoroethane		10.00					12.50
Temperature (°C.)	65	65	65	65	65	65	65
Pressure (p.s.i.g.)	5,000	5,000	5,000	5,000	5,000	5,000	5,000
Inherent viscosity (Nsp/C)	1.24	1.08	1.26	0.82	4.12	0.20	4.32

[1] Methocel solution, a 4.95% by weight, aqueous solution of a methoxylated propoxylated cellulose (Methocel 65-HG-50).

For purposes of comparison, the percent conversion after 3 hours of the vinyl fluoride monomer to the polymer employing the various polymerization systems are set forth below.

Polymerization aid	Percent conversion after 3 hrs.	Inherent viscosity (Nsp/C)
Isopropanol	58.0	1.24
Cyclohexane	75.0	1.26
Toluene	4.5	0.82
Benzene	28.0	4.12
Carbontetrachloride	30.3	0.20
1,1,2-trichlorotrifluorethane	90.0	4.32
None		5.20

As can be seen from a comparison of this data, the polyvinyl fluoride prepared by the polymerization system containing 1,1,2-trichlorotrifluoroethane is produced at a faster rate of conversion than the other systems without a decrease in its molecular weight.

Directly Coalescible Organosols

L.E. Wolinski; U.S. Patent 3,492,259; January 27, 1970; assigned to E.I. du Pont de Nemours and Company describes the polymerization of vinyl fluoride in mixtures of dimethyl sulfoxide/water to form directly coalescible organosol. Polyvinyl fluoride is a known thermoplastic polymeric material described in, for example, U.S. Patent 2,419,010. It also is known to prepare shaped structures of polyvinyl fluoride such as, for example, self-supporting films and coatings by solution casting techniques wherein the polyvinyl fluoride is first dissolved in a heated solvent such as, for example, cyclohexanone, dimethylformamide and tetramethylene sulfone, in a manner also described in U.S. Patent 2,419,010 or by the use of dispersions of polyvinyl fluoride as described in U.S. Patent 3,096,299.

A process for preparing directly coalescible organosols of polyvinyl fluoride has been found which comprises polymerizing vinyl fluoride in an aqueous medium comprising dimethyl sulfoxide and up to about 85% by weight, based upon the total weight of the aqueous medium, of water, at a temperature of between about 55° and 140°C. and a pressure of between about 1,000 and 15,000 psi thereby to obtain a directly coalescible organosol of polyvinyl fluoride. The polymerized vinyl fluoride is preferably coalesced and extruded in an aqueous medium having a weight ratio of dimethyl sulfoxide-to-water of between about 88/12 and 97/3. In a preferred example, the directly coalescible organosol of polyvinyl fluoride is heated to coalesce the polyvinyl fluoride, followed by extruding the coalesced polyvinyl fluoride into a shaped structure, preferably a self-supporting film structure.

The process is illustrated by the following examples. The inherent viscosity of the polymer systems in the examples were determined at 30°C. using the 0.5% solutions in hexamethylphosphoramide. The weathering test, designated FS/O$_3$, employed in the examples was performed by exposing test film samples to an intense ultraviolet light at 60°C. in a test device containing a rotating bank of twenty Westinghouse FS20T12 Fluorescent Sun Lamps and sealed except for inlet and outlet bleeds controlling passage therethrough of 200 cubic feet per minute of fresh air containing 250 ppm of ozone.

Example 1: A 1 l. reactor vessel equipped with an agitator was charged with 135 cc of deionized water that had been boiled under nitrogen to remove the oxygen therein. 367 cc of dimethyl sulfoxide (distilled from 5 A. molecular sieves under nitrogen) was added to the reactor to provide a 75/25 mixture of dimethyl sulfoxide and water. The reactor was purged twice with nitrogen under a pressure of 800 psig. The initiator line was purged twice with water and once with initiator. The agitator was set into operation at 500 rpm and the pressure in the reactor vessel was pressurized with vinyl fluoride at 4,000 to 4,250 psig and the temperature maintained at about 80°C. A solution of 0.1% by weight, based upon the total solution weight, of a water-soluble polymerization initiator of 2,2'-diguanyl-2,2'-azopropane dihydrochloride was injected into the reactor vessel. Thereafter, addition of vinyl fluoride monomer to the reactor was commenced in a continual manner at 2 minute intervals and continued until 1,200 g. of vinyl fluoride monomer had been added.

After a total reaction time of 21 minutes, the reaction vessel was found to be filled with a thick dispersion of polyvinyl fluoride weighing 590 g. and of 18.6% solids and the solids having an inherent viscosity of 1.72. The dispersion was placed into a vacuum flask equipped with an agitator and a distillation column. The vacuum flask was maintained at a temperature of 50° to 80°C. and under a vacuum of 2 to 5 mm. Hg. Water and dimethyl sulfoxide were removed to yield a dispersion of 22% solids having 71% dimethyl sulfoxide and 7% water corresponding to a weight ratio of dimethyl sulfoxide and 7% water corresponding to a weight ratio of dimethyl sulfoxide-to-water of 91/9. The dispersion was coalesced and then extruded at 150°C. onto a quench drum maintained at 15°C. The extruded film was reheated to 90°C. for 10 seconds and then stretched at that temperature 2.5 times by 2.5 times in both of its planar directions, followed by heating at 165°C. for 1 minute under restraint to remove dimethyl sulfoxide and water and a 2 mil thick film was obtained. The resulting 2 mil stretched film exhibited the following properties: Elongation 275%, Tenacity 20,000 psi and Modulus 300,000 psi.

The film had outstanding clarity with no surface or internal haze and its color was water white. An accelerated weathering test (FS/O3) decreased the elongation-at-break at a rate predicted to yield a 9 1/2 year life to 10% elongation-at-break.

Examples 2 through 11:   The procedure described in Example 1 was repeated except that the organosols were concentrated and cast on ferrotype plates, doctored to the desired thickness, coalesced by infrared heating to self-supporting solvent-containing film, and quenched by immersing the plates in cold water. After stripping from the plate, each film was clamped in a rectangular frame and dried free of volatiles by exposure for five minutes at 150°C. in a forced air draft oven. The results and pertinent data regarding the process conditions and the film structures obtained are summarized in Table 1.

TABLE 1:   POLYMERIZATION OF VINYL FLUORIDE IN AQUEOUS DIMETHYL SULFOXIDE SYSTEM

	Polymerization conditions								Film properties		
	Solvent system		Temp., °C.	Pressure, p.s.i.×10⁻²	Time, min.	η inh.	Percent solubility	Percent solids	Ten., p.s.i. ×10⁻³	Elong., percent I/W	Mod., p.s.i. ×10⁻⁵
Example	Type	Wt. ratio									
2........	DMS......	100	80	45	31	1.40	100	12.9	4.39	17/1	2.28
3..........	DMS/H₂O..	85/15	80	45-47	32	1.36	100	15.7	4.45	18/3	2.38
4..........	DMS/H₂O..	70/30	80	40-42	21	1.92	100	19.0	3.37	16/4	2.10
5..........	DMS/H₂O..	65/35	80	40-42	28	1.90	100	19.0	4.47	60/20	2.29
6..........	DMS/H₂O..	60/40	80	40-42	18	2.05	100	18.8	4.09	50/23	2.32
7..........	DMS/H₂O..	55/45	80	40-42	10	2.99	100	15.7	4.75	49/15	2.48
8..........	DMS/H₂O..	52.5/47.5	80	45-47	21	3.20	100	21.8	4.21	¹ 140/49/10	2.41
9..........	DMS/H₂O..	50/50	80	40-42	12	3.00	100	16.2	4.49	144/43	1.96
10..........	DMS/H₂O..	45/55	80	40-42	11	3.52	100	13.6	4.36	156/52	2.21
11..........	DMS/H₂O..	45/55	98	40-42	22	1.54	100	9.8	4.69	20/6	2.32

¹ Weathered 630 hours in accelerated FS/O₃ test.
OMS = Dimethyl Sulfoxide; I = Initial Elongation; W = Weathered, after 315 hrs. Accelerated FS/O₃ test.

COPOLYMERS

Methacrylate Copolymers

H.F. Reinhardt; U.S. Patent 3,407,247; October 22, 1968; assigned to E.I. du Pont de Nemours and Company describes a two-stage process for preparing fluoroolefin copolymers which consists essentially of:

A. A first stage of polymerizing at least one of the monomers selected from the group of methyl methacrylate, t-butly methacrylate, methacrylonitrile, methacrylamide, and phenyl methacrylate in an aqueous emulsion in the presence of a polymerization initiator which produces active reaction sites on the resulting prepolymer, and

B. A second stage of reacting the resulting prepolymer with a fluoroolefin compound by adding the fluoroolefin compound to the reaction mixture of the first stage and maintaining a mixture of these in an aqueous emulsion at a temperature of about 0° to 125°C. to produce a fluoroolefin copolymer.

This process yields a finely dispersed form of fluoroolefin copolymers which have the attractive properties of excellent formability, thermal stability and film formation. Such polymers are useful in making durable and formable paints, lacquers and similar finishes.

In a preferred form of the process, a prepolymer, polymethyl methacrylate, is prepared using conventional emulsion polymerization conditions and reacted in pressure equipment with vinyl fluoride or vinylidene fluoride to form a copolymer of methyl methacrylate and vinyl fluoride or vinylidene fluoride. These copolymers are more economical to produce than vinyl fluoride or vinylidene fluoride homopolymers and yet possess many of the properties of these homopolymers such as good formability, thermal stability and film formation.

The formability of films cast from the various vinyl fluoride and vinylidene fluoride copolymers decreases with increasing amounts of methyl methacrylate. This is illustrated in the following table which gives the results of a bump test for vinyl fluoride (VF) and vinylidene fluoride (VF_2)/methyl methacrylate (MMA) copolymers produced by reacting varying amounts by weight of the monomers.

The bump test is a laboratory method for determining the relative formabilities of polymers. A coating of the polymer is applied to a depth of approximately 1 mil on a 25 mil aluminum sheet to form a sample. A hemispherical plunger is placed with its curved surface upon the sample either on the coated side or the uncoated side as desired. A guided metal block weighing 1,454 g. is then dropped from a height of 20.5 inches so that it lands on the center of the plunger to produce a concave or convex impression in the coating depending upon whether the plunger is on the coated or uncoated side. The impressions are compared with standards for cracking and adhesion of the coating and graded from one to ten, one representing a complete loss of adhesion with extreme cracking and brittleness and ten representing the absence of visible failure of adhesion and visible cracking. The table on the following page gives the results of the bump test.

Polymer Weight Ratio	Concave	Convex
75 VF/25 MMA	10	10
50 VF/50 MMA	8	9
25 VF/75 MMA	5	5
75 VF₂/25 MMA	10	10
50 VF₂/50 MMA	9	9
25 VF₂/75 MMA	5	4

Example 1: The following ingredients are charged to a three-necked round bottom flask fitted with a thermometer, nitrogen inlet, glass paddle stirrer and a reflux condenser:

	Parts
Methyl methacrylate (0.1% hydroquinone)	25
Potassium persulfate	0.6
Sodium metabisulfite	0.2
Sodium lauryl sulfate	1.0
Dibasic sodium phosphate (7 waters of hydration)	1.5
Deoxygenated water	250

The mixture is heated to 85°C. with stirring (approximately 450 rpm) under a slow stream of nitrogen over a period of 15 minutes. The cloudy mixture is stirred at 85°C. for 45 minutes and then cooled to about 50°C. and transferred to a shaker tube mounted with an injection system. The injection system and shaker tube are flushed thoroughly with nitrogen and then the shaker tube is heated to 73°C. At this point, 75 parts of vinyl fluoride is injected and the pressure adjusted to 1,500 to 1,850 psi by water injection. The pressure is held constant by additional water injection until no further pressure drop occurs and the temperature is held at 70° to 80°C. The solid polymer obtained is filtered and then slurried three times with 250 parts of distilled water in a blender, filtered, rinsed with methanol and dried.

About 94 parts of a white polymer is obtained with a fluorine analysis of 29.46% (weight ratio of approximately 2.6 vinyl fluoride units/1.0 methyl methacrylate units). The white polymer is insoluble in methyl ethyl ketone, butyrolactone and toluene. A film cast from a 15% solids dispersion in butyrolactone on "Alodine" aluminum (6 mil blade, baked for 3 minutes at 210°C.) is colorless, has good flow on baking and excellent formability (bump test: concave 10, convex 10). "Alodine" aluminum is manufactured by the Amchem Products Co., Inc. and is a specially prepared aluminum that has undergone a treating process to provide a surface having improved adhesion characteristics and corrosion resistance. The inherent viscosity of this copolymer measured in a 0.5% hexamethylphosphoramide solution at 25°C. is 1.88.

Example 2: A charge is used consisting of:

	Parts
Methyl methacrylate (uninhibited)	50
Potassium persulfate	0.6
Sodium metabisulfite	0.2
Dibasic sodium phosphate (7 waters of hydration)	1.0
Deoxygenated water	250

The methyl methacrylate prepolymer is prepared as in Example 1 and then transferred to a shaker tube mounted with an injection system. The injection system and shaker tube are flushed thoroughly with nitrogen and the prepolymer is heated to 85°C. Fifty parts of vinylidene fluoride (distilled) is injected and the pressure is raised to approximately 6,000 psi by water injection and the temperature and pressure are maintained until no further pressure drop is observed. The main pressure drop occurs in the first 2 hours, but the mixture is left in the shaker tube where it is maintained at 85°C. for a total of 6 hours. A very fine white dispersion which runs through a fine filter is obtained. The aqueous medium is evaporated and the product copolymer dried in an air-circulation oven at 50°C. Yield is 94 parts of copolymer. The copolymer is slurried with 400 parts of distilled water in a blender, filtered and dried. Its fluorine analysis is 28.41% (weight ratio of app. 1.0 vinylidene fluoride units/1.0 methyl methacrylate units). It is soluble in cyclohexanone, partially soluble in methyl ethyl ketone and partially soluble in butyrolactone. A film cast from a 15% solids cyclohexanone solution on "Alodine" aluminum (6 mil blade, baked for 3 min. at 180°C.) has good formability (bump test: concave 9, convex 9).

In related work H.F. Reinhardt; U.S. Patent 3,438,934; April 15, 1969; assigned to E.I. du Pont de Nemours & Co. describes the use of calcium solicylate to stabilize the coating vehicles described above (U.S. Patent 3,407,247).

Copolymers with Bicyclo[2.2.1]Hept-2-ene

R.D. Lundberg and F.E. Bailey, Jr.; U.S. Patent 3,294,767; December 27, 1966; assigned to Union Carbide Corporation have found that when vinyl fluoride is copolymerized with bicyclo[2.2.1]hept-2-ene there is obtained a copolymer which has improved thermal stability as well as a reduced melting point, and melt viscosity thus permitting the melt processing of a vinyl fluoride polymer without undue danger of thermal decomposition. In addition these copolymers retain the

Vinyl Fluoride

excellent toughness, flexibility and resistance to ultraviolet light characteristic of poly(vinyl fluoride). For example a vinyl fluoride/bicyclo[2.2.1]hept-2-ene copolymer containing about 13 weight percent (6.8 mol percent) polymerized bicyclo[2.2.1]hept-2-ene can be molded at 220° to 230°C. for 1 minute, with little or no discoloration, to form a clear, tough plaque. In addition, this copolymer can be exposed to ultraviolet light for over 710 hours at 50°C. with only a slight yellowing.

It has been found that, if vinyl fluoride is copolymerized with a monomer containing an N-unsubstituted carboxamide group, a polymer having greatly improved thermal stability over poly(vinyl fluoride) or a vinyl fluoride/bicyclo[2.2.1]-hept-2-ene copolymer is obtained. For example a vinyl fluoride/bicyclo[2.2.1]hept-2-ene/bicyclo[2.2.1]hept-2-en-5-yl carboxamide terpolymer containing about 2.8 weight percent (1 mol percent) bicyclo[2.2.1]hept-2-en-5-yl carboxamide and 3.7 weight percent (2 mol percent) bicyclo[2.2.1]hept-2-ene, can be heated up to 275°C. for up to 1 min. with no yellowing.

It is believed that the improved processability and thermal stability obtained by the copolymerization of vinyl fluoride with bicyclo[2.2.1]hept-2-ene or its derivatives is due to the presence of the 2,3-bicyclo[2.2.1]heptanylene nucleus:

in the polymer chains. The bulky nature of this group disrupts the crystalline structure present in poly(vinyl fluoride), thus causing a reduction in melting point. In addition, the presence of this group inhibits the "unzipping" of the polymerized vinyl fluoride, thus providing a certain degree of thermal stability.

The vinyl fluoride/bicyclo[2.2.1]hept-2-ene copolymers and the vinyl fluoride/amide copolymers are readily produced by conventional polymerization techniques, as by a free-radical catalyzed process employing solvent, bulk, suspension or emulsion processes, either batchwise or continuously. The following examples are illustrative. In these examples the following tests were employed in evaluating the copolymers produced thereby:

(1) Melt index—ASTM D1238—52T at 190°C. and 43.1 psig
(2) Flow rate—ASTM D1238—52T at 190°C. and 206 psig
(3) Density—ASTM 1505-57
(4) Stiffness modules—ASTM D638-56T
(5) Tensile strength—ASTM D638-56T
(6) Percent elongation—ASTM D638-56T
(7) Reduced viscosity—Determined at 140°C. from a solution of 0.2 g. of copolymer in 100 ml. of cyclohexanone according to the equation:

$$I_R = \frac{I - I_o}{I_o C}$$

where: I_R = reduced viscosity
I = viscosity of the solution
I_o = viscosity of cyclohexanone
C = concentration of the polymer in the solution in g./100 ml. and is 0.2 g./100 ml.

Example 1: A 2 l., stainless steel, stirrer-equipped autoclave was charged with 50 ml. of oxygen-free isooctane, 5 g. of bicyclo[2.2.1]hept-2-ene, and 4 g. of diisopropylperoxydicarbonate. The autoclave was then sealed, cooled to -80°C. and 500 g. of gaseous vinyl fluoride were added. The autoclave was heated to about 25°C. and was maintained at 25 ± 5°C. for 18 hours. The vinyl fluoride pressure during the reaction was about 400 to 450 psig. The unreacted vinyl fluoride was vented to the atmosphere and the autoclave contents were mixed with isopropanol and a vinyl fluoride/bicyclo[2.2.1]hept-2-ene copolymer precipitated. After washing the polymer and drying, there were obtained 243 g. of a white vinyl fluoride/bicyclo[2.2.1]hept-2-ene copolymer having a reduced viscosity of 0.75. Employing a portion of this copolymer, a clear, brittle plaque was molded at 195° to 200°C. with very little decomposition. Poly (vinyl fluoride) produced in a similar manner decomposes under these conditions.

Examples 2 through 7: A 1.5 l., stainless steel, stirrer-equipped autoclave was charged with 475 g. of benzene, 35 g. of bicyclo[2.2.1]hept-2-ene and 2.5 ml. of a 5 weight percent solution of di-tert-butyl peroxide in benzene. The autoclave was flushed twice with nitrogen and then once with vinyl fluoride by pressuring the autoclave to 50 psig and then slowly venting to the atmosphere. The autoclave was then sealed and pressured to 1,000 psig with vinyl fluoride. Agitation of the mixture was begun, the autoclave was heated to 157°C., and the vinyl fluoride pressure was increased to about 15,000 psig. The reaction was continued at 14,000 to 15,000 psig and 157° to 161°C. for 2 hours and 24 minutes. At the end of this time, the autoclave was cooled to 50°C. and unreacted vinyl fluoride was vented.

The reaction product was mixed with methanol to precipitate the vinyl fluoride/bicyclo[2.2.1]hept-2-ene copolymer produced. After washing twice with methanol and oven-drying overnight at 50°C. the white, powdery copolymer weighed 125 g. The copolymer had a melt index of 0.82 dg./min., flow rate of 8.52 dg./min., a stiffness modulus of 149,885 psi, a tensile strength of 3,832 psi and an elongation of 432%. The copolymer was found to contain 87.2 weight percent polymerized vinyl fluoride by fluorine analysis. The copolymer had a density of 1.330 g./cc and a reduced viscosity of 0.73. The copolymer could be heated at 220° to 230°C. for 1 minute without significant discoloration, whereas poly-(vinyl fluoride) starts to decompose at about 200°C. after 1 minute. There was no apparent deterioration of a plaque of the copolymer after over 700 hours exposure to ultraviolet light.

Employing apparatus and procedures similar to those described above, five additional vinyl fluoride/bicyclo[2.2.1]hept-2-ene copolymers were produced. The date for these polymerizations are summarized in the following table with the data for the above experiment being included as Example 2.

Example	2	3	4	5	6	7
Charge:						
Bicyclo[2.2.1]-hept-2-ene, grams	35	6	12	8	59	112
Benzene, grams	475	475	475	475	475	475
Di-tert.-butyl peroxide solution, ml.	2.5	2.5	2.5	2.5	2.6	2.75
Conditions:						
Maximum Temperature, °C.	161	162	165	167	160	166
Pressure, p.s.i.$\times 10^{-3}$	15	15	15	15	15	15
Time, hours	2.4	4.0	2.4	1.8	2.3	4.0
Product:						
Weight, grams	125	137.5	163	155	137	111
Reduced viscosity	0.73	1.21	1.26	0.78	0.21	.....
Melt Index, dgm./min.	0.828	0	0	0	0.21	14.7
Flow Rate, dgm./min.	8.52	0.107	0.021	0	0.46	88
Stiffness, p.s.i.$\times 10^{-4}$	15.0	11.7	12.4	13.7	15.6	14.7
Tensile, p.s.i.$\times 10^{-3}$	3.8	4.2	4.4	4.1	4.5	3.5
Elongation, percent	432	56	307	112.5	114	8
Yield Stress, p.s.i.$\times 10^{-3}$	5.2	5.0	4.5	5.2	4.9	.....
Melting Temperature, °C.	160	192	194	183	151	.....
Composition:						
Vinyl fluoride, wt. percent	87.2	93.4	91.6	89.8	80.2	70.7
Bicyclo[2.2.1]hept-2-ene, wt. percent	12.8	6.6	8.4	10.2	19.8	29.3

Chlorinated Polyvinyl Fluoride

R. Bacskai; U.S. Patent 3,520,864; July 21, 1970; assigned to Chevron Research Corporation has found that the processability of polyvinyl fluoride may be enhanced by chlorinating the polymer to a chlorine content of about 0.5 to 60%, preferably 1 to 30%, by weight. The enhancement results from a significant lowering of the Vicat softening point. Additionally, it was found that chlorinated essentially linear polyvinyl fluoride having a chlorine content in the above range has significantly improved ductility and impact strength.

The chlorinated polyvinyl fluorides of this process are colorless solids having number average molecular weights of about 10,000 and above. Their Vicat softening points are significantly lower than those of their unchlorinated precursors. The extent of lowering is proportional to the amounts of chlorine incorporated in them. The chlorination increases with solubility of the polymer in conventional solvents. The reduced specific viscosities (measured in N,N-dimethylacetamide at 0.1 g./100 ml.) of the chlorinated essentially linear polyvinyl fluorides of this process will usually be 0.6 to 1.1 dl./g.

F^{19} nuclear magnetic resonance analysis of a chlorinated, essentially linear polyvinyl fluoride containing 38.9% by weight chlorine showed absorption maxima at 42 and 104 ppm upfield from the standard, trifluoracetic acid. This analysis indicates the distribution of chlorine atoms with respect to the α and β carbon atoms of the vinyl fluoride monomer unit is about random. The following examples illustrate the process.

Examples 1 through 3: A 10 g. portion of an essentially insoluble polyvinyl fluoride polymer (Du Pont 5097 about 90% insoluble in dimethyl formamide) was dispersed in 150 ml. carbon tetrachloride in a vessel. While illuminating the dispersion from 2 inches with a 275 watt G.E. sun lamp, chlorine gas at 170 ml./min. was bubbled through this dispersion at ambient temperature for 5 minutes. After this time, the reaction product was poured into 500 ml. methanol, and the resulting slurry was filtered. The polymer residue obtained in this way was dried. The polymer was analyzed and found to contain approximately 0.5 weight percent chlorine. Two grams of this chlorinated polymer with 0.04 g. of a heat stabilizer (Ferro 203) were molded in a 2 inch diameter circular die at 360° to 380°F., under 15 tons of pressure for 1 min. This molded sample was used to determine Vicat melting point. In a similar manner other chlorinated polyvinyl fluorides were made using longer chlorination times and the resulting polymers were molded and tested. The results along with the physical data on an unchlorinated polyvinyl fluoride sample are reported in the following table.

Example	Chlorination time, min.	Percent Cl.	Softening point (Vicat) °F.[1]
1	5	c.a. 0.5	327
2	20	10–12	290
3	40	25–28	241
Unchlorinated polyvinyl fluoride.	0	0	343

[1] ASTM D 1525–58T.

TRIFLUOROCHLOROETHYLENE

GENERAL PROCESSES

Continuous Production from Trifluorotrichloroethane

J.W. Jewell; U.S. Patent 3,014,015; December 19, 1961; assigned to Minnesota Mining and Manufacturing Company describes a continuous process for the production of solid polymers of trifluorochloroethylene from trifluorotrichloroethane. The process involves effecting temperature sensitive reactions under controlled conditions in angularly disposed elongated tubular reactors having a high surface to volume ratio. Separation of produced polymer from monomer is effected by passing the monomer-polymer slurry into an elongated tubular drying zone, where heat is applied, thereby vaporizing the monomer and leaving the solid polymer as a dispersed powder in the monomer vapor.

Figure 4.1 diagrammatically illustrates a suitable arrangement of apparatus in elevation which is used for the production of normally solid thermoplastic polymers of trifluorochloroethylene. The principal pieces of equipment for effecting the process are shown in Figure 4.1. They are dechlorination reactor 14, settler 17, polymerization reactor 82 and flash dryer 92. The process will be described for the polymerization of trifluorochloroethylene to produce the homopolymer.

According to Figure 4.1, trifluorotrichloroethane, commercially available as Freon 113, is passed through conduit 11 into preheater 12. A slurry of metallic zinc in methanol is simultaneously introduced into preheater 12 through conduit 11. Although zinc is a preferred dechlorinating agent other metals may be used such as tin, manganese, magnesium and iron. Methyl alcohol is a preferred solvent although other suitable solvents, such as ethanol, butanol, propanol, glycerol, dioxane and the cellosolves may be used. The function of preheater 12 is to mix the reactants and to bring them to reaction temperature so as to initiate the reaction; preheater 12 is maintained at a temperature between 75° and 250°C., preferably between 90° and 180°C. In a preferred method of operation the reactants pass from preheater 12, which is maintained at a temperature of at least about 10°C. above the temperature of the dechlorination reactor preferably about 30°C. above the temperature of the dechlorination reactor through conduit 13 into dechlorination reactor 14 where the dechlorination is effected with accurate temperature control.

Dechlorination reactor 14 consists of either a jacketed pipe or a spiral coil in a bath with a fairly steep downflow pitch throughout its length. The steep downflow pitch is necessary to maintain the metallic dehalogenating agent, which has quite a high settling rate, in suspension. From the horizontal, the angle of the reactor is maintained at about a minimum of 5° to a maximum of 90°. Preferably, the angle is maintained between 20° and 60° from the horizontal.

Velocity of the reactants through the reactor is maintained between 0.5 and 10 feet per second preferably between 2 and 6 feet per second. At a velocity less than about 0.5 foot per second the zinc settles and the reaction cannot be properly controlled. Both the time and temperature control in this reactor are exceptionally good particularly when the reactor is used in its preferred form, that is as a sectionally jacketed pipe since a sectional construction permits a change in the degree of cooling for each section.

Reaction temperature is maintained within a desired range between 0° and 200°C. with pressure controlled as desired so as to maintain liquid phase between about atmospheric and about 300 psig. Preferred temperature is between 40° and 175°C. with a particularly suitable temperature of between 50° and 150°C. Residence time in the reactor is between about 3 seconds and about 4 minutes. Produced monomer, zinc chloride, methanol and any other products of the reaction are removed from reactor 14 through conduit 16 to settler 17.

Heat exchanger 18 is located on conduit 16 to regulate the temperature of the stream into the settler. Temperature in the settler is adjusted, i.e., between 75° and 130°C. preferably between 90° and 120°C. so that monomer will pass overhead through conduit 19 to cooler 20 where it is condensed and accumulated in accumulator 22. The unreacted zinc concentrate

FIGURE 4.1: CONTINUOUS PRODUCTION OF TRIFLUOROCHLOROETHYLENE FROM TRIFLUOROTRICHLOROETHANE

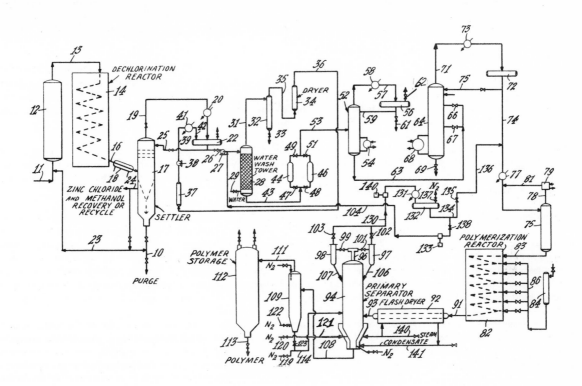

Source: J.W. Jewell; U.S. Patent 3,014,015; December 19, 1961

suspended in methanol and containing zinc chloride is collected at the bottom of settler 17 and may be recycled through conduit 23 to preheater 12. Complete elimination of monomer from this stream while desirable, is not essential, a small fraction of this stream may be withdrawn through line 10 to control the concentration of heavy impurities; zinc so withdrawn may be recovered. Zinc chloride and methanol are removed through conduit 24 where they may be recovered or recycled. A portion of the monomer contained in accumulator 22 is recycled through conduit 25 as to control the quality of the product.

The monomer contained in accumulator 22 is contaminated with small quantities of methanol and other impurities from the dechlorination zone, such as trifluoroethylene, difluorovinyl chloride, etc. In order to remove these impurities, the monomer is passed through conduit 26 by means of pump 27 into the top of water-wash tower 28. In the water-wash tower, the monomer flows downward through an upward moving stream of water which scrubs the monomer free of water-soluble impurities such as alcohol. Water containing methanol and a small amount of monomer is removed from the top of the water-wash tower through conduit 31 and is passed into the top portion of separator 32. Water and methyl alcohol are removed from the bottom of separator 32 through conduit 33 and may be either discarded or circulated to an alcohol recovery system.

Monomer containing water is passed to dryer 34 through conduit 35. Dryer 34 contains a conventional drying agent, such as calcium sulfate. The dried monomer is then recycled through conduit 36 to absorber 37 then to compressor 38 where it is compressed and passed through conduit 39 through cooler 41 into accumulator 22 through conduit 42. Absorber 37 serves to remove corrosive materials which might damage compressor 38. The water-washed monomer, free of water-soluble impurities is removed from the bottom of tower 28 through conduit 43 to dryers 44 and 46. Dryers 44 and 46 contain a conventional drying agent such as calcium sulfate. Valves 47, 48, 49 and 51 are used to close down either of the dryers for cleaning purposes.

The dry monomer containing low and high boiling impurities from the dechlorination zone is passed to distillation column 52 through conduit 53. The temperature of distillation column 52 is maintained at about 50°C. by means of reboiler 54 so that low boiling impurities are removed overhead in vapor phase. The low boiling impurities which are removed overhead as vapor are passed through conduit 57 to accumulator 56. Cooler 58 is a conventional cooler. A portion of the low boiling impurities is recycled to the top of distillation column 52 through conduit 59 as reflux. The low boiling impurities may be recovered for other use or discarded as liquid through conduit 61 or as vapor through valve 62 which valve is used to control the pressure in the system. The monomer free of low boiling impurities is removed as a liquid from the bottom of distillation column 52 through conduit 63 and is passed to the center portion of distillation column 64. The monomer is introduced into column 64 by means of valves 66 and 67. The temperature in distillation column 64 is so adjusted as to remove monomer as a vaporous fraction overhead leaving the high boiling impurities such as difluorovinyl chloride as a bottoms product.

Temperature control is effected by reboiler 68 which is maintained at about 52°C. The high boiling impurities are removed from column 64 as a bottoms product through conduit 69 and may be either recovered or discarded. Substantially pure monomer is removed as a vapor from the top of column 64 through conduit 71 and is collected in accumulator 72. Cooler 73 is a conventional cooler and is used to condense vaporized monomer. A portion of the pure monomer contained in accumulator 72 is recycled through conduit 73 to the top of column 64 as reflux.

Purified monomer is passed from accumulator 72 through conduit 74 to accumulator 75 where it is held prior to introduction into the polymerization zone. Exchanger 77 is a conventional unit which is used to bring the monomer to proper temperature for the desired pressure in accumulator 75. In order to maintain this equilibrium vaporized monomer leaves accumulator 75 through conduit 78 and enters condenser 79 where it is condensed and recycled through conduit 81.

Condenser 79 and connecting lines maintain monomer in accumulator 75 under its vapor pressure at the operating temperature. Monomer is withdrawn as required from accumulator 75 and is introduced into polymerization reactor 82 through conduit 83. Polymerization reactor 82 consists of a series of coils situated in an annular tank containing any suitable liquid for the purpose of temperature control. A refrigeration coil not shown in the diagram is used to maintain the desired temperature of the liquid in which the coil is immersed and forced circulation of this liquid is used to assure uniform temperature and predictable transfer rates. A suitable promoter such as trichloroacetyl peroxide is introduced into the reactor from promoter storage tank 84 through conduit 86.

The promoter is introduced in at least one point preferably a plurality of points along the length of the reaction coil. The continuous injection of promoter at a plurality of points in a small cross section flowing stream gives a much better assurance that all the monomer has an immediate opportunity to contact the promoter than is possible when the promoter is injected into a large volume of monomer in an autoclave. This procedure provides an added method of controlling the quality, that is the molecular weight distribution of the produced polymer since promoter concentration affects the quality of the polymer. Obviously, this type of control is not possible in an autoclave type reactor.

From the horizontal, the angle of the reactor coils is maintained between a minimum of about 1° and a maximum of about 30°, preferably between 5° and 15°. The velocity of the reactants passing through the coil is between 0.1 and 10 feet per second with a particularly suitable velocity of between 0.1 and 5 feet per second, and a preferred velocity between 0.3 and 4 feet per second.

Temperature in the polymerization zone is maintained between –20° and 150°C. depending on the promoter employed and the dimension of the polymerization zone and the velocity of the reactants. Preferred temperature is between 10° and 75°C. Generally higher temperatures are employed when the reactants are moved at high velocities through reaction zones of short length.

While approximately 25% of the monomer may be converted to polymer within the polymerization reactor, it is preferred to keep the concentration of polymer suspended in monomer, between 3 and 12%. The monomer stream containing suspended particles of polymer is passed from polymerization reactor 82 through conduit 91 to flash dryer 92. Flash dryer 92 consists of an elongated tubular heater contained in a jacket through which by means of lines 140 and 141, a heating medium such as steam is circulated so as to maintain a temperature between 60° and 250°C. preferably between 80° and 120°C. The slurry of polymer in liquid monomer is admitted to the dryer through pressure reducing and control valve 91. With the reduction in pressure and application of heat the monomer is vaporized very rapidly. The consequent increase in volume results in a rapid increase in velocity which prevents accumulation of polymer on the heating surface. The relative increase in velocity is from 0.5 to 10 feet per second at the inlet to about 200 feet per second at the outlet.

Conditions should be adjusted, i.e., by using a higher temperature or a smaller pipe diameter so that the minimum outlet velocity is about 50 feet per second, preferably between about 100 and 300 feet per second and still more preferably between 150 and 250 feet per second. When the outlet velocity is maintained at less than about 50 feet per second the polymer powder tends to accumulate on the heating surface and plug the line. The high turbulence under these conditions results in the rapid and uniform evaporation of substantially all of the monomer leaving a polymer powder dry enough to fluidize readily. The mixture is discharged through conduit 93, connected either radially or tangentially to primary separator 94.

The powder settles to the bottom of 94 which is heated sufficiently to maintain a temperature that will vaporize any residual monomer occluded or adsorbed in the polymer powder. A stream of aeration gas is introduced to the bottom of the vessel 94 through valve 120 on line 121 to maintain a fluid bed and to act as a stripping medium to remove vaporized monomer. The use of nitrogen for this purpose is shown although any other gas which is nonreactive in the system may be used. In order to insure a powder particle size distribution that will fluidize readily and to prevent accumulation of polymer on the walls of the separator, a portion of finished polymer which has been extruded and reground to coarse particle size is introduced into the coarse polymer hopper 109. It flows from the hopper through bottom valve 123 into line 114 where it is picked up in a stream of nitrogen admitted to the line through valve 119 and is carried to a point in primary separator 94 such that it will thoroughly mix with the powder discharged from the flash dryer.

A portion of the polymer product is allowed to mix with the coarse powder in hopper 109 and is recycled to the primary hopper 94. This permits control of the rate of powder flow in the bottom of 94 and thus assists in the ability to maintain a good fluid condition. The net polymer product is elutriated from hopper 109 in a stream of nitrogen or other suitable gas introduced into the bottom part of hopper 109 through valve 122 in sufficient quantity to provide the required elutriating velocity in that vessel and discharged through line 111 to storage hopper 112.

Vaporized monomer passes overhead from separator 94 through line 96 to filters 97 and 98. Valves 99, 101, 102 and 103 are used to shut down either of the filters for cleaning or repair. Vaporized monomer free of polymer but containing nitrogen flows through line 130, compressor 140 and condenser 131 into accumulator 132. This accumulator is held under sufficient pressure, by nitrogen flow through valve 137, to flow through line 134, valve 135 and line 136 to line 74 and so back to the reactor feed accumulator. To control the concentration of contaminants in the recycle monomer a part of the recycle is bled off through valve 138 and is pumped by pump 133 to accumulator 22 for monomer purification. Primary separator 94 and polymer hopper 109 are maintained under slight pressure with nitrogen gas. The nitrogen is removed from the equipment, is recovered and recycled by means not shown. The following example illustrates the process.

The dehalogenation reactor is made of a 90 foot length of 1 inch pipe positioned as an angular downward sloping unit so that gravity will assist a velocity of from 2 to 6 feet per second to carry the zinc powder through the line. The reactor is situated in a temperature controlled bath which is maintained at about 100°C. Using this size reactor the fresh feed consists of:

Material	Pounds/Hour
Freon 113	274
Zinc	192
Methanol	500

In order to get 100% conversion of the Freon 113 to monomer et al a considerable excess of zinc, about 10%, must be available, and to provide this condition zinc is recycled from the product separator to the reactor heater as a slurry suspension in a liquid mixture of zinc chloride and methanol. An excess, about 5%, of monomer is preferably added to make up for losses. Methanol is added as required. In order to maintain the desired velocity, one-tenth of the total hourly quantity of reactants and the total quantity of methanol are passed through the reactor in one pass. Ten passes per hour making up the required hourly input and output. The feed per pass consists of:

Material	Pounds/Pass
Freon 113	27.4
Zinc	19.2
Methanol	50.0

From the reactor the mixture flows to the separator-settler tower where the monomer product about 163 pounds plus a small fraction of methanol is flashed overhead. A conventional overhead reflux system serves to control the quantity of methanol in the overhead stream. The liquid slurry bottoms from the separator section flows into the settler section where the unreacted zinc settles in more concentrated slurry with liquid zinc chloride and methanol and is pumped as recycle to the reactor heater.

To control the concentration of zinc chloride in the circulating stream a portion mixed with methanol is withdrawn as a decanted liquid from the top of the settler. The methanol is recovered from the zinc chloride and the latter may be disposed of as such or further treated to recover zinc. There are minor quantities of side reaction products, about 1 pound, in the product and recycle streams which are considered in the recovery and purification equipment described above. The temperature control throughout the system is such that the zinc chloride is held in the liquid phase. Substantially pure monomer is accumulated in accumulator 72.

Polymerization of the accumulated monomer, in this instance, is effected in a polymerization reactor which comprises a 1 1/4 inch standard pipe approximately 16,540 feet long, arranged in a helix or coil approximately 31 feet in diameter. The coils are so disposed that the velocity of the reactants in the polymerization reactor is between 0.3 and 4 feet per second preferably about 0.8 foot per second. The 163 pounds per hour of produced monomer, is added to about 2,600 lbs.

per hour of recycled monomer in an accumulator 75. 2,763 pounds per hour of monomer are passed through the polymerization reactor together with about 8 pounds of trichloroacetyl peroxide promoter dissolved in a suitable solvent such as trichloromonofluoromethane which is introduced in at least one point along the polymerization reactor. In this instance, conditions are adjusted so as to effect a 6% conversion of the monomer to polymer. Thus, about 160 pounds per hour of polymer suspended in approximately 2,600 pounds per hour of monomer leave the polymerization reactor and are passed to flash dryer 92.

In the process of this example, the flash dryer is a 1 inch pipe approximately 40 feet in length. The monomer-polymer slurry is introduced into the flash dryer at a velocity sufficient to avoid settling. A pressure reduction between the polymerization reactor and the flash dryer controls the flow rate. Temperature of the flash dryer is maintained at about 100°C. by means of low pressure steam. The polymer leaves the flash dryer in the form of a fluidized powder dispersed in monomer vapor, at a velocity of about 200 feet per second and is passed into primary separator 94 where a major portion of the monomer is separated from the polymer and flow overhead through filters to the recycle system.

The polymer product settles out and is maintained by a heat source at a temperature high enough to evaporate any residual monomer. A stripping medium such as nitrogen is used to maintain an efficient stripping velocity in the fluid bed. A recycling stream of coarse ground polymer product is mixed with the reactor product in the primary separator to assist in maintaining a particle size distribution suitable for good fluidization in the bottom bed. This quantity of coarse ground polymer may be as much or more than the polymer product i.e. between about 0.5 and 2 times. The mixture of powder flows through a control valve from the bottom of the primary settler, is picked up in a stream of nitrogen and carried to the coarse polymer hopper where the fine polymer product is carried overhead in suspension in nitrogen and the coarse polymer with some fines settles out to be recycled to the primary separator.

Perfluorochloro Carboxylic Acid Dispersing Agent

A process described by A.N. Bolstad and F.W. West; U.S. Patent 3,006,881; October 31, 1961; assigned to Minnesota Mining and Manufacturing Company involves the emulsion polymerization of fluorine-containing olefins such as chlorotrifluoroethylene using a fluorochloro dispersing agent. The dispersing agent, an aliphatic perfluorochloro acid produced by hydrolyzing a telomer of chlorotrifluoroethylene, or a water-soluble inorganic salt of such an acid to produce a solid polymerization product in the form of a dispersion. Acids produced from chlorotrifluoroethylene telomers by hydrolysis may be represented by the generic formula $Z(CF_2CFCl)_{n-1}CF_2COOH$ where n is an integer from 2 to 16 and where Z is a monofunctional radical of the group consisting of a perhalogenated radical having a total atomic weight of not higher than 146.5 and a halogen atom selected from the group consisting of fluorine, chlorine and bromine atoms.

These aliphatic perfluorochloro acids may be prepared by the hydrolysis with fuming sulfuric acid of a bromotrichloromethane telomer of chlorotrifluoroethylene or a sulfuryl chloride telomer of chlorotrifluoroethylene. The preferred telomer acids are those containing from 6 to 12 carbon atoms. A suitable perfluorochloro acid may be prepared by the treatment with fuming sulfuric acid of a bromotrichloromethane telomer of chlorotrifluoroethylene. The telomer, having the formula $CCl_3(CF_2-CFCl)_nBr$, where n is an integer from 2 to 16, preferably from 3 to 6, may be prepared in a specific example by dissolving 3.5 parts of benzoyl peroxide in 408 parts of bromotrichloromethane and charging this solution to a pressure vessel along with 300 parts of chlorotrifluoroethylene. The system is heated for about 4 hours at about 100°C. with agitation to produce high yields of relatively low molecular weight polymers having the above formula.

These polymers may be distilled to produce individual compounds of the above formula where n is a particular integer. The mixed polymer is treated with fuming sulfuric acid containing less than about 20% excess sulfur trioxide at a temperature between about 125° and 175°C. for a period of time between 10 and 25 hours to produce monocarboxylic acids having an odd number of carbon atoms in the aliphatic chain with the structure $CCl_3(CF_2-CFCl)_{n-1}CF_2COOH$. A detailed description of this method of preparing perfluorochloro acids may be found in U.S. Patent 2,806,865.

The polymerization of the monomers requires an aqueous medium, a dispersing agent of the class discussed above, and a polymerization initiator, such as a water-soluble inorganic polymerization initiator, e.g., the ammonium and alkali metal persulfates, perborate or percarbonates, or an organic peroxide, such as cumene hydroperoxide. The preferred polymerization initiators or promoters are ammonium persulfate and alkali metal persulfates, particularly sodium or potassium persulfate.

A particular advantage of the product dispersions obtained by the method when chlorotrifluoroethylene is polymerized, is that the latex product may be used directly as a coating dispersion, since it contains nothing which would be detrimental in a coating or film. Polymerization initiators, etc. are present in extremely small concentrations and normally would not affect the characteristics of the coating or film. The perfluorochloro acids and salts used as dispersing agents may be decarboxylated at film-forming temperatures to produce normally liquid perfluorochlorinated olefins which are very similar in structure to the polychlorotrifluoroethylene resin, and which, therefore, acts as a plasticizer. This is in contrast to the perfluoro acids previously used as dispersing agents, which when decarboxylated, produce products insoluble in the polymer, with a resulting lack of homogeneity in the films. The following examples illustrates the process.

Example: Emulsion Polymerization of Chlorotrifluoroethylene — In this example four polymerization runs were carried out in aqueous dispersion using fluorochloro telomer acid as a dispersing agent. Each polymerization was carried out in a 100 ml. heavy walled, clean and nitrogen flushed Pyrex polymerization tube, using the indicated ingredients charged in the following order.

C_8-perfluorochloro telomer acid, water, a 2 1/2% solution of sodium metabisulfite ($Na_2S_2O_5$), a 2 1/2 solution of potassium persulfate ($K_2S_2O_8$), a 1.25% solution of ferrous sulfate ($FeSO_4 \cdot 7H_2O$), and chlorotrifluoroethylene. The C_8 telomer acid, referred to above, is a monocarboxylic acid derived from the treatment with fuming sulfuric acid of a C_8 telomer prepared with sulfuryl chloride as the telogen and chlorotrifluoroethylene as the monomer. The acid has the formula $Cl(CF_2CFCl)_3CF_2COOH$.

The polymerizations were conducted at +5°C. for 20 hours with rocking of the tube. The polymers obtained were then oven dried overnight at 190°C. The results of these experiments are shown in the table below.

Experiment Number	Pts. by Wt. CF2=CFCl	Pts. by Wt. Water	Initial pH*	Pts. by Wt. K2S2O8	Pts. by Wt. Na2S2O5	Pts. per Million (ppm) of Monomer FeSO4·7H2O	Pts. by Wt. Cotelomer Acid	Percent Yield
1	100	300	2.0	2.0	1.1	300	None	6.4
2	100	300	2.0	2.0	1.1	30	0.4	12.0
3	100	300	2.0	2.0	1.1	100	0.4	12.5
4	100	300	2.0	2.0	1.1	300	0.4	14.4

*pH was adjusted using a 5% solution of KOH.

Note: 2.0 parts by weight of $K_2S_2O_8$ = 40.0 ml. of 2.5% $K_2S_2O_8$.
2.2 parts by weight of $Na_2S_2O_5$ = 20.0 ml. of 2.5% $Na_2S_2O_5$.
100 ppm of $FeSO_4 \cdot 7H_2O$ = 1.0 ml. of 1.25% $FeSO_4 \cdot 7H_2O$.
0.4 parts by weight of C_8-acid = 0.15 g. of acid.
100 parts by weight of monomer = 41.7 g. of CF_2=$CFCl$.

Carbon Disulfide Moderator

A process described by R.M. Mantell and J.M. Hoyt; U.S. Patent 3,043,823; July 10, 1962; assigned to Minnesota Mining and Manufacturing Company involves polymerizing a monoolefin, particularly a halogenated monoolefin which is at least half fluorinated and which contains from 2 to 5, preferably from 2 to 3 carbon atoms, in an aqueous medium which contains carbon disulfide and a dispersing agent. When the monoolefin is fluorinated, the preferred dispersing agents are aliphatic halocarbon acids, particularly aliphatic fluorocarbon acids, such as produced by hydrolyzing a telomer of chlorotrifluoroethylene, or a water-soluble inorganic salt of such an acid, to produce a solid high molecular weight polymerization product in the form of a dispersion.

The polymerization process is preferably carried out at high pH, such as pH 7 to 14, preferably at a pH of 8 to 11, and at temperatures from 0° to 100°C. usually 15° to 35°C., for solid polymers. The polymerization temperatures can be varied within the above range depending on the molecular weight of the product desired. Pressures used are preferably autogenous and may go as high as 50 atmospheres. With carbon disulfide used as an "activator" the control of pH within the above ranges is essential if the full effect of the carbon disulfide on the yield is to be obtained.

The concentration of monoolefin in the aqueous dispersion may vary from 5 to 50% by weight of the water present. The dispersing agent may be present in concentrations from 0.1 to 6% by weight of the water present. The polymerization initiator or promoter may be present in concentrations from 0.001 to 5%, based on the weight of monomer used. The amount of carbon disulfide used can vary from 1 to 50 parts by weight of monomer, the preferred range being between 5 and 25 parts by weight of monomer. The salt of the variable valence metal is usually present, when used, in an amount to provide from 0.2 to 600 parts per million of ferrous ions based on the weight of the aqueous medium; and the reductant, when used, is usually present in an amount from 1 to 6 parts by weight per 100 parts of total monomer, usually in equal amounts with the promoter.

An advantage of the high yield product dispersions obtained by the method is that the latex product may be used directly as a coating dispersion, since the carbon disulfide volatilizes off at film-forming temperatures. Thus the addition of the carbon disulfide does not in any way alter the characteristics of the coating or film. This is particularly advantageous where such coating is subjected to severe conditions of temperature and chemical attack. The following examples illustrate the process.

Example 1: Emulsion polymerization of trifluorochloroethylene was conducted in sealed 250 ml. glass polymerization tubes at 25°C. for 20 hours with shaking, using the following emulsion recipe:

$CF_2{=}CFCl$	41.2 g.
2.5 wt. percent aqueous solution of $K_2S_2O_8$	40.0 ml.
4.0 wt. percent aqueous solution of $Cl(CF_2CFCl)_3CF_2COOH$ alkalized to pH 9 to 10 with KOH	46.8 ml.
5.25 wt. percent aqueous solution of $Na_2HPO_4 \cdot 7H_2O$	38.2 ml.

At the completion of the 20 hours the sealed tubes were vented and the polymer coagulated by rapid cooling in Dry Ice. The coagulate was then filtered on a Büchner filter and washed with hot water until free from soap bubbles, after which the washed coagulate was vacuum dried overnight at 70°C. and about 100 mm. Hg pressure. Trifluorochloroethylene polymer was produced in a 60% yield with a dilute solution viscosity of about 1.3 cs. (0.75 wt. percent solution in dichlorobenzotrifluoride at 266°F.).

The above experiment was repeated with the addition of 4.2 g. of CS_2 to the emulsion recipe. The resulting trifluorochloroethylene polymer had a ZST (zero strength time) of about 150 seconds and a dilute solution viscosity (0.75% solution in dichlorobenzotrifluoride at 266°F.) of about 0.90 cs. A yield of 95% was produced, indicating the effect of carbon disulfide addition to the polymerization recipe. No sulfur could be detected in the polymer by the usual bomb combustion-barium sulfate analysis.

Example 2: Five recipes as shown in Table 1 below were sealed in glass ampules and shaken at 25°C. for 7 hours. After venting the ampules each polymer sample was coagulated by rapid cooling in a Dry Ice-acetone bath and filtered with a sintered Buchner filter, the filtrates being collected for further testing. Each coagulate or precipitated polymer was wetted with 50 to 75 cc of methanol, and 100 to 300 cc of distilled water was added, after which the liquid was drawn through the coagulate by suction. This washing procedure was repeated 9 times. A final wash with methanol completed the wash operation. The precipitates were vacuum oven dried in glass dishes at 70°C. and about 100 mm. Hg pressure over a period of 24 hours and then weighed. Test results on the original filtrate and yield data appear in Table 2 below.

TABLE 1

Run No.	CS_2, g.	$CF_2{=}CFCl$, g.	2.5 Wt. Percent Aqueous Solution, ml.[a]	4.0 Wt. Percent Aqueous Solution, ml.[b]	5.25 Wt. Percent Aqueous Solution, ml.[c]
A	–	41.2	40	46.8	38.2
B	4.2	41.2	40	46.8	38.2
C	8.4	41.2	40	46.8	38.2
D	4.2	41.2	40	46.8[d]	38.2
E	–	41.2	40	46.8[d]	38.2

[a] $K_2S_2O_8$
[b] $Cl(CF_2CFCl)_3CF_2COOH$ alkalized with KOH to pH 9 to 10
[c] $Na_2HPO_4 \cdot 7H_2O$
[d] H_2O

TABLE 2

	Polymer Yield		Polymer Properties		Final Polymer Latex Properties		
Run No.	Gram	Percent	Dilute Solution Viscosity Centistokes[1]	NST, °C.[2]	$S_2O_8^=$ titer, ml.[3]	Surface Tension, dynes/cm.[2]	pH
A	10.5	15.5	1.222	304	2.5	39.0	1.55
B	16.2	39.3	1.034	287	2.5	38.7	1.55
C	17.7	43.0	0.965	277	2.5	38.6	1.55
D	0.49	1.2	–	–	–	–	–
E	0.54	1.3	–	–	–	–	–

[1] 0.75 wt. percent solution in dichlorobenzotrifluoride at 266°F.
[2] No strength temperature
[3] Ml. 0.1 N thiosulfate to equivalence for 5 ml. aliquot

Surface tension was determined on the Du Nuoy tensiometer, the pH by using a Beckman pH meter with a glass reference electrode, and the final $S_2O_8^=$ concentration by adding excess KI to a 5.0 mol aliquot and titrating with 0.1 N thiosulfate solution to the starch-iodide equivalence point. From the results obtained in the runs it was concluded that the addition of CS_2 causes no detectable change in pH, surface tension, or $S_2O_8^=$ concentration in the filtrate. However, the yields of B and C, which contain CS_2, were 54% and 69% higher than A, which contained no CS_2.

An 8 g. sample of polymer from each of runs A, B and C was cold pressed at 5,000 psi, then at 220° to 230°C. and 5,000 psi and quenched. Clear, bubble-free sheets resulted, the samples from each of the runs being indistinguishable in transparency, appearance and flexibility. Samples D and E were polymerized without the use of a fluorinated acid salt. Sample D contained CS_2, while sample E contained no CS_2. The percent yields, both with and without CS_2, are low in the absence of the fluorinated acid salt.

Chain Transfer Agents for Molecular Weight Control

R.L. Herbst and B.F. Landrum; U.S. Patent 3,024,224; March 6, 1962; assigned to Minnesota Mining and Manufacturing Company describe a process for controlling the molecular weight of normally solid polymers of trifluorochloroethylene. In the process, the polymerization of fluorine-containing olefins is effected in the presence of a polymerization modifier of the formula:

$$\begin{matrix} F \\ \\ F \end{matrix} \hspace{-4pt} {\Large \diagdown} C = C {\Large \diagup} \hspace{-4pt} \begin{matrix} H \\ \\ X \end{matrix}$$

where X is fluorine, chlorine, bromine or iodine. Between 0.01 and 10 mol percent of the above modifier based on monomer is used during polymerization, the exact concentration depending upon the molecular weight of polymer desired.

This process applies generally to the polymerization of perhalogenated olefins containing fluorine to produce both solid homopolymers and copolymers. Since molecular weight determinations for the polymer are difficult and time consuming, the polymers are classified usually by the no strength temperature (NST degrees centigrade) which is proportional to the molecular weight of the product. In general, the thermoplastic polymers have a minimum NST of about 220°C. and a maximum of about 350°C. which correspond to a minimum molecular weight of about 50,000 and a maximum molecular weight of about 200,000.

The polymerization is carried out in the presence of a specific and definite amount of the modifier of the above formula to obtain a predetermined molecular weight product (NST product). It has been found that the NST of the product may be controlled within plus or minus 3°C. by regulating the concentration of promoter and modifer according to the following equation:

$$(1) \quad NST \ = \ 324 - 277(Y - 0.03) - 202(Y - 0.05)(Z - 0.02) - (2 \times 10^6)X^3$$

in which X is the concentration in weight percent of promoter based on monomer, Y is the concentration in mol percent of $CF_2=CHCl$ and Z is the concentration in mol percent of $CF_2=CHF$. The above equation takes into account the cumulative effect of two of the most important modifiers. In some instances it may be desired to use only one modifier and the following equations provide the required concentrations of a single modifier to produce an NST product within plus or minus 3°C.

$$(2) \qquad NST \ = \ 205.6 + \frac{35.66}{Y + 0.29} - (2 \times 10^6)X^3$$

$$(3) \qquad NST \ = \ 314 - 3.8(Z - 0.02) - (2 \times 10^6)X^3$$

The variables in Equations 2 and 3 have the same definitions as the variables in Equation 1. The most convenient method of carrying out the process is by introducing the modifier or modifiers in the required predetermined concentration into the monomer feed passed to the polymerization zone. However, the modifiers may be added separately from the feed and separately from each other without departing from the scope of this process. In a batch type system the modifiers are added initially in the desired concentration without further additions. In continuous operations, continuous additions of monomer are made during the polymerization either with the feed or independent of the feed. Even in the batch system continuous introduction of modifier may be employed during the polymerization.

Example 1: This example illustrates the effect of varying concentrations of 1-chloro-2,2-difluoroethylene as a modifier upon the yield and NST of solid polymeric trifluorochloroethylene. This monomer ($CF_2=CFCl$) feed used contained 0.02 mol percent $CF_2=CHF$ and 0.03 mol percent $CF_2=CHCl$ as determined by mass spectrometer analysis. The $CF=CHCl$ to be added in addition to that present in the feed was prepared by the debromochlorination of $CF_2Br-CHCl_2$ using zinc-methanol and analyzed 99.1 mol percent purity. For the latter examples, $CF_2=CHF$ was prepared from

CF₂Br—CHClF in the same manner. Polymerizations were carried out in sealed glass tubes (200 ml. capacity) agitated by mechanical shaking. The tubes were thoroughly degreased with trichloroethylene and evacuated before charging. Bis-trichloroacetyl peroxide promoter in Freon 11 solution was measured by pipetting and the Freon 11 stripped off under vacuum before charging the monomer. Trifluorochloroethylene was charged in liquid phase.

The $CF_2=CHCl$ was charged as a vapor and its initial concentration in the reaction mixture based on monomer was determined by filling evacuated flasks of known volume and temperature to a precalculated pressure of $CF_2=CHCl$ vapor. The amount of trifluorochloroethylene charged was determined by weighing. The amount of polymer formed was determined by a weighing procedure which did not involve any polymer handling. The data obtained are summarized in the table below for a 20 hour polymerization at 5°C. using 0.01 weight percent bis-trichloroacetyl peroxide as the promoter.

Mol Percent $CF_2=CHCl$ Added	Percent Yield of Polymer	NST of Product
None	5.3	316
0.05	3.7	311
0.1	3.8	286
0.2	4.6	282
0.3	4.3	259
0.5	4.0	262
1.0	3.8	228
2.0	2.5	222

It can be seen from this example that the NST of the product depends upon the concentration of $CF_2=CHCl$ modifier. The general range of use of this modifier is between 0.04 and 2 mol percent based on monomer. The polymer product obtained is a thermoplastic homopolymer having a first order transition temperature of about 210°C. It is chemically and physically inert and can be fabricated into various articles of manufacture, such as valve diaphragms, gaskets, insulators, etc., by molding at elevated temperatures above the transition temperature. It can be formed into continuous and homogeneous films from dispersions or solutions, or by extrusion, which films serve as excellent protective surfaces. The fabricated polymer is nonporous and water repellant. The use of a modifier improves the embrittlement characteristics of the polymer.

Example 2: This example illustrates the effect of promoter concentration upon the yield and NST of polymeric trifluorochloroethylene using 0.1 mol percent $CF_2=CHCl$ as a modifier. In this example, the same procedure was employed as in Example 1, except that 0.1 mol percent $CF_2=CHCl$ was added and the promoter concentration varied. The data are summarized below.

Weight Percent Promoter	Percent Yield of Polymer	NST of Product
0.013	4.3	283
0.015	5.2	274
0.018	5.4	274
0.021	9.1	271
0.025	9.7	269

The promoter concentration is important in determining yield and has but little effect on the NST of the polymer, whereas from Example 1 it can be seen that the concentration of modifier is the critical factor in the determination of the NST of the polymer and has a lesser effect on the yield of polymer.

Alkyl Aluminum Halides, Metal Halide and Carbon Tetrachloride

A process described by G.H. Crawford; U.S. Patent 3,084,144; April 2, 1963; assigned to Minnesota Mining and Manufacturing Company comprises reacting a fluorine-containing organic compound having at least one ethylenic carbon-to-carbon double bond in the presence of a catalyst composition comprising a chlorine-containing organic compound, an organic compound of a metal of group IIIa, and a halide of a transition heavy metal to produce a polymer. The process is carried out at a temperature from -100° to +200°C., and usually at a temperature between -20° and 100°C. The process is particularly advantageous for polymerizing fluorine-containing monoolefins and polyolefins including both partially halogenated and perhalogenated fluorine-containing compounds to normally solid polymer products.

The preferred organic metallo compounds which are used to prepare the catalysts are the compounds of aluminum, indium and gallium in which the metal is bonded to at least one alkyl group such as, for example, in the alkyl aluminum halides, alkyl aluminum hydrides and trialkyl aluminum compounds. Of the halides of a transition heavy metal to be used as a starting material in the preparation of the above catalyst composition, the halides, and particularly the chlorides, of metals of groups IVb and VIII are preferred. The chlorine-containing organic compound used as a starting material in the preparation of the catalyst compositions include aliphatic and aromatic chlorine-containing compounds, preferably having only chlorine or a combination of chlorine and hydrogen bonded to carbon. The catalyst compositions are prepared by independently adding the organic metallo compound and the metal halide to the chlorine-containing organic compound to produce a liquid composition having a characteristic red color, the red color being attributed to the formation of a complex between the organic metallo compound, the metal halides and the chlorine-containing organic compound.

When carbon tetrachloride is employed as the chlorine-containing organic compound starting material, the liquid catalyst composition obtained is stable, that is, no precipitation of solid material is observed after prolonged storage or during its use to effect polymerization of the fluorine-containing olefins. When the other chlorine-containing organic compounds used in accordance with this process, such as chloroform, for example, are used to prepare the catalyst compositions, some precipitation of solid material occurs upon prolonged standing, which precipitation can be avoided to yield stable liquid catalysts by the addition of n-hexane to the chloroform before or at the same time that either the organic metallo compound or metal halide are added.

The advantages obtained by using the liquid catalyst compositions to effect polymerization of fluorine-containing vinyl type compounds are numerous, one very important advantage being that polymerization is capable of being effected in a homogeneous medium, i.e., in a medium in which the monomer to be polymerized and the catalyst are present in the same phase. This homogeneity is realized not only at the beginning of the polymerization reaction but also throughout the entire reaction, since the catalyst remains in a completely liquid and dissolved state from the beginning to the conclusion of any one polymerization reaction.

Due to the homogeneity of the reaction medium, the polymerization process is easily controlled and represents a generally simplified process which is capable of being conducted in simple equipment and is readily adapted to continuous operation. Since the active catalyst species is in a stable liquid state, the handling of the catalyst is facilitated. A further advantage of the stable liquid catalyst compositions is that when they are used to effect polymerization of various fluorine-containing vinyl type compounds such as trifluorochloroethylene, for example, the polymer product which forms cannot become occluded within the catalyst and the catalyst in turn, which remains in the liquid state, does not become occluded within the polymer to any appreciable or serious degree. The work-up of the polymer product is thereby facilitated and the efficiency of the process is at a maximum. The polymer product is freed of substantially all color by a simple washing procedure with carbon tetrachloride, for example.

In the following examples, the organic metallo compounds were prepared by conventional procedures. For example, the diethyl aluminum bromide (boiling point 56°C. at 0.3 mm. Hg pressure) was prepared by reaction of magnalium alloy (2:1 by weight of aluminum:magnesium) with ethyl bromide in the presence of a catalytic amount of a magnesium Grignard reagent at a temperature of about 56°C. In each of the following examples normal precautions were taken to exclude air and moisture, i.e., the chlorine-containing organic compounds and n-hexane were dried over sodium prior to use; the preparation of the catalysts and the polymerization reactions were conducted in a nitrogen atmosphere; and the reaction vessels were dried and flushed with nitrogen prior to introduction of the starting materials.

Example 1:   This example is intended to illustrate the preparation of the liquid catalyst compositions used in the process. To a 200 ml. glass flask there were added 100 ml. of anhydrous carbon tetrachloride followed by the independent addition thereto of 2 ml. (0.016 mol) of diethyl aluminum bromide and 0.1 ml. (0.0009 mol) of titanium tetrachloride at room temperature (25°C.), the addition of the titanium tetrachloride being accompanied by stirring. A clear wine-red liquid resulted which was filtered in order to remove any trace amounts of small particles. The filtering was accomplished by transferring the solution to a filtering apparatus in a nitrogen atmosphere. The liquid was then passed through a sintered glass fine grade filter disc into a glass container, and the red colored mother liquor was stored under nitrogen until ready for use. No precipitation of solid material from this liquid catalyst was observed after prolonged standing for many days.

Examples 2 to 4:   These examples illustrate the polymerization of trifluorochloroethylene in the presence of the liquid catalyst composition. A series of experiments were conducted in which various aliquots of the stable liquid catalyst prepared as described in accordance with Example 1 above were added to 300 ml. glass polymerization tubes which had previously been flushed with nitrogen. The contents of each tube were then frozen at liquid nitrogen temperature and 100 g. of trifluorochloroethylene were condensed into each of the tubes. In each case the tube was sealed and the tubes were allowed to stand at 25°C. for a period of 7 days.

At the end of this time, solid polymer particles were observed in each tube and the tubes were vented to atmospheric pressure to allow unreacted monomer to escape. In each case carbon tetrachloride (200 ml.) was added to the contents of the tubes with agitation. The reaction mixture was then filtered and the solid polymer was again washed with carbon tetrachloride until essentially all of the red color was removed from the polymer product. In each experiment the solid white polymer was dried for 8 hours at 190°C. The amount of liquid catalyst employed in each experiment as well as the physical properties of the white resinous polytrifluorochloroethylene homopolymer product are indicated in the following table. In each run the conversion of monomer employed to polymer product was good, e.g., in Example 2 a 45% conversion was obtained.

Trifluorochloroethylene

Example	Volume of Liquid Catalyst (ml.)*	Dilute Solution Viscosity (cs.)**	ZST (seconds)
2	40	–	–
3	2	1.812	818
4	10	1.204	312

*Liquid catalyst prepared as described in Example 1, i.e., in
which the volume ratio of $CCl_4 : (C_2H_5)_2AlBr : TiCl_4$ is $100 : 2 : 0.1$.
**As determined in 0.75% solution of dichlorobenzotrifluoride at 266°F.

Alkali Metal Fluoride Catalysts

F.N. Teumac and L.W. Harriman; U.S. Patent 3,244,684; April 5, 1966; assigned to The Dow Chemical Company have found that chlorotrifluoroethylene can readily be polymerized by contacting the monomer with an alkali metal fluoride or an alkali metal silicofluoride at elevated temperatures to form products which range from liquid dimers to greases to normally solid polymers.

The catalyst materials can be an alkali metal fluoride such as sodium fluoride, potassium fluoride, cesium fluoride, rubidium fluoride, cesium silicofluoride, rubidium silicofluoride, potassium silicofluoride or sodium silicofluoride or mixtures of any two or more of such alkali metal or alkali metal silicofluorides. The catalyst materials are preferably employed in an amount corresponding to from 0.8 to 1.2 times the weight of the chlorotrifluoroethylene to be reacted. The following examples illustrate the process.

Example 1:   A charge of granular cesium fluoride was enclosed in a nickel screen cage and was suspended in a 500 ml. capacity stainless steel autoclave. The autoclave was evacuated to remove air, then was filled with chlorotrifluoroethylene vapors under pressure and was heated at a temperature of 130°C. and a maximum pressure of 790 psig pressure for a period of about 72 hours. Thereafter, the autoclave was cooled, was vented and the products recovered. The conversion was 81.8% based on the monochlorotrifluoroethylene initially used. The products were separated. They consisted of liquid polymer, grease polymer and solid polymer. The liquid polymer was recovered by distillation. It was perfluoro-1,2-dichlorocyclobutane. The grease polymer was recovered by dissolving it in perchloroethylene, filtering to separate the solution from the insoluble solid polymer and evaporating the perchloroethylene solvent to recover the grease polymer as residue. The yield of perchloro-1,2-dichlorocyclobutane was 14.5%, based on the total polymeric product obtained. The yield of grease polymer was 18.8%. The yield of solid polymer was 66.7%.

Example 2:   Vapors of chlorotrifluoroethylene were fed at a rate of 36 g./hr. to a reaction zone and into contact with a bed of granular cesium fluoride at a temperature of 300°C. for a contact time of about 1.8 seconds, then were passed through successive cold traps and were cooled to condense and separate liquid product. The experiment was carried out in continuous manner for a period of 24 hours. The conversion at the start of the experiment was about 48%, but decreased with time and was about 6% at the end of the 24 hour test period, because of gradual depletion of the cesium fluoride catalyst material during the polymerization reaction. There were obtained 202 g. of polymeric products. The products were separated by distillation. They consisted of 71% by weight of liquid polymer, and 29% of grease polymers. The liquid product consisted of:

> 1,1,3-trichloroheptafluorobutane,
> 3,6-dichlorodecafluorobutane,
> 1,1,4,4-tetrachlorodecafluorohexane,
> C_8F_{16}, $C_{10}F_{10}Cl_2$ isomers of $C_{10}F_{12}Cl_3$, $C_8F_{13}Cl_3$,
> $C_{10}F_{16}Cl_4$ and higher polymers.

Stabilization with Chlorine Gas

R.R. Divis; U.S. Patent 3,045,000; July 17, 1962 has found that discoloration and molecular weight degradation of polychlorotrifluoroethylene during fabrication can be eliminated for all practical purposes by a preliminary treatment with elemental chlorine gas.

In a preferred example of the process, polychlorotrifluoroethylene, in a form exposing a large surface area, such as powder or porous lumps, is maintained in an atmosphere of chlorine gas at a temperature of 60° to 80°F. and atmospheric pressure for 1 hour. The treated polymer is then heated at 300°F. for 48 hours to remove excess absorbed but uncombined and unreacted chlorine. The product may then be fabricated in the usual manner. The process is illustrated by the following examples.

Example 1:   Into a 500 ml., round-bottom flask connected to a pressure indicator, vacuum pump and chlorine source, and immersed in a constant temperature bath at 77°F., was charged 82 g. of powdered polychlorotrifluoroethylene. The system was evacuated to a pressure of about 0.1 mm. of mercury, then chlorine was added to a pressure of 760 mm. of mercury. A pressure drop of 170 mm. of mercury was observed over a contact period of 80 minutes, although most of this pressure drop occurred within the initial 5 minutes of contact time. The polymer was transferred to a beaker and heated at 300°F. for 67 hours for removal of excess absorbed chlorine. A test sheet was molded for spectrophotometric color observation. At a wave length of 320 millimicrons, this sheet gave a percent transmittance value of 63.8%. A sheet molded from untreated polychlorotrifluoroethylene from the same lot and in the same manner gave a percent transmittance value of 39.6%.

Example 2:   Into a 500 ml., three-necked, round-bottom flask fitted with a thermometer, vacuum pump, pressure indicator and chlorine source, and immersed in a constant temperature bath, was placed 75 g. of pulverized, porous polychlorotrifluoroethylene. The system was evacuated to a pressure of about 0.1 mm. of mercury, then chlorine was added to a pressure of 560 mm. of mercury. The polymer temperature before chlorine addition was 75°F. and after a contact time of 5 minutes was 77°F. Then the polymer temperature slowly dropped to 75°F. over an additional contact time of 145 minutes. The change of pressure with contact time was:

Contact Time (minutes)	Pressure, mm. Hg
0	560
5	414
30	364
60	346
90	338
120	330
150	322

The treated polymer was transferred to a beaker and heated at 300°F. for 7 days for removal of excess absorbed chlorine. The flow index of the polychlorotrifluoroethylene before treatment was 3.0 mg./min. and after treatment was 3.6 mg./min. The flow index after heating at 570°F. for one hour of untreated polychlorotrifluoroethylene was 104 mg./min. and that of treated polymer was 46 mg./min. The untreated polychlorotrifluoroethylene had a brown color after heating at 570°F. for one hour, but the treated polymer showed no discoloration.

COPOLYMERS WITH VINYLIDENE FLUORIDE

Alkaline pH Conditions

S. Bandes; U.S. Patent 3,018,276; January 23, 1962; assigned to Allied Chemical Corporation has found that chlorotrifluoroethylene and mixtures containing chlorotrifluoroethylene and up to about 50% by weight of vinylidene fluoride may be polymerized in aqueous suspension medium in the presence of a redox catalyst comprising a water-soluble perphosphate and a water-soluble reducing agent, while maintaining the aqueous suspension medium at an approximately neutral to slightly alkaline pH of 6.5 to 8.5, to produce excellent polymer yields; i.e., about 80% up to almost 100% yields of polymer are obtained.

In the redox catalyst of the process the water-soluble perphosphate acts as a promoter and the water-soluble reducing agent as an activator. Although any water-soluble perphosphate may be employed, such as the alkali metal (e.g. sodium and potassium) and ammonium perphosphates. It is preferred to use an alkali metal perphosphate, and particularly, potassium perphosphate.

Any suitable water-soluble reducing agent may be used in preparing the redox catalyst. For example, bisulfates (including metabisulfites), hydrosulfites, sulfides, thiosulfates, hydrazine and hydroxylamine may be employed. It has been found convenient to employ as reducing agent bisulfites or metabisulfites, and particularly the alkali metal salts. When a metabisulfite is used, it hydrolyzes immediately in the aqueous medium to the corresponding bisulfite.

The pH of the aqueous suspension medium must be maintained within the range of 6.5 to 8.5 in order to attain the desired results of the process. In order to achieve a pH within the indicated range when a substantially pure perphosphate is employed, it is generally necessary, depending upon the nature of the reducing agent used, to add a suitable basic material such as potassium hydroxide or a suitable acidic material such as phosphoric acid to the aqueous suspension medium.

For example, when an acidic material such as a bisulfite or metabisulfite is employed as reducing agent with a substantially pure perphosphate, addition of a basic material is usually required to achieve the desired pH. However, when the potassium perphosphate-containing electrolyte solution is used in conjunction with a bisulfite or metabisulfite, it is found that the solution, as produced, contains sufficient basic material, i.e., dipotassium perphosphate, to obtain the desired pH. The dipotassium perphosphate also acts as a buffering agent in aiding maintenance of the initially produced pH. In each of the following examples the potassium perphosphate used was in the form of electrolyte solution produced by the process of Fichter and Gutzwiller. The pH of this solution ranged from 9.5 to 11.0.

Examples 1 to 3: Chlorotrifluoroethylene was polymerized in the following manner. Distilled water and a redox catalyst comprising potassium perphosphate and sodium metabisulfite were charged to a stainless steel closed reactor. The reactor was then quickly immersed in a bath of Dry Ice and acetone (at about −65°C.) and evacuated. The chlorotrifluoroethylene was condensed into the evacuated reactor, and the reactor was placed in an end-over-end agitator which was housed in a thermostated water bath.

After the polymerization reaction was completed, chlorotrifluoroethylene polymer was recovered by venting off unreacted chlorotrifluoroethylene and discharging the polymer in water to a filter where the polymer was washed several times with hot water and filtered off. The polymer was then dried at about 130°C. in a filtered air circulating oven. The operating conditions and results of typical runs are set forth in the table below.

Example	Chlorotrifluoro-ethylene, Parts	H_2O, Parts	Potassium Perphosphate, Parts	Sodium Metabisulfite, Parts	pH	Time Hours	Temperature °C.	Yield,* Percent	NST,** °C.
1	99	250	0.74	0.50	7.1 to 7.2	18	35	92	225
2	105	300	0.74	2.0	6.9 to 7.35	18	26	83	325
3	100	250	0.74	0.50	7.3 to 7.9	18	33	82	300

*Parts of polymer divided by parts of monomer fed.
**NST (no strength temperature) — a test used to provide an indication of relative molecular weight in accordance with the procedure outlined on pages 636, 638 and 641 of "Preparation, Properties and Technology of Fluorine and Organic Fluorine Compounds" by Slesser and Schram.

Examples 4 to 7: Mixtures of chlorotrifluoroethylene and vinylidene fluoride were polymerized as follows. Distilled water in which was dissolved a catalyst comprising potassium perphosphate and sodium metabisulfite was charged to a stainless steel, water-jacketed closed reactor supplied with an agitator having vertical motion. The reactor was then evacuated. The chlorotrifluoroethylene and vinylidene fluoride, in that order, were then condensed in a cylinder immersed in a bath (at −65°C.) of Dry Ice and acetone. The cylinder was then heated to about 60°C. and its contents charged to the evacuated reactor.

After the polymerization reaction was completed, copolymer of chlorotrifluoroethylene and vinylidene fluoride was recovered by venting off unreacted chlorotrifluoroethylene and vinylidene fluoride and discharging the copolymer in water to a filter where the copolymer was washed several times with hot water and filtered off. The copolymer was then dried at about 100°C. in a filtered air circulating oven. The operating conditions and results of typical runs are set forth in the table below. All examples include 1,500 parts water.

Example	Chlorotrifluoro-ethylene, Parts	Vinylidene Fluoride, Parts	Potassium Perphosphate, Parts	Sodium Metabisulfite, Parts	pH	Time, Hours	Temp. °C.	Yield, Percent	NST, °C.
4	386	20	2.8	1.9	about 6.8	18	35	92	266
5	385	20	3.0	4.0	6.6 to 7.1	19	25	99	302
6	376	20	2.0	1.1	7.3 to 7.4	18	35	92	266
7	426	22.4	3.0	4.0	7.1 to 7.8	3 1/2	26	82	280

Example 8: Chlorotrifluoroethylene was polymerized by the process described in Examples 1 to 3, while maintaining a highly acidic pH. The operating conditions and results are given below.

Example	Chlorotrifluoro-ethylene, Parts	H_2O, Parts	Potassium Perphosphate, Parts	Sodium Metabisulfite, Parts	pH	Time, Hours	Temperature °C.	Yield, Percent
8	100	250	0.74	0.40	3.0 to 2.7*	18	25	1

*Phosphoric acid was initially added to the aqueous suspension medium in this run until a pH of 3.0 was obtained.

Thus, it is seen that when a highly acidic pH such as described by the prior art was used in the process for producing chlorotrifluoroethylene polymer, practically no yield was obtained.

Chain Transfer Agents

A.N. Bolstad and F.J. Honn; U.S. Patent 2,956,048; October 11, 1960; assigned to Minnesota Mining and Manufacturing Company have found that resinous thermoplastic copolymers of trifluorochloroethylene and vinylidene fluoride, having superior chemical inertness and corrosion-resistant properties to oils, hydrocarbon fuels and various powerful reagents and which are also more easily brought into solution in various commercial solvents, can be prepared by copolymerizing the trifluorochloroethylene monomer with the vinylidene fluoride monomer, under polymerization conditions, in the presence of polymerization modifiers.

It is found that the trifluorochloroethylene monomer-content in the finished copolymer, can be increased as a result of employing the modifiers in the polymerization system, without impairing the ease with which the finished copolymer is brought into solution in various solvents, for application to surfaces as protective coatings. The process, therefore, results in producing resinous thermoplastic copolymers of trifluorochloroethylene and vinylidene fluoride suitable for application to various surfaces.

The polymerization modifiers are selected from the group consisting of tertiary mercaptans, chlorinated alkanes and brominated alkanes. These tertiary mercaptans, preferably have from 4 to 20 carbon atoms per molecule. An example of such tertiary mercaptans is dodecyl mercaptan ($C_{12}H_{25}SH$). The chlorinated and brominated alkanes, may possess hydrogen and other halogen substitution, for example, fluorine, in addition to the chlorine and/or bromine substituents. The following examples illustrate the process.

Example	Modifier	Parts of Modifier	Charge [1] Moles $CF_2=CFCl$/ $CF_2=CH_2$	Polymerization Time (hrs.)	Percent Conv.	0.5% Soln.[2] Viscosity	Mole Percent $CF_2=CFCl$ in Polymer
1.........	CHCl₃............	10	90/10	7.5	6	0.55	83
		10	80/20	7.5	54	0.53	71
		10	86/14	7.5	49	0.53	78
		5	85/15	4.8	55	0.57	78
		1	85/15	4.8	89	0.74	81.5
2.........	CCl₄............	10	90/10	7.5	63	0.70	86.5
		10	86/14	7.5	76	0.68	80
		10	80/20	7.5	66	0.61	74
3.........	CH₂Cl₂	10	85/15	5.2	60	.62	81.5
4.........	CH₃Cl	10	85/15	3.3	46	.92	79.5
5.........	CBr₃H	2	80/20	6	39	0.45	76.5
		2	86/14	7	46	0.46	81.5
		1	85/15	6.5	59	0.49	80
		1	80/20	7.5	62	0.48	78
6.........	CBrCl₃	4	86/14	7	8	.38	91
		2	86/14	7	8	.39	90
		0.5	85/15	15	95	.50	86.5
		0.5	80/20	15	98	.57	87
7.........	CFCl₃	10	85/15	2.6	61	.96	81
		5	86/14	7	100	.98	84
		5	90/10	7	77	.86	89.5
		10	86/14	7	76	.99	78
8.........	CF₂ClCFCl₂	10	85/15	4.2	82	.80	80
		10	80/20	3.2	82	.87	74.5
		5	85/15	3.2	71	.83	84
		5	80/20	2.7	87	.96	73
9.........	CCl₂CF₂CFClBr	5	85/15	7.1	27	.42	86
10.........	CCl₃O (O, Cl)	2.5	86/14	7	99	.97	75
		2.5	90/10	7	95	.84	89
11.........	CCl₃O	1	85/15	2	86	.87	82
12.........	Dodecyl Mercaptan	0.3	86/14	7	10	0.42	75
		0.1	86/14	7	12	0.44	81
		0.1	90/10	7	13	0.43	82
13.........	None	---------	86/14	7.5	99	1.18	80
		---------	80/20	7.5	95	1.23	80
		---------	80/20	4	7	1.58	72
		---------	75/25	4	99	1.19	72

[1] Recipe (parts by wt.): Water 200; Monomers 100; $(NH_4)_2S_2O_8$ 1.0; $Na_2S_2O_3$ 0.4; $FeSO_4.7H_2O$ 0.05. Temperature=20° C.
[2] In dichlorobenzotrifluoride at 266° F.

It will be seen from the examples, 1 through 12, that the presence of the polymerization modifiers of the process in the polymerization recipe, results in obtaining a resinous thermoplastic copolymer, which, when dissolved in the solvent, results in obtaining a coating composition possessing a low solution viscosity. It will also be seen that there can also be obtained an increase in the trifluorochloroethylene content in the finished resinous copolymer over that previously obtained by reason of the modifier being present.

For comparative purposes, the polymerization reaction shown in Example 13, in which no modifier was present in the polymerization recipe, results in obtaining a resinous copolymer, which, when dissolved in the solvent, produces a mixture having a solution viscosity so high as to make it impractical to employ these coating compositions for commercial applications.

Interpolymer Compositions

F.J. Honn and J.M. Hoyt; U.S. Patent 3,053,818; September 11, 1962; assigned to Minnesota Mining and Manufacturing Company describe trifluorochloroethylene interpolymers. The interpolymeric compositions are produced by interpolymerizing

131

trifluorochloroethylene, vinylidene fluoride and another fluoroolefin. Examples of the other fluoroolefin which is inter-polymerized with the trifluorochloroethylene and the vinylidene fluoride monomers are vinyl fluoride, 1,1-chlorofluoro-ethylene, 4,6,7-trichloroperfluoroheptene-1, trifluoroethylene, perfluoropropene, perfluoroisobutene, tetrafluoroethylene, 2-chloropentafluoropropene, bromotrifluoroethylene, perfluorocyclobutene and 1,1,1,4,4,4-hexafluorobutene-2.

The interpolymers thus produced are valuable macromolecules and are adaptable to a wide variety of commercial uses. They are chemically and physically stable, resistant to oil and hydrocarbon fuels, selectively soluble in various commer-cial solvents and can be molded by conventional techniques to yield a wide variety of useful articles. They also serve as durable, flexible, protective coatings on surfaces which are subjected to environmental conditions in which they may come into contact with various corrosive substances, such as oils, fuels and strong chemical reagents.

The most useful interpolymers of the process are produced from monomeric mixtures comprising the trifluorochloroethylene monomer present in an amount of at least 30 mol percent, the vinylidene monomer present in an amount of at least 30 mol percent and the other fluoroolefin monomer present in an amount of at least 10 mol percent. The interpolymers are preferably prepared by carrying out the polymerization reaction in the presence of a free-radical promoter. For this pur-pose, the polymerization reaction is carried out by employing a water-soluble peroxy type initiator in a water-suspension type recipe or an organic peroxide initiator in a bulk-type system. The water-suspension type recipe system is preferred. The following example illustrates the process.

Example: A stainless steel reaction vessel was flushed with nitrogen and then charged with 60 ml. of deionized water and 10 ml. of an aqueous solution containing 2.0 g. of sodium metabisulfite in 100 ml. of solution. The contents of the vessel were then frozen, and the vessel was next charged with 20 ml. of an aqueous solution containing 2.5 g. of ammonium per-sulfate dissolved in 100 ml. of water. The contents of the vessel were next refrozen, and there was then charged to the vessel 10 ml. of an aqueous solution containing 0.5 g. of $FeSO_4 \cdot 7H_2O$ in 100 ml. of solution.

The vessel was connected to a vacuum-transfer system and evacuated at liquid nitrogen temperature. To the frozen con-tents of the vessel were added, by distillation, 11.5 g. of trifluorochloroethylene, 27.5 g. of vinylidene fluoride, and 10.7 g. 1,1-chlorofluoroethylene, to make a total monomer charge containing 15 mol percent trifluorochloroethylene, 65 mol percent vinylidene fluoride and 20 mol percent 1,1-chlorofluoroethylene. After the contents of the reaction vessel had been refrozen at liquid nitrogen temperature, the vessel was evacuated, closed and rocked at room temperature for a period of 5 hours. At the end of this time, the product was collected, washed with hot water and dried to constant weight in vacuo at 35°C. A rubbery interpolymeric product of trifluorochloroethylene, vinylidene fluoride and 1,1-chloro-fluoroethylene was obtained in an amount corresponding to an 86% conversion.

A sample of the raw interpolymer was compression molded at 300°F. for approximately 10 minutes. After molding, the sample was firm, flexible and retained its rubbery characteristics. A volume increase of 20% was observed in the molded sample when tested by ASTM designation, D-471-49-T, in ASTM type II fuel, consisting of isooctane (60% by volume), benzene (5% by volume), toluene (20% by volume) and xylene (15% by volume). Gehman stiffness of the molded sample of raw interpolymer, determined according to ASTM designation D-1053-49-T, was as follows: $T_2 = 8.5$°C.; $T_5 = -1.5$°C.; $T_{10} = -6.5$°C.; $T_{100} = -21.0$°C.

2,5-Dichlorobenzotrifluoride Addition

D.A. Rausch; U.S. Patent 3,072,590; January 8, 1963; assigned to The Dow Chemical Company describes an emulsion polymerization which is conducted in the presence of a small but effective amount of certain readily available aryl halides and substituted aryl halides. Specifically, the aryl halide is one of the group consisting of ortho-dichlorobenzene and 2,5-dichlorobenzotrifluoride.

These materials are particularly useful in the preparation of latexes of chlorotrifluoroethylene polymers and of copolymers of chlorotrifluoroethylene with vinylidene fluoride. They are simply added to the mixture of water, monomer, dispersing agent and catalyst before emulsification and polymerization are carried out in the known manner. Advantageously, these useful aryl halides are employed in an amount equal to from 1.0% by weight to 10.0%, based on the weight of monomeric material.

The latexes resulting from the use of these modifying agents in the emulsification polymerization of fluoroolefins are stable materials well adapted to the formation of continuous films. In general, it will be found, when ortho-dichlorobenzene is em-ployed, that the modifying agent will separate from the latex on standing and may readily be removed without impairing the stability of the latex. The 2,5-dichlorobenzotrifluoride, on the other hand, usually remains incorporated in the latex. The process is illustrated by the following examples.

Example 1: A mixture of 50 parts of oxygen-free distilled water, 0.3 part of sodium bisulfite, 0.015 part of ferrous sul-fate heptahydrate and 2 parts of perfluorooctanoic acid was adjusted to a pH of 3.0. This mixture was then charged into a bottle containing 2.0 parts of 2,5-dichlorobenzotrifluoride and the contents of the bottle were frozen. One part of potassium persulfate and 20 parts of chlorotrifluoroethylene were added, and the bottle was tightly sealed. The bottle was then agitated for 20 hours in a constant temperature bath maintained at 25°C. There was formed a stable, fluid latex having a solids content of 28% in which the average particle size was about 0.05 micron.

Example 2: A mixture of 100 parts of distilled, oxygen-free water, 0.5 part of sodium bisulfite, 0.05 part of ferrous sulfate heptahydrate and 4.0 parts of perfluorooctanoic acid was adjusted to a pH of 3.0. This mixture was then charged into a stainless steel pressure vessel. 6 parts of 2,5-dichlorobenzotrifluoride were added and the contents of the vessel were frozen. The air was then evacuated therefrom. One and one-half parts of potassium persulfate dissolved in 20 parts of distilled water was then charged into the vessel. The contents were again frozen and the vessel was reevacuated.

Chlorotrifluoroethylene (CTFE) and vinylidene fluoride (VF_2) were then distilled into the vessel in a ratio of 80 mols of the former to 20 mols of the latter. The vessel was sealed and placed in a water bath where it was agitated at 20°C. for 20 hours. A stable, fluid latex resulted which had an average particle size of 0.05 micron. When cast onto cellophane the latex gave, when dried, a clear continuous coating.

Example 3: Following the method of Example 2, latexes were prepared from several monomer mixtures of varying compositions. The compositions employed are indicated in the table below.

	Mol Ratio, CTFE/VF_2	2,5-Dichlorobenzotri-fluoride as Percent of Monomer	Solids Percent	Observation
(2)	80/20	4.0	48	Stable, fluid, latex.
(1)	81/19	0.0	49	Coagulation.
(4)	89/11	4.0	31	Stable, fluid, latex.
(5)	89/11	4.0	48	Stable, fluid, latex.
(3)	84/16	4.0	48	Stable, fluid, latex.
(6)	62/38	4.0	37	Stable, fluid, latex.

MISCELLANEOUS

Phosphorus-Containing Polymers

In a process described by R.H. Wade; U.S. Patent 3,054,785; September 18, 1962; assigned to Minnesota Mining and Manufacturing Company a halogen-containing monomer, or monomers, is polymerized with a phosphorus halide in the presence of a catalyst, to produce an open chain polymer. The phosphorus halide provides the terminal groups, tends to control polymerization and gradually modifies free radical reaction. By varying the amount of halide used, the molecular weight of the polymeric material may be varied to produce oils, greases, waxes or solids. The phosphorus halide is selected from a group of phosphorus halides of the formula PX_n where X is a halogen selected from the group consisting of chlorine, iodine and bromine, and n is either 3 or 5. Typical examples of these phosphorus halides are phosphorus tribromide, phosphorus pentabromide, phosphorus trichloride, phosphorus pentachloride and phosphorus triiodide.

The reaction occurs essentially as shown below in a typical equation using chlorotrifluoroethylene and phosphorus tribromide as an example:

$$n(CF_2{=}CFCl) + PBr_3 \longrightarrow Br_2P(CF_2{-}CFCl)Br$$

where n is an integer from 2 to below about 30 for the distillable oils, greases and waxes. These polymeric phosphorus-containing halocarbons may be separated by distillation under anhydrous conditions.

Haloethylene monomers polymerized in the presence of phosphorus trihalides contain a $-PX_2$ terminal group which is subject to hydrolysis under conventional conditions to form an acidic polymer, i.e., a substituted phosphorus acid. Derivatives of these acids, such as amides, esters and salts, may also be obtained. Not only may the polymer be used as a starting material to form the above recited chemical combinations, but it is contemplated that it may also be used in further polymerization reactions.

The hydrolyzed terminal group of the polymer acts as a typical acid, forming metallic salts with metallic hydroxides or with free metals. If the polymer chain is not too long, i.e., not more than about 10 monomeric units, the polymer molecule may readily be brought into solution in an aqueous or aqueous alkaline media. On the other hand, the solubility of the polymer decreases as the molecular weight or chain length of the polymer increases. As the magnitude of the polymer chain becomes sufficiently great, the organic portion remains insoluble while the terminal group dissolves, and an emulsion results.

The acidic polymer is also capable of emulsifying other polymers with which it is compatible. The polymer to be emulsified is finely divided and added to an emulsion of the acidic phosphorus-containing polymer or the solvent is added to a mixture of the polymers. The acidic terminal group of the phosphorus-containing polymer remains dissolved, while the finely divided polymer particles adhere to the monomer portion of the phosphorus-containing polymer. An emulsifier for haloolefinic polymers may be used in the polymerization of haloolefins in aqueous media, in which adequate contact of the various components is assured. It may also be used in compounding waxes or paints in which an aqueous or alkaline solvent is desired. Or it may be used to facilitate application of a polymer to an object through spraying or dipping.

The ability of the polar terminal group to combine with metals also provides a means of securing the phosphorus-containing polymer to a metal surface. The polymer is spread over the metal surface and fused or pressed. Another polymer, which is compatible with the phosphorus-containing polymer, such as a solid homopolymer or trifluorochloroethylene, may then be spread over the surface and the entire polymeric film fused or pressed, thus cementing the film to the metal.

Polymerization in the presence of phosphorus halides adds not only a terminal $-PX_2$ or $-PX_4$ group to the polymer but also a halogen terminal group to the other end of the polymer chain. The presence of a chlorine terminal group is particularly advantageous when it is desirable to prepare phosphorus-containing perfluorochloro polymers. The direct addition of the chlorine during polymerization eliminates the necessity of stabilizing the polymer by further chlorination or fluorination. The following examples illustrate the process.

Example 1: A charge of 465 g. (4 mols) of trifluorochloroethylene, 550 g. (4 mols) of phosphorus trichloride and 24 g. of benzoyl peroxide was placed in a stainless steel bomb and reacted at a temperature of 100°C. for 5 hours, shaking the bomb to facilitate reaction. The pressure initially rose to 275 psi, then decreased until a pressure of 130 psi was reached at the completion of the run. 105.5 g. of unreacted trifluorochloroethylene monomer was recovered. Additional phosphorus trichloride (375 g.) was added to the remaining solid material in the bomb and allowed to stand overnight at room temperature, after which the bomb was heated on a steam bath. The solid polymeric product was recovered, corresponding to a conversion of 77% of the original monomer charge, was largely insoluble in CCl_4, benzene, ether, ethyl alcohol and acetone. Hot 5% sodium hydroxide partially disintegrated the solid, but a test of the filtrate for chloride ion was negative.

Example 2: A charge of 116 g. (1 mol) of trifluorochloroethylene, 208 g. (1 mol) of phosphorus pentachloride, and 6 g. of benzoyl peroxide (approximately 5 weight percent of monomer) was added to a stainless steel bomb and reacted at 100°C. for 5 hours with shaking of the bomb. Pressure initially rose to 260 psi then decreased to a final value of 190 psi at the end of the run. Unreacted monomer (83 g.) was recovered. The polymeric product produced corresponded to a 28% conversion of the total monomer charge.

Polymer Blends to Aid Processability

S. Gates and D.H. Mullins; U.S. Patent 2,944,997; July 12, 1960; assigned to Union Carbide Corporation have found a way to mold, mill or extrude the polymers of chlorotrifluoroethylene without resort to the high processing temperatures leading to severe molecular degradation. This is accomplished by blending a copolymer of chlorotrifluoroethylene and vinylidene fluoride with the chlorotrifluoroethylene polymer prior to mechanical working of the copolymer. Such a blend may be sheeted on a two-roll mill for example, at a temperature of 185°C. whereas the chlorotrifluoroethylene polymer cannot be fluxed at this temperature. The incorporation of the copolymer as a fluxing aid does not impair the desirable physical properties of the base polymers, and because lower processing temperatures are required, the degradation problem is minimized. The following example shows one method of mixing the homopolymer and copolymer.

Example: A mechanical mixture consisting of equal parts of a chlorotrifluoroethylene-vinylidene fluoride copolymer which contained 6.25% vinylidene fluoride, by analysis, and a chlorotrifluoroethylene homopolymer (melt viscosity 55) fluxed readily on a laboratory two-roll mill heated to 180°C. Since the homopolymer could not be fluxed at this temperature, the value of the copolymer as a processing aid is apparent. The following data demonstrate that the improvements obtained by blending do not impair the physical properties of the base homopolymer.

Resin	Blend	Homopolymer
Tensile strength,[1] psi	3,700	4,200
Elongation,[1] percent	50	25
Stiffness modulus,[2] psi	122,000	140,000
T_F, °C.[3]	32	47
T_4, °C.[3]	104	119
Izod impact,[4] ft. lbs. per inch of notch at -50°C.	0.9	1.3

[1]As determined on a Scott L-6 Tensile Tester operating at an ambient temperature of 25°C. and at a constant rate of elongation of 48 in./min.
[2]ASTM Method D-747-50.
[3]ASTM Method D-1043-51. T_F is the point corresponding to 135,000 psi and T_4 to 10,000 psi on the stiffness-temperature curve.
[4]Tests run of annealed samples. The other tests are on quenched samples.

Trifluorochloroethylene

Conversion of Powdered Polymers to Granular Form

M.H. Nickerson; U.S. Patent 3,058,967; October 16, 1962 has found that the powdered polychlorotrifluoroethylene produced by the method described in U.S. Patent 2,851,407, and by other methods which produce the polymer in similar form may be converted into a much more convenient form if it is exposed in the form of a relatively thin layer to a temperature between 450° and 575°F. Under these conditions the material partially melts or sinters to a loosely formed cake which, when subsequently ground up in a mechanical chopping device fitted with a coarse screen, yields a material having the general appearance of coarse sand or fine gravel. This treatment completely eliminates the original dusty character, together with the tendency of the powder to pack and "bridge" in the hopper of an extruder or injection molding machine.

The sand-like material obtained by sintering flows freely and behaves in a highly desirable manner in the screw of an extruder. Care must be exercised that the polymer is not exposed to temperatures above 575°F. for an excessive period of time since the effect of very high temperatures upon this polymer is to cause a degradation of molecular weight. When the sintering operation is carried out properly as described, the degradation is slight and the granular product will be found to have substantially the same molecular weight as the original material. Any slight loss in molecular weight is more than compensated for by the improved handling characteristics of the polymer. If desired, and it is most convenient to do so, the polymer recovered in moist form from the water dispersion in which it was polymerized may be dried and sintered in the same operation.

The following exemplifies one manner of carrying out the conversion of the polymer to an improved form for handling. The washed and dried powdered polymer, having an NST value of 300, was spread out uniformly on trays to a depth of 3/8 of an inch, approximately 7 lbs. of the polymer covering a tray area of approximately 20 inches by 40 inches. The trays used were, and preferably are, coated with a fused continuous film of polytetrafluoroethylene to permit easy separation of the sintered polymer from the surface.

The trays were then inserted between two electrically heated metal plates, approximately 2 to 3 inches apart. These plates were essentially the same size as the trays and were maintained at a temperature of 500°F. After 20 to 25 minutes the trays were withdrawn, and the loosely sintered cake broken up into smaller pieces for feeding to a chopper or other mechanical disintegrator, which reduced the cake to a free-flowing granular material free of dust.

Crystalline Ethylene Copolymers

A process described by M. Ragazzini and D. Carcano; U.S. Patent 3,501,446; March 17, 1970; assigned to Montecatini Edison SpA, Italy provides ethylene-monochlorotrifluoroethylene copolymers which are crystalline in structure, have a melting point considerably higher than the melting points of polyethylene and polymonochlorotrifluoroethylene, and a molar ratio of ethylene to monochlorotrifluoroethylene between 1 and 2. The melting point is a function of the monochlorotrifluoroethylene content of the copolymer, the maximum corresponding to a molar ratio of monochlorotrifluoroethylene to ethylene equal to 1. The copolymers are produced by contacting the monomers with a catalyst selected from the group consisting of boron alkyls, boron hydrides, alkyl boron hydrides and their complexes in the presence of a substance yielding oxygen. The following examples illustrate the process.

Example 1: In a 500 cc flask, fitted with a perfectly tight propeller stirrer, immersed in a trichloroethylene bath and Dry Ice at -78°C., 200 cc of chlorotrifluoroethylene are condensed in nitrogen current. Through a dipping tube there is then introduced the quantity of ethylene necessary for obtaining a molar composition of the liquid phase chlorotrifluoroethylene/ethylene = 75/25. By means of a hypodermic syringe there are introduced in this order: 1 cc ethylic ether, 100 cc oxygen and 10 cc of a solution at 10% by volume of triethyl boron in trichlorotrifluoroethane ($CCl_2F-CClF_2$). The polymerization is conducted for 5 hours and 150 g. of a copolymer containing 24.6% by weight of chlorine, corresponding to a molar ratio chlorotrifluoroethylene/ethylene equal to 1/1, are thus obtained.

The infrared spectrum shows that it is an actual copolymer. The x-ray spectrum is characteristic for a crystalline product. In this copolymer the two comonomers regularly alternate along the chain and such an arrangement gives the product peculiar characteristics, such as, for instance, the melting point, which is 260°C.

Example 2: In a 500 cc flask, immersed in a bath of trichloroethylene cooled to -40°C. with Dry Ice and fitted with a perfectly tight propeller stirrer, a copolymerization test is conducted for 4 hours, following the same procedures as those of Example 1. 180 g. of a copolymer are obtained that will contain 24.6% by weight of chlorine, corresponding to a molar ratio chlorotrifluoroethylene/ethylene = 1/1. The melting point of the product will be about 235°C. and the structural analysis will yield the same results as those obtained for the copolymers prepared at -78°C.

FLUORODIENES

FLUORINATED BUTADIENES

2-Fluorobutadiene

In a process described by A.N. Bolstad and E.S. Lo; U.S. Patent 2,951,063; August 30, 1960; assigned to Minnesota Mining and Manufacturing Company the fluoroprene (2-fluorobutadiene) monomer is copolymerized with a polymerizable straight-chain diene containing a terminal carbon atom having two fluorine substituents. Examples of these latter polymerizable straight-chain dienes that can be polymerized with the fluoroprene monomer are 1,1,2-trifluorobutadiene; 1,1,3-trifluorobutadiene; 1,1,-difluorobutadiene; 1,1,2,4-tetrafluorobutadiene; 1,1,2,4,4-pentafluorobutadiene; and perfluorobutadiene. The following example illustrates the copolymerization of fluoroprene and 1,1,2-trifluorobutadiene to produce an elastomeric copolymer.

Example: A heavy-walled glass polymerization tube was flushed with nitrogen and then charged with 9 ml. of a soap solution comprising 6 g. of the ammonium salt of perfluorooctanoic acid ($C_7F_{15}COONH_4$), as an emulsifier; 0.4 g. of sodium metabisulfite; 0.5 g. borax; and 0.1 g. dodecyl mercaptan dissolved in 180 ml. of water. The tube was then immersed in a liquid nitrogen freezing bath.

When the contents of the tube were frozen solid, 1 ml. of an aqueous solution containing 1.0 g. of potassium persulfate dissolved in 20 ml. of water were added. The contents of the tube were then refrozen, and the tube was then connected to a gas-transfer system and evacuated at liquid nitrogen temperature. To the frozen contents of the tube were added, by distillation, 2.94 g. of 1,1,2-trifluoro-1,3-butadiene and 2.06 g. of fluoroprene, which comprised a 50/50 molar ratio.

The polymerization tube was sealed and rotated end-over-end in a temperature-regulated bath at 50°C. under autogenous pressure for a period of 22 hours. At the end of this time, the contents of the tube were coagulated by freezing at liquid nitrogen temperature. The coagulated product was then removed from the tube, washed with hot water until entirely free of soap and dried to constant weight in vacuo at 35°C. An extensible rubbery copolymer was obtained and then found, upon analysis, to comprise approximately 58 mol percent fluoroprene and the remaining major constituent being 1,1,2-trifluoro-1,3-butadiene. The copolymer was obtained in an amount representing a 100% conversion.

Fluoroprene-Vinylidene Chloride Copolymers

J.M. Hoyt; U.S. Patent 2,915,510; December 1, 1959; assigned to Minnesota Mining and Manufacturing Company found that the copolymerization of fluoroprene and a chlorine-containing monoolefin having from two to three carbon atoms per molecule, such as vinylidene chloride, vinyl chloride or 2-chloropentafluoropropene produces an elastomeric polymeric fluoroprene composition possessing good chemical and physical stability and resistance to oils, fuels, and various strong chemical reagents.

The elastomeric polymeric compositions of the process are preferably prepared by carrying out the polymerization reaction in the presence of a free-radical-forming promoter. For this purpose, the polymerization reaction is carried out by employing a water-soluble peroxy-type initiator in a water-suspension-type recipe or an organic peroxide initiator in a bulk-type system. The water-suspension-type recipe is preferred. The following examples illustrate the process.

Example 1: A heavy-walled glass polymerization tube of about 20 ml. capacity was flushed with nitrogen and then charged with 5 ml. of an aqueous solution prepared by dissolving 5 g. of potassium stearate in 100 ml. of water, and adjusted to a pH of 11.8 by the addition of potassium hydroxide. The soap solution in the glass polymerization tube was then frozen. To the frozen contents of the tube were then charged 4 ml. of an aqueous solution, prepared by dissolving 0.75 g. of potassium persulfate in 80 ml. of water. The contents of the tube were then refrozen. Thereafter, 1 ml. of an aqueous solution, prepared by dissolving 0.4 g. of sodium metabisulfite and 0.5 g. of borax in 20 ml. of water, was added to the tube. The contents of the tube were once more frozen in liquid nitrogen. The tube was then connected to a gas-transfer system and evacuated at liquid nitrogen temperature. To the frozen contents of the tube were added, by distillation, 2.13 g. of fluoroprene and 2.87 g. of vinylidene chloride, which comprised a comonomer mixture containing 50 mol percent fluoroprene and 50 mol percent vinylidene chloride. The polymerization tube was then sealed under vacuum and agitated in a temperature-regulated water bath at 50°C. for a period of 24 hours.

At the end of this time, the contents of the tube were coagulated by freezing. The coagulated product was then removed from the tube, washed with hot water and then dried to constant weight in vacuo at 35°C. A copolymeric elastomeric product was obtained, which was found, upon analysis, to comprise approximately 70 mol percent fluoroprene, and the remaining major constituent, vinylidene chloride, being present in an amount of approximately 30 mol percent. The copolymer was obtained in an amount corresponding to a 53% conversion.

Example 2: Employing the procedure set forth in Example 1 and the same polymerization system, the tube was charged with 2.68 g. of fluoroprene and 2.32 g. of vinyl chloride, which comprised a comonomer mixture containing 50 mol percent fluoroprene and 50 mol percent vinyl chloride. The copolymerization reaction was carried out at a temperature of 50°C. for a period of 23 hours. The resultant elastomeric product was worked up in accordance with the same procedure as set forth in Example 1. A soft rubbery product was obtained, and upon analysis, was found to comprise approximately 96 mol percent fluoroprene, and the remaining major constituent, vinyl chloride, being present in an amount of approximately 4 mol percent. The copolymer was obtained in an amount corresponding to an 8% conversion.

Trifluorobutadiene

E.S. Lo; U.S. Patent 2,986,556; May 30, 1961; assigned to Minnesota Mining and Manufacturing Company has found that the copolymerization of trifluorobutadiene and a fluorinated ethylene produces an elastomeric copolymeric composition possessing good chemical and physical stability, and good resistance to oils, fuels and various strong chemical reagents.

The trifluorobutadienes which are copolymerized with the fluorinated ethylene include 1,1,2-trifluorobutadiene and 1,1,3-trifluorobutadiene. The fluorinated ethylene comonomers, which are copolymerized with the trifluorobutadienes, include 1,1-dichloro-2,2-difluoroethylene, 1-chloro-1-fluoroethylene, perfluoroethylene, 2-chloro-1,1-difluoroethylene, trifluoroethylene, vinyl fluoride and vinylidene fluoride.

The 1,1,2-trifluorobutadiene monomer is obtained by adding dibromofluoromethane to 1-fluoropropene to produce the adduct $CF_2BrCHFCHBrCH_3$, which, upon dehydrobromination, yields $CF_2{=}CFCH{=}CH_2$, BP 4.8° to 8.0°C. The 1,1,3-trifluorobutadiene monomer is obtained by adding dibromodifluoromethane to 2-fluoropropene to produce the adduct, $CF_2BrCH_2CFBrCH_3$, which is then dehydrobrominated at about 150°C. using tri-n-butyl amine to yield $CF_2{=}CH{-}CF{=}CH_2$, BP 17.5° to 19.4°C.

In general, the copolymeric compositions of the process are produced from the polymerization of monomeric mixtures containing the trifluorobutadiene and the fluorinated ethylene at temperatures between –20° and 120°C., with intermediate temperature ranges being selected with reference to the specific polymerization system employed. The preferred elastomeric copolymeric compositions are copolymers produced from monomeric mixtures containing between 25 and 60 mol percent of the trifluorobutadiene and the remaining major constituent being any of the above fluorinated ethylene comonomers.

Example: A heavy-walled glass polymerization tube of about 20 ml. capacity was flushed with nitrogen and then charged with 5 cc of a solution prepared by dissolving 0.1 g. of dodecyl mercaptan and 6 g. of the ammonium salt of perfluorooctanoic acid in 100 cc of water. The contents of the tube were then frozen, and the tube was then charged with 1 cc of a promoter solution prepared by dissolving 0.4 g. of sodium metabisulfite and 0.5 g. of borax in 20 cc of water. To the contents of the tube were next charged 4 cc of a solution prepared by dissolving 1 g. of potassium persulfate in 80 cc of water.

The contents of the tube were then refrozen, and the tube was next connected to a gas-transfer system and evacuated at liquid nitrogen temperature. To the frozen contents of the tube were added, by distillation, 1.02 g. of 1,1,3-trifluorobutadiene and 1.66 g. of 1,1-dichloro-2,2-difluoroethylene, which comprised a comonomeric mixture containing 45 mol percent 1,1,3-trifluorobutadiene and 55 mol percent 1,1-dichloro-2,2-difluoroethylene. After the contents of the tube were thoroughly frozen with liquid nitrogen, the tube was evacuated and sealed.

The polymerization tube and its contents were agitated in a temperature-regulated water bath at 50°C. for a period of 72 hours. At the end of this time, the contents of the tube were coagulated by freezing. The coagulated product was then removed from the tube, washed with hot water and then dried to constant weight in vacuo at 35°C. A copolymeric rubbery product was obtained which was found, upon analysis, to comprise approximately 74 mol percent 1,1,3-trifluorobutadiene and the remaining major constituent, 1,1-dichloro-2,2-difluoroethylene, being present in an amount of approximately 15 mol percent. The copolymer was obtained in an amount corresponding to 31% conversion.

Fluorodienes

A sample of the raw copolymer was compression molded at 350°F. for a period of about 10 minutes. After molding, the sample remained flexible, retaining its rubbery characteristics. A volume increase of 30.2% was observed in the molded sample when tested by ASTM Designation D-471-49T, in ASTM Type II Fuel, consisting of isooctane (60% by volume), benzene (5% by volume), toluene (20% by volume) and xylene (15% by volume.) Gehman stiffness of the molded sample of raw copolymer determined according to ASTM Designation D-1053-49T, was as follows: T_2 = -10°C.; T_5 = -19.5°C.; T_{10} = -24.5°C.; T_{100} = below -50°C.

A.N. Bolstad and J.M. Hoyt; U.S. Patent 3,398,128; August 20, 1968; assigned to Minnesota Mining and Manufacturing Company describe copolymers of 1,1,2-trifluorobutadiene-1,3 and another fluorinated 1,3-diene having from 4 to 5 carbon atoms per molecule containing two fluorine atoms on a terminal carbon atom and at least one hydrogen atom. The following example illustrates the copolymerization of 1,1,2-trifluorobutadiene and 1,1,3-trifluorobutadiene to produce an elastomeric copolymer.

Example: A heavy-walled glass polymerization tube was flushed with nitrogen and was then charged with 7 ml. of a 0.75% aqueous solution of the C_8-telomer acid derived from chlorotrifluoroethylene, $Cl(CF_2CFCl)_3CF_2COOH$, which had been adjusted to pH 9.5 with KOH solution. The potassium C_8-telomerate, $Cl(CF_2CFCl)_3CF_2COOK$, functions as an emulsifier. The stoppered tube was then placed in a Dry Ice-acetone freezing bath. After the contents of the tube were frozen solid, the tube was charged with 3 ml. of a 1% aqueous solution of potassium persulfate. In a separate experiment it was found that the final pH is 7.0 when the aforesaid solutions, in the amounts stated, are mixed without freezing.

The contents of the tube were then refrozen, and the tube was connected to a gas-transfer system and evacuated at liquid nitrogen temperature. Thereafter 3.75 g. of 1,1,3-trifluoro-1,3-butadiene and 1.25 g. of 1,1,2-trifluoro-1,3-butadiene were distilled into the tube to make up a total monomer charge comprising 75 mol percent of 1,1,3-trifluoro-1,3-butadiene and 25 mol percent of 1,1,2-trifluoro-1,3-butadiene. The monomers are prepared by the procedure as is disclosed in the Journal of the American Chemical Society, vol. 77, page 2,786 (May 20, 1955).

The polymerization tube was then sealed and rotated end-over-end in a temperature-regulated water bath at 50°C. The polymerization was conducted under autogenous pressure at 50°C. for a period of 24 hours. The polymer latex thus obtained was coagulated by freezing at liquid nitrogen temperature. The coagulated product was collected, washed with hot water, and dried to constant weight in vacuo at 35°C. A relatively soft, snappy rubber was obtained which was found, upon analysis, to comprise approximately 35 mol percent of 1,1,2-trifluoro-1,3-butadiene, and the remaining major constituent 1,1,3-trifluoro-1,3-butadiene. The copolymer was obtained in an amount corresponding to a 63% conversion.

A sample of the raw copolymer was compression molded at 250°F. for 5 minutes. After molding, the sample remained as a snappy rubber. A volume increase of only 18.5% was observed in the molded sample, when tested according to ASTM Designation D-471-49T, in ASTM Type II Fuel, which consists of isooctane (60% by volume), benzene (5% by volume), toluene (20% by volume) and xylene (15% by volume), Gehman stiffness of the molded sample of raw copolymer, determined according to ASTM Designation D-1053-49T is as follows: T_2 = -19°C.; T_5 = -21°C.; T_{10} = -26°C.; T_{100} = -30°C.

J.T. Barr; U.S. Patent 3,379,773; April 23, 1968; assigned to Pennsalt Chemicals Corporation describes the copolymerization of 1,1,2-trifluorobutadiene-1,3 with the comonomers hexafluorobutadiene-1,3; 3,4-dichloro-3,4,4-trifluorobutene-1; 2,2,2-trifluoroethyl vinyl ether; vinyl chloride; styrene; 1,1,2-trifluorobutene-1; and 1,1,4,4-tetrafluorobutadiene-1,3. 1,1,2-trifluorobutadiene-1,3 is a relatively new composition and its preparation is described in U.S. Patent 3,308,175. The process is illustrated by the following examples.

Example 1: A pressure reactor was charged with 50 parts of 1,1,2-trifluorobutadiene-1,3 and 50 parts of hexafluorobutadiene-1,3, and the mixture allowed to stand at room temperature for 3 days. On evaporating off the unreacted monomers a very elastic, rubbery solid, quite different from the homopolymer of either 1,1,2-trifluorobutadiene-1,3 or hexafluorobutadiene-1,3 was obtained.

Example 2: A pressure reactor was charged with 50 parts of 1,1,2-trifluorobutadiene-1,3, 50 parts of 3,4-dichloro-3,4,4-trifluorobutene-1, and 0.5 part of benzoyl peroxide and the mixture reacted at 70°C. for 40 hours. 26.8 parts of a very rubbery solid were obtained.

Example 3: A pressure reactor was charged with 6 parts of 1,1,2-trifluorobutadiene-1,3, 94 parts of methyl acrylate, 150 parts of water, 3.0 parts of Triton X-200, and 0.5 part of potassium persulfate, and agitated at 50°C. for 17 hours. 96 parts of a brittle, crumbly polymer were obtained. 30 parts of this polymer were cured at 144°C. and 500 psi pressure with 12 parts of Philblack A (carbon black), 0.3 part of zinc oxide, 0.12 part of sulfur, and 0.105 part of Altax (benzothiazyl disulfide). The resultant rubber had a tensile strength of 2,380 psi and 425% elongation at its break point.

Example 4: A pressure reactor was charged with 50 parts of 1,1,2-trifluorobutadiene-1,3, 50 parts of methyl methacrylate, 150 parts of water, 3.0 parts of Triton X-200, and 0.5 part of potassium persulfate, and agitated at 50°C. for 17 hours. 93.3 parts of a hard, rubbery polymer were obtained. This polymer had a tensile strength of 3,000 psi and 325% elongation at its break point without having been cured.

2-Trifluoromethyl-1,3-Butadiene Copolymers

A process described by E.S. Lo and G.H. Crawford, Jr.; U.S. Patent 2,951,065; August 30, 1960; assigned to Minnesota Mining and Manufacturing Company comprises polymerizing 2-trifluoromethyl-butadiene with an ethylenically monounsaturated hydrocarbon in which one or more hydrogen atoms is substituted only with a corresponding number of fluorine atoms, in the presence of a polymerization promoter. The polymerization reaction is carried out at temperatures between 15° and 150°C. and preferably at a temperature between 15° and 75°C. The preferred polymerization catalyst system is an aqueous system comprising a peroxy compound and an emulsifier. The copolymers thus obtained are valuable macromolecules having particularly good low temperature flexibility, elasticity, and resilience.

In addition, these compositions are chemically and thermally stable, selectively soluble in various organic solvents and can be molded by conventional techniques to yield a wide variety of useful end products, and are also useful as durable, flexible, protective coatings on surfaces which are subject to environmental conditions in which they may come into contact with temperatures as low as -60°F. and strong chemical reagents.

The preferred comonomers have not more than 6 carbon atoms per molecule and have at least one carbon atom bearing two fluorine substituents. Of this preferred class, the fluoromonoolefins having at least one fluorine atom for every carbon atom, that is, at least as many fluorine substituents as there are carbon atoms, are particularly preferred. Examples of suitable comonomers which are polymerized with 2-trifluoromethyl-butadiene in accordance with the process are vinylidene fluoride, trifluoroethylene, tetrafluoroethylene, 3,3,3-trifluoropropene, 2,3,3,3-tetrafluoropropene, 1,1,3,3,3-pentafluoropropene, hexafluoropropene, hexafluoroisobutene and 1,1,1-trifluoro-3-trifluoromethyl-butene-2.

In order to obtain copolymers of 2-trifluoromethyl-butadiene having a good combination of the above mentioned desirable characteristics, at least 10 mol percent of 2-trifluoromethyl-butadiene is incorporated into the copolymer product. A copolymer containing less than 10 mol percent of 2-trifluoromethyl-butadiene and correspondingly more than about 90 mol percent of the fluoro-monoolefin, does not exhibit appreciably improved low temperature flexibility. The most useful copolymers produced in the process are the elastomers containing between about 10 mol percent and about 95 mol percent of combined 2-trifluoromethyl-butadiene, the remaining major constituent being any one of the above fluoro-monoolefins. Of these copolymers those containing between 40 and 90 mol percent of 2-trifluoromethyl-butadiene are particularly preferred. The following examples illustrate the process.

Example 1: This example illustrates the copolymerization of 2-trifluoromethyl-butadiene with vinylidene fluoride. A heavy-walled glass polymerization tube was flushed with nitrogen and was then charged with 2 ml. of a 2% by weight aqueous solution of sodium metabisulfite. The stoppered tube was then placed in a liquid nitrogen freezing bath. After the contents of the tube were frozen solid, the tube was charged with 5 ml. of a 10% by weight aqueous solution of potassium stearate, having a pH adjusted to about 11 by the addition thereto of an aqueous potassium hydroxide solution.

The contents of the tube were refrozen and the tube was then charged with 3 ml. of a 3.5% by weight aqueous solution of potassium persulfate. The contents of the tube were then refrozen and the tube was connected to a gas transfer system and evacuated at liquid nitrogen temperature. Thereafter the tube was charged with 3.47 g. of 2-trifluoromethyl-1,3-butadiene and 1.76 g. of vinylidene fluoride to make up a total monomer charge containing 50 mol percent of each monomer. The polymerization tube was then sealed and rotated end-over-end in a temperature regulated bath at 25°C. The polymerization reaction was conducted under autogenous pressure at 25°C. for a period of 24 hours. The polymer latex obtained was coagulated by freezing it at liquid nitrogen temperature. The coagulated product was collected, washed with hot water to remove residual salts and dried to constant weight in vacuo at 35°C.

A rubbery material was obtained which, upon analysis for fluorine content, was found to comprise approximately 61 mol percent of combined 2-trifluoromethyl-butadiene, the remaining major constituent being vinylidene fluoride monomer units. The polymer product of this example was obtained in an amount corresponding to a 39% conversion.

A sample of this raw copolymer was compression molded at 120°F. for 10 minutes. After molding the sample remained as a rubbery material. The 2-trifluoromethyl-butadiene copolymer of this example possesses good physical and mechanical properties and retains its flexibility at temperatures as low as -32.5°C. and lower. It is particularly suited for the manufacture of end products such as resilient gaskets which are to be used at relatively low temperatures where retention of rubbery properties is a prime requisite.

Example 2: This example illustrates the copolymerization of 2-trifluoromethyl-butadiene with tetrafluoroethylene. Employing the procedure set forth in Example 1, and the same aqueous emulsion polymerization system, the tube was charged with 2.75 g. of 2-trifluoromethyl-1,3-butadiene and 2.25 g. of tetrafluoroethylene, to make up a total monomer charge containing 50 mol percent of each of the monomers. The polymerization reaction was carried out under autogenous pressure at a temperature of 25°C. for a period of 24 hours. The resultant polymer latex was worked up in accordance with the same procedure set forth in Example 1. A tough, snappy rubber was obtained which, upon analysis for fluorine content, was found to comprise approximately 67 mol percent of combined 2-trifluoromethylbutadiene, the remaining constituent being combined tetrafluoroethylene monomer units. The copolymer was obtained in an amount corresponding to a 22% conversion. A sample of this raw copolymer was compression molded at 250°F. for 10 minutes. After molding, the sample remained as a tough, snappy rubber having a torsional modulus of 12.6 psi. The raw copolymer milled easily at 25°C. in a conventional rubber mill and remained as a tough rubber. The raw copolymer retains its flexibility at temperatures as low as −26°C. and lower without marked evidence of embrittlement.

In related work, E.S. Lo; U.S. Patent 2,979,489; April 11, 1961; assigned to Minnesota Mining and Manufacturing Company has found that the incorporation of 2-chloropentafluoropropene into the polymer of 2-trifluoromethyl-1,3-butadiene, improves the properties of the 2-trifluoromethyl-1,3-butadiene homopolymer to such an extent that excess solvent-swell is appreciably prevented and that the 2-trifluoromethyl-1,3-butadiene homopolymer is not attacked, thereby resulting in a useful copolymeric composition. These copolymeric compositions of 2-trifluoromethyl-1,3-butadiene and 2-chloropentafluoropropene, an addition to being highly resistant to excess solvent-swell, exhibit increased resistance to oil and hydrocarbon fuels, improved flow properties and improved chemical and physical stability. The following example illustrates the process.

Example: A heavy-walled glass polymerization tube of about 20 ml. capacity was flushed with nitrogen and then charged with 5 ml. of a catalyst solution prepared by dissolving 0.75 g. of the C_8-telomer acid of trifluorochloroethylene, viz., $Cl(CF_2CFCl)_3CF_2COOH$, in 100 ml. of water. This solution in the polymerization tube was then frozen. Thereafter, there was added to the polymerization tube 4 ml. of a solution prepared by dissolving 1 g. of potassium persulfate in 80 ml. of water. The contents of the tube were then refrozen.

Thereafter, there was added to the frozen contents of the tube 1 ml. of a solution prepared by dissolving 0.4 g. of sodium metabisulfite in 20 ml. of water. The contents of the tube were once more frozen. The tube was next connected to a gas-transfer system and evacuated at liquid nitrogen temperature. To the frozen contents of the tube were added, by distillation, 1.85 g. of 2-chloropentafluoropropene and 3.15 g. of 2-trifluoromethyl-1,3-butadiene, which comprised a total monomeric charge containing 30 mol percent of 2-chloropentafluoropropene and 70 mol percent of 2-trifluoromethyl-1,3-butadiene.

The polymerization tube was next sealed under vacuum and agitated in a temperature-regulated water bath at 50°C. for a period of 17 hours. At the end of this time, the contents of the tube were coagulated by freezing at liquid nitrogen temperature. The coagulated product was then removed from the tube, washed with hot water and then dried to constant weight in vacuo at 35°C. An elastomeric copolymeric product was obtained, which was found to comprise 2-chloropentafluoropropene and 2-trifluoromethyl-1,3-butadiene in an amount of 2.1 g. This amount corresponded to a 42% conversion. This copolymer when subjected to the action of ASTM Type II fuel, consisting of isooctane (60% by volume), benzene (5% by volume), toluene (20% by volume) and xylene (15% by volume), is found to show no appreciable solvent-swell increase.

E.S. Lo; U.S. Patent 2,951,064; August 30, 1960; assigned to Minnesota Mining and Manufacturing Company also describes copolymers which contain 2-chloro-3,3,3-trifluoropropene and 2-fluorobutadiene or 2-trifluoromethylbutadiene in varying comonomer ratios. The most valuable polymeric compositions produced in the process are those containing between 5 and 40 mol percent of 2-chloro-3,3,3-trifluoropropene, the remaining major constituent being 2-fluorobutadiene or 2-trifluoromethylbutadiene. Such copolymers are produced by employing an initial monomer mixture containing between 20 and 60 mol percent of 2-chloro-3,3,3-trifluoropropene and correspondingly between 80 and 40 mol percent of one of the above comonomers.

For the production of high molecular polymers, it is preferably to employ a monomer mixture containing less than 60 mol percent of 2-chloro-3,3,3-trifluoropropene. The particularly preferred copolymers contain between about 10 and 30 mol percent of 2-chloro-3,3,3-trifluoropropene and are obtained by charging an initial monomer feed containing between 25 and 50 mol percent of the chlorofluoropropene to the polymerization zone. The polymeric compositions of the process are stable elastomers ranging from soft to hard rubbery materials including tough, snappy rubbers and retain their rubbery characteristics at temperatures as low as −70°F.

2,3-Bis(Trifluoromethyl)-1,3-Butadiene

B.C. McKusick and R.E. Putnam; U.S. Patent 3,035,034; May 15, 1962; assigned to E.I. du Pont de Nemours and Company describe the preparation of monomeric 2,3-bis(trifluoromethyl)-1,3-butadiene and the polymerization of this compound to a solid polymer by free-radical catalysis. The polymers obtained are inert, but contrary to expectations, are not rubbery. Monomeric 2,3-bis(trifluoromethyl)-1,3-butadiene is obtained by the following reactions:

(1) $CF_3-C{\equiv}C-CF_3 + CH_2{=}CH-CH{=}CH_2 \longrightarrow$

Fluorodienes

(2)   [structure: cyclohexadiene with CF₃ groups] $\xrightarrow[\text{PtO}_2]{\text{H}_2}$ [structure: cyclohexene with CF₃ groups]

(3)   [structure: 1,2-bis(trifluoromethyl)cyclohexene] $\xrightarrow{\Delta}$ $CH_2{=}C(CF_3){-}C(CF_3){=}CH_2$

The following examples illustrate the preparation and polymerization of 2,3-bis(trifluoromethyl)-1,3-butadiene.

Example 1: Preparation of 2,3-Bis(Trifluoromethyl)-1,3-Butadiene — A pyrolysis apparatus, which consisted of a dropping funnel and vertical quartz tube 12 inches in length and 3/4 inch in diameter and packed with quartz rings, was attached to a Dry Ice trap and liquid nitrogen cooled trap in series and evacuated to 3 mm. The tube was heated to 750°C. by means of a standard tube furnace and 15 g. of 1,2-bis(trifluoromethyl)cyclohexene was admitted dropwise. There was obtained 14 g. of light yellow condensate of BP greater than 25°C. Distillation gave 6 g. (40%) of 2,3-bis(trifluoromethyl)-1,3-butadiene of BP 58° to 60°C., $n_D^{25} = 1.3236$. Small amounts of trifluoromethyl benzene and 1,2-bis(trifluoromethyl)benzene were also isolated. Analysis — Calcd. for $C_8H_4F_6$: C, 37.9; H, 2.1; F, 60.0. Found: C, 38.91; H, 2.41; F, 59.54.

Example 2: Polymerization of 2,3-Bis(Trifluoromethyl)-1,3-Butadiene — In a Carius tube were charged 0.005 g. of azodiisobutyronitrile, 1.5 g. of 2,3-bis(trifluoromethyl)butadiene and 3 cc of benzene. The tube was cooled, evacuated and sealed. It was heated to 90°C. for 4 hours. The tube was cooled, opened and the benzene was removed from the product by evaporation. There remained 0.1 g. (7%) of poly-2,3-bis(trifluoromethyl)butadiene as a white powder.

Example 3: Polymerization of 2,3-Bis(Trifluoromethyl)-1,3-Butadiene — In a platinum tube, 6 inches x 0.5 inch in diameter and sealed at one end, were placed 3.0 g. of bis(trifluoromethyl)butadiene and 0.01 g. of azodiisobutyronitrile. The tube was cooled to -80°C., flushed with dry nitrogen, and sealed. It was then placed in a 200 cc stainless steel high pressure bomb, which was pressured to 2,000 atmospheres with nitrogen, and was heated to 80°C. for 16 hours.

There was obtained 3.0 g. (100%) of white opaque polymer as a solid plug. The polymer could be pressed to a stiff film at 325°C. and 15,000 lbs. ram pressure. The polymer burned but did not support combustion. On heating on a block it melted and decomposed slowly at 355°C. It was completely unaffected by boiling dimethylformamide, boiling aniline, boiling n-butylamine and boiling concentrated nitric acid.

Tetra- and Pentafluorobutadienes

E.S. Lo and G.H. Crawford, Jr.; U.S. Patent 2,992,211; July 11, 1961; assigned to Minnesota Mining and Manufacturing Company have found that the copolymerization of tetrafluorobutadiene and fluorine-containing monoolefins produces copolymeric compositions possessing good chemical and physical stability and good resistance to oils, fuels and various strong chemical reagents. Examples of the fluorine-containing monoolefins which can be copolymerized with the tetrafluorobutadiene are trifluorochloroethylene, 1,1-dichloro-2,2-difluoroethylene, trifluoroethylene or vinylidene fluoride.

In general, the copolymeric compositions are produced from the polymerization of monomeric mixtures containing the tetrafluorobutadiene and the fluorine-containing monoolefin at temperatures between -20° and 150°C. with intermediate temperature ranges being selected with reference to the specific polymerization system employed. The preferred copolymeric compositions are copolymers produced from monomeric mixtures containing between 25 and 75 mol percent of the tetrafluorobutadiene and the remaining major constituent being the fluorine-containing monoolefin. The following example illustrates the process.

Example: A heavy-walled glass polymerization tube of about 20 ml. capacity is flushed with nitrogen and then charged with 5 cc of a solution prepared by dissolving 5 g. of potassium stearate in 100 cc of water. This solution is next adjusted to a pH of 11 by the addition of potassium hydroxide. The contents of the tube are then frozen, and the tube is next charged with 4 cc of a solution prepared by dissolving 0.75 g. of potassium persulfate in 80 cc of water. The contents of the tube are next refrozen, and the tube is then charged with 1 cc of a solution prepared by dissolving 0.4 g. of sodium metabisulfite in 20 cc of water. The tube and its contents are then refrozen. To the frozen contents of the tube are then added, by distillation, 2.6 g. of 1,1,2,4-tetrafluorobutadiene and 2.4 g. of trifluorochloroethylene, comprising a comonomeric mixture containing 50 mol percent of each monomer.

The polymerization reaction is carried out at a temperature of 50°C. for a period of 70 hours. At the end of this time, the contents of the tube are coagulated by freezing. The coagulated product is then removed from the tube, washed with hot

141

water and then dried to constant weight in vacuo at 35°C. A white, powdery copolymeric product is obtained comprising 82.5 mol percent 1,1,2,4-tetrafluorobutadiene and 17.5 mol percent trifluorochloroethylene, in a good yield.

H. Iserson, F.E. Lawlor and M. Hauptschein; U.S. Patent 3,062,794; November 6, 1962; assigned to Pennsalt Chemicals Corporation have found that the fluorinated butadiene 1,1,2,3-tetrafluorobutadiene-1,3 is unique among the fluorinated butadienes in that, although having a high proportion of fluorine, it polymerizes with ease to high molecular weight polymers having excellent elastomeric properties and other outstanding properties including high melting point, high decomposition temperature, excellent resistance to attack by solvent and other chemical reagents, and good low temperature properties. The following examples illustrate the preparation of the homopolymer.

Example 1: 20 parts by weight of deoxygenated water, 0.03 part by weight of potassium persulfate, 0.5 part of sodium lauryl sulfate and 9.8 parts of 1,1,2,3-tetrafluorobutadiene-1,3 are introduced into a pressure reaction vessel which is sealed under vacuum. The vessel and contents is heated at 60°C. for 16 1/2 hours, cooled and vented. The polymer is washed thoroughly with water and dried in a vacuum oven at 60° to 80°C. for 3 hours. The polymer, obtained in 60% conversion, is a strong, white elastomer.

Example 2: This example illustrates the polymerization of 1,1,2,3-tetrafluorobutadiene-1,3 by the use of ultraviolet light. 1.43 g. of the diene monomer is introduced into a Vycor tube 9" in length and having an outside diameter of 9 mm. which is then sealed in vacuo. The tube is placed 1" away from an ultraviolet source (Hanovia lamp type S-100) and exposed for 40 hours. The polymer is removed from the tube and placed in a vacuum oven at 85°C. for 3 hours. The polymer product, obtained in 86% conversion is a firm white elastomer.

The polymers obtained as described above are the result essentially of the 1,4-addition polymerization of the monomer as shown by the fact that the only observable band in the C=C stretching vibration region of the infrared spectrum is at 5.8μ. These homopolymers accordingly are made up essentially of repeating:

$$\left[CF_2-CF=CF-CH_2\right]$$

units as distinguished from:

$$\left[\begin{array}{c}CF_2-CF-\\ |\\ CF\\ \|\\ CH_2\end{array}\right]$$

or

$$\left[\begin{array}{c}CF-CH_2-\\ |\\ CF\\ \|\\ CF_2\end{array}\right]$$

units with their pendant —CF=CH2 or —CF=CF2 units resulting from 1,2- or 3,4-addition polymerization respectively. The lack of any substantial proportion of such pendant groups is highly advantageous since such groups detract considerably from the thermal and chemical stability of the polymer. In the case of repeating:

$$\left[CF_2-CF=CF-CH_2\right]$$

units derived by 1,4-addition, it will be noted that the hydrogens are flanked on each side by fluorine atoms which tend to protect the hydrogens against chemical attack, and the double bond is perfluorinated.

The homopolymers of $CF_2=CF-CF=CH_2$ prepared as described above are high molecular weight polymers as evidenced by their insolubility in organic solvents and by their high melting points. These polymers have melting ranges from 225° to 300°C., the melting range being determined by placing a small sample of the elastomer between two glass plates which are in turn placed on a melting point block and heated slowly, maintaining slight pressure on the sample between the two glass plates. As illustrated by the foregoing examples, the diene $CF_2=CF-CF=CH_2$ may be polymerized to homopolymers having high softening points under mild polymerization conditions. The preferred homopolymers of the process are those having softening points of at least 150°C. and especially those having softening points of at least 200°C.

The homopolymers of $CF_2=CF-CF=CH_2$ are further characterized by their excellent resistance to organic solvents, the homopolymers prepared in the above examples being substantially insoluble in solvents such as acetone, 1,1,2-trichlorotrifluoroethane, carbon tetrachloride, chloroform and 2,4-dichlorobenzotrifluoride. The high resistance of the homopolymer to attack by solvents is a highly valuable characteristic and further evidence of the high molecular weight of the polymer. A further characteristic of the homopolymer of $CF_2=CF-CF=CH_2$ is its excellent thermal stability. The polymers prepared as above described showed no decomposition throughout their melting point range, that is up to about 300°C. The polymers furthermore, have good low temperature properties, retaining their elastomeric properties at temperatures below 0°C. They also show high resistance to attack by chemical reagents in general including, for example, to attack by strong acids and alkalis and the like in contrast to hydrocarbon rubbers and less highly fluorinated rubbers.

Example 3: Copolymer of $CF_2{=}CF{-}CF{=}CH_2$ and Vinylidene Fluoride — 100 parts by weight of deoxygenated water, 0.4 part by weight of ammonium perfluorooctanoate and 0.5 part by weight of di-tert-butyl peroxide are introduced into a 300 ml. 316 stainless steel reaction vessel. The vessel is cooled in liquid nitrogen, evacuated and 26.6 parts by weight of 1,1,2,3-tetrafluoro-1,3-butadiene and 13.3 parts by weight of vinylidene fluoride are charged by gas transfer in vacuo. The vessel and contents are heated at 80°C. for 16 hours. It is then cooled and vented. The resulting copolymer is an elastomer having good low temperature properties.

Example 4: Copolymer of $CF_2{=}CF{-}CF{=}CH_2$ and Tetrafluoroethylene — A Vycor tube is charged with 1.33 g. of $CF_2{=}CF{-}CF{=}CH_2$ and 1.18 g. of $CF_2{=}CF_2$ and is then sealed under vacuum and irradiated by ultraviolet light for 46 hours. The tube is then cooled and vented after which the polymer is removed and heated in a vacuum oven at about 70°C. for 4 hours. The product is a solid containing about 31.5% carbon and about 47.5 mol percent of the tetrafluoro-butadiene monomer. The preparation of 1,1,2,3-tetrafluorobutadiene-1,3 is given in detail in the patent.

The preparation of 1,1,4,4-tetrafluorobutadiene and related copolymers is described by J.L. Anderson and K.L. Berry; U.S. Patent 3,218,303; November 16, 1965; assigned to E.I. du Pont de Nemours and Company. In the process, a fluoro-cyclobutene having, on the singly bonded carbons of the cyclobutene ring, only halogen atoms of which at least two are fluorine atoms, having hydrogen on at least one of the doubly bonded carbons of the cyclobutene ring, and having, as any remaining substituent halogen, hydrocarbon or halogenated hydrocarbon is pyrolyzed at a temperature within the range of 350° to 900°C. and the resulting fluorobutadiene is isolated.

The fluorocyclobutene starting materials can be prepared in situ, if desired, by pyrolysis of corresponding cyclobutanes having substituents which are converted to fluorocyclobutenes of the type defined above under the reaction conditions of this process, i.e., at temperatures of 350° to 900°C. For example, 1-acetoxy-1-methyl-2,2,3,3-tetrafluorocyclobutane is convertible to 1-methyl-3,3,4,4-tetrafluorocyclobutene at a temperature of 600°C. and the resulting cyclobutene is in turn converted to 2-methyl-1,1,4,4-tetrafluorobutadiene at 500° to 725°C. The following examples illustrate the process.

Example 1: Preparation of 1,1,4,4-Tetrafluoro-1,3-Butadiene — One part of 3,3,4,4-tetrafluorocyclobutene is distilled through a cylindrical quartz tube one inch in diameter and 12 inches long filled with quartz packing and heated by an electric heater at 400°C. The reaction gases are led through a trap cooled by liquid nitrogen. The pressure, measured between the cold trap and the vacuum pump, is maintained at 1 micron of mercury. The reaction products are isolated in the cold trap at a temperature of −196°C. The reaction products contain a significant amount of 1,1,4,4-tetrafluoro-1,3-butadiene, as indicated by infrared absorption analysis. The presence of strong absorption bands at 5.85, 7.55 and 7.60, 8.52 and 8.58, and 10.80 and 10.90 microns in the infrared absorption spectrum of 1,1,4,4-tetrafluoro-1,3-butadiene is consistent only with the structure of this compound. Pyrolysis of 3,3,4,4-tetrafluorocyclobutene by the same method used above with the exception that the temperature of pyrolysis is 500°, 600° and 700°C., respectively, give increasing conversions of the fluorocyclobutene to the corresponding fluorobutadiene as the temperature is increased.

Example 2: Preparation of Thermal Polymers of 1,1,4,4-Tetrafluoro-1,3-Butadiene — A mixture of 900 parts of 1,1,4,4-tetrafluoro-1,3-butadiene and 3 parts of a mixture of terpene hydrocarbons known to inhibit the polymerization of other fluorine-containing olefins is heated at 200°C. for 16 hours in a closed stainless steel reaction vessel at the autogenous pressure developed under these conditions. After cooling, the reaction vessel is opened and the reaction product is fractionally distilled. There are obtained 103 parts of a dimer of 1,1,4,4-tetrafluorobutadiene, believed to be cis-1,2-bis(2,2-difluorovinyl)-3,3,4,4-tetrafluorocyclobutane, boiling at 99° to 100°C. and having a refractive index, n_D^{25} of 1.3448; 280 parts of a trimer of tetrafluorobutadiene, believed to be cis-1-(2,2,3,3,6,6-hexafluoro-4-cyclohexenyl)-2-(2,2-di-fluorovinyl)-3,3,4,4-tetrafluorocyclobutane, boiling at 47° to 48°C./0.2 mm. and melting at about 50°C.; 100 parts of a sublimable solid tetramer of tetrafluorobutadiene; and a nonvolatile residue amounting to 190 parts which is an ether-soluble polymer of 1,1,4,4-tetrafluorobutadiene having a molecular weight of 6,300 (determined ebullioscopically).

Analysis of nonvolatile polymer calculated for $(C_4H_2F_4)_x$: C, 38.1%; H, 1.59%; F, 60.2%. Found: C, 38.61%; H, 1.82%; F, 59.2%. These thermal polymers are useful as solvents for halogenated compounds, as stable compounds for heat transfer substances, and as chemical intermediates. For example they can be hydrogenated and halogenated to saturated products which show markedly increased thermal and chemical stability, and they can be oxidized to polybasic acids of value in condensation polymerization.

Example 3: Preparation of Linear Addition Polymer of 1,1,4,4-Tetrafluoro-1,3-Butadiene — 500 parts of 1,1,4,4-tetra-fluoro-3-butadiene is held at 75°C. for 10 days and at room temperature (20° to 30°C.) for 7 days in a stainless steel con-tainer. After removal of excess monomer from the reaction mixture, there is obtained 12.5 parts of very high molecular weight polymer of 1,1,4,4-tetrafluoro-3-butadiene. This polymer is pressed into a film 6 mils thick at 360°C. at about 2,000 psi pressure during a period of 2 minutes. This film is cut into strips 2 to 3 mm. wide and the strips are cold drawn to fine, highly oriented fibers.

Fluorodienes

In a related process described by E.S. Lo; U.S. Patent 2,959,575; November 8, 1960; assigned to Minnesota Mining and Manufacturing Company pentafluorobutadiene is copolymerized with a fluorohalomonoolefin. Of the latter, the fluorochloromonoolefins are preferred, and particularly the fluorochloroethylenes, e.g., trifluorochloroethylene or 1,1-dichloro-2,2-difluoroethylene. The following example illustrates the process.

Example: A heavy-walled glass polymerization tube of about 20 ml. capacity is flushed with nitrogen and then charged with 5 cc of a solution prepared by dissolving 0.1 g. of dodecyl mercaptan and 1 g. of the ammonium salt of perfluorooctanoic acid in 100 cc of water. The contents of the tube are then frozen, and the tube is next charged with 1 cc of a promoter solution prepared by dissolving 0.4 g. of sodium metabisulfite and 0.5 g. of borax in 20 cc of water. To the contents of the tube are next charged 4 cc of a solution prepared by dissolving 1 g. of potassium persulfate in 80 cc of water.

The contents of the tube are then refrozen, and the tube is next connected to a gas-transfer system and evacuated at liquid nitrogen temperature. To the frozen contents of the tube are added, by distillation, 2.77 g. of 1,1,2,4,4-pentafluorobutadiene and 2.23 g. of trifluorochloroethylene, comprising a comonomeric mixture containing 50 mol percent of each monomer. After the contents of the tube are thoroughly frozen with liquid nitrogen, the tube is evacuated and sealed. 1-chloro-1,2-dibromo-1,2,2-trifluoroethane, $CF_2BrCFClBr$, is added to vinylidene fluoride to yield $CF_2BrCFClCH_2CF_2Br$ which is then dehydrobrominated using potassium hydroxide, followed by debromochlorination using zinc to yield $CF_2=CF-CH=CF_2$, BP 15.0° to 15.5°C.

The polymerization tube and its contents are next agitated in a temperature-regulated water bath at 50°C. for a period of 24 hours. At the end of this time, the contents of the tube are coagulated by freezing. The coagulated product is then removed from the tube, washed with hot water and then dried to constant weight in vacuo at 35°C. A copolymeric product is obtained which is found, upon analysis to comprise approximately 92 mol percent 1,1,2,4,4-pentafluorobutadiene and the remaining major constituent, trifluorochloroethylene, being present in an amount of approximately 8 mol percent. The copolymer is obtained in an amount corresponding to an 11% conversion. In a manner similar to that described above, 1,1,2,3,4-pentafluorobutadiene is polymerized with trifluorochloroethylene to produce an elastomeric copolymer of these two monomers.

Copolymers with Styrene

In a process described by F.J. Honn; U.S. Patent 2,949,446; August 16, 1960 and U.S. Patent 2,962,484; August 29, 1960; both assigned to Minnesota Mining and Manufacturing Company copolymers of fluorodienes, which are preferably fluorobutadienes, containing at least two fluorine atoms, are produced by copolymerizing the fluorodiene with styrene to produce elastomeric and thermoplastic copolymers which are moldable at temperature between 200° and 450°F. into tough, flexible and clear sheets which are particularly suitable for applications requiring materials possessing a high degree of thermal stability and resistance to strong and corrosive chemicals. The following example illustrates the process.

Example: A polymerization tube was charged with: (1) 180 parts of a soap solution prepared by dissolving 5.0 parts of potassium fatty acid soap in 180 parts of deionized water by stirring and heating at a temperature not in excess of 50°C. When solution was complete, the solution was cooled to room temperature (22°C.) and the pH was adjusted to 10.2; 0.3 part of tertiary-dodecyl mercaptan was added. (2) 20.0 parts of a solution containing 0.3 part of potassium persulfate. (3) 50 parts of 1,1-difluoro-3-methyl butadiene; and (4) 50 parts of styrene. The mixture was frozen after each addition in a freezing bath consisting of a slush of solid carbon dioxide and trichloroethylene.

The polymerization tube was sealed in vacuo at the temperature of liquid nitrogen and placed in a water bath maintained at a temperature of 50°C., and the tube was then shaken for a period of 24 hours. At the end of this period, the tube was frozen, at the temperature of liquid nitrogen, to reduce the vapor pressure of the unreacted monomers and to coagulate the polymer. The contents of the tube were then removed and washed with distilled water until free of soap. The washed contents were then dried to constant weight in vacuo at a temperature of 35°C. A white solid polymeric material was obtained in a quantity representing a 5% conversion and upon analysis was found to contain 34 mol percent of combined 1,1-difluoro-3-methyl butadiene and 66 mol percent of combined styrene.

Copolymers with Chloroethylene

A.N. Bolstad and J.M. Hoyt; U.S. Patent 2,996,487; August 15, 1961; assigned to Minnesota Mining and Manufacturing Company describe the copolymerization of fluorobutadienes and a chloroethylene. Preferred fluorobutadienes include 1,1-difluorobutadiene; 1,1,2-trifluorobutadiene; 1,1,3-trifluorobutadiene; and 1,1,2,4,4-pentafluorobutadiene. Preferred chloroethylenes include vinyl chloride and vinylidene chloride.

The copolymers of the process are particularly useful for the fabrication of a wide variety of materials having high desirable physical and chemical properties. The copolymers possess important utility in the fabrication of resilient gaskets, seals, valve-diaphragms, films and various other commercial applications. Another important use of the copolymers is in the form of durable, flexible, protective coatings on surfaces which are subjected to distortion in normal uses, e.g., fabric surfaces. For these purposes, the copolymers may be dissolved in various commercial solvents. Particularly useful solvents comprise the aliphatic and aromatic esters, ketones and halogenated hydrocarbons. The following examples illustrate the process.

Example 1: A heavy-walled glass polymerization tube of about 20 ml. capacity was flushed with nitrogen and then charged with 9 cc of a catalyst solution prepared by dissolving 25 g. of potassium stearate and 1.5 g. of dodecyl mercaptan in 900 cc of water, adjusted to a pH of 10 by the addition of potassium hydroxide. The contents of the tube were then frozen and the tube was then charged with 1 cc of a catalyst solution prepared by dissolving 1.5 g. of potassium persulfate in 100 cc of water. The contents of the tube were next refrozen, and the tube was then connected to a gas-transfer system and evacuated at liquid nitrogen temperature. To the frozen contents of the tube were added, by distillation, 1.78 g. of 1,1-difluorobutadiene and 3.22 g. of vinyl chloride, comprising a comonomeric mixture containing 50 mol percent of each monomer. After the contents of the tube were thoroughly frozen with liquid nitrogen, the tube was evacuated and sealed.

The polymerization tube and its contents were next agitated in a temperature-regulated water bath at 50°C. for a period of 24 hours. At the end of this time, the contents of the tube were coagulated by freezing. The coagulated product was then removed from the tube, washed with hot water and then dried to constant weight in vacuo at 35°C. A rubbery co-polymeric product was obtained which was found, upon analysis, to comprise approximately 79 mol percent 1,1-difluoro-butadiene and the remaining major constituent, vinyl chloride, being present in an amount of approximately 21 mol percent. The copolymer was obtained in an amount corresponding to a 27% conversion, and was found to have a 59% volume increase for the raw copolymer, in ASTM Fuel Type II, consisting of isooctane (60% by volume), benzene (5% by volume), toluene (20% by volume) and xylene (15% by volume).

Example 2: Employing the procedure set forth in Example 1 and the same polymerization system, the tube was charged with 2.25 g. of 1,1-difluorobutadiene and 2.43 g. of vinylidene chloride, comprising a comonomeric mixture containing 50 mol percent of each monomer. The polymerization reaction was carried out at a temperature of 50°C. for a period of 22 hours. The resultant elastomeric product was worked-up in accordance with the same procedure as set forth in Example 1. A rubbery product was obtained and, upon analysis, was found to compromise approximately 66 mol percent 1,1-difluorobutadiene and the remaining major constituent, vinylidene chloride, being present in an amount of approximately 34%. The copolymer was obtained in an amount corresponding to a 43% conversion.

Chlorine-Substituted Fluorodienes

J.T. Barr; U.S. Patent 3,020,267; February 6, 1962 and U.S. Patent 3,148,175; September 8, 1964; both assigned to Pennsalt Chemicals Corporation has found that certain highly fluorinated dienes in which the fluorine is present in alternating $-CF_2-$ and $=CF-CH=$ groupings can be modified by the substitution of chlorine for a portion of the hydrogen and fluorine in the molecule to form a diene with a molecular structure which has greatly improved stability and resistance to oxidation. Furthermore, monomeric dienes with this modified structure can be polymerized readily into rubber-like polymers with correspondingly enhanced properties and at lower cost. Among the advantages resulting from such monomers and polymers the following are outstanding:

(1) increased chemical and physical stability and oxidation resistance of the monomers and polymers;
(2) decreased tendency for spontaneous polymerization among some of the unsymmetrical partially-fluorinated dienes, resulting in simplified storage and handling of the monomers, and
(3) reduced crystallinity of the rubber-like products which can be formed, resulting in enhanced physical properties and a wider usable temperature range of the polymers.

Some specific examples of starting butenes and the chlorine-substituted fluorodiolefins derived from them by the process are the following:

Butene:
  1,2-dichloro-1,1,2-trifluorobutene-3
Butadiene:
  3-chloro-1,1,2-trifluorobutadiene-1,3
  3,4-dichloro-1,1,2-trifluorobutadiene-1,3

Butene:
  1-bromo-2,2-dichloro-1,1-difluorobutene-3
Butadiene:
  2,3-dichloro-1,1-difluorobutadiene-1,3
  2,3,4-trichloro-1,1-difluorobutadiene-1,3

Example 1: Preparation of 3-Chloro-1,1,2-Trifluorobutadiene-1,3 — Chlorine was passed into a solution of 1-bromo-2-chloro-1,1,2-trifluorobutene-3 in an equal part of carbon tetrachloride until the weight increase of the solution indicated that the theoretical amount of chlorine required just to saturate the double bond had been absorbed. Distillation gave a 90% yield of 1-bromo-2,3,4-trichloro-1,1,2-trifluorobutane (BP 180° to 183°C., n_D^{25} 1.4540), and an 8% yield of a mixture of 1-bromo-2,3,4,4-tetrachloro-1,1,2-trifluorobutane and 1-bromo-2,3,3,4-tetrachloro-1,1,2-trifluorobutane but predominantly the former (BP 96° to 99°C. at 12 mm. n_D^{27} 1.4718).

Dehydrochlorination of the 1-bromo-2,3,4-trichloro-1,1,2-trifluorobutane by KOH in alcohol at 0° to 10°C. gave an 83% yield of 1-bromo-2,3-dichloro-1,1,2-trifluorobutene-3 (BP 132° to 134°C., n_D^{28} 1.4330). The 1-bromo-2,3-dichloro-1,1,2-trifluorobutene-3 was then dehalogenated by dropping it into a refluxing suspension of zinc dust in alcohol containing a little t-butyl-catechol. The mixture was refluxed 1/3 hour, then cooled and filtered. Addition of 2 volumes of water and a little HCl to the filtrate resulted in the formation of a lower layer which was separated, washed, dried and distilled. The product was 3-chloro-1,1,2-trifluorobutadiene-1,3 (BP 54°C., n_D^{28} 1.3835). A conversion of 43.5% was obtained.

The 3-chloro-1,1,2-trifluorobutadiene-1,3 was homopolymerized in 90% conversion to form a soft elastomer which was vulcanized into a strong rubbery material. It was copolymerized in 3:1 ratio with styrene to 73% conversion. It was also copolymerized with trifluoroethyl vinyl ether in 1:1.36 ratio to 63% conversion. The high boiling mixture of 1-bromo-2,3,4,4-tetrachloro-1,1,2-trifluorobutane and 1-bromo-2,3,3,4-tetrachloro-1,1,2-trifluorobutane was dehalogenated directly and the isomer 4-chloro-1,1,2-trifluorobutadiene was isolated. This diene was polymerized to a soft elastomer which vulcanized to a strong rubbery material.

Example 2: A pressure reactor was charged with 100 parts 3-chloro-1,1,2-trifluorobutadiene-1,3, 180 parts water, 5 parts Dupanol WA, 4 parts borax, 4 parts sodium bisulfite, 4 parts potassium persulfate and 1 part tertiary dodecyl mercaptan. The void space was purged with nitrogen, and the reactor was sealed. The reactor was heated at 50°C. and rotated at 29 rpm for 24 hours. 90 parts of soft white rubbery products were obtained. 100 parts of product were compounded and milled with 1 part paraffin, 40 parts Philblack O (carbon black), 0.5 part 2-mercaptoimidiazol, and 1 part benzothiazyl disulfide. The product was cured at 310°F. and 500 psig for 30 minutes to give a strong rubbery sheet possessing good physical properties.

2-Chloro-1,1-Difluorobutadiene-1,3

A process described by J.T. Barr; U.S. Patent 3,053,815; September 11, 1962 and U.S. Patent 3,308,175; March 7, 1967; both assigned to Pennsalt Chemicals Corporation involves 2-chloro-1,1-difluorobutadiene-1,3 and elastomeric compositions based on its copolymerization with unsaturated fluoroesters.

The compound 2-chloro-1,1-difluorobutadiene-1,3 is formed by the dehalogenation of 1-bromo-2,2-chloro-1,1-difluorobutadiene-3, e.g., with zinc dust in ethanol, to remove the bromine atom and one of the chlorine atoms. 2-chloro-1,1-difluorobutadiene-1,3 is a monomeric liquid which boils at 45° to 47°C. and has a refractive index n_D^{25} 1.3960. The above halogenated butene-3 compound is prepared by the method described in U.S. Patent 3,308,175. The method of its preparation is illustrated by the following equations:

$$CF_2BrCCl_2I + CH_2{=}CH_2 \longrightarrow CF_2BrCCl_2CH_2CH_2I \xrightarrow[\text{alcohol}]{\text{KOH}} CF_2BrCCl_2CH{=}CH_2$$

2-chloro-1,1-difluorobutadiene-1,3 is readily polymerized to a homopolymer by use of standard polymerization procedures, for example, emulsion polymerization in the presence of peroxide catalysts and emulsifying agents. Depending on the degree of polymerization, the homopolymer may be made with properties ranging from a viscous sticky liquid to a hard rubbery solid.

The unsaturated fluoroesters of 2,2,2-trifluoroethanol which copolymerize with 2-chloro-1,1-difluorobutadiene-1,3 according to the process are represented by the following compounds: 2,2,2-trifluoroethyl acrylate, 2,2,2-trifluoroethyl-2-chloroacrylate, 2,2,2-trifluoroethyl-2-fluoroacrylate, bis(2,2,2-trifluoroethyl) maleate, and bis(2,2,2-trifluoroethyl) fumarate. The preferred ester is 2,2,2-trifluoroethyl acrylate.

The proportions of 2-chloro-1,1-difluorobutadiene-1,3 and of unsaturated fluoroester used may be varied over a wide range depending on the properties desired in the copolymer. For example, useful copolymers based on as little as 5% of one monomer and 95% of the other may be prepared according to the process, but at least 15% of 2-chloro-1,1-difluorobutadiene-1,3 is desirably used. As the amount of 2-chloro-1,1-difluorobutadiene-1,3 in the copolymer is increased, the copolymer becomes more elastic. Copolymers containing about 50 to 75% of 2-chloro-1,1-difluorobutadiene are especially preferred.

Rubbers prepared by curing the homopolymer or copolymers of this process have outstanding resistance to solvents, oils, oxygen, sunlight, heat, aging and chemicals, and are particularly useful where resistance to these is necessary, as for example, in the chemical process and allied industries. The following examples illustrate the process.

Example 1: A pressure reactor is charged with 75 parts 2-chloro-1,1-difluorobutadiene-1,3, 25 parts 2,2,2-trifluoroethyl acrylate, 144 parts water, 4 parts Dupanol WA, 0.8 part tertiary dodecyl mercaptan, 3.2 parts sodium bisulfite, 3.2 parts borax and 3.2 parts potassium persulfate and is agitated at 5°C. for 44 hours. About 55 parts of a weak, rubbery, tacky polymer are obtained.

Example 2: 100 parts of the copolymer of Example 1 is compounded with a standard polyamine recipe consisting of 35 parts of high abrasion furnace black, 1 part of sulfur, 1 part of paraffin, and 1 part of triethyl tetramine ("Teta") and milled on a rubber mill. The mass is then cured for 30 minutes at 310°F. at 500 psi pressure. The cured sheet of the product has the following properties:

Tensile strength, psi	2,300
Percent elongation	440
Shore A hardness	82

Cured polymer made as above described was immersed in Esso Turbo Oil 15 at about 390°F. for 46 hours with no indication of swelling being observed. The same cured polymer was immersed in carbon tetrachloride for 20 hours with no indication of swelling being observed.

Acyloxy-1,3-Butadienes

W.J. Middleton; U.S. Patent 3,424,732; January 28, 1969; assigned to E.I. du Pont de Nemours and Company describe the preparation and polymerization of perfluoroalkyl and ω-chloroperfluoroalkyl-3-acyloxy-1,3-butadienes having the general formula:

$$\begin{array}{c} R_1 \\ \diagdown \\ \diagup \\ R_2 \end{array} C=CH-\underset{\underset{\underset{\underset{R_3}{C=O}}{|}}{O}}{\overset{}{C}}=CH_2$$

R$_1$ and R$_2$ taken separately can be the same or different and represent perfluoroalkyl or ω-chloroperfluoroalkyl containing 1 to 4 carbon atoms.

R$_1$ and R$_2$ taken together represent perfluoroalkylene containing 3 to 5 carbon atoms.

R$_3$ is a saturated aliphatic or aromatic hydrocarbyl group selected from the group consisting of alkyl containing 1 to 10 carbon atoms, aralkyl containing 7 to 14 carbon atoms, aryl containing 6 to 14 carbon atoms, and alkylaryl containing 7 to 14 carbon atoms.

These monomers can be homopolymerized and copolymerized with copolymerizable monomers containing ethylenic unsaturation.

Ziegler-Type Catalyst

A process described by G.H. Crawford, Jr.; U.S. Patent 3,089,866; May 14, 1963; assigned to Minnesota Mining and Manufacturing Company comprises polymerizing a fluorine-containing olefin in the presence of a polymerization activator comprising a Ziegler-type catalyst to produce a fluorine-containing polymer.

Typical examples of suitable Ziegler-type catalysts to be used are aluminum trihydride; beryllium hydride; dimethyl magnesium; diethyl zinc; dimethyl aluminum hydride; ethyl aluminum dihydride; n-hexyl aluminum dihydride; beryllium diethyl; zinc dimethyl; triethyl aluminum; trimethyl aluminum and tripropyl aluminum.

The polymerization activator may consist essentially of one of the above mentioned Ziegler catalysts, or it may be used in conjunction with cocatalysts comprising various derivatives of a transition metal of groups IV, V, VI and VIII of the periodic system. Thus, various derivatives of titanium, zirconium, hafnium, thorium, vanadium, columbium, tantalum, chromium, molybdenum, tungsten, iron, cobalt, nickel, and palladium may be used.

Of the cocatalysts to be employed, the halide derivatives are preferred and are exemplified by titanium tetrachloride, zirconium tetrachloride, ferric chloride, ferrous chloride, nickel chloride, palladium dichloride, manganese dichloride, chromium dichloride and tungsten hexachloride.

Among the advantages realized by using such compounds as cocatalysts in conjunction with the above described compounds of metals of groups II and III are: the rate of reaction is considerably faster; the polymerization may be carried out at a lower temperature; and the conversions of monomer(s) to polymer product are significantly increased. When a normally solid material is desired as a product of the process, the use of such cocatalysts is recommended.

The concentration of the Ziegler-type catalyst with respect to the concentration of the cocatalyst may vary over a relatively wide range. Thus, they may be used in a ratio of between 0.05 to 2.0 mols of cocatalyst per mol of Ziegler catalyst and preferably are used in a mol ratio of between 0.1 and 0.5 per mol of Ziegler catalyst. The following examples illustrate the process.

Fluorodienes

Example 1: After flushing a 10 ml. glass polymerization tube with nitrogen, the tube was evacuated and 0.10 g. of di-ethyl aluminum bromide was added. The contents of the tube were then frozen at liquid nitrogen temperature followed by the addition of 0.5 g. of titanium tetrachloride. After refreezing the contents of the tube, 5 g. of 2-trifluoromethyl buta-diene were then condensed into the tube in the absence of air and moisture. The tube was then sealed and maintained at a temperature of -20°C. for about 48 hours. At the end of this time, the contents of the tube were washed with 100 ml. acetone followed in sequence by 100 ml. of water, 100 ml. of 10% HCl and 100 ml. of water. The washed polymer was then vacuum dried at about 50°C. A rubbery polymer, namely poly-2-trifluoromethyl butadiene homopolymer was obtained in about a 20% conversion of total monomer employed to polymer product.

Example 2: After flushing a 10 ml. glass polymerization tube with nitrogen, the tube was evacuated and 0.10 g. of di-ethyl aluminum bromide was added. The contents of the tube were then frozen at liquid nitrogen temperature followed by the addition of 0.05 g. of titanium tetrachloride. After refreezing the contents of the tube, 5 g. of 1,1,3-trifluorobuta-diene were then condensed into the tube in the absence of air and moisture. The polymerization reaction occurred very rapidly going to completion in less than 2 minutes. The product was worked up following the same procedure as described in Example 1. Approximately 5 g. of a tough, snappy rubbery homopolymer of 1,1,3-trifluorobutadiene were obtained. The polymer product of this example possesses good low temperature flexibility and is particularly useful in the fabrication of resilient gaskets, pump diaphragms and O-rings.

OTHER FLUORINATED DIENES

3,4,5,5,5-Pentafluoropentadiene

E.E. Frisch and O.W. Steward; U.S. Patent 3,202,643; August 24, 1965; assigned to Dow Corning Corporation de-scribe the preparation and homopolymerization of 3,4,5,5,5-pentafluoropentadiene. The following examples illustrate the process.

Example 1: In a 1.4 liter stainless steel Aminco autoclave were placed 1-bromo-1,1,2,3,3,3-hexafluoropropane (870 g. 3.6 mols) and t-butylperoxide (35.5 g., 4% by weight). The autoclave was heated to 115°C. and ethylene was introduced continuously from a cylinder at a pressure of 180 psi. The reaction was carried out over a period of 24 hours while the tem-perature was maintained at 115°C., and the autoclave was rocked.

A second run was carried out similarly with 1-bromo-1,1,2,3,3,3-hexafluoropropane (880 g., 3.8 mols) and t-butylper-oxide (35.2 g.). The contents from the autoclave from both runs were combined, washed with water, and dried over an-hydrous calcium sulfate. Fractional distillation gave 5-bromo-1,1,1,2,3,3-hexafluoropentane (330 g., 1.28 mols), BP 120°C. (739 ml.) a 17% yield.

Potassium hydroxide, 85% minimum (64 g., about 1.0 mol) was dissolved in a solution of dimethyl Carbitol (180 ml.) and water (40 ml.). On dissolving, two layers resulted. This mixture was added over a period of 15 minutes to 5-bromo-1,1,1,2,3,3-hexafluoropentane (141 g., 0.5 mol) and 4-t-butyl catechol (1 g.). During the addition, the reaction tem-perature rose to 70°C. After stirring for one-half hour, the reaction mixture was basic. After refluxing for 3 hours, the reaction mixture was only weakly basic. The material boiling below 100°C. was removed through a Vigreaux column, the distillate was dried over anhydrous calcium sulfate, and 4-t-butyl catechol (1 g.) was added to prevent polymerization of the diene. Fractional distillation gave 3,4,5,5,5-pentafluoro-1,3-pentadiene, $CF_3CF{=}CFCH{=}CH_2$ (41 g., 0.26 mol), BP 40° to 41°C. (746 ml.) 52% yield.

Example 2: The product of Example 1 was irradiated with ultraviolet light for 4 days at room temperature. A tough, strong, flexible polymer resulted, having a softening point of about 170°C.

Copolymers of Octafluorocyclohexadiene

W. Hopkin and A.K. Barbour; U.S. Patent 2,958,683; November 1, 1960; assigned to The National Smelting Company Limited, England have found that two compounds, which may conveniently be referred to as the polyfluoro cyclic olefins, exhibit extraordinary reactivity in polymerization systems. Their structures have been elucidated as follows:

Octafluoro-cyclo-hexa-1:3-diene

Fluorodienes

Octafluoro-cyclo-hexa-1:4-diene

The range of comonomers with which these polyfluoroolefins copolymerize is wide and embraces a number of types which may be said to be typical of the majority of monomers capable of homopolymerization. Thus, ethylene and substituted ethylenes are tangibly represented, by ethylene, vinyl chloride, vinyl acetate and vinylidene fluoride. Conjugated unsaturated monomers are similarly represented by the straight chain 1:3-butadiene for its many derivatives, by styrene for vinyl aromatic compounds and by ethyl acrylate, methyl methacrylate and acrylonitrile for α-β unsaturated acid derivatives. The following examples illustrate the process.

Example 1: 53 parts by weight of octafluoro-cyclo-hexa-1:3-diene and 75 parts by weight of 1:3-butadiene were reacted in a sealed glass tube at 50°C. for 18 hours with continuous agitation in the presence of 180 parts by weight of water, 5 parts by weight of sodium stearate, 0.5 part by weight of n-dodecyl mercaptan and 0.3 part by weight of sodium persulfate.

At the end of this period the tube was opened and unreacted 1:3-butadiene (50 parts by weight) was discharged. The tube contained a white solid elastomeric copolymer together with some latex. Virtually all of the octafluoro-cyclo-hexa-1:3-diene charged to the reactor was found to have been consumed in the copolymerization reaction. The elastomeric product was separated by precipitation with dilute hydrochloric acid, washed with water and dried in vacuo to give 76 parts by weight of copolymeric product which contained 43% fluorine as determined by a sodium fusion method.

Example 2: 224 parts by weight of octafluoro-cyclo-hexa-1:3-diene and 35 parts by weight of 1:3-butadiene were reacted in a sealed glass tube in the presence of the same mixture as in Example 1, and for the same time at the same temperature. After separation and drying there remained 190 parts by weight of a white elastomeric copolymer containing about 50% fluorine by analysis by sodium fusion. No unreacted 1:3-butadiene was left and the analysis of the copolymer corresponds closely to that required for a 1:1 molar copolymer.

Example 3: A solution of 0.741 part by weight of benzoyl peroxide in 4.718 parts by weight of methyl methacrylate was prepared. 0.477 part by weight of this solution was mixed with 5.476 parts by weight of octafluoro-cyclo-hexa-1:3-diene in a glass tube which was then cooled to -180°C., evacuated and sealed.

On melting the frozen liquid after sealing, a lump of white solid polymer presumably polymethylmethacrylate was observed. On heating at 80°C., a polymeric material rapidly separated and was deposited as a glassy solid on the walls of the tube. After 22 hours the tube was cooled. 4.6 parts by weight of liquid, nearly pure octafluoro-cyclo-hexa-1:3 diene and 0.835 part by weight of solid polymeric material were recovered.

Part of the polymeric material was the homopolymer mentioned which was a white granular solid, and the remainder was a glassy flexible solid. Of the remainder, 0.819 part by weight was taken up in 15 parts by weight of chloroform and added with stirring to 450 parts by weight of methylated spirit to deposit 0.308 part by weight of flocculent precipitate of solid high polymer. Evaporation of the liquors after separation of the solid high polymer by filtration yielded 0.372 part by weight of gummy material. Elemental analysis yielded the following results: High polymer F = 11.3%. Gummy material F = ca. 26%.

Examination of Copolymer from Example 2: (a) Thermal Stability — A sample of copolymer which had been pressed to a thin sheet was heated at 210°C. while a current of air (0.5 l./min.) was passed over it. After an initial loss in weight of ca. 12% due to volatile materials not completely removed by drying, the sample stabilized at 83% of its original weight within a few minutes, and remained substantially unchanged in weight after 4 hours.

A further sample was heated in high vacuum for 62 hours at 148°C. It was substantially unchanged in weight and properties except for a slight darkening. In another experiment, exhaustively dried samples were heated in high vacuum while suspended from a quartz spiral spring, and the change in weight noted after one hour at several temperatures. It is apparent that they stabilize at 60 to 66% loss in weight at temperatures of 500° to 800°C.

(b) Resistance to Red Fuming Nitric Acid — A small sample was warmed for 1.5 hours with red fuming nitric acid, but was apparently not seriously affected. A further sample withstood immersion in boiling acid for a few minutes.

(c) Resistance to Solvents — A sample was immersed in benzene for one day. The extent of swelling was estimated approximately by weighing the sticky swollen mass, and found to be ca. 130% of the original volume. A further sample was

149

seriously swollen in acetone to a slimy jelly. A "dumb bell" specimen was made and stretched at room temperature (ca. 23°C.). Its ultimate breaking stress was greater than 1,000 psi while the elongation was to 830% of the original length of a section of the specimen.

4-Chloroperfluoroheptadiene-1,6

L.A. Wall and J.E. Fearn; U.S. Patent 3,211,637; October 12, 1965; assigned to the U.S. Secretary of the Navy describe the preparation and polymerization of 4-chloroperfluoroheptadiene-1,6. Efforts to polymerize this material through use of azobisisobutyronitrile at 60°C. and ultraviolet radiation at room temperature have been unsuccessful. It appears that a temperature of 100°C. or above is essential for a rapid polymerization reaction to take place. The upper limit of the temperature variable would appear to be the decomposition temperature of the monomer starting material.

A more successful polymerization of the 4-chloroperfluoroheptadiene-1,6 was obtained by gamma irradiation of the monomer at 100°C. under autogenous pressure. The speed of the reaction is a function of the dosage of radiation. To produce suitable reaction rates, a dosage of 0.01 to one megarad per hour is sufficient. A suitable source of actinic radiation is cobalt-60. The monomer can be dissolved in any suitable solvent which may or may not be a solvent for the polymer. The following examples are illustrative of the polymerization.

Example 1: 33 g. (0.1 mol) of 4-chloroperfluoroheptadiene-1,6 were placed in a glass tube attached to a vacuum, frozen with liquid nitrogen and the tube evacuated. The heptadiene was thoroughly degassed by the repeated freezing, pumping, thawing technique and the tube was then sealed. This tube and contents were irradiated for 71 hours by being subjected to a cobalt-60 source at 0.2 megarad per hour and at a temperature of 150°C.

A brittle, glassy polymer was obtained in 100% conversion. A sample of this material was dissolved in hexafluorobenzene, precipitated with xylene, redissolved in hexafluorobenzene and put through the freeze dry technique. This material was found to have an intrinsic viscosity of 0.09 when hexafluorobenzene is used as a solvent. A KBr pellet of this polymer indicated the absence of a carbon-carbon double bond.

Example 2: 10 g. (0.03 mol) of 4-chloroperfluoroheptadiene-1,6 was treated in the same fashion described above excepting that the temperature of reaction was lowered to 100°C. 8 g. of polymer was obtained or 80% conversion. This polymer upon purification as described above and dissolved in hexafluorobenzene showed an intrinsic viscosity of 0.05.

Fluorinated Diene-Diol Reaction Products

E.W. Cook; U.S. Patent 3,391,118; July 2, 1968; assigned to FMC Corporation has found that terminal diols which are completely or almost completely perfluorinated will react with highly fluorinated terminal dienes in the presence of KOH or other alkaline condensing agents, to produce elastomeric polymers which can be identified by proton nuclear magnetic resonance (NMR) as having a structure in which diene moieties alternate with diol moieties, the polymers terminating with a group derived either from the diol or from the diene and which are characterized by extremely good low temperature properties. The following examples illustrate the process.

Examples 1 to 5: 2,2,3,3,4,4-Hexafluoropentanediol-1,5 and 4-Chloroperfluoroheptadiene-1,6 — The diol used, a known compound, has recently become available on the market; the 4-chloroperfluoroheptadiene-1,6 can be prepared from 4,6,7-trichloroperfluoroheptene-1 by dechlorination with zinc at low temperatures (reflux in isopropanol or tetrahydrofuran).

The precursor 4,6,7-trichloroperfluoroheptene-1 (TCPFH) can be prepared from the sodium salt of $Cl(CF_2CFCl)_3CF_2COOH$, by decarboxylation under vacuum. In a typical preparation, 267 g. of this sodium salt $Cl(CF_2CFCl)_3CF_2—COONa$ were placed in a 1 liter vacuum flask, which was evacuated through a Dry Ice-cooled trap to about 1 mm. pressure, and then heated to 150°C. The precursor TCPFH distilled into the trap, was washed with water, 10% aqueous sodium carbonate, and water again, and dried. Distillation of the crude dried product gave 159 g. of pure TCPFH boiling at 161° to 162°C., a 75% yield. Its infrared spectrum checked that of known material. The 4-chloroperfluoroheptadiene-1,6 was prepared as follows:

To a stirred refluxing mixture of 16 g. (0.25 mol) of zinc dust, 1 g. of zinc chloride, and 25 ml. of isopropanol was added 39.9 g. (0.1 mol) of 4,6,7-trichloroperfluoro-1-heptene over a 6 hour period. After refluxing an additional 0.5 hour, the mixture was filtered from unreacted zinc and the filtrate diluted with several volumes of water. The organic layer was taken up in methylene chloride, water washed several times, and dried over magnesium sulfate. Distillation gave a 11.3 g. of product, BP 97° to 112°C./761 mm., n_D^{25} 1.3311. The purity by gas chromatography was 67.5%.

The product can be further purified by redistillation, or it can be used in crude forms, using the crude in correct proportion to give the indicated amount of reactant. The following experiment is typical of those carried out. All are tabulated in the table which follows the description of procedure.

Fluorodienes

To 11 g. of the pentanediol in 20 ml. acetone was added 3.0 g. potassium hydroxide. This mixture was then rapidly poured into 16 g. of the heptadiene contained in a 250 ml. Erlenmeyer flask and the resultant mixture stirred magnetically. The light yellow color which formed immediately slowly darkened and, after 72 hours, tan to brown polymer was found in the flask along with a viscous oil. The oil was dissolved in acetone, the solution decanted, and the polymer washed with acetone. The polymer was leached with acetone in a Soxhlet extractor overnight and then dried in a vacuum oven.

Using this procedure the yield of elastomer is generally about 20%, while the lower molecular weight oils are obtained in about 50% yield. The elastomer has extremely good low temperature properties. The value of the glass transition point (T_g) appears to be in the neighborhood of −50°C.

Copolymerization of 4-Chloroperfluoroheptadiene-1,6 with 2,2,3,3,4,4-Hexafluoropentanediol-1,5

Example	Amt. Diol (g.)	Amt. Diene (g.)	KOH (g.)	Solvent Amt. (ml.)	Oil Amt.[1] (g.)	Polymer[2] Amt. (g.)	T_g, °C.
1	11	16.0	3.0	20 acetone	–	6	−58
2	5.5	8.0	1.5	10 acetone	5.7	8.1[3]	–
3	5.5	8.2	1.5	10 acetone	5.5	2.1	−48
4	5.5	8.0	1.7	10 acetone	8	3.5	−47
5	11	16.0	3.0	20 acetone	13.4	5.7	−57

[1] There are mechanical losses.
[2] Constants determined with Differential Scanning Calorimeter.
[3] Lower molecular weight products not extracted from polymer.

FLUOROETHERS

VINYL ETHERS

Trifluoroethyl Vinyl Ether

Various copolymers of 2,2,2-trifluoroethyl vinyl ether are described in U.S. Patent 2,851,449. In general, 2,2,2-trifluoroethyl vinyl ether can be prepared by vinylation of 2,2,2-trifluoroethanol with acetylene in the presence of the corresponding alkali metal alcoholate, for example, potassium 2,2,2-trifluoroethanolate. A more detailed description may be found in U.S. Patent 2,830,007.

The homopolymerization of 2,2,2-trifluoroethyl vinyl ether has been described in U.S. Patent 2,820,025. This compound homopolymerizes to form a high molecular weight, rubberlike, form-stable, nontacky product only with great difficulty. Such homopolymers are obtained by cationic polymerizations in the presence of certain activating chlorinated solvents at low temperatures.

It has been found by C.E. Schildknecht; U.S. Patent 2,991,278; July 4, 1961; assigned to Air Reduction Company, Inc. that copolymers of 2,2,2-trifluoroethyl vinyl ether and haloolefins can be prepared smoothly and rapidly, under certain conditions, by free radical means. The copolymer rubbers prepared in the process may be cured or vulcanized by conventional methods such as used with polymers of butadiene and chloroprene. For example, treatment of the copolymer with sulfur, sulfur-containing compounds, or magnesium or zinc oxides may result in crosslinking of the copolymer. The properties of the copolymer may also be modified by incorporating a small proportion, e.g., less than 10%, of a bifunctional monomer into the reaction mixture, and conducting the copolymerization under conditions designed to give fusible partial copolymers containing residual reactive double bonds. The following examples illustrate the process.

Example 1: Copolymers of 2,2,2-Trifluoroethyl Vinyl Ether and Tetrafluoroethylene —A high pressure reactor is flushed with nitrogen and charged with 200 parts of deoxygenated water, 0.15 part ammonium persulfate, and 1.0 part of sodium pyrophosphate. The reactor is closed, evacuated and cooled to -78°C. Then, the reactor is charged with 10 parts 2,2,2-trifluoroethyl vinyl ether and 60 parts tetrafluoroethylene. The reaction mixture is agitated at 55° to 65°C. for 15 hours at a pressure of 300 to 500 atmospheres. The copolymer formed is washed with water and dried. The product is characterized by less opacity and greater plasticity on heating than shown by tetrafluoroethylene homopolymers.

Example 2: Copolymers of 2,2,2-Trifluoroethyl Vinyl Ether and Vinyl Chloride — A pressure reaction vessel was charged with a mixture of:

	Grams
Water	100
Ammonium persulfate	2.0
Sodium bisulfite	1.0
Sodium lauryl sulfate	1.0
Formic acid	12 drops

The vessel was then swept with nitrogen and then with carbon dioxide to remove oxygen. The aqueous phase was then cooled until frozen and a mixture of 37.6 g. of vinyl chloride and 12.6 g. of 2,2,2-trifluoroethyl vinyl ether was added. The temperature was raised to 40° to 50°C. The batch was maintained within that temperature range for 2.5 hours. After that time, a copolymer latex had formed. The latex was coagulated by freezing with Dry Ice, and then allowed to thaw slowly. A solution of 4 g. of urea in 174 g. of methanol was stirred into the copolymer slurry and it was allowed to stand for 1/2 hour. The batch was then filtered and washed with methanol containing 1% of urea. The copolymer mass was voluminous, granular, and white. The copolymer was dried in vacuum at a temperature of 50°C.

Yield: 33.0 g.; conversion to copolymer 66%. Analysis showed 3.7% F in the copolymer or 8.2% by weight of combined trifluoroethyl vinyl ether. There was 14.3 mol percent of trifluoroethyl vinyl ether in the monomer mixture, and there was 4.3 mol percent of trifluoroethyl vinyl ether combined in the copolymer.

1,1,2,2-Tetrafluoroethyl Vinyl Ether

G.H. Crawford and E.S. Lo; U.S. Patent 2,975,164; March 14, 1961; assigned to Minnesota Mining and Manufacturing Company describe the polymerization of 1,1,2,2-tetrafluoroethyl vinyl ether with a polymerizable ethylenically unsaturated organic compound in the presence of a polymerization promoter. The polymerization may be effected in the presence of a free radical forming promoter or an ionic promoter, and may be carried out in aqueous or nonaqueous media. When the polymerization reaction is carried out in an aqueous polymerization catalyst system, the pH of the system should be no lower than 6, and preferably is 7 or above. The process is carried out at a temperature between about −30° and 150°C. under autogenous conditions of pressure, or at superimposed pressures up to about 1,200 psig.

The physical nature of the polymers of 1,1,2,2-tetrafluoroethyl vinyl ether produced by the process ranges from oils, greases and waxes to resinous thermoplastic and elastomeric materials. The molecular weight of the polymers ranges between about 5,000 and 100,000, or higher, depending upon the particular comonomer and the polymerization conditions employed.

The monomer, 1,1,2,2-tetrafluoroethyl vinyl ether, is prepared by the addition of ethanol to tetrafluoroethylene at a temperature of about 50°C. yielding 1,1,2,2-tetrafluoroethyl ethyl ether. This fluoro-ether is then chlorinated at a temperature of about 25°C. and dehydrochlorinated in the presence of triethanolamine and powdered sodium hydroxide in xylene to yield the desired 1,1,2,2-tetrafluoroethyl vinyl ether which has a boiling point of 39°C. and a density of 1.395 at about −78°C. The following examples illustrate the process.

Example 1: Copolymer of 1,1,2,2-Tetrafluoroethyl Vinyl Ether and Trifluorochloroethylene — The following neutral emulsion polymerization system was employed in carrying out the reaction of this example.

	Parts by Weight
Water	200.0
Total monomers	100.0
Potassium persulfate	1.0
Potassium 3,5,7,8-tetrachloroundeca-fluorooctanoate	0.75

After flushing a glass polymerization tube with nitrogen, the tube was charged with 150 grams of water containing 0.75 gram of dissolved potassium 3,5,7,8-tetrachloroundecafluorooctanoate. The polymerization tube was then placed in a solid carbon dioxide-trichloroethylene bath. When the contents of the tube were frozen solid, the tube was charged with 50 grams of water containing 1 gram of dissolved potassium persulfate. The pH of the resulting polymerization medium was adjusted to 7 using a 5% aqueous potassium hydroxide solution.

The contents of the tube were refrozen and the tube was charged with 55 grams of freshly distilled 1,1,2,2-tetrafluoroethyl vinyl ether and 45 grams of trifluorochloroethylene to make up a total monomer charge containing 50 mol percent of each monomer. The tube was then sealed and rotated end-over-end in a water bath at a temperature of 30°C. for a period of 24 hours. The polymerization was conducted under autogenous pressure. The contents of the tube were then frozen in a liquid nitrogen bath to coagulate the polymer latex. The coagulated product was collected, thoroughly washed with hot water and dried in vacuo at a temperature of 35°C.

The product was a tough inelastic resinous solid and was obtained in a 97% conversion based on the total monomers charged. Analysis for chlorine and fluorine content showed the product to contain 50 mol percent of combined 1,1,2,2-tetrafluoroethyl vinyl ether and 50 mol percent of combined trifluoroethylene. This copolymer was found to become rubbery at 100°C. and was compression molded at 150°C. at a pressure of 10,000 psi to form a continuous, clear, flexible and tough film.

The above procedure was repeated except that a temperature of 50°C. was maintained over a period of 24 hours yielding a tough inelastic resinous product in 99.1% conversion. Analysis of this product showed it to contain 50 mol percent of combined 1,1,2,2-tetrafluoroethyl vinyl ether and 50 mol percent of combined trifluorochloroethylene. These copolymers of 1,1,2,2-tetrafluoroethyl vinyl ether and trifluorochloroethylene can be molded into a variety of useful articles such as O-rings and gaskets, and are also valuable as protective linings for polymerization reactors, tanks, and the like.

Example 2: Copolymer of 1,1,2,2-Tetrafluoroethyl Vinyl Ether and Vinylidene Fluoride — A polymerization tube was charged with the same neutral emulsion catalyst system used in Example 1 above. Thereafter the tube was charged with

69.2 grams of 1,1,2,2-tetrafluoroethyl vinyl ether and 30.8 grams of vinylidene fluoride to make up a total monomer charge containing 50 mol percent of each monomer. The polymerization reaction was conducted for a period of 24 hours at a temperature of 25°C. under autogenous pressure. The polymer latex thereby obtained was coagulated by freezing and the coagulated product was treated in the same manner as described in Example 1.

The product was a tough and slightly rubbery product obtained in 99% conversion based on the total monomers charged. Analysis for fluorine content showed the product to contain 50 mol percent of combined 1,1,2,2-tetrafluoroethyl vinyl ether and 50 mol percent of combined vinylidene fluoride. Thus, product milled easily at 25°C. and was found to have a torsional modulus of 704.2 The percent volume increase of the raw polymer as determined in ASTM Fuel, Type II, which consists of isooctane (60% by volume), benzene (5% by volume), toluene (20% by volume) and xylene (15% by volume) was only 9.8%. The copolymer is useful as a substitute for tough natural rubbers. The fluorine content and high thermal stability of this copolymer makes it particularly suited for use in the tropics where natural rubber is subject to degradation from biological attack.

Trifluoromethyl Vinyl Ether

In a process described by P.E. Aldrich; U.S. Patent 3,162,622; December 22, 1964; assigned to E.I. du Pont de Nemours and Company trifluoromethyl vinyl ether, $CF_3-O-CH=CH_2$, is prepared by dehydrohalogenation of a 2-chloro- (or 2-bromo-) ethyl trifluoromethyl ether by means of an alkali metal hydroxide. The dehydrohalogenation is conveniently carried out by contacting the 2-chloro- (or 2-bromo-) ethyl trifluoromethyl ether with at least an equimolar quantity of an alkali metal hydroxide, e.g., potassium hydroxide. While the use of an inert reaction medium is not essential in this dehydrohalogenation process, it is preferred that one be employed. Absolute ethyl alcohol or denatured alcohol are suitable as they dissolve the alkali metal hydroxide and the dehydrohalogenation is conveniently carried out at the reflux temperature of the mixture. A dispersion of powdered alkali metal hydroxide in a high boiling hydrocarbon can also be employed.

The 2-chloroethyl and 2-bromoethyl trifluoromethyl ether starting materials for the dehydrohalogenation process can be prepared by known methods. For example, 2-chloroethanol or 2-bromoethanol can be heated at 100° to 125°C. with carbonyl fluoride and the reaction product, after removal of excess carbonyl fluoride, can then be treated with sulfur tetrafluoride at temperature of from 100° to 175°C. The reaction mixture is then treated with a slurry of powdered sodium fluoride in xylene and the filtrate is distilled to obtain the 2-haloethyl trifluoromethyl ether.

Copolymerization of trifluoromethyl vinyl ether with one or more other copolymerizable ethylenically unsaturated monomers is accomplished by bulk or solution methods in the presence of free radical liberating initiators. Thus, trifluoromethyl vinyl ether can be copolymerized with tetrafluoroethylene in hexafluoropropylene dimer as a solvent and dinitrogen difluoride as the initiator, or in a solvent such as octafluoro-1,2-dithiane. The bulk polymerization of trifluoromethyl vinyl ether with other ethylenically unsaturated compounds, e.g., hexafluoropropylene, can be accomplished with dinitrogen difluoride as initiator. Emulsion copolymerization between trifluoromethyl vinyl ether and other ethylenically unsaturated monomers, e.g., tetrafluoroethylene, can be carried out in the same manner as emulsion homopolymerization. The following examples illustrate the process.

Example 1: Dehydrochlorination of 2-Chloroethyl Trifluoromethyl Ether — (A) A glass reaction vessel having three necks is fitted with a reflux condenser, a dropping funnel, and a magnetic stirrer. Provision for collecting gaseous products is made by connecting the top of the condenser to a trap cooled by a mixture of acetone and solid carbon dioxide. The vessel is charged with 56 g. (1 mol) of potassium hydroxide in 210 ml. of 2B denatured alcohol. The reaction vessel is heated until the alcohol refluxes and there is then added dropwise, during a period of 60 minutes, 36.6 g (0.246 mol) of 2-chloroethyl trifluoromethyl ether. At the end of the reaction there are 15 to 16 ml. of condensate in the cold trap. Distillation of this material in a low temperature still gives 18.8 g. of crude trifluoromethyl vinyl ether boiling at −18° to −14°C.

Analysis of this product by vapor phase chromatography shows that it is a mixture of two components in the ratio 76:24. The retention time of the smaller peak is in agreement with that of an authentic sample of vinyl chloride (boiling point, −12°C.). Elemental analysis of the mixture shows that it contains 10.55% fluorine. On the assumption that the mixture is free of other impurities, this fluorine analysis indicates that the ratio of trifluoromethyl vinyl ether and vinyl chloride in the mixture is 71:29. An analytical sample of trifluoromethyl vinyl ether is separated from the mixture by preparative gas chromatography. Analysis: Calculated for $C_3H_3F_3O$: F, 50.87%. Found: F, 50.91%.

(B) A mixture of 32 g. of 2-chloroethyl trifluoromethyl ether and 30 ml. of absolute ethyl alcohol is heated to reflux in a reaction vessel of the type described above. To this mixture is added in a slow but steady stream a solution of 18 g. of 85% potassium hydroxide in 120 ml. of absolute ethyl alcohol during a period of about 20 minutes. After the addition is completed the mixture is heated at reflux for another 30 minutes. The crude product is transferred by distillation directly from the trap to a receiver, 10 g. of distillate being obtained. This contains trifluoromethyl vinyl ether and can be used without further purification for subsequent reactions.

Example 2: A platinum tube is charged with 30 ml. of gaseous trifluoromethyl vinyl ether and 1.5 ml. of gaseous dinitrogen difluoride and then sealed. The tube is heated at 70°C. for 4 hours under an external pressure of 1,000 psi. After the tube is cooled and opened, there is obtained a polymer of trifluoromethyl vinyl ether in the form of a viscous oil.

Example 3: A platinum tube of the type used in the preceding experiments is charged with 1.5 g. of trifluoromethyl vinyl ether, 1.5 g. of tetrafluoroethylene, 2 ml. of gaseous dinitrogen difluoride and 2 ml. of the saturated dimer of hexafluoropropylene (as solvent). The tube is sealed and heated at 75°C. for 4 hours under an external pressure of 100 atmospheres. There is obtained 2.18 g. of sticky, soft, solid polymer with properties similar to trifluoromethyl vinyl ether homopolymer.

The nuclear magnetic resonance spectrum of the copolymer in acetone solution shows a fluorine peak corresponding closely in position to the trifluoromethoxy fluorine peak of the trifluoromethyl vinyl ether homopolymer. Two other closely adjacent peaks are also present and they are of nearly equal area. The polymer is entirely soluble in acetone and this indicates it is not a mixture of homopolymers. Elemental analysis (fluorine, 62.04%) of the polymer indicates it to be a copolymer consisting of 52.7% of units derived from trifluoromethyl vinyl ether and 47.3% of units derived from tetrafluoroethylene.

Perfluoroalkoxy Perfluorovinyl Ethers

In a process described by J.F. Harris, Jr. and D.I. McCane; U.S. Patent 3,132,123; May 5, 1964; assigned to E.I. du Pont de Nemours and Company high molecular weight polymers are obtained by the polymerization of perfluoro-vinyl ethers having the general formula:

$$CF_2=CF-OR$$

where R is a perfluoroalkyl radical. The perfluorovinyl ethers may further be copolymerized with halogenated ethylenes, and particularly with tetrafluoroethylene, vinyl fluoride, vinylidene fluoride and hexafluoropropylene to give rise to high molecular weight solid copolymers. Examples of the perfluoroalkyl perfluorovinyl ethers are perfluoromethyl perfluorovinyl ether, perfluoropropyl perfluorovinyl ether, etc.

Perfluorovinyl ethers are prepared by the electrolytic fluorination of 2-alkoxy propionic acids using established techniques, followed by the decarboxylation and defluorination of the sodium salt of the perfluorinated 2-alkoxypropionic acid which leads to the formation of the perfluoroalkyl perfluorovinyl ether. The preparation of the perfluorovinyl ethers employed in this process is illustrated by the following experimental procedures.

Into a glass vessel was charged 30 g. of cesium fluoride and 75 ml. of diethylene glycol dimethyl ether. The vessel was cooled to –80°C., evacuated, and 66 g. of carbonyl fluoride and 83 g. of hexafluoropropylene epoxide were charged to the vessel. The vessel was heated to 75°C. for 4 hours. Low temperature distillation of the resulting product afforded 82 g. of perfluoro-2-methoxypropionyl fluoride, BP 10° to 12°C. The propionyl fluoride was dehalocarbonylated by passage through a bed of dry potassium sulfate pellets at 300°C. for a contact time of 10 minutes. A 60% yield of perfluoromethyl perfluorovinyl ether, BP -22°C., was obtained.

The polymerization may be carried out in bulk or in the presence of an inert diluent such as water or a perfluorinated solvent. The copolymerization of the perfluorovinyl ethers with tetrafluoroethylene can be carried out in an aqueous phase employing polymerization procedures such as described in U.S. Patent 2,559,752, or in a perfluorinated solvent phase employing polymerization procedures such as described in U.S. Patent 2,952,669. The preparation of the perfluorovinyl ether polymers is illustrated by the following examples.

Example 1: A 100 ml. stainless steel autoclave fitted with a magnetically driven stirring blade was flushed with nitrogen and evacuated. A solution of 10 g. (0.06 mol) of perfluoromethyl perfluorovinyl ether in 64 ml. of perfluorodimethyl-cyclobutane was admitted to the autoclave. The solution was heated to 60°C., and then tetrafluoroethylene was pressured into the autoclave until a pressure of 300 psig was attained. Approximately 10^{-4} mol of N_2F_2 diluted with N_2 was added to the rapidly stirred mixture. The contents of the autoclave were heated with stirring for 45 minutes at 60°C., and then cooled to room temperature and vented to atmospheric pressure. Solid polymer, weighing 11.4 g., was obtained. The melt viscosity of the copolymer at 380°C. was 16×10^4 poises. Infrared analysis of films of the resin pressed at 350°C. and 25,000 pounds platen pressure indicated that the copolymer contained 11.3 wt. percent perfluoromethyl perfluorovinyl ether. The films of the copolymer were tough, transparent and colorless.

Example 2: Into an evacuated stainless steel 100 ml. autoclave fitted with a magnetically driven stirrer was placed a solution of 9 g. of perfluoropropyl perfluorovinyl ether (0.034 mol) in 64 ml. of perfluorodimethylcyclobutane. The solution was heated to 60°C.; tetrafluoroethylene was admitted to the autoclave until a pressure of 268 psig was attained. To the rapidly stirred mixture was added approximately 10^{-4} mol of N_2F_2 diluted with nitrogen. The contents of the autoclave were heated and stirred for 1 hour at 60°C., and then cooled and gaseous materials vented off.

Fluoroethers

The solid polymer obtained weighed 15.0 g. and had a melt viscosity at 380°C. of 3.6 x 10^4 poises. Films of the co-polymer pressed at 350°C. and 20,000 pounds platen pressure were tough, clear and colorless. Infrared analysis of the resin indicated the presence of 9.7 wt. percent perfluoropropyl perfluorovinyl ether.

Trifluoroethyl Vinyl Ether-Vinyl Glycidyl Ether Copolymers

In a process described by H. Sorkin; U.S. Patent 3,414,634; December 3, 1968; assigned to Air Reduction Co., Inc. vinyl glycidyl ether of the formula

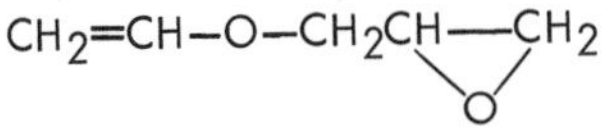

$$CH_2=CH-O-CH_2CH-CH_2$$
$$O$$

is copolymerized with trifluoroethyl vinyl ether to form a copolymer which may be cured. The following examples illustrate the process.

Example 1: A mixture of 860 g. (10 mols) iso-propyl vinyl ether, 100 g. (1.35 mols) glycidol, and 25 g. mercuric acetate was stirred and refluxed for 16 hours. The condenser was then replaced by a simple Claisen head and 72.9 g. liquid boiling at 121° to 131°C. was collected. This was fractionated in a small Vigreux column and four fractions were obtained:

Fraction	BP (°C.)	Weight (g.)
A	50 to 86	3.8
B	90 to 120	4.2
C	120 to 135	34
D	High boiling residues	

The fraction boiling at 125° to 135°C. was collected and subjected to a final fractionation in a 55-plate column to give 17.4 g. glycidyl vinyl ether, BP 133.5°C., n_D^{25} 1.4356. $C_5H_8O_2$ Calculated: C, 59.97%; H, 8.08%. Found: C, 60.23%; H, 8.09%.

Example 2: A solution of 100 g. trifluoroethyl vinyl ether, 1 g. glycidyl vinyl ether, and 1 g. azo-bis-iso-butyro-nitrile was refluxed for 26.5 hours. The solution was poured into hexane, the precipated polymer washed with hexane, and dried to yield 16 g. of a clear, colorless polymer. $(C_4H_5F_3O)_9(C_5H_8O_2)$ Calculated: C, 39.8%; H, 4.29%; F, 41.60%. Found: C, 39.15%; H, 4.25%; F, 41.60%.

Perfluorinated Divinyl Ethers

C.G. Fritz; U.S. Patent 3,310,606; March 21, 1967; assigned to E.I. du Pont de Nemours and Company describes the homopolymerization of perfluorodivinyl ethers having the formula

$$CF_2=CF-O-C_nF_{2n}-O-CF=CF_2$$

where n is from 2 to 24 or the copolymerization of such divinyl ethers with other perfluorodivinyl ethers, perfluorovinyl ethers or perfluoroolefins. Specific examples of fluorocarbon divinyl ethers employed in the process are:

perfluorodimethylene-bis(perfluorovinyl ether)
perfluorotrimethylene-bis(perfluorovinyl ether)
perfluorotetramethylene-bis(perfluorovinyl ether)
perfluorohexamethylene-bis(perfluorovinyl ether)
perfluorododecamethylene-bis(perfluorovinyl ether)
perfluorooctadecamethylene-bis(perfluorovinyl ether)

The process is illustrated by the following examples.

Example 1: A glass tube sealed at one end was cooled to -80°C. and evacuated. It was then charged with 0.04 ml. of 2,3-bis(difluoroamino)perfluoro-2-butene and 9 ml. of perfluorodimethylene-bis(perfluorovinyl ether). The tube was sealed and placed in a 50°C. bath. After 30 minutes, solidification had occurred. Heating at 50°C. was continued for 4 hours and followed by 8 hours at 100°C. The thermosetting resin obtained was colorless, clear, stiff and quite tough. It was insoluble in all common solvents and could not be pressed to a film at 300°C. and 30,000 psi. The polymer had a flex modulus of 239,000 psi at 23°C. and 4,600 psi at 100°C. It had a heat distortion temperature of 100°C. at 264 psi. After heating at 250°C. for several hours, the flex modulus at 100°C. was improved to 143,000 psi indicating further curing of the polymer at 250°C. The polymer was completely unaffected by boiling concentrated sulfuric acid, nitric acid, and 20% aqueous potassium hydroxide.

Example 2: The procedure of Example 1 was repeated except that a viscous syrup was isolated after 15 minutes heating. This syrup was placed in a mold in an air oven and cured at 100°C. for 5 hours. A thermoset resin with the same properties as that of Example 1 was obtained.

Example 3: The syrup prepared by the process of Example 2 was coated on one surface of a 3/4" stainless steel nut. A second 3/4" stainless steel nut was placed in contact with the first at finger pressure. The assembly was then heated in an air oven for 5 hours at 100°C. under no pressure. The adhesive bond formed in this way required an 8 kg. weight for breaking.

Example 4: Into a small glass polymer tube was charged 0.5 g. of perfluorotrimethylene-bis(perfluorovinyl ether). The tube was cooled to -196°C. and evacuated. There was then introduced 1 cc of the initiator of Example 1 in the gaseous state. The tube was reevacuated, sealed, and slowly heated in a water bath. After 4 hours heating at 100°C. the contents had solidified in part. Heating at 100°C. was continued for an additional 48 hours. The solid plug so attained was hard, brittle, and completely insoluble in common solvents. A differential thermal analysis indicated a transition at 147°C.

Example 5: The procedure of Example 4 was used. The tube was charged with 1.1 cc of perfluorotetramethylene-bis-(perfluorovinyl ether) and 0.03 cc of liquid 2,3-bis(difluoroamino)perfluoro-2-butene. Heating of the tube at 50°C. was carried out for 6 hours. The temperature was then raised to 70°C. and heating continued for 7 hours. The temperature was then raised to 100°C. and heating was continued for 12 hours. The resin so attained was hard, brittle, and insoluble in common solvents.

In related work R.A. Darby; U.S. Patent 3,418,302; December 24, 1968; assigned to E.I. du Pont de Nemours and Company has found that perfluorodimethylene-bis(perfluorovinyl ether) when polymerized in certain ways gives rise to polymers having structures of the class consisting of

and no measurable unsaturation as determined by infrared spectroscopy. The two isomeric cyclic structures result through a bond being formed between the carbon atom of one of the vinyl groups which is attached to an oxygen atom with either carbon atom of the other vinyl group. Both structures are formed during polymerization.

The polymerization of perfluorodimethylene-bis(perfluorovinyl ether) can be controlled to result in the intramolecular cyclization by polymerizing the divinyl ether in dilute concentrations, e.g., in the presence of a liquid medium. The intramolecular cyclization also occurs when the divinyl ether is copolymerized with other monovinyl compounds in the presence of an inert liquid diluent. The following examples illustrate the process.

Example 1: In a glass carius tube were charged 2 g. of perfluorodimethylene-bis(perfluorovinyl ether) and 18 g. of perfluorodimethylcyclobutane. The tube was cooled to -80°C., evacuated and charged with 3 mol percent N_2F_2 based on monomer. The tube was sealed and allowed to stand at 25°C. for 24 hours. The gel so formed was dried in vacuo at 100°C. The resulting granular polymer could be pressed to a clear, stiff, tough film at 250°C. and 30,000 psi. The polymer was found to contain no unsaturation as measured by infrared analysis.

Example 2: In a platinum tube sealed at one end were placed 0.5 ml. of perfluorodimethylene-bis(perfluorovinyl ether), 0.27 g. of tetrafluoroethylene, 4 ml. perfluorodimethylcyclobutane and 0.02 ml. of 2,3-bis(difluoroamino)perfluoro-2-butene. The charge consisted of equimolar amounts of tetrafluoroethylene and perfluorodimethylene-bis(perfluorovinyl ether). The platinum tube was sealed, placed in a shaker tube, and pressured to 900 atmospheres with nitrogen. It was then heated to 75°C. for 3 hours. A quantitative conversion to a copolymer of approximately 1/1 composition was obtained. This polymer could be pressed to a clear, stiff, tough film at 250°C. and 30,000 psi. The polymer was free from unsaturation.

Fluorocarbon Vinyl Ether Polymers Containing Sulfonic Acid Groups

D.J. Connolly and W.F. Gresham; U.S. Patent 3,282,875; November 1, 1966; assigned to E.I. du Pont de Nemours and Company describe fluorocarbon vinyl ethers which have the formula

$$MSO_2CFR_fCF_2O(CFYCF_2O)_nCF\!\!=\!\!CF_2$$

R_f is a radical selected from the class consisting of fluorine and perfluoroalkyl radicals having from 1 to 10 carbon atoms.

Y is a radical selected from the class consisting of fluorine and the trifluoromethyl radical.

157

n is an integer of 1 to 3 inclusive.

M is a radical selected from the class consisting of fluorine, the hydroxy radical, the amino radical and the radicals having the formula —OMe, where Me is a radical selected from the class consisting of alkali metals and quaternary ammonium radicals.

The vinyl ethers are readily homopolymerized or copolymerized with ethylene or halogenated ethylenes. In these polymers it is generally preferred that at least one of the additional monomers be ethylene or a halogenated ethylene such as vinylidene fluoride, tetrafluoroethylene, or chlorotrifluoroethylene, while the other additional monomer is a perfluoro alpha-olefin such as hexafluoropropylene or a perfluoro (alkyl vinyl ether) of the type $CF_2=CF—O—(CF_2)_n—CF_3$ where n is 0 to 5, inclusive.

The concentration of fluorocarbon vinyl ethers which contain sulfonic acid groups is chosen in relation to the degree of cross-linkability desired for the copolymer product. For economic reasons, 5 mol percent, based on total monomer incorporated into the copolymer, is usually all that is used to produce high modulus vulcanized products, while 0.2 mol percent is about the minimum which will produce a satisfactory degree of crosslinking. For example, when copolymers are prepared from vinylidene fluoride, hexafluoropropylene and a perfluorovinyl ether of the structure

$$CF_2=CF—O—CF_2—CF(CF_3)—O—CF_2CF_2—SO_2F$$

good elastomers are obtained when the molar ratio of vinylidene fluoride to hexafluoropropylene lies within the range of 51:49 to 85:15 and the proportion of the fluorocarbon vinyl ether is present in the range of about 0.2 to 5 mol percent of the total monomer units present in the copolymer.

In the preparation of elastomeric copolymers from tetrafluoroethylene, perfluoro(methyl vinyl ether) and a perfluorovinyl ether of the structure

$$CF_2=CF—O—CF_2—CF—(CF_3)—O—CF_2—CF_2—SO_2F$$

a preferred range of molar ratios is 1.5 to 2.0 mols of tetrafluoroethylene per mol of perfluoro(methyl vinyl ether) with 0.5 to 4 mol percent of the total composition of the sulfonyl fluoride monomer.

The vinyl ethers of this process are prepared by the pyrolysis of compounds having the following formulas:

$$FSO_2CFR_fO(CFYCF_2O)_nCF(CF_3)COF \quad \text{and} \quad FSO_2CFR_fO(CFYCF_2O)_nCF(CF_3)COOX$$

where R_f, Y and n have the same meaning as previously given, and X is an alkali metal. The pyrolysis is carried out at temperatures of 200° to 600°C. In the case of the acid fluoride, a metal oxide such as zinc oxide or silica is preferably employed as a solid catalyst for the gas phase reaction. The acid fluoride employed in the pyrolysis is obtained by the reaction of hexafluoropropylene epoxide with a fluorosulfonyl fluoroacyl fluoride having the formula FSO_2CFR_fCOF. The alkali metal salt of the carboxylic acid is formed from the corresponding acid fluoride by reaction with an alkali metal salt of a weak acid, such as carbonic acid. The vinyl ethers of the process are preferably polymerized in a perfluorocarbon solvent using a perfluorinated free radical initiator. The process is illustrated by the following examples.

Example 1: Into a rotary evaporator were charged 200 g. of

$$FSO_2CF_2CF_2OCF(CF_3)CF_2OCF(CF_3)CO_2Na$$

The evaporator was heated to 180°C. until no further gas evolution was observed. The off-gases from the reaction were condensed in a cold trap. On distillation, there was obtained 48 g. of perfluoro[2-(2-fluorosulfonylethoxy)propyl vinyl ether], BP 118°C. The infrared and NMR spectra of the product were consistent with the structure of the ether. Analysis — Calculated for $C_7F_{14}O_4S$: C, 18.84%; F, 59.62%; S, 7.18%. Found: C, 19.11%; F, 59.13%; S, 7.11%.

Example 2: Into a rotary evaporator were charged 150 g. of

$$FSO_2CF_2CF_2O[CF(CF_3)CF_2O]_2CF(CF_3)CO_2Na$$

The evaporator was heated to 200°C. until no further gas evolution was observed. The off-gases from the reaction were condensed in cold traps. On distillation of the reaction product there were obtained 35 g. of

$$FSO_2CF_2CF_2O[CF(CF_3)CF_2O]_2CF=CF_2$$

having a boiling point at 159°C. Infrared and NMR spectra was consistent with the indicated structure.

Example 3: The fluorosulfonyl fluoroacyl fluoride, having the formula

$$FSO_2CF_2OCF(CF_3)CF_2OCF(CF_3)COF$$

was passed through a one inch stainless steel column packed with 1/4 inch pellets of ZnO and heated to a temperature of 285°C. The fluorosulfonyl fluoroacyl fluoride, 85 g., was vaporized by dripping on a flash evaporator in a nitrogen stream of 400 ml./min. The nitrogen stream was then passed through the column and multiple cold traps in which the reaction product was collected. Upon separation, there were obtained 60 g. of the perfluoro[2-(2-fluorosulfonylethoxy)-propyl vinyl ether].

Example 4: Into an evacuated 320 ml. stainless steel shaker tube were charged 40 g. of the vinyl ether, having the formula

$$FSO_2CF_2CF_2OCF(CF_3)CF_2OCF=CF_2$$

along with 40 g. of tetrafluoroethylene and 200 ml. of perfluorodimethylcyclobutane. While cold, a 30 ml. jumper line was pressured to 20 psig with 2.4 volume percent difluorodiazine in nitrogen. This catalyst was pressured into the shaker tube with 800 psi of nitrogen. The mixture was shaken and the temperature raised slowly to 80°C. and maintained there for one hour. On cooling and discharging, 28 g. of 9 wt. percent vinyl ether copolymer having a melt viscosity above 1×10^4 poises was obtained. The polymer could be compression molded into clear tough film.

Example 5: Into an evacuated 320 ml. stainless steel shaker tube were charged 30 g. of purified vinyl ether having the formula

$$NaSO_3CF_2CF_2OCF(CF_3)CF_2OCF=CF_2$$

along with 200 ml. of deoxygenated, distilled water, about 30 g. of tetrafluoroethylene and 1.0 g. of ammonium persulfate. The reaction mixture was heated under autogenous pressure to 68° to 70°C. for 2 hours. On discharging, 48 g. of gelatinous copolymer were obtained which after drying could be made into films of good stiffness and contained 14 wt. percent of the vinyl ether.

Example 6: A 400 ml. Hastelloy shaker bomb is swept with nitrogen and charged with 200 ml. of deoxygenated distilled water, 3.0 g. (11.0 mM) of disodium phosphate heptahydrate, 0.55 g. (2.4 mM) of sodium bisulfite, 0.15 g. (0.3 mM) of ammonium perfluorooctanoate, and 5.0 g. (11.0 mM) of perfluoro[2-(2-vinyloxy-1-methylethoxy)-ethane sulfonyl]-fluoride. The bomb is closed, cooled to -80°C., and purged of oxygen by evacuating to 1 ml. pressure of mercury. With the interior under reduced pressure, 18.1 g. (0.12 mol) of hexafluoropropene and 28.0 g. of vinylidene fluoride (0.44 mol) are introduced. The bomb is shaken and the temperature inside the reaction chamber is increased to 60°C. and held there for two hours. The bomb is then cooled to room temperature, and excess gaseous reactants vented to the atmosphere.

The partially coagulated product is removed and coagulation is completed by freezing. The polymer is isolated by filtration, washed thoroughly with water, and dried overnight at 70°C. in a vacuum oven. The dry, white polymer weighed 37.6 g. Analysis for carbon, hydrogen, fluorine and sulfur showed that the product contains 32.7% C, 2.2% H, 63.5% F and 0.23% S.

The product is compounded on a two-roll rubber mill to contain the following:

	Parts by weight
Terpolymer	100
Carbon black, medium thermal	20
MgO	12
PbO	3

This compound stock is vulcanized by pressing sheets in a mold for 30 minutes at 150°C., followed by removing the sheets and heating them in an air oven for 24 hours at 204°C. The following physical properties were measured at 21.1°C.

Tensile strength	1,625 psi
Elongation at break	340 %
Stress at 200% elongation	1,175 psi

Perfluoroalkanesulfonamido Vinyl Ethers

R.F. Heine; U.S. Patent 3,078,245; February 19, 1963; assigned to Minnesota Mining and Manufacturing Company describes fluorocarbon vinyl ethers which are perfluoroalkanesulfonamido, including perfluorocycloalkanesulfonamido, alkyl vinyl ethers of the formula

$$R_fSO_2N(R)R'OCH{=\!=\!}CH_2$$

R is hydrogen, or an alkyl group having from 1 to 12 carbon atoms, preferably from 1 to 4 carbon atoms.
R_f is aliphatic C_nF_{2n+1} or cycloaliphatic C_nF_{2n-1}.
n is an integer from 1 to 18, preferably from 3 to 12.
R' is an alkylene bridging group having from 1 to 12, preferably from 1 to 8, carbon atoms. R' can be branched or straight chain.

Illustrative of these compounds are: perfluoroethanesulfonamidoethyl vinyl ether, perfluorooctanesulfonamidoethyl vinyl ether, perfluorododecanesulfonamidoethyl vinyl ether, N-propyl perfluorooctanesulfonamidoethyl vinyl ether, N-ethyl perfluorooctanesulfonamidoethyl vinyl ether, and N-hexadecylperfluorooctanesulfonamidoethyl vinyl ether. These compounds are reactive monomers and can be polymerized to yield homopolymers as well as copolymers with another polymerizable ethylenically unsaturated monomer, particularly the polymerizable vinyl compounds capable of vinyl addition reactions.

Fluorocarbon vinyl esters, methacrylates, acrylates and the fluorinated monoolefins having from 2 to 3 carbon atoms are particularly preferred as comonomers. Vulcanizable copolymers can be made. Since the side chains can vary as to length and to chemical type, the production of various polymers having different physical properties is possible, including high molecular weight elastomers and thermoplastics and also lower molecular weight oils, greases and waxes.

The perfluoroalkanesulfonamido alkyl vinyl ethers of this process are prepared by reacting the sodium salt of a perfluoroalkanesulfonamide, e.g., as described in U.S. Patent 2,732,398, with a chloroalkyl vinyl ether in a solvent such as ethylene glycol, acetone, or dimethyl formamide, according to the following equation:

$$R_fSO_2NHR + Cl(R')OCH{=\!=\!}CH_2 \xrightarrow{\text{NaOCH}_3,\ \text{ethylene glycol}} NaCl + R_fSO_2N(R)R'OCH{=\!=\!}CH_2$$

with R_f, R' and R as defined above. Other N-substituted perfluoroalkanesulfonamides can be prepared by reacting various perfluoroalkanesulfonyl fluorides, which are described in U.S. Patent 2,732,398, with diverse amines. Some of these perfluoroalkanesulfonamides are, for example, N-methyl perfluorooctanesulfonamide, N-isopropyl perfluorooctanesulfonamide, and N-ethyl perfluorododecanesulfonamide.

The perfluoroalkanesulfonamidoalkyl vinyl ether polymers of this process are resistant to hydrolysis or loss of the perfluoroalkyl group, even when in contact with fatty oils at elevated temperatures. This is particularly important when such polymers come into contact with foodstuff, as is the case when they are used to treat packaging materials, and is in contrast to polymers which contain relatively easily hydrolyzable vinyl ester groups. Various fabrics, synthetic and natural, as well as paper, leather and diverse other products, may be rendered both oil and water repellent with these polymers using known techniques, e.g., spraying, roll coating, brushing, padding, dipping, etc. The following examples illustrate the process.

Example 1: This example illustrates the homopolymerization of the perfluorooctanesulfonamidoalkyl vinyl ethers. The following charge was made into a glass ampule:

2.0 grams 2-(N-propyl perfluorooctanesulfonamide) ethyl vinyl ether
4.0 grams distilled xylene hexafluoride
2 drops 47% $BF_3 \cdot (C_2H_5)_2O$

The $BF_3 \cdot (C_2H_5)_2O$ was added to the frozen monomer solution (cooled in liquid air). The ampule was frozen, evacuated, sealed and placed in a -78°C. bath. The bath was allowed to warm slowly to room temperature and was maintained at room temperature for 3 hours. As the solution thawed it became viscous and a dark green color appeared, finally turning brown. The ampule was then opened and the contents were poured into methanol containing a small amount of ammonia. A brown, viscous, low molecular weight polymer layer separated and was recovered. The brown homopolymer was insoluble in acetone.

Example 2: This example illustrates the copolymerization of a perfluoroalkanesulfonamidoalkyl vinyl ether and a fluorinated acrylate. The following recipe was charged to a polymerization tube, after which the tube was sealed and tumbled in a 50°C. water bath.

	Parts by weight
Monomers	100
Water	126
Acetone	54
$C_8F_{17}SO_2NHC_3H_6N(CH_3)_2 \cdot HCl$	5
$K_2S_2O_8$	0.2

The monomer charge consisted of 50 wt. percent of 2-(N-ethyl perfluorooctanesulfonamide) ethyl vinyl ether and 50 wt. percent of 2-(N-propyl perfluorooctanesulfonamide) ethyl acrylate. After about 18 hours the tube was removed from the bath and a stable, clear blue latex was observed. The coagulated copolymer was a clear flexible plastic.

Example 3:   To a polymerization ampule was charged:

	Parts by weight
2-(N-propyl perfluorooctanesulfonamido) ethyl vinyl ether	8.4
5% aqueous solution of Brij 35 (a reaction product of ethylene oxide and $C_{11}H_{23}CH_2OH$, Atlas Co.)	5.0
Dioxane	6.3
5% aqueous solution of potassium acid phosphate	1.0
5% aqueous solution of ammonium persulfate	1.0
Perfluoropropene	1.6
Distilled water	5.0

The ampule was charged with all materials and then evacuated while the contents were frozen. The ampule was maintained at 50°C. for 20 hours, and a clear, stable latex, together with 9.4 parts of coagulum, was found. Overall conversion to the copolymer was 89%. A portion of the latex was coagulated, and a clear soft copolymer film was prepared.

Example 4:   This example indicates the use of the copolymers in paper treatment. The stable latex of Example 3 was diluted with water to 1% solids concentration. Kraft paper was treated with various concentrations of this copolymeric latex. Pickup was assumed to be 100%, and samples were dried at 230°F. To determine oil repellency of both creased and uncreased samples, SAE 10 oil at 80°F. was applied to the surface of the specimens, and the underside of the specimens was observed for penetration. The time required for oil penetration represents a measure of oil repellency. As indicated below, the resistance to oil penetration of the treated paper was excellent.

Solids concentration	Uncreased, days	Creased, days
0.3%	40+	5 to 20
0.2%	40+	3 to 5
0.1%	15	2

ALLYL ETHERS

2-Chloro-1,1,2-Trifluoroethyl Allyl Ether

J.G. Abramo and R.H. Reinhard; U.S. Patent 2,975,161; March 14, 1961; assigned to Monsanto Chemical Company have found that ethers of the type such as 2-chloro-1,1,2-trifluoroethyl allyl ether may be polymerized to yield useful plastic masses. The homopolymers of the fluorinated allyl and methallyl ethers are in general low molecular weight liquid or semisolid products. When copolymerized with other vinyl monomers, particularly with from 5 to 95% of the vinyl monomer, solid polymers are obtained. The following examples illustrate the process.

Example 1:   2-chloro-1,1,2-trifluoroethyl allyl ether was prepared by bubbling chlorotrifluoroethylene at room temperature through allyl alcohol in which a catalytic amount of KOH had been dissolved. After absorption of the olefin had ceased, the reaction mixture was washed with water to remove the catalyst, unreacted allyl alcohol, and any dissolved unreacted chlorotrifluoroethylene. The wash water was decanted, the organic material was dried with anhydrous calcium chloride, and then was distilled to yield the ether product having a boiling point of 109°C. at 771 mm. of Hg. Approximately 15 g. of the 2-chloro-1,1,2-trifluoroethyl allyl ether and 0.45 g. of benzoyl peroxide were charged to a tube which was then sealed and placed in a constant temperature oil bath maintained at 85°C. for a period of 25 hours. At the end of this time, the tube was opened and its contents were discharged into a 25 cc round-bottomed flask and subjected to heating under vacuum to remove all volatile components boiling below 150°C. at 200 mm. of Hg. The polymeric product recovered was a pale, greenish-yellow viscous liquid at room temperature. This homopolymer was somewhat sticky, extremely resistant to wetting with water, and soluble in a 1:1 mixture of absolute ethanol and carbon tetrachloride.

Example 2: 60 parts by weight of vinyl acetate, 40 parts by weight of 2-chloro-1,1,2-trifluoroethyl allyl ether, and 3 parts by weight of benzoyl peroxide in a sealed tube were heated at 75°C. for 25 hours in a constant temperature oil bath. A semisolid polymer was produced which is soluble in ethyl acetate.

Example 3: A monomer-catalyst mixture consisting of 70 parts by weight of styrene, 30 parts by weight of 2-chloro-1,1,2-trifluoroethyl allyl ether, and 3 parts by weight of benzoyl peroxide was heated in a sealed tube at 75°C. for 25 hours. The semisolid copolymer was dissolved in ethyl benzene, then precipitated in methanol and dried overnight at 40°C. in a vacuum oven. The recovered copolymer was a white amorphous powder.

Haloalkyl Perfluoroallyl Ethers

E.S. Lo; U.S. Patent 2,975,163; March 14, 1961; assigned to Minnesota Mining and Manufacturing Company describes the preparation of a class of fluorine-containing organic compounds, namely haloalkyl perfluoroallyl ethers. These ethers find particular utility as chemical intermediates for the production of other valuable materials. The fluorine-containing organic ethers possess, as one of the groups bonded to the ether oxygen atom, a wholly fluorinated allyl group, and as the second group which is bonded to the ether oxygen, a saturated halocarbon radical. These ethers have the general formula:

$$R—O—CF_2CF\!\!=\!\!CF_2 \qquad\qquad (1)$$

where R may be a partially halogenated or perhalogenated alkyl group having from 2 to 12 carbon atoms per radical, and preferably not more than 8 carbon atoms per molecule. The preferred ethers are the 1,1-dihydroperfluoroalkyl perfluoroallyl ethers having the general formula:

$$C_nF_{2n+1}CH_2OCF_2CF\!\!=\!\!CF_2 \qquad\qquad (2)$$

where n is an integer from 1 to 11. These compounds are prepared by interacting 3-chloropentafluoropropene with a 1,1-dihydroperfluoroalcohol or an alkali metal derivative in an alkaline medium. It has been found that the best yields of desired ether are obtained by effecting this reaction in a strongly alkaline medium containing an alkali metal hydroxide such as sodium hydroxide and potassium hydroxide, although milder alkaline reagents may be used.

When the 1,1-dihydroperfluoroalkyl perfluoroallyl ethers are subjected to the polymerization conditions, they homopolymerize to relatively low molecular weight oils such as the dimer and trimer. On the other hand, it has been found that the perfluoroallyl ethers of this process copolymerize with various monomers such as olefins containing at least one fluorine atom to yield polymers having a molecular weight from about 1,000 to 100,000 or higher depending upon the particular comonomer and the polymerization conditions employed. The polymers have particularly good adhesion properties. The lower molecular weight polymers in the oil, grease and wax range are particularly useful, for example, as bonding agents, lubricants, plasticizers, and as ingredients of polish compositions. The higher molecular weight polymers such as the thermoplastics and elastomers are of value as protective coatings on metal and fabric surfaces in producing adhesives, laminates, films, and other molded articles. The following examples illustrate the process.

Example 1: To a 250 cc round bottom, one-neck flask containing 100 g. of 1,1-dihydroperfluoroethanol, there were added 11.2 g. (about 0.2 mol) of potassium hydroxide pellets. The resulting solution was then frozen and there were then added 20 cc (about 33.5 g., 0.2 mol) of 3-chloropentafluoropropene. A reflux condenser was then placed in the neck of the flask using a mixture of Dry Ice and acetone as the cooling medium. The reaction mixture was allowed to warm to room temperature (about 22°C.) over a period of 1.5 hours, during which time a large amount of white precipitate formation was observed. The flask was then warmed in a hot water bath at 80°C. for about 4 hours. After cooling the reaction mixture to room temperature, the pH of the reaction mixture was found to be 6, indicating that all of the potassium hydroxide had reacted.

The reaction mixture was then subjected to fractional distillation to yield the following fractions: Fraction (1) boiling point 56° to 71°C.; fraction (2) boiling point 74°C. Mass spectrometer analysis of fraction (1) showed this fraction to contain about 20 mol percent of $CF_3\!\!=\!\!CFCF_2OCH_2CF_3$ and mass spectrometer analysis of fraction (2) indicated this fraction to contain a substantial amount of the same compound. Further purification was carried out as follows. Fractions (1) and (2) were combined and washed with water to remove CF_3CH_2OH, dried on magnesium sulfate, and fractionated. About 22 g. of colorless liquid boiling between 55° and 56°C. were collected. Infrared analysis of this fraction showed the presence of —CH, —CF, —CF=CF—, and —C—O—C— groups.

Example 2: To a three-necked glass flask there were added 23 g. (0.23 mol) of 1,1-dihydroperfluoroethanol and

43.8 g. (0.263 mol) of 3-chloropentafluoropropene. The solution was stirred with magnetic stirrers for about 2 hours at 22°C., but under these conditions essentially no reaction occurred. A reflux condenser was then fitted to the flask using a Dry Ice-acetone mixture as the cooling medium. About 17 cc of water containing about 15 g. of dissolved potassium hydroxide was then slowly added to the contents of the reaction flask. After complete addition of the aqueous potassium hydroxide solution, the reaction mixture was refluxed for a period of about 2 hours.

The solution was then cooled and washed several times with water. The lower layer was dried with Drierite and 45.5 g. of a colorless liquid was obtained, which liquid was then subjected to fractional distillation to obtain the following fractions: Fraction (1), boiling point 53° to 55°C.; fraction (2), boiling point 56°C.; fraction (3), boiling point 59° to 88°C.; and fraction (4), boiling point 91°C. A very viscous residue remained in the distillation flask and is believed to be compounds of higher molecular weight than $CF_2=CFCF_2OCH_2CF_3$. Mass spectrometer analysis of fraction (2) above showed this fraction to contain at least 90 mol percent $CF_2=CFCF_2OCH_2CF_3$, and fluorine analysis showed this fraction to contain 64.8% fluorine (calculated fluorine percent for $C_5F_8H_2O$ is 66.0%).

Example 3: A stainless steel polymerization bomb having a volume capacity of about 45 cc was flushed with nitrogen and was then charged with 12 cc of a solution prepared by dissolving 0.25 g. of perfluorooctanoic acid and 1 g. of disodium hydrogen phosphate heptahydrate in 30 cc of water. After freezing the contents of the bomb, the bomb was then charged with 8 cc of a solution prepared by dissolving 0.25 g. of potassium persulfate in 20 cc of water. The contents of the bomb were then frozen at liquid nitrogen temperature and the bomb was evacuated. Thereafter, 6.98 g. of 1,1-dihydroperfluoroethyl perfluoroallyl ether, prepared as described in the above examples, and 2.92 g. of vinylidene fluoride were charged to the bomb to make up a total monomer charge containing 40 mol percent of the perfluoroallyl ether and 60 mol percent of vinylidene fluoride.

The bomb was then closed and rocked in an electric rocker maintained at a constant temperature of 50°C. for a period of 71 hours. The bomb was then vented to atmospheric pressure followed by freezing of the reaction product at a temperature of about -5°C. The coagulated product was cooled, thoroughly washed with hot water, and dried in vacuo at a temperature of 35°C. A sticky solid copolymer product was obtained in about a 25% conversion. This copolymer is particularly useful as an adhesive for bonding halogenated polymer surfaces, for example, to metal surfaces such as chromium, steel and aluminum.

Example 4: After flushing a polymerization glass tube having a volume capacity of 20 cc with nitrogen, there were added 6 cc of a solution prepared by dissolving 1.0 g. of perfluorooctanoic acid and 4 g. of disodium hydrogen phosphate heptahydrate in 120 cc of water. After freezing the contents of the tube, there were then added 4 cc of a solution prepared by dissolving 1 g. of potassium persulfate in 80 cc of water. After refreezing the contents of the tube, there were then added 2.75 g. of 1,1-dihydroperfluoroethyl perfluoroallyl ether and 2.25 g. of 1,1-chlorofluoroethylene to make up a total monomer charge containing 30 mol percent of the ether and 70 mol percent of the 1,1-chlorofluoroethylene.

The bomb was then closed and rocked for a period of 23 hours in a constant temperature water bath maintained at 50°C. The contents of the bomb were then frozen at about -5°C. to coagulate the resulting polymer latex. The coagulated product was cooled, thoroughly washed with hot water, and dried in vacuo at a temperature of 35°C. The copolymer product obtained was a grease and is particularly useful as a bonding agent and as a high temperature lubricant.

MISCELLANEOUS

Fluorocarbon Polyethers

R.A. Darby; U.S. Patent 3,450,684; June 17, 1969; assigned to E.I. du Pont de Nemours and Company describes fluorocarbon polyethers which have the general formulas:

$$XCF_2CF_2(OCFXCF_2)_nOCF=CF_2 \quad and \quad (CF_2)_m \begin{array}{c} (OCFXCF_2)_lOCF=CF_2 \\ (OCFXCF_2)_pOCF=CF_2 \end{array}$$

X is a member of the class consisting of fluorine, chlorine, hydrogen, the difluoromethyl group, the chlorodifluoromethyl group and the perfluoromethyl group.
n is an integer of at least one.
m is an integer of at least two.
The sum of l and p is an integer of at least one, the total number of carbon atoms in the vinyl ether not to exceed 24.

163

Fluoroethers

The fluorocarbon compounds of the process can be prepared in several ways which are illustrated by the following reaction equations, where X and n have the same meaning as given before and where MeOH is an alkali metal hydroxide.

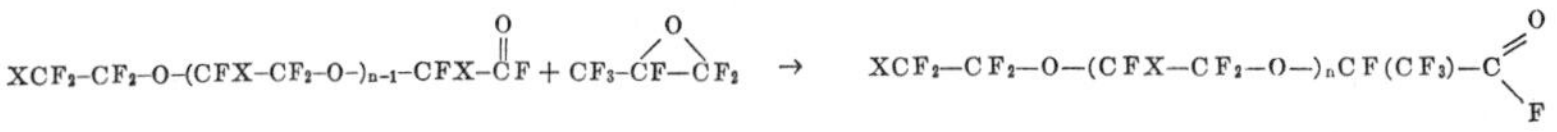

$$XCF_2-CF_2-O-(CFX-CF_2-O-)_{n-1}-CFX-\overset{\overset{O}{\parallel}}{C}F + CF_2-\overset{O}{\overset{\diagdown}{C}F}-CF_2 \rightarrow XCF_2-CF_2-O-(CFX-CF_2-O-)_nCF(CF_3)-\overset{\overset{O}{\diagup}}{\underset{F}{C}\diagdown} \tag{1}$$

$$n+2CF_2-\overset{O}{\overset{\diagup\diagdown}{C}F}-CF_2 \longrightarrow CF_3-CF_2-CF_2-O-[CF(CF_3)-CF_2-O]_nCF(CF_3)\overset{\overset{O}{\diagup}}{\underset{F}{C}\diagdown} \tag{2}$$

$$XCF_2-CF_2-O-(CFX-CF_2-O)_n-CF(CF_3)-\overset{O}{\overset{\diagup}{C}}\underset{F}{\diagdown} \xrightarrow{H_2O} XCF_2CF_2O-\left(CFX-CF_2O\right)_n-\overset{CF_3}{\overset{\mid}{C}}F-\overset{\overset{O}{\diagup}}{C}\underset{OH}{\diagdown}$$

$$\downarrow MeOH$$

$$XCF_2-CF_2-O-\left(CFX-CF_2-O\right)_n-\overset{CF_3}{\overset{\mid}{C}}F-\overset{\overset{O}{\diagup}}{C}\underset{OMe}{\diagdown}$$

heat + activator $\qquad$ $\downarrow$ heat

$$XCF_2-CF_2-O-\left(CFX-CF_2-O\right)_n-CF=CF_2 \tag{3}$$

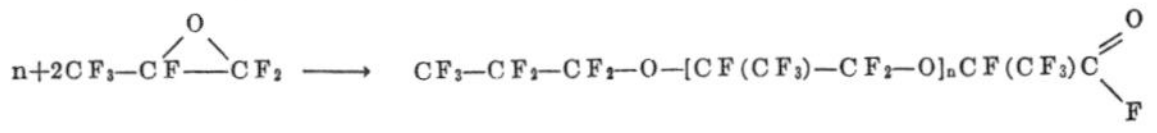

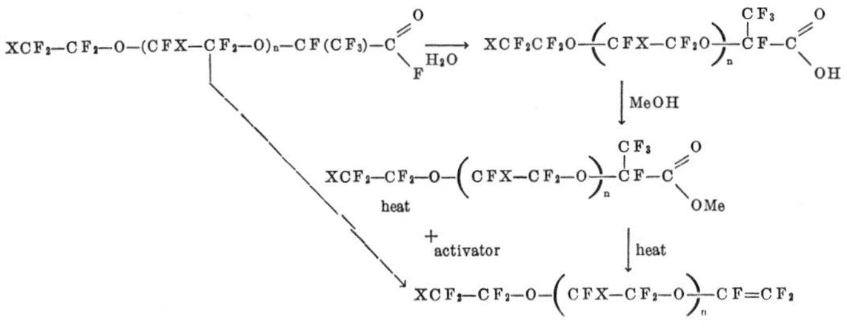

The fluorocarbon ether employed in reaction (1) is obtained by the polymerization of epoxides of tetrafluoroethylene, hexafluoropropylene, chlorotrifluoroethylene, etc. This ether is reacted with hexafluoropropylene epoxide to result in a polyether of a higher degree of polymerization which contains a pendent trifluoromethyl group on the carbon atom in the alpha position to the carbonyl group. Where X is a trifluoromethyl group, the same product can also be obtained by the direct polymerization of hexafluoropropylene epoxide as illustrated in reaction Equation 2.

The acid fluoride obtained by reactions (1) and (2) is hydrolyzed to the acid, converted to a monovalent metal salt, and then heated to result in the perfluorovinyl ether. The acid fluoride may also be directly pyrolyzed to the perfluorovinyl ether in the presence of an activator such as zinc oxide or silica (Equation 3). The acid fluoride may also be directly hydrolyzed to the salt.

The reaction of the hexafluoropropylene epoxide with the monoacid fluoride (reaction Equation 1) or with the diacid difluoride is carried out in bulk using active carbon as the catalyst, or by reaction in a polar solvent using an alkali metal or quaternary ammonium fluoride catalyst. The solvents suitable in the preparation of the alpha-substituted acid fluoride by the latter technique are nitriles and polyalkyl ethers liquid at reaction temperatures. Examples of these solvents are the dimethyl ether or ethylene glycol, the dimethyl ether of diethylene glycol, benzonitrile, acetonitrile, etc. Other highly polar solvents which have no active hydrogen are also useful in the process. These solvents include dimethyl sulfoxide and N-methyl pyrrolidone (reaction Equations 1 and 2).

The pyrolysis may be carried out with the acid fluoride directly or the acid fluoride can be converted into a monovalent salt such as the alkali metal salt of the acid and then pyrolyzed to the vinyl ether. The hydrolysis is carried out by contacting the acid fluoride with water. The formation of the alkali metal salt, for example, is accomplished by carrying out the hydrolysis in the presence of alkali metal base, such as KOH. The acid fluoride is generally pyrolyzed in gaseous form by passage through a reaction zone maintained at temperatures of 300° to 600°C. In the presence of certain activators, such as oxygen-containing sodium salts, the pyrolysis may be carried out at somewhat lower temperatures. The pyrolysis of the alkali metal salts is carried out at temperatures of 170° to 250°C. In the presence of the solvents described above, the pyrolysis may be carried out at temperatures as low as 100°C.

The resulting vinyl ethers are useful as monomers which are capable of homopolymerizing as well as capable of forming high molecular weight copolymers with tetrafluoroethylene, chlorotrifluoroethylene, vinylidene fluoride and similar vinyl monomers. In this respect, the perfluorovinyl ethers are different from perfluoroolefins of similar molecular weight and carbon numbers which have not been homopolymerized. The vinyl ethers are polymerized through addition to the double bond.

The homopolymers of the perfluorovinyl ethers of the process have the general formulas shown on the following page.

Fluoroethers

$$-\!\!\left(\!CF\!-\!CF_2\!-\!\right)_d \atop \left(\!-O\!-\!CF_2CFX\!-\!\right)_n\!-\!O\!-\!CF_2\!-\!CF_2X$$ and $$-\!\!\left(\!CF\!-\!CF_2\!-\!\right)_d \atop (OCF_2\!-\!CFX\!-\!)_l(CF_2\!-\!)_m(OCFX\!-\!CF_2\!-\!)_pOCF\!=\!CF_2}$$

where d is the degree of polymerization and the remaining symbols have the previously indicated meanings. In the polymerization of the vinyl ether, polymer formation readily extends to the second vinyl group to give rise to insoluble network type polymers. The following examples illustrate the process.

Example 1:   To 1,000 ml. of water were added 596 g. of the trimer of hexafluoropropylene epoxide in acid form having the formula:

$$CF_3\!-\!CF_2\!-\!CF_2\!-\!O\!-\!CF(CF_3\!-\!)CF_2\!-\!O\!-\!CF(CF_3)\!-\!\overset{\overset{O}{\parallel}}{C}\!-\!OH$$

The resulting mixture was titrated with 20% aqueous NaOH solution until a pH of 10 was obtained. The sodium salt was isolated by evaporation and carefully dried. A total of 508 g. of the dry salt was obtained. The resulting salt was pyrolyzed at 200° to 275°C. in 150 g. batches in a one liter flask at 2 to 4 mm. Hg pressure. The resulting perfluorovinyl ether was distilled. There was obtained 160 g. of the perfluorovinyl ether having the formula

$$CF_3\!-\!CF_2\!-\!CF_2\!-\!O\!-\!CF(CF_3)\!-\!CF_2\!-\!O\!-\!CF\!=\!CF_2$$

and a boiling point of 103°C.

Example 2:   A 100 ml. stainless steel autoclave was flushed with nitrogen and evacuated. A solution of 3.3 grams (0.0076 mol) of the perfluorovinyl ether of Example 1, having the formula:

$$CF_3CF_2\!-\!CF_2\!-\!O\!-\!CF(CF_3)\!-\!CF_2\!-\!O\!-\!CF\!=\!CF_2$$

dissolved in 76 ml. of perfluorodimethylcyclobutane was placed in the autoclave. The contents of the autoclave were heated to 75°C. with vigorous agitation, whereupon tetrafluoroethylene was admitted to the autoclave until a pressure of 205 psi was attained. Then, approximately 10^{-3} mols of dinitrogen difluoride, N_2F_2, diluted with nitrogen was added. Tetrafluoroethylene was periodically added to maintain the initial pressure. After 45 minutes the autoclave was cooled to room temperature and vented.

Solid white copolymer weighing 13.7 g. after drying was obtained. The resin had a melt viscosity of 9.9×10^4 poises at 380°C. The resin could be compression molded into clear, tough, colorless films at 340°C. and 20,000 psi platen pressure. Infrared analysis of the films indicated that 5.0% by weight of the perfluorovinyl ether was incorporated in the copolymer.

Example 3:   Into a Carius tube were placed 8 g. of the perfluorovinyl ether described in Example 1 and 10 mg. of perfluoroazobutane, $C_4F_9N\!=\!NC_4F_9$. The contents of the tube were cooled in Dry Ice, flushed with nitrogen and evacuated. The tube was then sealed and the vinyl ether was irradiated by means of a mercury ultraviolet lamp powered by a 7,500 v. step-up transformer for seven days. The resulting reaction mixture was heated at a reduced pressure of 2 to 4 mm. Hg, until no distillate was evolved. The residual viscous colorless oil weighed 4 grams. Analysis by nuclear magnetic resonance and infrared spectroscopy verified the structure of the oil to be a homopolymer having the following general formula:

$$-\!\!\left(\!CF\!-\!CF_2\!-\!\right)_d \atop O\!-\!CF_2\!-\!(CF_3)CF\!-\!O\!-\!C_3F_7}$$

where d is the degree of polymerization.

Example 4:   Potassium perfluoro-2-(2-n-propoxy)propoxypropionate, 98.5 g., was pyrolyzed at about 2 mm. Hg pressure for 25 hours at 212°C. The product, perfluoro-2-n-propoxypropyl perfluorovinyl ether, weighed 76 g. The infrared band showed the characteristic vinyl ether absorption band at 5.48 microns and the complete absence of hydrogen-containing impurities.

Example 5:   Using the procedure of Example 1, perfluoro-2-ethoxyethyl perfluorovinyl ether is prepared from the reaction product of the dimer of tetrafluoroethylene epoxide, $CF_3CF_2OCF_2COF$, and hexafluoropropylene epoxide.

Example 6:   Into a stainless steel reaction vessel of 80 ml. capacity were charged 15 g. of tetrafluoroethylene, 5 g. of perfluoro-2-n-propoxypropyl perfluorovinyl ether and 40 ml. of a 0.25% aqueous ammonium persulfate solution. The vessel was heated under autogenous pressure at 85°C. for one hour.

A solid polymer, which could be molded at 380°C. and 35,000 psi into tough films, weighing 20 g. was obtained.

165

Fluoroethers

Perfluoro(2-Methylene-4-Methyl-1,3-Dioxolane)

S. Selman and E.N. Squire; U.S. Patent 3,308,107; March 7, 1967; assigned to E.I. du Pont de Nemours and Company describe the preparation and polymerization of perfluoro(2-methylene-4-methyl-1,3-dioxolane) which has the structural formula:

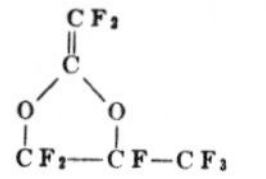

The compound is a clear, colorless liquid boiling at 44.8°C. at atmospheric pressure. It is prepared by contacting gaseous perfluoro(2,4-dimethyl-2-fluoroformyl-1,3-dioxolane) of the formula

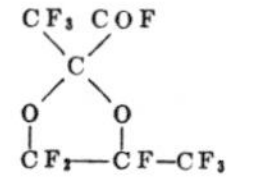

in an inert atmosphere and at a temperature of about 150°C. to 400°C. with an anhydrous alkali or alkaline earth metal oxygen-containing salt, or an oxide of a metal from Groups II-A, II-B, III-A or IV-A of the Periodic Table for sufficient time to obtain the new compound of this process. Generally a contact time of 1 to 100 seconds is sufficient. The contacting step can be carried out by passing the starting material through the salt or oxide contained in an inert organic solvent such as the diethyl ether of diethylene glycol, or the salt or oxide can be in the form of granules to form a solid or permeable base over or through, respectively, which the starting material is passed.

The compound, perfluoro(2-methylene-4-methyl-1,3-dioxolane), is readily homopolymerized in conventional fashion; namely, in solution, emulsion or in bulk, at temperatures ranging from –40° to +200°C., but preferably at room temperature, in a sealed container with an inert atmosphere and in the presence of a free radical initiator.

The homopolymer of perfluoro(2-methylene-4-methyl-1,3-dioxolane) obtained by conventional polymerization is characterized by the repeat unit

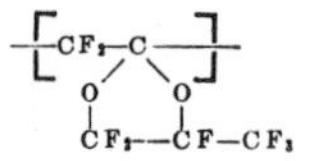

and is normally solid and of high molecular weight as measured by a high inherent viscosity and the ability to be formed into films, which when plasticized exhibit considerable toughness. The homopolymer can also be formed into fibers. The homopolymer of perfluoro(2-methylene-4-methyl-1,3-dioxolane) is soluble in certain fluorinated organic solvents such as the perfluoro(alkyl furanes) and 2,3-dichloro-hexafluorobutene-2. Such solutions are useful as adhesives, paints and for dip-coating. Films of the homopolymer have unusual gas permeability properties and can be employed for the separation of many gases. The following examples illustrate the process.

Example 1: Perfluoro(2,4-dimethyl-2-fluoroformyl-1,3-dioxolane) was passed through a loosely packed bed of 8-14 mesh dried sodium carbonate in an 18" x 3/4" diameter vertical Pyrex tube maintained at a temperature of 295°C. in a current of nitrogen. The perfluoro(2,4-dimethyl-2-fluoroformyl-1,3-dioxolane) was fed at a rate of 0.02 mol per hour and the nitrogen was fed at a rate of 100 ml. per minute. The contact time was thus about 35 seconds. The product gas collected in a cold trap and consisted of perfluoro(2-methylene-4-methyl-1,3-dioxolane), boiling point 44.8°C. The conversion was 100% and the yield was 96 to 99.4% theoretical. Analysis — Calculated for $C_5F_8O_2$: C, 24.6%; F, 62.3%. Found: C, 24.9%; F, 62.3%.

The infrared and nuclear magnetic resonance spectra of the compound are in accord with the structure:

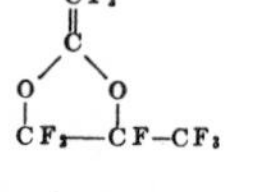

The infrared spectrum is characterized by major absorption bands at 8.0 and 8.1 (doublet), 7.5, 8.6, 8.9, 9.75 microns ascribable to C—C, C—F, and C—O bonds. There is a weak absorption at 5.4 microns probably due to the $C=CF_2$ group. An additional major absorption at 13.7 microns can be ascribed to the perfluoromethyl group. The infrared absorption bands normally ascribed to C—H vibrations are not present in the spectra. The NMR F^{19} spectrum at 56.4 mc. per second indicates the presence of $-CF_3$, $-CF_2O$, $=CF_2$, and $-CF-$ groups by absorptions respectively at 82.8, 87.3 (4 line pattern), 128.4 and 129.9 ppm high field of standard CCl_3F. The area ratios of 3:2:2:1 of the peaks is in accord with the above assignments.

Example 2:   A glass Carius tube was evacuated, cooled to -78°C., and charged with 32.9 g. of perfluoro(2-methylene-4-methyl-1,3-dioxolane) and 0.003 g. of nitrogen fluoride (N_2F_2) as a mixture of the cis- and trans-isomers. The tube was sealed, and upon warming to room temperature an exotherm was observed with a temperature rise to 32°C. The polymerization was allowed to proceed at room temperature for 16 hours. The tube was opened and there were obtained 31.0 grams of polymer. The polymer was found to be soluble in a fluorocarbon solvent sold under the trade name FC-75 and believed to be consisting predominantly of perfluorinated alkylfuranes. The inherent viscosity of 0.5% solution in FC-75 was found to be 1.22 at 29°C.   The polymer could be compression molded at 275°C. and 600 psi into a clear, colorless, transparent film having a tensile strength at 23°C. of 2,830 psi and an elongation of 1.7%. Analysis — Calculated for $(C_5F_8O_2)_n$:   C, 24.6%; F, 62.3%.   Found: C, 24.7%; F, 62.2%.

The compound of this process can be copolymerized with a wide variety of one or more monomers which contain ethylenic unsaturation, preferably terminal unsaturation, and which are polymerizable with a free radical initiator. Copolymerization of the compound with monomers, such as ethylene and tetrafluoroethylene, which homopolymerize to crystalline homopolymers, can reduce the crystallinity of such polymers. The compound also contributes increased thermoplasticity and stiffness to crystalline homopolymers. Generally, at least 0.05% by weight of the copolymer should consist of units derived from perfluoro(2-methylene-4-methyl-1,3-dioxolane) in order for the effect of this compound to be noticeable.

Hexafluoro-2-(Pentafluorophenoxy)-1-(Trifluorovinyloxy) Propane

D.B. Pattison; U.S. Patent 3,467,638; September 16, 1969; assigned to E.I. du Pont de Nemours and Company describes compounds prepared by the reaction of cesium or potassium perfluorophenoxides, with from 2 to 3 mols of hexafluoropropylene epoxide in the presence of an inert aprotic polar organic solvent, which can be described by the formula:

$$\text{F}_5\text{C}_6\text{—O—}\left(\begin{matrix}\text{CF}_3\\|\\\text{CF—CF}_2\text{—O}\end{matrix}\right)_n\begin{matrix}\text{CF}_3\ \text{O}\\|\ \ \ \|\\\text{CF—C—F}\end{matrix}$$

n = 0, 1 or 2

The intermediates can be converted readily to other derivatives such as the acid or alkali metal salts. Pyrolysis of the intermediates at 100° to 200°C. results in monomers having the formula:

$$\text{F}_5\text{C}_6\text{—O—}\left(\begin{matrix}\text{CF}_3\\|\\\text{CF—CF}_2\text{—O}\end{matrix}\right)_n\text{CF=CF}_2$$

where n has the same meaning as above. These monomers can be copolymerized by conventional methods with fluoride-containing vinyl monomers such as vinylidene fluoride, hexafluoropropene, tetrafluoroethylene and perfluoro(alkyl vinyl ethers) to form useful, vulcanizable plastics and elastomers.

Perfluorophenol and its alkali metal salts are known compounds and are described, for example, by Wall et al in the Journal of Research of the National Bureau of Standards A. Physics and Chemistry, vol. 67A, pp. 481-497 (1963). Hexafluoropropylene epoxide is also a known compound and can be made by the action of alkaline hydrogen peroxide on perfluoropropylene at a temperature of about 30°C. The solvents employed to react the cesium perfluorophenoxide and hexafluoropropylene epoxide are aprotic polar organic solvents of which the lower alkyl ethers of ethylene glycol, diethylene glycol, triethylene glycol, and tetraethylene glycol are preferred. The process is illustrated by the following examples.

Example 1:   Preparation of Tetrafluoro-2-(Pentafluorophenoxy) Propionyl Fluoride — Cesium perfluorophenoxide is prepared by the reaction of pentafluorophenol and cesium carbonate by the following procedure. In a 1,000 ml., 4-necked flask fitted with a mechanical stirrer, a reflux condenser, dropping funnel, under nitrogen, there are put 178 g. of cesium carbonate and 330 ml. of bis(2-methoxyethyl)ether. The mixture is heated to 90°C. with stirring and a solution of 74.8 g. of pentafluorophenol in 80 ml. bis(2-methoxyethyl)ether is added dropwise over 30 minutes. The mixture is heated three hours at 90° to 100°C. and cooled to room temperature. Solids are separated by filtration under nitrogen and the solids rinsed with 100 ml. of bis(2-methoxyethyl)ether. The combined filtrate is evaporated to dryness in a spinning flask, finishing half an hour at 150°C. under 0.5 mm. pressure. The yield of cesium perfluorophenoxide is 118 g. (92% of the theoretical yield).

Tetrafluoro-2-(pentafluorophenoxy) propionyl fluoride is prepared by the reaction of cesium perfluorophenoxide with hexafluoropropylene oxide. In a 500 ml., 3-necked flask fitted with a Dry Ice cooled reflux condenser, dropping funnel, mechanical stirrer, and thermometer, under nitrogen, and connected to a weighed cylinder containing hexafluoropropylene oxide, there are put 118 g. of cesium perfluorophenoxide and 150 ml. of 2,5,8,11,14-pentaoxapentadecane, purified by redistillation at 1 mm. pressure from lithium aluminum hydride. The mixture is heated to about 50°C. until all solids dissolve and then cooled to 10°C. With stirring at 10° ± 2°C., hexafluoropropylene oxide gas is added continuously at a rate to maintain moderate reflux. During 2.2 hours, 89.2 g. of hexafluoropropylene oxide (1.45 mols per mol of cesium) are added. At this point addition of hexafluoropropylene is stopped. After sitting for half an hour the mixture is distilled under reduced pressure. The yield is 136 g. (90% of the theoretical yield), mostly boiling at 48° to 55°C. under 3 mm. pressure, but with some higher boiling material. Distillation is stopped when 2,5,8,11,14-pentaoxapentadecane comes over as an upper layer, separating from the product. The theoretical yield is calculated on the basis that the product is 55% tetrafluoro-2-(pentafluorophenoxy) propionyl fluoride and 45% tetrafluoro-2-[hexafluoro-2-(pentafluorophenoxy)propoxy] propionyl fluoride, based on the use of 1.45 mols of hexafluoropropylene oxide.

The mixed fluorides show the following analyses. Calculated for $C_9F_{10}O_2$: C, 32.8%; H, 0.0%; F, 57.6%; mol. wt. 330. Calculated for $C_{12}F_{16}O$: C, 29.0%; H, 0.0%; F, 61.3%; mol. wt. 496. Found: C, 31.9%; H, 0.2%; F, 57.2%; mol. wt. 4,000 by IR spectra.

Example 2:   Preparation of Perfluorophenyl Perfluorovinyl Ether — Perfluorophenyl perfluorovinyl ether is prepared by the reaction of tetrafluoro-2-(pentafluorophenoxy) propionyl fluoride with sodium carbonate. In a 1,000 ml., 4-necked flask fitted with a mechanical stirrer, a dropping funnel with a tip extended near the middle of the flask, a distilling head, and thermometers to measure both vapor and liquid temperatures, there are put 125 g. of sodium carbonate. This is heated while sweeping with nitrogen to remove water. After cooling, 300 ml. of bis(2-ethoxyethyl)ether is added. The mixture is heated, and about 100 ml. of distillate is collected and discarded. Next, 130.6 g. of the mixed perfluorophenyl propionyl fluorides made in Example 1 are added dropwise over 70 minutes, maintaining 175° to 180°C. pot temperature and 165° to 170°C. head temperature. Heating is continued one-half hour longer until the pot temperature is 190°C. and the head temperature is 187°C.

To the crude product, 150 ml. of water are added and the mixture is distilled at atmospheric pressure, returning the upper water layer to the pot and collecting the lower layer as product. Two fractions are obtained, cut 1, BP 90° to 95°C., wt. 42 g., and cut 2, BP 95° to 100°C., wt. 37 g. By vapor phase chromatography (VPC), cut 1 is 93.8% perfluorophenyl perfluorovinyl ether and 1.6% bis(2-ethoxyethyl)ether. Cut 2 is a mixture of 54.6% perfluorophenyl perfluorovinyl ether, 37.2% hexafluoro-2-(pentafluorophenoxy)-1-(trifluorovinyloxy)propane, and 6.9% bis(2-ethoxyethyl)ether. Cut 1 shows the following analysis. Calculated for C_8F_8O: C, 36.3%; H, 0.0%; F, 57.6%. Found: C, 36.1%; H, 0.2%; F, 58.3%.

Example 3:   Preparation of Hexafluoro-2-(Pentafluorophenoxy)-1-(trifluorovinyloxy) Propane — This ether is prepared by fractional distillation of cut 2 of Example 2. Fractions from two runs are combined and distilled five times with an excess of water through a 2-ft. Vigreaux column. The product boiling at 98°C. as a water azeotrope is mostly hexafluoro-2-(pentafluorophenoxy)-1-trifluorovinyloxy propane as determined by VPC. At this stage the bis(2-ethoxyethyl)-ether content is 0.4% by VPC. Redistillation of 20.9 g. of the ether (without water) and analysis by VPC shows the following:

Calc'd for—	$C_{14}F_{20}O_3$	Cut 1 (discarded)	Cut 2	Cut 3
Boiling pt. ° C. at 16 mm. press.		73–76	76	76
Weight g.		1.3	9.2	8.6
Percent C	28.2	30.6	30.5	30.3
Percent H	0	0.1	0.1	0.1
Percent F	63.8	62.2	63.4	63.3
Percent $C_6F_5OC_3F_6OCF_2CF_2H$			19.5	21.8
Percent $C_6F_5OC_3F_6OCF=CF_2$			77.2	77.2
Percent unidentified			3.4	1.0

NOTE.—Cuts 2 and 3 are mainly hexafluoro-2-(pentafluorophenoxy)-1-(trifluorovinyloxy) ether, with hexafluoro-2-(pentafluorophenoxy)-1-(2H-tetrafluoroethoxy) propane as the major impurity.

Example 4:   Preparation of a Copolymer of Hexafluoro-2-(Pentafluorophenoxy)-1-(Trifluorovinyloxy) Propane, Perfluoromethyl Perfluorovinyl Ether, and Tetrafluoroethylene —This example shows that this ether copolymerizes readily with perfluoromethyl perfluorovinyl ether and tetrafluoroethylene. For use in polymerization, the perfluorophenoxypropyl perfluorovinyl ether is washed twice with an equal volume of water to remove any traces of 2,5,8,11-tetraoxadodecane. The material is redistilled through a Vigreaux column, and low boiling material, containing less than 50% of the desired vinyl ether by VPC, is rejected. The remainder, typically containing about 58% hexafluoro-2-(pentafluorophenoxy)-1-(trifluorovinyloxy) propane, and 34% of hexafluoro-2-(pentafluorophenoxy)-1-(2H-tetrafluoroethoxy) propane, and no perfluorophenyl perfluorovinyl ether, is used as described below.

In a 1 liter stainless steel autoclave thoroughly flushed with nitrogen and then with gaseous perfluoromethyl perfluorovinyl ether is charged 346 ml. of phosphate solution (prepared by dissolving 76 g. of $Na_2HPO_4 \cdot 7H_2O$ and 4 g. of $NaH_2PO_4 \cdot H_2O$ in 1,930 ml. H_2O); 1,554 ml. of distilled and deaerated water; 15 g. of ammonium perfluorooctanoate; and 2 g. of ammonium persulfate. The contents of the autoclave are stirred and maintained at 50°C. and 140 psig pressure during a five hour run. Pressure is first raised to 140 psig by metering over a 30-minute period perfluoromethyl perfluorovinyl ether and tetrafluoroethylene at rates of 234 g./hr. and 94 g./hr., respectively, as gaseous monomers. Liquid hexafluoro-2-(pentafluorophenoxy)-1-(trifluorovinyloxy) propane, 7.6 g., is introduced during the same period by a screw injector. Near the start of the run, 50 ml. of sulfite solution (prepared by dissolving 0.07 grams of $CuSO_4 \cdot 5H_2O$ and 9 grams of Na_2SO_3 in 900 ml. distilled deaerated H_2O) is pumped into the autoclave, and during the run 10 ml. of the sulfite solution are added each hour. The pressure is then maintained at 140 psig by adjusting the gas flow rates or by addition of the sulfite solution. Typically, during the run perfluoromethyl perfluorovinyl ether is charged at the rate of 50 g./hr., tetrafluoroethylene at the rate of 45 g./hr., and the ether at the rate of 3.23 g./hr. During the 5 hours at 140 psig, 432 g. of monomers are added. After the run there are 14.0 liters of off-gas at 25°C. and atmospheric pressure. The copolymer latex weighs 2,404 g. and has a pH of 6. The solid copolymer is coagulated by freezing the latex in a Dry Ice-acetone mixture, warmed to room temperature, filtered from the liquid, and washed twice with water in a Waring Blendor. The polymer is put in a 5-liter, 3-necked flask with a mechanical stirrer, water separator, and reflux condenser, and is heated at reflux with about 2 liters of water until no more lower layer comes over. The yield of liquid water-insoluble distillate is 2.2 g.

The polymer was filtered, washed with water, and dried to constant weight at room temperature and on a 2-roll rubber mill at 100°C. The yield of solid elastomer is 353 g. It has an inherent viscosity of 0.78 measured at 30°C. in 2,3-dichloroperfluorobutane at a concentration of 0.1 g. per 100 ml. of solvent. A thin pressed film shows a strong infrared absorption band at 6.55 microns (characteristic of the perfluorophenoxy group), a weaker infrared absorption band at 10.00 microns (characteristic of the perfluoropropoxy ether group),

$$-OCF_2\overset{\overset{\displaystyle CF_3}{|}}{C}F-O-$$

and a strong infrared absorption band at 11.25 microns (characteristic of the perfluoromethoxy group).

The copolymer of this example is compounded on a 2-roll rubber mill according to the following weight proportions:

	A	B
Copolymer	100	100
Medium thermal carbon black	20	20
Magnesium oxide	15	15
p-Phenylene diamine	1	
Tetraethylene pentamine		1

The compositions are molded under pressure for 30 minutes at 160°C. by conventional compression molding techniques into sheets and pellets which are suitable for use in physical testing. The sheets are post cured in an air oven for 24 hours at 204°C. to complete vulcanization, and the pellets are post cured 1 day at 120°C., 1 day at 140°C., 1 day at 160°C., 3 days at 180°C., and 1 day at 204°C.

Properties measured at 25° C.	A	B
Tensile strength, p.s.i.	1,600	1,950
Elongation at break, percent	205	280
Hardness, Durometer A	82	80
Percent volume increase in 2,3-dichloroperfluorobutane (7 days)	326	161
Compression set, 70 hrs. 121° C., percent	55	59

The unvulcanized elastomer is completely soluble in 2,3-dichloroperfluorobutane. The vulcanized elastomers swell but do not dissolve in this solvent.

FLUORINATED NITROSO POLYMERS

TRIFLUORONITROSOMETHANE

Tetrafluoroethylene

In a process described by G.H. Crawford, Jr.; U.S. Patent 3,072,592, January 8, 1963; U.S. Patent 3,213,009, October 19, 1965; and U.S. Patent 3,399,180, August 27, 1968; all assigned to Minnesota Mining and Manufacturing Company, a fluorine-containing nitrosoalkane is polymerized with one or more unsaturated comonomers while the monomers are in solution in fluorinated organic solvent to produce directly a high molecular weight solid copolymer. The powerful tendency toward alternation in most of these copolymer systems renders practicable and reproducible the introduction of a third monomer to produce terpolymers having special and unique properties. The solid polymers of the process have average molecular weights above 50,000 and generally above 100,000, and as high as 150,000 and 200,000 or higher, and generally speaking, inherent viscosities between about 0.3 and 1.5. Many of the elastomeric and thermoplastic copolymers are insoluble in hydrocarbon solvents. Many of the polymers of this process are thermally stable up to about 200°C. The proportion of the monomeric units in the final polymer varies between about 25 and 75 mol percent for each of the components. Usually the copolymer is a 1:1 copolymer.

Typical examples of the fluorine-containing nitrosoalkanes which can be copolymerized in the process include trifluorodinitrosoethane, pentafluoronitrosoethane, tetrafluorodinitrosoethane, tetrafluoronitronitrosoethane, trifluorochlorodinitrosoethane, heptafluoronitrosopropane, hexafluorodinitrosopropane, hexafluoronitronitrosopropane, mononitrosoperfluorobutane and octafluorodinitrosobutane.

The mononitrosoalkanes are typically prepared by reacting a fluorine-containing alkyl halide of less than 13 carbon atoms, such as an alkyl bromide or an alkyl iodide, with nitric oxide in approximately equal molar ratios in the presence of mercury and ultraviolet light for about 24 hours at ambient temperature to produce the corresponding mononitrosoalkane. The use of the bromide is preferred because it is much cheaper than the iodide. For example, trifluoromethylbromide or iodide is reacted with nitric oxide to produce trifluoronitrosomethane; pentafluoroethylbromide or iodide is reacted with nitric oxide to produce pentafluoronitrosoethane; and heptafluoropropylbromide or iodide is reacted with nitric oxide to produce heptafluoronitrosopropane. Also, the chlorofluoronitrosoalkanes can be prepared from chlorotrifluoroethylene telomers of trichlorobromomethane in a similar manner.

A convenient empirical formula for representing the mononitroso compounds is R_f—NO where R_f is a perhalogenated alkyl radical containing fluorine on the carbon atom adjacent to the nitroso group and in which the other halogens are selected from the group consisting of chlorine and fluorine. Preferably, the alkyl radical has not more than 6 carbon atoms.

The nitroso-containing adducts of olefins are prepared by reacting nitric oxide with a fluorine-containing olefin to produce a compound containing a carbon chain of at least two carbon atoms having an oxide of nitrogen group attached to each of two adjacent carbon atoms in the chain in which at least one of the oxide of nitrogen groups is a nitroso group. Preferably, the olefin has at least one halogen atom attached to at least one of the carbon atoms of the double bond and therefore the resultant adduct also contains at least one halogen atom on at least one of the carbon atoms of the former double bond. The reaction between the nitric oxide and the olefin is carried out in either the liquid or vapor phase within a wide range of conditions. The temperature of reaction is usually between about 10° and 100°C.

Preferably, the reaction is carried out at ambient temperature conditions. In some cases, an energy source, such as infrared light or ultraviolet light, can be employed. Under such conditions, the reaction is carried out generally in the vapor phase but sufficient pressure may be utilized to cause either or both of the reactants to be present in the reaction zone in the liquid phase without departing from the scope of this process. The reaction time is usually between about one-half hour and 30 hours. Preferably, the reaction is carried out with an excess of olefin being present at the reaction site. The total charge of reactants is a mol ratio of nitric oxide to olefin of approximately 2:1.

The olefin employed in the reaction with the nitric oxide is preferably an acyclic monoolefin-1 containing not more than 12 carbon atoms per molecule, generally not more than 6 carbon atoms, and preferably contains at least one halogen atom attached to at least one of the carbon atoms of the double bond. Preferably, these halogens are normally gaseous halogens. The resulting adducts are the 1,2-nitronitrosoalkanes and 1,2-dinitrosoalkanes, such as 1,2-nitronitrosotetrafluoroethane and 1,2-dinitrosotrifluorochloroethane. The preferred class of olefins is the perhalogenated olefins in which at least one fluorine atom is attached to each of the carbon atoms of the double bond. Examples of perhalogenated olefins are tetrafluoroethylene, trifluorochloroethylene, perfluoropropene, and perfluorobutene-1, unsymmetrical difluorodichloroethylene, 2-chloropentafluoropropene-1, and 1,1-dichlorotetrafluoropropene-1.

The solid high molecular weight polymers of the process are useful as sealants, adhesives and surface coatings such as for metal and glass surfaces. The polymer can be coated on various surfaces directly from the solution produced in the polymerization system. In the case of using the solution for coating of a surface, the deposited polymer after evaporation of the solvent medium forms a continuous homogeneous nonporous film on the surface with satisfactory adhesion. The solid rubbery copolymers of this process may be preformed at temperatures above 150°C. into various articles, such as gaskets and O-rings, and vulcanized to produce stiffer and harder articles. The following examples illustrate the process.

Example 1: A 50/50 mol ratio charge of CF_3Br (74.5 grams) and NO (15.0 grams) was agitated in the presence of mercury and ultraviolet light (2537 A.) for 24 hours. The pressure was maintained at about one atmosphere by intermittently charging NO as the pressure decreased. The product was distilled in a 35 inch long reflux column having 70 theoretical plates using aluminum turnings as packings and at a reflux temperature of about -84°C. to produce a 60% yield of trifluoronitrosomethane substantially free from CF_3Br (less than one weight percent).

Example 2: 5 grams of CF_3NO (made and purified as above) and 5 grams of C_2F_4 were charged to a 30 ml. Pyrex ampule and agitated therein in the absence of a catalyst for 24 hours at -20°C. An 85% conversion was obtained based on the C_2F_4 charged. The product was a rubbery high molecular weight polymer having an inherent viscosity of about 0.85 corresponding to an estimated average molecular weight above 100,000. The polymer in the glass reactor was dissolved in perfluorocyclobutane and removed from the reactor in solution (no insoluble residue). The copolymer product of the above run had the following physical and chemical properties.

CF_3NO/C_2F_4 gum — properties:

$\langle n \rangle$ inherent viscosity 0.85 gum rubber
Analysis (C, F, N) — indicates 1:1 comonomer ratio
Infrared — shows disappearance of N=O bond
NMR — linear structure $(-N-O-CF_2-CF_2-)_n$
$|$
CF_3
Tg (by n^d) — -51°C.
24 hours swell, ASTM fluid II — (gum)
Gehman T_{10} — -38°C. (vulcanizing stock)
Torsional modulus 40 psi (vulcanized stock)
Thermally stable to 200°C.

Example 3: Equimolar quantities of NO and C_2F_4 were charged into a 12.5 liter 3-necked flask to a pressure totaling one atmosphere or 0.25 mol each. The vessel was irradiated 16 hours with infrared light. At the end of this period a pressure drop of 0.5 atmosphere was noted. The vessel contained a blue gas. This was forced out of the flask through a series of three water scrubbers by introducing water into the flask. The gas was collected, then fractionated by distillation. The distillation was carried out in a screen-saddle packed column of 10 theoretical plates. To the distillation pot was added an equal volume of water for the absorption or hydrolysis of any remaining impurities. 21 grams of a deep blue liquid boiling 21.5 to 22.5°C., accounting for 65 weight percent of the starting materials was obtained. The infrared spectrum showed a strong band at 6.25μ corresponding to the —N=O grouping. The compound analyzed 11.9 weight percent carbon. The Dumas molecular weight of the gas averaged 179. Nuclear magnetic resonance determinations and the above analytical information indicated the gas to comprise essentially the nitro nitroso adduct ($ONCF_2CF_2NO_2$).

Example 4: Into a 12.5 liter 3-necked flask were introduced 2/3 atmosphere of NO and 1/3 atmosphere CF_2—CFCl. After five days the pressure had dropped to 0.78 atmosphere and the flask contained a blue gas. The contents of the flask were pumped out and condensed in Dry Ice. 2.1 grams of blue liquid were obtained after water washing to remove oxides of nitrogen and hydrolyzable impurities. This liquid was distilled under vacuum. Its boiling point, corrected to one atmosphere, was 64°C. Nuclear magnetic resonance analysis and molecular weight determinations appeared to indicate the structure to be:

$$CF_2-CFCl$$
$$||$$
$$NONO$$

Example 5: 3 grams of CF_3NO (purified) and 3 grams of C_2F_4 were dissolved in perfluorotributylamine in a concentration of 30 weight percent. The solution was then placed in a polymerization Pyrex ampule and polymerized therein for 24 hours at -20°C. A 90% conversion was obtained based upon C_2F_4 charged. The polymer was recovered from the solvent by distillation of the solvent. The polymer thus recovered had the following properties and structure:

$$(NO-CF_2CF_2)_n$$
$$|$$
$$CF_3$$

T_g (by n^d)	-51°C.
ASTM #2 fuel swell, 24 hours	3%
Inherent viscosity	1.05
Percent crystallinity	0
Weight loss, 24 hours at 200°C.	0%
Weight loss, 24 hours at 250°C.	20%

Example 6: The following table shows the solubility of the copolymers in various solvents at room temperature. The polymers had an inherent viscosity of about 0.8.

Solvent	Solubility
Copolymer of CF_3NO/C_3F_4:	
Acetone	Totally insoluble.
Methyl ethyl ketone	
Dimethyl sulfoxide	
Tetrahydrofuran	
Carbon disulfide	
Benzene	Very slight swell.
Dichloromethane	Slight swell.
Carbon tetrachloride	Slight swell.
Fluorotrichloromethane (Freon 11)	Slightly soluble.
1,1,2-trichlorotrifluoroethane (Freon 113)	Soluble to about 30%.
Mixed perfluorocyclic ethers $(C_4F_9)_3N$	Miscible, all proportions.
Perfluoropropene	
Perfluorocyclobutane	
Copolymer of $ONCF_2CF_3NO_2/C_2F_4$:	
Hydrocarbon solvents $(C_4F_9)_3N$	Insoluble 2%.
1,1,2-trichlorotrifluoroethane	Miscible, all proportions.
Xylene hexafluoride	

Example 7: The following table shows the results of copolymerizations similarly carried out in accordance with the conditions of Example 5 with different solvents. The monomers were CF_3NO and C_2F_4. A polymer with an inherent viscosity $<n>$ below about 0.3 was considered as being too low for satisfactory elastomeric properties.

Run No.	Solvent	Time, Hrs.	Conversion, percent	$<n>$
1	CCl_3F	18	30	0.21
2	Cl_3FCCF_2Cl	18	42	0.24
3	C_4F_7H	46	35	0.20
4	$n-C_7H_{16}$	17	45	0.10
5	$C_8F_{16}O$	18	56	0.46
8	$(C_4F_9)_3N$	22	78	1.04

Example 8: A copolymer of CF_3NO and C_2F_4 ($n = 0.825$) produced in accordance with the examples was milled with a vulcanization composition and then cured at about 250°F. for 30 minutes and then post cured at 240°F. for 36 hours. The vulcanization recipe is shown below based on 100 parts by weight of copolymer.

	Parts
ZnO	5.0
Sulfur	2.0
HMDA	3.0
Filler	None

The results of the cure is shown on the following page.

F break (psi)	225
Elongation (break)	670%
24 hour swell in 70% iso-octane/30% toluene	3%
Set at break	0%
Gehman T_{10}	−41°C.

Example 9: The following are some of the copolymers that have been produced in accordance with the process.

Nitroso Monomer	Olefin Monomer	Polymerization			Nature of Polymer
		Time, Hrs.	Temp., °C.	Conver., percent	
C_2F_5NO	C_2F_4	24	−20	100	Rubber linear.
C_3F_7NO	C_2F_4	24	−16	100	Do.
$C_8F_{17}NO$	C_2F_4	24	−20	50	Do.
$NO_2CF_2CF_2NO$	C_2F_4	24	−20	85	Solid, linear.
$NO_2CF_2CF_2NO$	CF_2CCl_2	24	−20	80	Thermoplastic, linear.
$NO_2CF_2CF_2NO$	CF_2CFCl	24	−20	85	Do.
CF_3NO	$CH_2=CF_2$	3 wks.	−15 to −20	37	Grease.
CF_3NO	CF_2CFCl	24	−15	100	Rubber, linear.
CF_3NO	CF_2CCl_2	24	−20	100	Short, tough, rubber, linear.
CF_3NO	$C_6H_5CH=CH_2$	24	−14	....	Oil.
CF_3NO	$CH=C(CH_3)COOC_4H_9$	24	−14	80	Do.
CF_3NO	$CF_2=CFH$	6	+20	85	Rubber, linear.
CF_3NO	$CClF=CF-O-CH_3$	6	−20	85	Glassy.
CF_3NO	$CF_2=CFCH=CH_2$	24	−20	60	Short rubber, linear.
CF_3NO	$CF_2CF_2CF=CH_2$	48	+40	5	Thermoplastic, linear.
$NO_2CF_2CF_2NO$	CF_2CFH	24	+20	40	Elastomeric, stiff.

A process described by J.B. Rose; U.S. Patent 3,065,214; November 20, 1962; assigned to Imperial Chemical Industries Limited, England provides normally solid copolymers of equimolar quantities of trifluoronitrosomethane and tetrafluoroethylene. The process for their production comprises reacting together in the absence of oxygen equimolar quantities of trifluoronitrosomethane and tetrafluoroethylene and holding the reaction mixture at a temperature above 20°C. until a solid product is formed. The following example illustrates the process.

Example: Equimolar quantities of trifluoronitrosomethane and tetrafluoroethylene were condensed, in vacuo, into a thick-walled glass tube having a volume such that the monomers were liquefied at the reaction temperature. The quantities of monomers used were accurately measured by means of a constant-volume, variable-pressure gas burette built into the vacuum apparatus. The tube was surrounded by a methyl chloride bath at −24°C. for a period of 15 hours. Reaction between the monomers occurred yielding a viscous oil. The unopened tube was then heated at 40°C. for 10 days and at the end of this period all volatile materials were pumped out of the tube. A solid rubbery polymer was obtained which was extremely inert and did not appear to be soluble in common, nonfluorinated solvents, though it was partially soluble in perfluoromethylcyclohexane. It was unaffected by prolonged heating at 200°C. in air.

Hexafluoropropylene

R.N. Haszeldine and C.J. Willis; U.S. Patent 3,114,741; December 17, 1963; assigned to National Research Development Corporation, England describe a method of manufacturing polymers comprising reacting hexafluoropropylene with a perfluorinated nitroso alkane at a pressure of at least 30 atmospheres and a temperature of at least 50°C. At pressures of from about 30 to 60 atmospheres, temperatures of the order of at least 100°C. are necessary, while at higher pressures temperatures of about 50°C. will suffice. The reaction conditions are preferably such that the pressure is between about 30 and 100 atmospheres while the temperature is of the order of 100°C. The relative proportions of the two reactants used is not critical, but it is preferred to use approximately equimolar quantities as better results are obtained and polymers of longer chain length are yielded. Thus, on reacting together hexafluoropropylene and trifluoronitroso-methane, the following reaction occurs:

$$CF_3-CF{=}CF_2 + CF_3-NO \longrightarrow \cdots N-O(C_3F_6)-N-O-$$
$$\qquad\qquad\qquad\qquad\qquad\qquad\quad |\qquad\qquad\quad |$$
$$\qquad\qquad\qquad\qquad\qquad\qquad\quad CF_3\qquad\qquad CF_3$$

The copolymer $\cdots-N-O-CF_2-CF-N-O-\cdots$ predominates.
(with CF_3, CF_3, CF_3 substituents)

The products range from a colorless transparent oil through a wax to a translucent elastomeric gel. They are insoluble in water and common organic solvents and are stable in vacuo up to at least 200°C. They are of use as coolants, heat transfer media, plasticizers for polymers such as Viton A, surface coatings with good temperature and/or chemical resistance, rubber coatings for sealing chemical plant and for coating chemical equipment subject to corrosive action, particularly by strong acid, for fabrication (as rubber) into O rings, flexible hosing for transportation of corrosive chemicals, lubricants for use with equipment where chemical corrosion is particularly prevalent, etc. The following example illustrates the process.

Example: Trifluoronitrosomethane (and its homologues) may be obtained by application of the method described by D.A. Barr and R.N. Haszeldine in a paper entitled "Perfluoroalkyl Derivatives of Nitrogen" (J. Chem. Soc., June 1955, pp. 1881-1889).

When trifluoronitrosomethane and hexafluoropropylene were heated in equimolar proportions at 30°C. in a stainless steel autoclave at an initial pressure of 10 atmospheres for 8 days, no reaction was detected and the reactants were recovered substantially unchanged. Equimolar proportions of hexafluoropropylene and trifluoronitrosomethane were then reacted in a steel autoclave at an initial pressure of 30 atmospheres and a temperature of 100°C. for 14 days. At the end of this time a 34% yield of a colorless viscous oil was obtained which analysis showed to be a 1:1 copolymer. Found: C, 19.2; H, 0.1; N, 5.6. (C_4F_9NO) requires: C, 19.3; H, nil; N, 5.6.

Trifluoroethylene

In a process described by R.N. Haszeldine and C.J. Willis; U.S. Patent 3,197,451; July 27, 1965; assigned to National Research Development Corporation, England, there is provided a polymer consisting essentially of polymeric chains of structure:

$$\left[\begin{array}{c} -N-O-C_2HF_3- \\ | \\ CF_2Z \end{array} \right]_n$$

in which Z represents a group selected from the groups consisting of a fluorine atom, a chlorine atom, and a halogeno-alkyl group selected from the group consisting of a fluoroalkyl and a fluoro- and chloroalkyl group, and n represents an integer from 3 to 5,000, said polymer having a molecular weight not greater than 2,000,000. The molecular weight limitation applies to the molecular weight when calculated by the viscosity method using a fluorocarbon solvent.

A particular feature of the polymers is that the polymeric chains can be cross-linked by treating the polymer with a source of free radicals. The free radicals extract hydrogen atoms from the polymeric chains, thus leaving polymer free radicals which intercombine to form the cross-linked polymer. The degree of cross-linking can be varied by variation of the temperature at which the polymer is treated, the limit of treatment and the concentration of the free radical source. For instance treatment with 2% by weight of dibenzoyl chloride at about 100°C. for about 30 minutes has been found to considerably harden and toughen the polymers and render them insoluble in all common organic solvents.

The process for the manufacture of a polymer comprises reacting trifluoroethylene with a fluorinated nitrosoalkane of structure $Z-CF_2-NO$ in which Z represents a group selected from the groups consisting of a fluorine atom, a chlorine atom, a fluorine substituted alkyl group and a fluorine and chlorine substituted alkyl group, at a temperature in the range of about −40° to 150°C. and a pressure in the range of about 0.1 to 200 atmospheres.

The ratio of reactants used is not critical but equimolar proportions are desirable. This leads to the formation of a copolymer of higher molecular weight than is the case if a large excess of one of the reactants is employed. Nevertheless use of an excess of one or other of the reagents still results in a strict 1:1 polymer. The following examples illustrate the process.

Example 1: Equimolar proportions of the trifluoronitrosomethane (12 parts) and trifluoroethylene (10 parts) are condensed in vacuo into a Pyrex tube which is then sealed and allowed to warm up to about 20°C. so that the reaction proceeds under a pressure of about 8 atmospheres. The blue color of the nitroso compound slowly fades and, after about 14 days, on the removal of the volatile products, an approximately 50% yield of a clear, colorless elastomer is obtained. Analysis of the elastomer obtained has shown it to be a 1:1 copolymer of trifluoronitrosomethane and trifluoroethylene. Found: C, 19.8; H, 0.8; N, 7.0%; C_3HNOF_6 requires C, 19.8; H, 0.6; N, 7.7%. When the polymer was dissolved in diethyl ketone and applied as a surface coating to wood, the wood then showed marked resistance to chemical attack by sulfuric acid, hydrochloric acid and potassium permanganate solution.

When the polymer was heated at 100°C. for 30 minutes with 2% by weight of dibenzoyl peroxide, the cross-link polymer was much harder and tougher, and was insoluble in ketone solvents and in all common organic solvents. Treatment of the polymer with concentrated nitric acid failed to cause chemical attack.

Example 2: Trifluoronitrosomethane (2.39 parts) and trifluoroethylene (1.98 parts) are kept at 70°C. in a sealed Pyrex tube at 5 to 10 atmospheres for two days. The copolymer is formed in 8% yield and analysis has shown it to be a 1:1 copolymer. Found: C, 19.8; H, 0.8; N, 7.9%; C_3HNOF_6 requires C, 19.9; H, 0.6; N, 7.7%. The oxazetidine

$$CF_3N-O-C_2HF_3 \rceil$$

(found: C, 19.7; H, 0.8; N, 0.8%; required: C, 19.9; H, 0.6; N, 7.7%) is formed in 80% yield, and the weight balance is made up by recoverable starting materials.

Terpolymers Containing Aliphatic Nitroso Monocarboxylic Acids

In a process described by G.H. Crawford, Jr. and D.E. Rice; U.S. Patent 3,321,454; May 23, 1967; assigned to Minnesota Mining and Manufacturing Company, an ethylenically unsaturated aliphatic monoolefin containing fluorine is copolymerized with a halogenated nitroso monocarboxylic acid at a substantially constant temperature to produce directly a high molecular weight, solid, essentially linear polymer. To control the number of carboxyl groups in the final polymer, a fluorine-containing nitrosoalkane is substituted for part of the nitroso carboxylic acid. The solid linear polymer has an average molecular weight above 100,000 and as high as 1,000,000 to 2,000,000 or higher.

The first component of the polymerization system is a nitroso aliphatic monocarboxylic acid. Preferably these carboxylic acids are saturated and perhalogenated, the halogens being chlorine or fluorine. Useful carboxylic acids include the omega nitroso saturated perfluorocarboxylic acids such as omega nitroso perfluorobutyric acid, omega nitroso perfluorovaleric acid and omega nitroso perfluorononanoic acid. In general, the nitroso carboxylic acids of this process contain from 2 to 10 carbon atoms per molecule. The nitroso carboxylic acids are prepared by reacting an aliphatic anhydride of a dicarboxylic acid containing halogen substitution with nitrogen sesquioxide to produce the corresponding halogen containing acyl dinitrite. The acyl dinitrite is then monodecarboxylated and hydrolyzed to produce the corresponding nitroso aliphatic carboxylic acid. The preparation of such carboxylic acids and the conditions of their preparation are described in U.S. Patent 3,192,247.

The second comonomer with which the nitroso carboxylic acid is copolymerized is a polymerizable aliphatic monoolefin having only ethylenic unsaturation and not more than 8 carbon atoms per molecule. Examples of the preferred fluorine-containing monoolefins include trifluoroethylene, difluoromonochloroethylene, tetrafluoroethylene, trifluorochloroethylene and unsymmetrical difluorodichloroethylene.

The fluorine-containing mononitrosoalkane monomeric material which is used as the third component is perhalogenated in which the halogens are normally gaseous halogens and preferably the nitrosoalkane contains less than 13 carbon atoms per molecule. Typical examples of the fluorine-containing mononitrosoalkanes include trifluoronitrosomethane, pentafluoronitrosomethane, heptafluoronitrosopropane, nitrosoperfluorobutane, nitrosoperfluorooctane, trifluorodichloronitrosomethane, 1-nitroso-1,3,5,7,7,7-hexachlorononafluoroheptane, and 1-nitroso-1,3,5,7,9,9,9-heptachlorododecafluorononane.

The mononitrosoalkanes are prepared by reacting a fluorine-containing alkyl halide of less than 13 carbon atoms, such as an alkyl bromide or an alkyl iodide, with nitric oxide in approximately equal molar ratios in the presence of mercury and ultraviolet light for about 24 hours to produce the corresponding nitrosoalkane.

The proportion of the reactants, nitrosoalkane to olefin to nitroso carboxylic acid, in the reaction mixture is usually in a mol ratio of about 1:0.5:0.01 to about 1:1.5:0.2. The preferred mol ratio of the nitroso-containing monomers to the monoolefin, is about 1:1, and the mol ratio of these reactants in the final product is approximately the same. In any event, sufficient carboxylic acid monomer should be used to assure at least three free carboxyl groups per polymer molecule. The terpolymer may be represented by the following linear structure which has been substantiated by chemical analysis and nuclear magnetic resonance determination:

$$\left[\left(\underset{\underset{R'}{\mid}}{N} - O - \underset{\underset{R''}{\mid}}{C} - \underset{\underset{X}{\mid}}{\overset{\overset{X}{\mid}}{C}} \right)_m \left(\underset{\underset{(CX_2)_p}{\mid}}{N} - O - \underset{\underset{COOH}{\mid}}{C} - \underset{\underset{X}{\mid}}{\overset{\overset{X}{\mid}}{C}} \underset{R''}{} \right) \right]_n$$

in which R' is the alkyl group of the nitrosoalkane and previously defined, and R" is an alkyl group derived from the monoolefin or halogen or hydrogen; X is a halogen or hydrogen, and preferably X is fluorine or chlorine; m is 0 to 500; p is 1 to 9; and n is generally 250 to 1,000. The solid high molecular weight terpolymers are useful as sealants, adhesives and surface coatings such as for metal and glass surfaces.

The solid rubbery terpolymers of this process may be reformed at temperatures above 50°C. into various articles, such as gaskets and O rings; and vulcanized to produce stiffer and harder articles. The elastomer may be vulcanized with conventional vulcanization or cross-linking agents under conventional vulcanization conditions. Examples of suitable cross-linking agents include the basic metal oxides and hydroxides, such as the metals magnesium, cadmium, manganese, calcium, zinc and strontium, the polyhydric alcohols such as ethylene glycol and the diepoxides such as the diglycidyl ether or a bisphenol. The following examples illustrate the process.

Example 1: Approximately 10 grams of N_2O_3 were condensed into a flask at Dry Ice temperature and 16 grams of perfluorosuccinic acid anhydride, previously cooled to 0°C., were added. The flask was connected to a Dry Ice reflux condenser and placed in an ice-salt bath at −5°C. After a few minutes, the reaction mixture had become lighter in color and a precipitate had started to form. After two hours, the reaction mixture consisted of a yellow solid perfluorosuccinyl dinitrite, which was dried for eight hours under vacuum at room temperature. The yield was 17.5 grams. The material quickly formed perfluorosuccinic acid upon exposure to moisture of the atmosphere, or reacted rapidly with water, and melted at 44° to 48°C. Analysis — Calculated for $C_4F_4N_2O_6$: C, 19.4; F, 30.7; N, 11.3. Found: C, 19.6; F, 31.7; N, 10.5.

The NMR spectrum showed a single peak at $\phi^{*14} = 126.0$. About 10 grams of perfluorosuccinyl dinitrite $(CF_2COONO)_2$ was placed in a 250 cc flask and connected through carbon dioxide-acetone and liquid nitrogen traps to a vacuum pump, used to maintain a pressure of ~1 mm. throughout the system. The reaction flask was then subject to ultraviolet irradiation (lamp ~6 in. from flask). After reaction, water (50 cc) was added to the carbon dioxide-acetone trap to convert the $ONOOCCF_2CF_2NO$ to $HOOCCF_2CF_2NO$. The aqueous solution was then extracted with ether and the ether evaporated to obtain omega nitroso perfluoro propionic acid $(HOOCCF_2CF_2NO)$.

Example 2: A 30 cc glass ampule was charged with 1.5 grams $HOOCCF_2CF_2NO$, cooled to liquid nitrogen temperature and 5.1 grams CF_3NO and 5.7 grams C_2F_4 condensed in. The ampule was sealed, warmed to −78°C., shaken briefly to obtain a homogeneous solution and then allowed to stand at −30°C. for three days. The product was a tacky elastomeric material similar in appearance to samples of low MW CF_3NO/C_2F_4 copolymer. The polymer was soluble in $CF_2ClCFCl_2$ and perfluorocarbons and insoluble in water, methanol and acetone. The IR and NMR spectra were consistent for a terpolymer of $CF_3NO/C_2F_4/HOOCCF_2CF_2NO$. The T_g was found to be −33°C. The Neutralization Equivalent and C, F, N, analyses indicated the following structure:

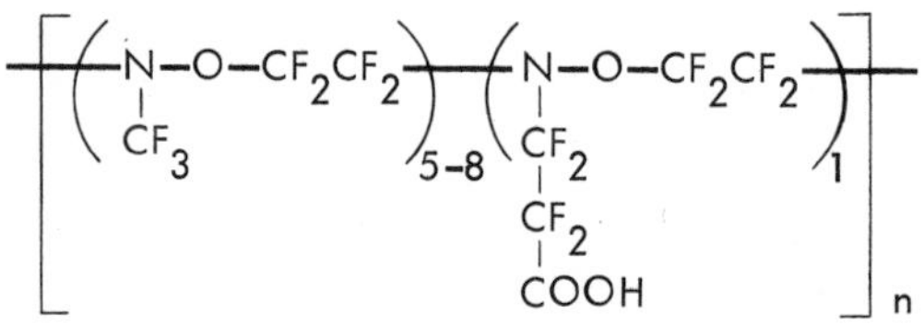

$$\left[\left(\underset{\underset{CF_3}{|}}{N}-O-CF_2CF_2\right)_{5-8}\left(\underset{\underset{\underset{\underset{COOH}{|}}{CF_2}}{\overset{|}{CF_2}}}{N}-O-CF_2CF_2\right)_1\right]_n$$

When a solution of 0.5 gram of the above terpolymer in 10 cc $CF_2ClCFCl_2$ was treated with 0.5 cc of a saturated $Ba(OH)_2$ solution, gel formation occurred instantly. The gel, after drying, exhibited no cold flow and could not be dissolved in $CF_2ClCFCl_2$ indicating cross-linking had occurred.

Example 3: Omega nitrosoperfluorocarboxylic acid was charged into a 60 cc Pyrex ampule at room temperature and the ampule was then cooled to liquid nitrogen temperature and the gaseous monomers (CF_3NO and C_2F_4 or CF_2CFCl) condensed in under vacuum. The ampule was then sealed, warmed to −65°C., and shaken until a homogeneous solution resulted. The ampule was then kept at a desired temperature until it appeared that a high conversion had been reached as evidenced by the viscosity and color of the reaction mixture. The polymer was freed of unreacted monomer by drying under vacuum at 70°C. The conditions of reaction are shown in the table below. The polymer produced was a high molecular weight elastomer which could be cross-linked readily with basic metal oxides or hydroxides, polyhydric alcohols and epoxides.

Run No.	Monomers	Mol Ratio Charged	Rx Temp. (°C.)	Rx Time, Days	Conver., Percent
1	$HOOCCF_2CF_2NO/CF_3NO/C_2F_4$	1.5/48.5/50	−65	30	91
2	$HOOCCF_2CF_2CF_2NO/CF_3NO/C_2F_4$	0.5/49.5/50	−65	25	97
3	$HOOCCF_2CF_2CF_2NO/CF_3NO/C_2F_4$	1/49/50	−65	25	97
4	$HOOCCF_2CF_2CF_2NO/CF_3NO/C_2F_4$	0.25/49.75/50	−65	25	97
5	$HOOCCF_2CF_2CF_2NO/CF_3NO/CF_2CFCl$	1.5/48.5/50	−25	3	85

Polyamine Cross-Linking Agents

A.M. Borders; U.S. Patent 3,072,625; January 8, 1963; assigned to Minnesota Mining and Manufacturing Company describes the use of polyfunctional amino compounds as linking agents for the copolymers discussed previously (U.S. Patent 3,072,592, page 170).

The alkylene polyamines which may be used as cross-linking agents include the aliphatic alkylene polyamines and the dialkylene arylene polyamines, such as ethylene diamine, diethylene triamine, triethylene tetramine, tetraethylene pentamine and hexamethylene diamine and meta-xylylene diamine. Examples of suitable arylene diamine cross-linking agents include benzidine and p,p'-diaminodiphenylmethane. The arylene diamines require somewhat more severe conditions of temperature and time for cross-linking. Compounds which yield alkylene polyamines by dissociation or decomposition or by reaction with other materials under the conditions of molding or vulcanization are also useful as cross-linking agents. The carbamic radical containing salts of acyclic primary and secondary polyamines are among the compounds of this class because upon heating under the conditions of vulcanization the alkylene polyamines are produced. Examples are the carbamic radical salts of hexamethylene diamine, triethylenetetramine, triethylenetriamine, etc. The following example illustrates the process.

Example: The copolymer of Example 2 in the previously mentioned U.S. Patent 3,072,592 was compounded and vulcanized in several runs with different vulcanization recipes as shown in Table 1 below. The physical results on the best cures are shown in detail in Table 2 below. Run No. J-1 is the same as run No. J except that the cure was oven treated at 300°F. for an additional 43 hours.

TABLE 1

CURING RECIPES

Run	A	B	C	D	E	F	G	H	I	J
Recipe	Quantities expressed in parts by weight									
CF₃NO/C₃F₄ Copolymer	100	100	100	100	100	100	100	100	100	100
Hexamethylene Diamine	4	3	3			2				
Hexamethylene Diamine Carbamate				4	6			3		
Triethylene Tetramine							1.25		1.5	1.5
CaO								6		
MgO				20	20					
Carbon Black		15		15	15	20	20		25	25

EXPERIMENTED CURING CYCLES

Temperature °F.	Time, hours									
Mold in press:										
at 300°F.	½	½	½	½	½	½	½		½	½
at 250°F.										10
Sample in Oven:										
at 140°F.								1		
at 180°F.								1	72	
at 212°F.				1						
at 250°F.							16			7
at 300°F.				1						
at 400°F.				1						
Obtained Cure	Fair	Fair	Fair	Fair	Fair	Fair	Good	Fair	Fair	Good

TABLE 2

Run	J	J-1
Ingredients:		
C₃F₄/ONCF₃, parts by wt.	100	100
Triethylene tetramine, parts by wt.	1.5	1.5
Carbon Black, parts by wt.	25	25
Curing Cycle:		
Mold in press—		
at 300°F., hr	½	½
at 250°F., hr	10	10
Post cure in Oven—		
at 250°F., hr	7	7
at 300°F., hr		43
Properties:		
Tensile Strength at break, p.s.i.	49.3	184
Tensile at 100% Elong., p.s.i.	23.7	99
Elongation, percent	425	150
Gehman T₁₀, °C.	−37	
Brittle Point, °C.	−37	
Tortional T₁₀ modulus, p.s.i.	7.74	
Volume Swell, at room temp.:		
in 70:30 (Iso. Oct. Toluene), percent	0	0
in Acetone, percent	0	0
in Heptane, percent	0	0
in Toluene, percent	0	0
in Carbon Tetrachloride, percent	0	0

In the above recipes, carbon black is a filler to impart body and abrasive resistance to the ultimate vulcanized product, and whether or not carbon black constitutes a part of the vulcanization recipe will depend upon the ultimate use of the vulcanized product. In addition to the physical properties of the vulcanized product, another indication of cure is insolubility of the treated polymer in a fluorocarbon solvent such as the cyclic ether, $C_8F_{16}O$. All of the vulcanized polymers were insoluble in a fluorocarbon solvent; whereas, those polymers that were treated and did not vulcanize continued to be soluble in fluorocarbon solvents.

Reaction of Nitric Oxide with Olefins

G.H. Crawford, Jr.; U.S. Patent 3,436,384; April 1, 1969; assigned to Minnesota Mining and Manufacturing Company describes the preparation of adducts of nitric oxide and a fluorine-containing olefin. The fluorine, oxygen, and nitrogen-containing compounds are prepared by reacting nitric oxide with a fluorine-containing olefin to produce an adduct containing a carbon chain of at least two carbon atoms having an oxide of nitrogen group attached to each of two adjacent carbon atoms in the chain in which at least one of the oxide of nitrogen groups is a nitroso group. A typical structural formula for representing the adducts of this process is:

$$ON-\underset{\underset{R}{|}}{\overset{\overset{R}{|}}{C}}-\underset{\underset{R}{|}}{\overset{\overset{R}{|}}{C}}-Y$$

in which R is an alkyl radical or hydrogen or halogen and at least one R is fluorine. Preferably, the R is a halogen, such as fluorine, and/or a perhalogenated or partially halogenated alkyl radical of not more than 5 carbon atoms, and preferably the halogens are fluorine and/or chlorine; and in which Y is an oxide of nitrogen group, such as NO. The following examples illustrate the process.

Example 1: Equimolar quantities of NO and C_2F_4 were charged into a 12.5 liter 3-necked flask to a pressure totaling one atmosphere or 0.25 mol each. The vessel was irradiated 16 hours with infrared light. At the end of this period a pressure drop of 0.5 atmosphere was noted. The vessel was repressured with NO and the above procedure repeated. The vessel contained a blue gas. This was forced out of the flask through a series of three water scrubbers by introducing water into the flask. The gas was collected, then fractionated by distillation. The distillation was carried out in a screen-saddle packed column of 10 theoretical plates. To the distillation pot was added an equal volume of water for the absorption or hydrolysis of any remaining impurities. 21 grams of a deep blue liquid boiling 21.5° to 22.5°C., accounting for 60 weight percent of the starting materials was obtained. The infrared spectrum showed a strong band at 6.25μ corresponding to the $-N=O$ grouping. The compound analyzed 11.9 weight percent carbon. The Dumas molecular weight of the gas averaged 179. Nuclear magnetic resonance determinations and the above analytical information indicated the gas to comprise essentially the nitro nitroso adduct ($ONCF_2CF_2NO_2$).

Example 2: Into a 30 ml. Pyrex ampule was condensed 2.4 grams of the NO adduct to tetrafluoroethylene which was prepared and purified in accordance with Example 1. 1.5 grams C_2F_4 was likewise admitted to the ampule. The ampule was held at −25°C. for eight hours. After the first three hours the contents of the tube had solidified into a gum. The ampule was opened and the polymer was dissolved in Freon 113 and removed from the ampule as a solution. Upon removal of solvent, 2.8 grams of a transparent elastomeric gum were obtained. The infrared spectrum still showed the characteristic −NO band. However, its intensity was reduced by an amount indicating consumption in the polymerization reaction. This polymer was found to contain F, 54.3 weight percent, and had a molecular weight of about 100,000. From the solubility of the gum, the analytical data and molecular weight, the elastomeric gum was thus identified as a copolymer having a linear structural formula of:

$$\underset{CF_2CF_2NO_2}{(-NO-CF_2CF_2-)_n}$$

SUMMARY OF RELATED PROCESSES

Copolymers of fluorinated nitrosobenzenes are described by R.N. Haszeldine, J.M. Birchall and J.H. Umfreville; U.S. Patent 3,310,543; March 21, 1967 and by J.A. Castellano and J. Green; U.S. Patent 3,428,672; February 18, 1962; assigned to Thiokol Chemical Corporation.

Nitrosyl fluoroacylates are described by J.D. Park and R.W. Rosser; U.S. Patent 3,160,660; December 8, 1964; assigned to Minnesota Mining and Manufacturing Company.

C.W. Taylor; U.S. Patent 3,342,874; September 19, 1967; assigned to Minnesota Mining and Manufacturing Company describes a method of preparation for nitrosyl trifluoroacetate which can be pyrolized to trifluoronitrosomethane.

R.N. Haszeldine, R.E. Banks, and H. Sutcliffe; U.S. Patent 3,231,555; January 25, 1966; assigned to National Research Development Corporation, England describe a process for preparing solid copolymers of trifluoronitrosomethane and substituted ethylenes.

The copolymerization of nitroso-t-butane and tetrafluoroethylene is described by R.N. Haszeldine, R.E. Banks and M.K. McCreath; U.S. Patent 3,223,689; December 14, 1965.

OTHER FLUORINATED MONOMERS AND POLYMERS

FLUORINATED CYCLICS

Halotrifluorocyclopropenes

A.E. Barkdoll and P.B. Sargeant; U.S. Patent 3,413,275; November 26, 1968; assigned to E.I. du Pont de Nemours and Company describe 3-halotrifluorocyclopropenes and their copolymers with polymerizable ethylenically unsaturated monomers. The monomers of this process are represented by the structural formula

where X is fluorine, chlorine or bromine. Preferably X is fluorine. The monomers can be prepared by two methods, one being the dehydrohalogenation of a cyclopropane of the formula

and the other, which is the preferred method, being the dehalogenation of a compound of the formula

In these formulas X is as defined previously and Y is chlorine or bromine. The dehydrohalogenation and dehalogenation processes are, in general, well-known reactions. The monomers are colorless gases or low-boiling liquids, e.g., tetrafluorocyclopropene boils at about −13°C., melts at about −60°C. and is stable for several hours at 100°C., and is stable indefinitely at room temperature.

Example 1: Tetrafluorocyclopropene and 3-Chlorotrifluorocyclopropene — Potassium hydroxide (20 g.) was heated by an oil bath to 160° to 165°C. in a 50 ml. 3-neck round-bottom flask equipped with an additional funnel, magnetic stirrer, and water condenser leading to a Dry Ice-acetone cooled trap. 1-chloro-1,2,3,3-tetrafluorocyclopropane (1.0 g., 0.067 mol) was added rapidly to the stirring molten potassium hydroxide. The reaction mixture turned dark and gas evolution occurred. Vapor phase chromatography of the gaseous product showed the presence of several products, two of which were identified as being tetrafluorocyclopropene and 3-chlorotrifluorocyclopropene.

Example 2: Perfluorocyclopropene —

Aqueous potassium hydroxide (200 g. in 400 ml. of H_2O) was heated to 95°C. in a 1-liter 3-neck round-bottom flask equipped with a mechanical stirrer, gas inlet tube reaching below the surface of the liquid, and a condenser leading to a 50 ml. trap cooled in Dry Ice-acetone. Pentafluorocyclopropene (38 g., 0.29 mol) was slowly bubbled through the stirring solution (6 hours). The final portion was passed through by sweeping with N_2 for 15 minutes. There was obtained 15.5 ml. of product (25 g.) which was found to consist of 24% perfluorocyclopropene and 86% pentafluoro-cyclopropane by vapor phase chromatography.

Example 3: Perfluorocyclopropene/Methyl Vinyl Ether Copolymer — Perfluorocyclopropene (16 mmols), methyl vinyl ether (16.7 mmols), and benzoyl peroxide (0.01 g., 4.5×10^{-5} mols) were sealed in a glass tube (200 mm. x 8 mm. i.d. x 10 mm. o.d.) and heated at 85°C. for 8 hours to provide a white solid copolymer (1.94 g., 83% yield) containing 41.6% fluorine. This corresponds to a ratio of 1.2:1 methyl vinyl ether:perfluorocyclopropene. The copolymer was soluble in diethyl ether, acetone, benzene, and tetrahydrofuran. Clear, self-supporting films were pressed at 200°C. and also cast from benzene solution. The copolymer had an inherent viscosity of 0.32 (0.1% benzene solution at 25°C.).

Examples 4 through 17: Other Copolymers of Perfluorocyclopropene — Equimolar quantities of perfluorocyclopropene and polymerizable ethylenically unsaturated monomers (see table below) were added to a glass tube (18 mm. x 4 mm. i.d.) with benzoyl peroxide (5 mg., 2×10^{-5} mols). The tube was degassed, sealed and heated at 80° to 85°C. The copolymeric product was characterized by its infrared spectrum, by differential thermal analysis, and in some cases by fluorine elemental analysis.

The infrared spectrum of each copolymer was different from that of the corresponding homopolymer of the ethylenically unsaturated monomer. Each copolymer had characteristic absorption at 1730–1700 (strong), 1600 (weak to medium), and 1200 (strong) cm.$^{-1}$ in addition to other new bands. None of these characteristic absorptions are shown by the homopolymers. Differential thermal analysis curves of the copolymers were different from those of the said corresponding homopolymers. Those copolymers containing considerable perfluorocyclopropene (comonomer:perfluorocyclopropene ratio <5:1) exhibited a sizeable endotherm around 300°C.

The copolymer of Example 8 was found to have a different infrared spectrum than polyvinylfluoride. It was an elastic copolymer whereas polyvinyl fluoride is not.

Copolymers of Perfluorocyclopropene

Example No.	Vinyl Monomer	mmols	Reaction (1) Time (hr.)	Weight Copolymer (mg.)	Percent F	Monomer/ Perfluoro-Cyclopropene (2)
4	$CH_2{=}CH_2$	(3)3.6	21	151	37.18	3.31
5	$CH_2{=}CH_2$	(3)2.1	8	48	38.30	3.09
6	$CH_2{=}C(CH_3)_2$	(3)3.6	20	78	33.68	2.02
7	$CF_2{=}CF_2$	(3)2.9	20.5	23	(4)	(4)
8	$CH_2{=}CHF$	(3)2.1	8	25	(4)	(4)
9	$CH_2{=}CF_2$	(3)1.4	8	(5)	(4)	(4)
10	$CH_2{=}CHCl$	(3)1.4	8	48	6.24	17.7
11	$CH_2{=}CHC_6H_5$	(6)1.3	8	125	2.87	24.2
12	$CH_2{=}CHCN$	(3)2.3	(7)	114	1.20	117
13	$CH_2{=}CHCN$	2.3	8	110	1.23	115
14	$CH_2{=}CHOCH_3$	2.1	8	200	39.95	1.35
15	$CH_2{=}CHOCOCH_3$	(8)1.5	8	191	38.30	3.09
16	trans-$CH_3CH{=}CHCH_3$	3.4	8	(5)	25.86	3.24
17	cis-$CH_3CH{=}CHCH_3$	3.4	8	(5)	33.78	2.02

(1) Heated at 80° to 85°C.
(2) Mol ratio calculated from elemental F analysis.
(3) Equimolar amounts of perfluorocyclopropene.
(4) No elemental F analysis.
(5) Not weighed.
(6) 1.3 mmols styrene, 2.1 mmols perfluorocyclopropene.
(7) Allowed to stand at room temperature for 4 days.
(8) 1.5 mmols vinyl acetate, 2.1 mmols perfluorocyclopropene.

1,1-Difluoro-2-Methylene-3-Difluoromethylenecyclobutane

A process described by W.H. Knoth, Jr.; U.S. Patent 2,964,507; December 13, 1960; assigned to E.I. du Pont de Nemours and Company involves 1,1-difluoro-2-methylene-3-difluoromethlenecyclobutane, and its polymers having a substantial proportion of recurring units of

$$-CH_2-C{=}\!\!{=}C-CF_2-$$
$$CF_2-CH_2$$

The 1,1-difluoro-2-methylene-3-difluoromethylene-cyclobutane can be prepared in various ways. One method is by heating a solution of 1,1-difluoroallene in an inert solvent, under the autogenous pressure developed by the reaction system. Heating of the reaction mixture should take place at a temperature of 50° to 150°C. Temperatures between 75° and 125°C. are preferred for ease and speed of the reaction.

The exact time of heating depends on the particular reaction temperature employed. At 95°C., reaction times of 16 to 20 hours are sufficient. At lower temperatures, longer times are required, while at higher temperatures, shorter times can be used. Preferably, the reaction vessel is swept out with an inert gas, e.g., nitrogen, and then is evacuated prior to carrying out the heating. The 1,1-difluoroallene used as starting material in the process can be prepared by the pyrolysis of 1-methylene-2,2,3,3-tetrafluorocyclobutane, as described in U.S. Patent 2,733,278.

Monomeric 1,1-difluoro-2-methylene-3-difluoromethylenecyclobutane polymerizes spontaneously at ordinary temperatures. If desired, the monomer can be stabilized by incorporating in it a polymerization inhibitor immediately after its preparation. Hydroquinone and quinone inhibit the polymerization of the monomer for a short time. However, if permenant stabilization is desired, phenothiazine is used. An amount of inhibitor ranging from 0.5 to 20% of the weight of the monomer is satisfactory, although these percentages are by no means critical.

Copolymers of 1,1-difluoro-2-methylene-3-difluoromethylenecyclobutane with other copolymerizable ethylenic compounds containing substantial proportions of the former monomer can be prepared by adding a free radical-liberating initiator, e.g., benzoyl peroxide, to a mixture of 1,1-difluoro-2-methylene-3-difluoromethylenecyclobutane and a substantial proportion (5 to 95% by weight) of another polymerizable ethylenic compound, e.g., tetrafluoroethylene or acrylonitrile.

Example 1: A solution of 11 g. of 1,1-difluoroallene in 95 g. of decane is placed in a vessel capable of withstanding pressure, and the free space in the vessel is then flushed out with nitrogen, the vessel closed and evacuated. The closed reaction vessel is heated to 95°C. until no further drop in pressure is observed, 16 hours being required. During the heating period, the pressure in the reaction vessel decreases from 22 lb./sq. in., gauge, to 3 lb./sq. in. After cooling, the reaction vessel is opened, and the reaction mixture is fractionally distilled. There is obtained 0.7 g. of 1,1-difluoro-2-methylene-3-difluoromethylenecyclobutane.

1,1-difluoro-2-methylene-3-difluoromethylenecyclobutane has a boiling point of 68° to 70°C. at 760 mm. The structure of this compound is characterized by nuclear magnetic resonance analysis, infrared absorption, reaction with bromine to form a dibromide of the expected composition, and by elemental analysis of its polymer. 1,1-difluoro-2-methylene-3-difluoromethylenecyclobutane polymerizes at room temperature unless inhibited, preferably with phenothiazine.

Example 2: 1,1-difluoro-2-methylene-3-difluoromethylenecyclobutane stabilized by phenothiazine (5% by weight) is subjected to distillation. One part of the inhibitor-free distillate is placed in a glass vessel cooled by a bath of ice water. After two hours at 0°C., the reaction vessel is removed from the ice water bath, and the product is found to be a hard, white, solid polymer of 1,1-difluoro-2-methylene-3-difluoromethylenecyclobutane. This polymer is insoluble in common organic solvents, e.g., benzene, petroleum ether, acetone, xylene, chloroform, carbon tetrachloride, ethyl ether, and dimethylformamide.

The polymer of 1,1-difluoro-2-methylene-3-difluoromethylenecyclobutane is pressed at 125° to 160°C. under 500 to 18,000 lb./sq. in. pressure to a clear, tough, flexible film. This film softens at temperatures of about 150°C., and the film is resistant to boiling nitric acid. Heating the polymer to about 165°C. converts it temporarily to a rubbery form. This rubbery polymer can be stretched and worked at room temperature, and this treatment converts it to a non-rubbery form which is fibrilated and has the appearance of an oriented, cold drawn crystalline polymer. The polymer as originally formed cannot be cold drawn.

Perfluorovinylcyclobutene

H. Iserson; U.S. Patent 3,046,261; July 24, 1962; assigned to Pennsalt Chemicals Corporation describes the preparation of a conjugated perfluorinated diene, namely perfluorovinylcyclobutene which polymerizes with ease to form valuable polymeric materials.

This diene is prepared by the dehalogenation of the compound 1-iodo-2-(1,2-dichloro-1,2,2-trifluoroethyl)perfluoro-cyclobutane. Loss of the two chlorines, the iodine atom, and the adjacent fluorine, produces the desired conjugated diene.

$$CF_2ClCFCl-CF\!\!-\!\!CFI \longrightarrow CF_2\!\!=\!\!CF-C\!\!=\!\!CF$$
$$CF_2\!\!-\!\!CF_2 \qquad\qquad CF_2\!\!-\!\!CF_2$$

The dehalogenation of the iodide to produce the diene is preferably carried out with the use of an excess of metallic zinc in a suitable medium, preferably acetic acid, or other mediums such as acetamide, ethanol, or dioxane. The reaction is preferably carried out at room temperature to the reflux temperature of the reaction mixture. The complete preparation of perfluorovinylcyclobutene is given in the patent. The conjugated perfluorinated diene readily homo-polymerizes or copolymerizes with ethylenically unsaturated compounds to produce polymeric materials having elasto-meric properties and high chemical and thermal stability. The ease of polymerization is evidenced by effecting poly-merization at atmospheric pressures by UV irradiation or in the presence of catalysts such as organic peroxides at relatively low pressures. The following examples illustrate the preparation of the homopolymer.

Example 1:   In to an 8 mm. Vycor (96% silica glass) tube there is introduced 0.5 gram of perfluorovinylcyclobutene after which the tube is sealed in a vacuum and placed about 3 centimeters from an ultraviolet light source. The tube is irradiated for 3 weeks at room temperature. The solid polymer is removed from the tube and heated in a vacuum oven (operating at about 29 inches H_2O) at a temperature of 55° to 70°C. for 4 hours. Analysis of the polymer shows that it contains 31.8% carbon as compared to a theoretical carbon content for $(C_6F_8)_n$ of 32.15%.

Example 2:   Thirty grams perfluorovinylcyclobutene, 0.4 gram $C_7F_{15}COONH_4$ and a solution consisting of 1 gram $K_2S_2O_8$, 1.5 gram $Na_4P_2O_7 \cdot 10H_2O$, 0.5 gram Na_2SO_3 in 80 ml. distilled and deoxygenated water are placed in a 7 ounce bottle which is then purged with nitrogen and capped. The bottle is rotated end over end in a water bath at 60°C. for 20 hours. The contents are removed, cooled at -15°C. for 15 hours and then warmed to room temperature. The precipitated polymer polyperfluorovinylcyclobutene is washed thoroughly with water and dried in vacuum at 75°C.

The diene readily forms copolymers of valuable properties with copolymerizable ethylenically unsaturated compounds. Preferred comonomers are ethylenically unsaturated compounds having from 2 to 10 carbon atoms, particularly halo-genated and more especially fluorinated compounds of this type. A particularly valuable group of comonomers are the haloethylenes such as $CH_2\!\!=\!\!CF_2$, $CF_2\!\!=\!\!CF_2$, $CF_2\!\!=\!\!CFCL$, $CF_2\!\!=\!\!CCl_2$, $CF_2\!\!=\!\!CHCl$, $CF_2\!\!=\!\!CFH$, $CH_2\!\!=\!\!CHCl$, and $CH_2\!\!=\!\!CHF$. Particularly preferred among the haloethylenes as comonomers are vinylidene fluoride, tetrafluoroethylene and chlorotrifluoroethylene.

Example 3:   A Vycor tube is cooled in liquid nitrogen and is then charged with approximately 0.3 gram of perfluoro-vinylcyclobutene and 0.3 gram of vinylidene fluoride by vacuum gaseous transfer. The tube is sealed and irradiated by ultraviolet light for 19 days. It is then cooled, opened, vented and the tube and contents heated at 55°C. for 4 hours. The copolymer product is a solid of elastomeric properties.

FLUORINATED OLEFINS

Perfluoropropylene Homopolymer

In a process described by H.S. Eleuterio; U.S. Patent 2,958,685; November 1, 1960; assigned to E.I. du Pont de Nemours and Company a high molecular weight polymer of perfluoropropylene is obtained by polymerizing perfluoro-propylene with a highly fluorinated initiator at a temperature of 0° to 300°C. and a pressure of at least 1,000 atmos-pheres. Perfluoropropylene, also referred to as hexafluoropropylene, has the general formula $CF_3CF\!\!=\!\!CF_2$ and can be prepared by pyrolysis of tetrafluoroethylene, polytetrafluoroethylene and other methods which have been described in the literature. The process is illustrated by the following example.

Example: Into a 200 ml. stainless steel autoclave was charged 75 ml. of perfluoro-1,3-dimethylcyclobutane and 0.2 g. of mercury bis-trifluoromethylmercaptide. The reaction mixture was heated to 225°C. and pressured with hexa-fluoropropylene to 3,000 atmospheres. The reaction mixture was agitated for a period of 14 hours at that temperature and pressure. The reaction vessel was then cooled to room temperature, and excess monomer was then vented off. On filtering the reaction mixture, 50 g. of a solid, white polyperfluoropropylene was obtained. The polymer was found to have a softening point in the range of 225° to 250°C.

A sample of the polymer was molded into transparent, tough, flexible films by pressing the polymer at 250°C. for 2 minutes at a pressure of 20,000 pounds. X-ray analysis of the film indicated the polymer to be amorphous. The film could be oriented by drawing at a temperature below the softening range. The polymer could be drawn into monofilaments by heating a sample to 250°C. and drawing it. The average Tinius Olsen stiffness of the polymer was found to be 187,000 psi at room temperature employing compression molded samples and measuring the stiffness according to the method described in ASTM D-747. Solid polyperfluoropropylene has properties which are in many ways similar to those of polytetrafluoroethylene. Thus this polymer has outstanding corrosion resistance, weatherability, and dielectric properties.

The high stiffness of the polymer and the retention of stiffness at elevated temperatures indicates outstanding mechanical properties. The polymer can be compression molded into articles and films. Polyperfluoropropylene is useful as a fiber-forming polymer since it can be oriented by drawing.

Perfluoropropylene — Transition Metals and Organometallic Catalysts

D. Sianese and G. Caporiccio; U.S. Patent 3,287,339; November 22, 1966; assigned to Montecatini Societa Generale per l'Industria Mineraria e Chimica, Italy have found that by operating in a temperature range between –30° and +150°C., preferably between 0° and 90°C., and under pressures ranging between atmospheric pressure and 60 atm., preferably at autogenic pressure, it is possible to obtain normally solid polymers of totally or partially fluorinated olefins by employing particular catalysts. These catalysts are the products of interaction between a metallorganic compound and a transition metal compound and are normally employed dissolved or dispersed in an organic liquid.

The transition metal compounds are preferably compounds such as titanium, vanadium and vanadyl alkoxides and acetyl-acetonates in which the transition metal can have its maximum valence or a valence lower than its maximum; or the transition metal compound can be a mixture of alkoxides and acetylacetonates of the same or a different transition metal. The selection of the two reactants forming the catalytic system also depends on the particular olefin to be polymerized or copolymerized. For example, while titanium compounds are preferred for the polymerization of perfluoroolefins, vanadium compounds are preferably employed with the partially fluorinated olefins.

According to the process, the whole reaction product derived from the components of the catalytic system has a molar ratio of metallorganic compound to transition metal compound higher than 0.1, more particularly a ratio between 1 and 6.

Example 1: A solution of 1.45 g. of titanium tetraisopropylate in 10 cc of methylene chloride is introduced into a 50 cc glass flask provided with an agitator, a dropping funnel and a reflux-condenser, and kept under dry nitrogen. 10 cc of methylene chloride containing in solution 0.02 mol of triisobutyl aluminum are then added dropwise to the solution within 10 minutes and at a temperature of 40°C.

17 cc of the resulting brown suspension are placed, after excluding air therefrom, in a glass vial which is sealed under nitrogen and is placed in a thermostatic bath at the temperature of 60°C. for a period of 15 minutes. After cooling, the contents of the vial are introduced into a previously evacuated stainless steel 330 cc oscillating autoclave. 230 g. of monomeric hexafluoropropylene are then introduced. The polymerization is continued for 15 days at the temperature of 30°C. and for 9 days at the temperature of 40°C.

After this period, the unreacted monomer is recovered and the contents of the autoclave are poured into an excess of methanol acidified with nitric acid. The precipitated polymer is isolated, washed with methanol and dried at 60°C. under a high vacuum. 3.10 g. of solid polyperfluoropropylene, in the form of a white powder, are obtained. The polymer is demonstrated to be crystalline under x-ray examination. The temperature at which the complete disappearance of the crystallinity takes place, as determined with the hot-stage polarizing microscope, is between 110° and 115°C.

One gram of polyperfluoropropylene, obtained as described above, is dissolved in 50 cc of hot carbon tetrachloride. The solution is filtered and poured, while agitating, into 100 cc of methanol. The precipitated polymer is filtered, washed with methanol and dried. It has a melting temperature between 130° and 135°C. The polymer thus obtained is subjected to extraction with boiling n-heptane and an insoluble fraction is separated. The final melting point of this fraction is between 170° and 175°C.

Example 2: 20 cc of CH_2Cl_2, containing in solution 10×10^{-3} mol of $Ti(O\ i\text{-}C_3H_7)_4$ to which has been added while agitating, 20 cc of CH_2Cl_2 containing 35×10^{-3} mols of $Al(i\text{-}C_4H_9)_2Cl$ are introduced into a 50 cc glass vial under dry nitrogen at room temperature. The vial is sealed and heated in a thermostatic bath to 60°C. for 30 minutes. The vial is then cooled to room temperature and opened under nitrogen. 4 cc of the resulting brown-red suspension are taken by means of a syringe and introduced into a 15 cc glass vial, dried and kept under nitrogen. In this vial, after cooling to –80°C., 6 g. of monomeric perfluoropropylene are condensed.

The vial is sealed and the polymerization reaction is operated for 210 hours at 50°C. At the end of the reaction, the vial is opened and the polymer formed is precipitated with an excess of methanol. The polymer is then washed with methanol, acidified with nitric acid and is finally vacuum dried to constant weight. 0.15 g. of a solid white polymer, soluble in chlorinated hydrocarbon (e.g., CCl_4) and reprecipitable from methanol, are obtained. The polymer can be molded at about 100°C. into transparent laminae.

Under x-ray examination it appears to have a high degree of crystallinity (more than 30 to 40% ca. of polymer is crystallized).

Solid Poly(3,3,3-Trifluoropropene)

A process described by E.M. Sullivan, E.W. Wise and F.P. Reding; U.S. Patent 3,110,705; November 12, 1963; assigned to Union Carbide Corporation involves the free-radical catalyzed polymerization of 3,3,3-trifluoropropene. Solid, high molecular weight poly(3,3,3-trifluoropropene) products are obtained.

The pressure employed in the polymerization process is of prime importance, and should be at least about 18,000 pounds per square inch if solid, high-molecular weight 3,3,3-trifluoropropene homopolymers are to be obtained as products. The maximum pressure which can be employed is restricted solely by the limitations imposed by the equipment utilized, and pressures of from about 18,000 pounds per square inch to about 125,000 pounds per square inch, or higher, can readily be employed. The preferred pressure range is from about 25,000 pounds per square inch to about 100,000 pounds per square inch. The following examples illustrate the process.

Example 1: A 10-milliliter steel alloy reactor was charged under a nitrogen atmosphere with 0.05 gram of azobisisobutyronitrile and filled with approximately 10 milliliters of cold 3,3,3-trifluoropropene at a temperature of -30°C. The charge was compressed to 50,000 pounds per square inch and maintained at this pressure by the injection of additional 3,3,3-trifluoropropene, and at a temperature of between 59° and 60°C., for a period of 12 1/3 hours. Conducted in this manner, the ensuing polymerization reaction produced 5.68 grams of a solid homopolymer of 3,3,3-trifluoropropene. Physical studies showed the homopolymeric product to be an amorphous, elastomeric resin having a relative viscosity of 1.05 and a glass transition temperature of 35°C.

Example 2: To the reactor and in the manner described in Example 1, there were charged 0.05 gram of azobisisobutyronitrile and 10 milliliters of cold 3,3,3-trifluoropropene at a temperature of -30°C. The charge was compressed to 67,200 pounds per square inch and maintained at a pressure of between 64,800 and 69,000 pounds per square inch by the injection of additional 3,3,3-trifluoropropene, and at a temperature of between 56° and 60°C., for a period of 9 1/2 hours. The polymerization reaction produced 2.78 grams of a solid homopolymer of 3,3,3-trifluoropropene having a relative viscosity of 1.09.

A 0.2 gram sample of the polymer was molded at a temperature of 65°C. and at a pressure of 1,000 pounds per square inch using a Buehler hydraulic press to produce a transparent, nonflammable flexible plaque having a thickness of approximately 8.5 mils.

2,3,3,3-Tetrafluoropropene

E.S. Lo; U.S. Patent 3,085,996; April 16, 1963 and U.S. Patent 2,970,988; February 7, 1961; both assigned to Minnesota Mining and Manufacturing Company describes copolymers containing between about 15 and about 85 mol % of 2,3,3,3-tetrafluoropropene, the remaining major constituent preferably being the fluorine-substituted ethylenes such as tetrafluoroethylene, trifluoroethylene and vinylidene fluoride. Some of the copolymers of the process are resinous thermoplastic materials at room temperature, but become elastomeric and rubbery when heated to temperatures above 50°C. An example of this type of copolymer is the product obtained by copolymerizing 2,3,3,3-tetrafluoropropene with trifluorochloroethylene. Such polymeric materials also are useful in applications where a high temperature rubber is required.

Particularly valuable low temperature and oil and fuel resistant polymers of 2,3,3,3-tetrafluoropropene are those containing between about 2 and about 45 mol percent of 2,3,3,3-tetrafluoropropene, the remaining major constituent being a 1,1-difluorobutadiene such as 1,1,2-trifluorobutadiene and 1,1,3-trifluorobutadiene. The following examples illustrate the process. The monomer, 2,3,3,3-tetrafluoropropene (CF_3—CF=CH_2), is prepared as described in the Journal of the American Chemical Society, volume 66, page 497 (March 1946).

Example 1: This example illustrates the homopolymerization of 2,3,3,3-tetrafluoropropene to produce a resinous thermoplastic material. A heavy-walled glass polymerization tube was flushed with nitrogen and was then charged with 5 ml. of a 0.75% by weight aqueous solution of the potassium salt of the C_8-telomer acid derived from the C_8-sulfuryl chloride telomer of trifluorochloroethylene, namely potassium 3,5,7,8-tetrachloroperfluorooctanoate, the pH of this solution having been adjusted to 12 by the addition of an aqueous potassium hydroxide solution. The potassium C_8-telomerate functions as an emulsifier. The stoppered tube was then placed in a liquid nitrogen freezing bath. After the contents of the tube were frozen solid, the tube was charged with 1 ml. of a 2% by weight aqueous solution of sodium metabisulfite, the contents were refrozen, and the tube was further charged with 4 ml. of a 1.25% by weight aqueous solution of potassium persulfate. In a separate experiment it was found that the final pH is about 7.0 when the solutions, in the amounts stated, are mixed without freezing. The contents of the tube were then refrozen, and the tube was connected to a gas-transfer system and evacuated at liquid nitrogen temperature. Thereafter 2.5 grams of 2,3,3,3-tetrafluoropropene were distilled into the tube. The polymerization tube was then sealed and rotated end-over-end in a temperature regulated bath at 50°C. The polymerization was conducted under autogenous pressure at 50°C. for a period of 24 hours. The polymer latex thus obtained was coagulated by freezing at liquid nitrogen temperature. The coagulated product was collected, washed with hot water to remove residual salts, and dried to constant weight in vacuo at 35°C.

A high molecular weight resinous thermoplastic material was obtained. The poly-2,3,3,3-tetrafluoropropene homopolymer of this example is useful as a protective lining for reactor vessels and tanks, the metal surfaces of which may come into contact with strong and corrosive chemicals.

Example 2: This example illustrates the copolymerization of 2,3,3,3-tetrafluoropropene with vinylidene fluoride. Employing the procedure set forth in Example 1 and the same aqueous emulsion polymerization system, the tube was charged with 3.2 grams of 2,3,3,3-tetrafluoropropene and 1.8 grams of vinylidene fluoride to make up a total monomer charge containing 50 mol percent of each monomer. The polymerization reaction was carried out under autogenous conditions of pressure at a temperature of 25°C. for a period of 71 hours. The resultant polymer latex was worked up in accordance with the same procedure set forth in Example 1. A white, slightly rubbery product was obtained and, upon analysis for fluorine content, was found to comprise approximately 20 mol percent of combined 2,3,3,3-tetrafluoropropene, the remaining major constituent being vinylidene fluoride, that is, about 80 mol percent. The copolymer was obtained in an amount corresponding to a 41% conversion.

When this 2,3,3,3-tetrafluoropropene:vinylidene fluoride copolymer product was heated to 50°C., the copolymer became a very rubbery material having good physical and mechanical properties. It is particularly useful as a high-temperature, acid resistant rubber suitable as a protective coating or lining, as a wire insulator, and in the manufacture of flexible films, sheets of varying thickness, gaskets and other such end products. The copolymer is also relatively resistant to diester type hydraulic fluids such as Esso Turbo Oil, 15, thereby making it useful in aircraft component parts.

A process for the preparation of 2,3,3,3-tetrafluoropropene is described by D.M. Marquis; U.S. Patent 2,931,840; April 5, 1960; assigned to E.I. du Pont de Nemours and Co.

3,3,3-Trifluoro-2-Trifluoromethylpropene

A process described by G.B. Sterling; U.S. Patent 3,240,757; March 15, 1966; assigned to Dow Chemical Co. involves copolymers of 3,3,3-trifluoro-2-trifluoromethylpropene and certain ethylenically unsaturated monomers. The 3,3,3-trifluoro-2-trifluoromethylpropene starting material has the empirical formula $(CF_3)_2C:CH_2$, and may alternatively be named 1,1,1,3,3,3-hexafluoroisobutene. It can be prepared by procedure described in J. Chem. Soc. (London), page 3567, 1953. In brief, 1,1,1,3,3,3-hexafluoro-2-methylpropane is mildly chlorinated to produce $(CF_3)_2CH \cdot CH_2Cl$, which compound is dehydrochlorinated with potassium hydroxide-ethyl alcohol solution or produce the 3,3,3-trifluoro-2-trifluoromethylpropene of the formula $(CF_3)_2C:CH_2$. The following examples illustrate the process.

Example 1: In each of a series of experiments, a mixture of 3,3,3-trifluoro-2-trifluoromethylpropene and vinyl propionate in proportions as stated in the following table, together with an aqueous solution, was sealed in a pressure resistant glass bottle and agitated to form an aqueous emulsion employing the recipe:

	Parts by Weight
Monomers	100
Water	110
Sodium lauryl sulfate	0.85
Sodium bicarbonate	1.0
Potassium persulfate	0.75

The emulsion was heated in the closed bottle with agitating at a temperature of 60°C. for a period of 72 hours. Thereafter, the bottle was cooled and opened. The copolymer latex was removed and was heated to temperatures between about 98° to 100°C. while bubbling steam through to remove unreacted monomer. The copolymer was recovered by coagulating the latex and separating, washing and drying the coagulum. The copolymer was analyzed to determine the proportion of 3,3,3-trifluoro-2-trifluoromethylpropene chemically combined.

Other portions of the copolymer were compression molded at temperatures of about 170°C. and about 200 pounds per square inch gauge pressure and the molded product observed for its physical properties. The copolymer was also tested for its solubility in a number of organic solvents. The table below identifies the experiments and gives the proportion of monomers in the starting materials. The table also given the proportions of the monomers chemically combined or interpolymerized in the copolymer product. In the table the empirical formula $C_4H_2F_6$ is employed to designate 3,3,3-trifluoro-2-trifluoromethylpropene, for brevity.

Run No.	Monomers			Copolymer Product		
	$C_4H_2F_6$, %	Vinyl Propionate, %	Con- version, %	$C_4H_2F_6$, %	Vinyl Propionate, %	Remarks
1	90	10	24	62	38	Clear, soft, flexible
2	75	25	56	16	84	Clear, hard, brittle
3	50	50	100	50	50	Clear, soft, flexible
4	10	90	100	14	86	Clear, very soft

All of the copolymers were soluble in carbon tetrachloride, toluene and methyl isobutyl ketone. They were insoluble in ethyl alcohol and hexane.

Example 2: In each of a series of experiments, a mixture of 3,3,3-trifluoro-2-trifluoromethylpropene and styrene in proportions as stated in the following table was polymerized employing the procedure and recipe employed in Example 1.

Run No.	Monomers			Copolymer Product		
	$C_4H_2F_6$, %	Styrene, %	Con- version, %	$C_2H_4F_6$, %	Styrene, %	Remarks
1	90	10	29	17	83	Clear, hard, stiff
2	50	50	56	13	87	Clear, hard, stiff
3	10	90	91	7	93	Clear, hard, stiff

All of the copolymers were soluble in carbon tetrachloride, methyl ethyl ketone and toluene, but were insoluble in ethyl alcohol.

Hexafluoropropylene and Fluoranil

W.J. Brehm and A.S. Milian; U.S. Patent 3,053,823; September 11, 1962; assigned to E.I. du Pont de Nemours and Company describe fluorocarbon polyethers in the form of copolymers of hexafluoropropylene and fluoranil. The copolymers are believed to have the following general formula

where n is the number of hexafluoropropylene units and m the number of fluoranil units in the polymer chain section having one hexafluoropropylene fluoranil bond, and x represents the number of polymer chain sections in the copolymer. The fluoranil employed as comonomer may be obtained by the method described by Wallenfels et al, Chemische Berichte, vol. 90, page 2819 (1957).

The molecular weight of the copolymers may be varied to result in products ranging from greases and waxes to high molecular weight resins. The greases and waxes find utility as lubricating materials, particularly in applications involving high temperatures and/or a corrosive environment. The high molecular weight resins are useful for fabrication into film and fiber having the outstanding noncorrosive properties of fluorocarbon polymers. The process is illustrated by the following examples.

Example 1: Into a thin, approximately 10 ml. platinum tube equipped with a sealing device was charged 3.0 g. of fluoranil, and 745 ml. of gaseous hexafluoropropylene at room temperature was then condensed into the tube. The tube was sealed and placed into an autoclave which was pressured to 4,000 atmospheres with nitrogen and heated to a temperature of 225°C. The polymerization was allowed to proceed for ten hours under these conditions. The tube was then removed from the autoclave and placed in a Dry Ice bath, opened and allowed to come to room temperature. From the tube there was then isolated a yellow solid, a sample of which was digested with chloroform to remove the unreacted fluoranil.

Combustion analysis of the dried residue yielded 33.5% carbon and 57.9% fluorine. The calculated values for $C_9F_{10}O_2$, the empirical formula of the 1:1 copolymer, are 32.7% carbon and 57.6% fluorine.

Example 2: Employing the equipment described in the preceding example, 249 cm.^{3} of hexafluoropropylene gas and 2 g. of fluoranil were polymerized in the presence of 4.0 ml. of perfluorodimethyl cyclobutane at a temperature of 225°C. and at a pressure of 3,000 atmospheres for a period of 8 hours. On digestion with chloroform, there was obtained a white copolymer of hexafluoropropylene and fluoranil having an inherent viscosity of 0.143. The copolymer was cast from a solution in symmetrical difluorotetrachloroethane to give rise to a stiff, clear and transparent, self-supporting film. The copolymer softened at temperatures above 100°C. and could be cold-drawn into fibers.

1,1-Difluoroisobutylene

J.J. Drysdale; U.S. Patent 2,956,988; October 18, 1960; assigned to E.I. du Pont de Nemours and Company describes copolymers of 1,1-difluoroisobutylene with fluoroethylenes containing 5% to 95% by weight of difluoroisobutylene. These copolymers are especially amenable to injection molding and have outstanding chemical and thermal stability.

The copolymers of this process are prepared by subjecting a mixture of 1,1-difluoroisobutylene and at least one fluoroethylene to addition polymerization. Thus, 1,1-difluoroisobutylene can be copolymerized with the fluoroethylenes by emulsion, solution, bulk or bead methods in the presence of an addition polymerization initiator. The particular polymerization method employed with any particular mixture of comonomers is dependent on the particular comonomer employed with the difluoroisobutylene. The addition polymerization initiator can be any free radical polymerization catalyst. These initiators are used in conventional quantities, amounts ranging from 0.01 to 10% or more, based on the weight of the comonomers, being operable.

The 1,1-difluoroisobutylene used in making the copolymers can be prepared by pyrolyzing 1,1-dimethyl-2,2,3,3-tetra-fluorocyclobutane (cf. U.S. Patent 2,462,345) at 600° to 1000°C. preferably at 750° to 900°C. and a pressure of less than 50 mm. of mercury with very rapid passage of the pyrolysis mixture through the pyrolysis zone. The following examples illustrate the process.

Example 1: A mixture of 30 parts of vinyl fluoride and 5 parts of 1,1-difluoroisobutylene with 15 parts of distilled water and 0.1 part of di-tertiary butyl peroxide and 0.1 part of disodium hydrogen phosphate heptahydrate is heated eight hours at 135°C. under 1,000 atmospheres water pressure in a silver-lined reaction vessel. At the end of this period there is isolated 32.4 parts of a white copolymer of 1,1-difluoroisobutylene and vinyl fluoride having a softening point of 104°C. and a melting point of 126°C. A film pressed from this copolymer at 140°C. under pressure of 10,000 pounds is clear, tough, and cold-drawable.

This copolymer is heat stable, there being no apparent decomposition after being heated at 210°C. for five minutes. This copolymer has a flow number of 36, which indicates that it is well suited for injection molding or extrusion application. This polymer does not fail in an ultraviolet light-ozone accelerated aging test until after 120 hours. The copolymer is soluble up to 20% by weight in either dioxane or cyclohexanone at room temperature.

The unusual properties of this copolymer of Example 1 are evident from the following properties possessed by a vinyl fluoride homopolymer. The vinyl fluoride homopolymer has a flow number of 20 (determined at 230°C.) partially decomposes after heating at 210°C. for five minutes, and fails in the ultraviolet light-ozone accelerated aging test after 100 to 120 hours.

Furthermore, a comparable copolymer of vinyl fluoride and isobutylene, as regards solubility in dioxane, has a softening point appreciably lower than 100°C. and fails in approximately 90 to 100 hours in the accelerated aging test. In the examples the melt-flow characteristics of a polymer are indicated by "flow numbers." These flow numbers are determined as the square of the diameter in inches of a film pressed from one gram of polymer in five minutes at 210°C. under 10,000 pounds pressure. Polymers having flow numbers above 15 in this test can be molded satisfactorily, and those having flow numbers above 25 are especially suitable for injection molding.

Example 2: A solution of 3 parts of 1,1-difluoroisobutylene and 0.05 part of α,α'-azobis-isobutyronitrile in 45 parts of perfluorodimethylcyclohexane is placed in a silver-lined reaction vessel capable of withstanding high pressure. The reaction vessel is then pressured to 450 lbs./sq. in. gauge at 80°C. for four hours with tetrafluoroethylene. At the end of this reaction period there is obtained 1.7 parts of a white copolymer of 1,1-difluoroisobutylene and tetrafluoroethylene having a softening point of approximately 110°C. and a melting point of 140° to 150°C. This copolymer can be pressed at 140°C. under 1,000 pounds pressure to a clear film. On analysis this copolymer is found to contain 26.1% of 1,1-difluoroisobutylene.

Polytetrafluoroethylene is highly stable and chemically inert. It does not show a true melting point but sinters on heating to a high temperature (360° to 370°C.). A significant flow number cannot be obtained for this homopolymer. Copolymers of tetrafluoroethylene with perfluoropropylene do have melting points, but copolymers even containing high percentages of perfluoropropylene have melting points above 275°C. Copolymers having lower percentages of perfluoropropylene have even higher melting points.

In contrast, the copolymer of difluoroisobutylene and tetrafluoroethylene of Example 2, while maintaining the stability and chemical inertness of polytetrafluoroethylene, can be molded at temperatures as low as 140°C.

Perhaloalkyl-Substituted Unsaturated Fluoroalkanes

B.M. Lichstein and C. Woolf; U.S. Patent 3,472,905; October 14, 1969; assigned to Allied Chemical Corporation describe fluoroolefins having the formula

$$\begin{array}{c} R_1 \quad R_3 \\ | \quad \ | \\ C=C-CH_2F \\ | \\ R_2 \end{array}$$

where R_1 and R_2 are perhalogenated alkyl groups in which the halogen atoms can be fluorine and chlorine with at least one fluorine atom attached to each carbon atom and R_3 can be hydrogen or alkyl groups. Preferably R_1 and R_2 each contain from 1 to 5 carbon atoms and R_3 when it is an alkyl group contains from 1 to 5 carbon atoms.

The fluoroolefins can be prepared by reacting an unsaturated fluorine-containing hydroxy compound having the formula

$$\begin{array}{c} R_1 \quad R_3 \\ | \quad \ | \\ HO-C-C=CH_2 \\ | \\ R_2 \end{array}$$

where R_1, R_2 and R_3 have the meanings given above, with sulfur tetrafluoride. The fluoroolefins can be polymerized to linear fluorine-containing polymers in conventional manner, and they can also be mixed with known monomers including compounds having vinyl unsaturation and copolymerized to prepare fluorine-containing copolymers useful as thermally stable elastomers and lubricants. The following examples illustrate the process.

Example 1: 31 parts by weight of 1,1-bis(perfluoromethyl)-1,3-propanediol were added to 20 parts by volume of concentrated sulfuric acid, maintained at a temperature of 250°C., over a period of about 4 hours. The product boiling at temperatures up to 300°C. was collected, dried over magnesium sulfate, filtered and redistilled. A first fraction boiling at 76°C. was 98% pure according to vapor phase chromatographic analysis. Elemental analysis was as follows: Calculated for $C_5H_4OF_6$: C, 30.94; H, 2.08. Found: C, 30.6; H, 2.16.

The product was confirmed as 1,1,1-trifluoro-2-trifluoromethyl-3-butene-2-ol by infrared and nuclear magnetic resonance analyses. 23 parts of 1,1,1-trifluoro-2-trifluoromethyl-3-butene-2-ol as prepared above and 89 parts of sulfur tetrafluoride (mol ratio 1:6.95) were condensed at -78°C. and transferred under vacuum to a predried glass pressure vessel. The mixture was stirred at room temperature for about 3 days during which period a maximum pressure of 227 psig developed in the vessel. The resultant product was distilled under vacuum through a -78°C. trap and redistilled at atmospheric pressure. The product having a boiling point of 47°C. was collected.

A 47.4% yield of 1,1,1-trifluoro-2-trifluoromethyl-4-fluoro-2-butene was obtained. It was 100% pure as determined by vapor phase chromatographic analysis. The structure was confirmed by infrared analysis. The results of elemental analysis were: Calculated for $C_5H_2F_7$: C, 30.6; H, 1.5. Found: C, 31.8; H, 1.7.

Example 2: 12 parts of 1,1,1-trifluoro-2-trifluoromethyl-3-butene-2-ol as prepared in Example 1 and 17.3 parts of sulfur tetrafluoride (mol ratio 1:2.59) were condensed under vacuum at -78°C. into a nickel autoclave and stirred together at room temperature for 17 hours. The maximum pressure reached was 89 psig. The product mixture was distilled under vacuum through a trap at -78°C. and redistilled at atmospheric pressure. A 58.8% yield of product boiling at 56° to 63°C. was obtained. Nuclear magnetic resonance analysis confirmed the structure of 1,1,1-trifluoro-2-trifluoromethyl-4-fluoro-2-butene. Elemental analysis found was C, 30.3; H, 1.5.

Other Fluorinated Monomers and Polymers

Perfluorobutyne-2

In a process described by <u>A.N. Bolstad and F.J. Honn; U.S. Patent 2,966,482; December 27, 1960; assigned to Minnesota Mining and Manufacturing Company</u> perfluorobutyne-2 is copolymerized with a fluorinated olefin, such as 1-chloro-1-fluoroethylene or 1,1-difluoro-3-methyl butadiene, to produce useful polymeric compositions. The following example illustrates the process.

Example: A heavy-walled glass polymerization tube of about 300 ml. capacity was flushed with nitrogen and then charged with 60 ml. of a soap solution, prepared by dissolving 15 grams of lauryl sulfonate in 600 ml. of water. The contents of the tube were then frozen, and the tube was then charged with 20 ml. of a 2.5% aqueous solution of ammonium persulfate. The contents of the tube were next refrozen, and 10 ml. of a 2% aqueous solution of sodium metabisulfite was added. The contents of the tube were next refrozen in liquid nitrogen, and thereafter were added 10 ml. of a 1% aqueous solution of ferrous sulfate heptahydrate. The contents of the tube were then refrozen in liquid nitrogen.

The tube was next connected to a gas-transfer system and evacuated at liquid nitrogen temperature. To the frozen contents of the tube were added, by distillation, 33.5 grams of perfluorobutyne-2 and 13.2 grams of 1-chloro-1-fluoroethylene, which comprised a 50/50 molar ratio. The polymerization tube was then sealed and rotated end-over-end in a temperature regulated water bath at 20°C. for a period of 24 hours. At the end of this time, the contents of the tube were coagulated by freezing at liquid nitrogen temperature. The coagulated product was then removed from the tube, washed with hot water and then dried to constant weight in vacuo at 35°C.

A copolymeric elastomeric product was obtained which was found, upon analysis, to comprise approximately 7 mol percent perfluorobutyne-2 and the remaining major constituent, 1-chloro-1-fluoroethylene, being present in an amount of approximately 93 mol percent. The copolymer was obtained in an amount corresponding to a 13% conversion.

Fluoroisoprenes

<u>P. Tarrant and R.P. Lutz; U.S. Patent 2,945,896; July 19, 1960; assigned to Research Corporation</u> have found that fluoroisoprenes of the formula

$$CF_2{=}C{-}CX{=}CX_2$$
$$|$$
$$CF_2$$

where the X's represent hydrogen or fluorine, may be polymerized by conventional methods to form homopolymers, or copolymers with such substances as butadiene, which are highly useful elastomers resistant to the commonly used fuels and retaining their elastic properties at low temperatures.

The fluoroisoprenes can be made simply, conveniently and in good yields, by condensing a perhalogenated pentafluoropropane of the formula

$$CF_2Y'{-}CClY''{-}CF_3$$

where Y' is bromine or chlorine and Y'' is iodine, bromine or chlorine, with an olefin of the formula

$$CHX{=}CX_2$$

where X is hydrogen or fluorine, for example, by heating the reactants in the presence of a carboxylic acid peroxide, to give an halogenated fluoroisopentane of the formula

$$CF_2Y'{-}CCl{-}CH{-}CX_2Y''$$
$$| \quad |$$
$$CF_3 \quad X$$

dehydrohalogenating the fluoroisopentane by heating with an alkali to give a halogenated fluoroisopentane of the formula

$$CF_2Y'{-}CCl{-}CX{=}CX_2$$
$$|$$
$$CF_3$$

and dehalogenating the fluoroisopentene by heating with metallic zinc to give the corresponding fluoroisoprene. The following examples are illustrative of the process.

Example 1: 215 g. of 1,2-dichloro-2-iodo-1,1,3,3-pentafluoropropane are charged to a 300 ml. stainless steel reaction vessel with fluoropropane, 53.5 g. of trifluoroethylene, and 5 g. of benzoyl peroxide. The vessel is sealed and heated to 85°C. for 3 hrs. After the vessel has cooled, unreacted gas is removed through a valve and the liquid contents are filtered and distilled. About 169 g. of the one-to-one addition product, shown below, is obtained at 67–69°C./45 mm.:

$$CF_2ClCCl(CF_3)CHFCF_2I$$

Other physical properties are n_D^{25} 1.4174, d_4^{25} 2.1504, MR_D 48.09 (found) MR_D 47.82 (calculated).

Example 2: Into a 500 ml. 3-necked round-bottom flask equipped with a stirrer, addition funnel, and short column with take-off head, is placed 233 g. of

$$CF_2Cl\text{—}CCl(CF_3)\text{—}CHF\text{—}CF_2I$$

The liquid is heated and 40 g. of potassium hydroxide, dissolved in the least amount of water, is added dropwise. The product is collected as it distills over with water in the range 75° to 95°C. When all the KOH has been added, the reaction is stirred and heated for 2 hours, during which time the reflux temperature is about 97°C. The material which has collected is washed with sodium thiosulfate solution until colorless, dried over calcium chloride and distilled. The olefin is obtained at 88° to 89°C. A considerable amount of high-boiling material remains in the pot and it is treated as before with KOH along with unreacted material which is recovered from the original reaction vessel. The olefinic product,

$$CF_2Cl\text{—}CCl(CF_3)\text{—}CF\text{=}CF_2$$

amounts to 65.5 g.; n_D^{25} 1.3394, d_4^{25} 1.727, MR_D 34.29 (found) 34.55 (calculated).

Example 3: Into a 300 ml. 3-necked round-bottom flask equipped with a stirrer, addition funnel, and short column with take-off head are placed 23 g. of powdered zinc, 3 g. of zinc chloride, and 50 cc of isopropyl alcohol. The mixture is stirred and heated and 65.5 g. of

$$CF_2Cl\text{—}CCl(CF_3)\text{—}CF\text{=}CF_2$$

is added dropwise. The product distills and is collected at about 30°C. It is then washed with ice water, dried over calcium chloride, and distilled. Perfluoroisoprene,

$$CF_2\text{=}C(CF_3)\text{—}CF\text{=}CF_2$$

amounting to about 15 g. comes over at 30° to 31°C.; n_D^0 1.3000, d_4^0 1.527, MR_D 25.97 (found), 24.36 (calculated).

Example 4: The procedure and equipment are identical to those used in Example 3. The materials used are

88 g. of $CF_2Cl\text{—}CCl(CF_3)\text{—}CH\text{=}CH_2$
52.5 g. of powdered zinc
5 g. of zinc chloride
150 cc of isopropyl alcohol.

The product is removed from the reaction as it distills in the range 34° to 45°C. It is washed and dried as before to yield 46.5 g. of the pentafluoroisoprene

$$CF_2\text{=}C(CF_3)\text{—}CH\text{=}CH_2$$

which boils at 32° to 32.5°C.; n_D^0 1.3372, d_4^0 1.310, MR_D 25.11 (found) 24.36 (calculated).

Example 5: A mixture of 100 parts of the product of Example 4, 180 parts of water, 3 parts of Aerosol OT, 0.1 part of $K_2S_2O_3$, 0.3 part of borax and 0.3 part of t—C_{12} mercaptan is stirred for 24 hours at 25°C. An elastomeric latex is formed which is worked up in the usual manner.

Fluorinated Styrenes and Organometallic Catalysts

G. Natta and D. Sianesi; U.S. Patent 3,420,806; January 7, 1969; assigned to Montecatini Edison SpA, Italy have found that styrenes substituted by fluorine atoms polymerize very readily in contact with the organometallic catalysts, in contrast to the chloro-substituted styrenes. Also it is found that, in general, and particularly in the case of o- and p-fluorostyrene, as well as p-fluorostyrene substituted in the ortho position by lower alkyl groups containing 1 to 5 carbon atoms (e.g., p-fluoro-o-methyl styrene or p-fluoro-o-ethyl styrene), the polymers obtained have, prevailingly the isotactic structure and consequently are highly crystalline when examined at the x-rays. When the monomer is m-fluorostyrene, the polymer obtained has a regular structure, but is not crystallizable. The mixtures of different fluorostyrene isomers can be polymerized in contact with the stereo-specific organometallic catalysts, such as the catalyst obtained by (a) starting with a highly crystalline halide of the transition metal in which the metal has a valency not higher than 3, e.g., $TiCl_3$, and (b) mixing such halide with the organometallic compound e.g., $Al(C_2H_5)_3$, to obtain crystallizable polymers. The molecular weights of the obtained polymers, as evaluated from viscometric measurements, are always considerably in excess of 100,000. The following examples are given to illustrate the process.

Example 1: 0.5 g. $TiCl_4$, dissolved in 20 cc benzene, are introduced under dry nitrogen into a 250 cc glass flask fitted with glass stirrer and dropping funnel. The flask is held in a bath regulated at 70°C., and maintained in an atmosphere of inert gas. During a time interval of about 2 minutes, 20 cc benzene containing 0.90 g. of dissolved triethyl aluminum are added dropwise and, immediately afterwards, 20 g. para-fluorostyrene are introduced into the flask.

The reaction is stopped after 24 hours by adding methanol in excess, which decomposes the catalyst present, and coagulates the polymer formed. The polymer, separated from the solution by filtration, is then washed with an additional quantity of methyl alcohol and freed from traces of inorganic salts through digestion with methanol and hydrochloric acid.

The polymer is then subjected to extraction with acetone. After the extraction, 1.5 g. amorphous polymer of low molecular weight can be recovered from the acetone through precipitation with methanol. As a residue after the acetone extraction, 14.1 g. crystalline poly-para-fluorostyrene are obtained which corresponds to a yield of 70% based on the amount of monomer introduced. This polymer is insoluble in methyl ethyl ketone, ether and benzene, and soluble in hot tetralin.

The intrinsic viscosity determined at 100°C. in a tetralin solution, is 1.2. The polymer appears highly crystalline when subjected to x-ray examination and the identity period ascertained along the fiber axis is of 8.3 A. The distances between lattice planes corresponding to the reflexes appearing in a powder spectrum taken with a Geiger counter, are d_{hkl} = 8.9; 6.0; 5.3; 4.9 and and 3.9A. The melting point (complete disappearance of crystallinity) determined with a polarizing microscope, is between 275° and 280°C.

Example 2: A solution of 1.0 g. of vanadium tetrachloride and 0.33 g. titanium tetrachloride in a 10 cc mixture of fluoro-ethylbenzene isomers (consisting of 66% ortho-isomer, 19% para-isomer and 15% meta-isomer) is introduced under nitrogen into a 250 cc glass flask. The equipment is fitted with a glass stirrer and a graduated dropping funnel for measuring and introduction of the reagents and is maintained in an atmosphere of inert gas and immersed in a thermostatic bath, regulated at 40°C.

To the above solution are added under agitation, 1.43 g. $Al(C_2H_5)_3$ dissolved in 60 cc of the product obtained through dehydrogenation (carried out at 500° to 550°C. upon a catalyst based on Fe_2O_3 in the presence of steam) of a fluoro-ethyl-benzene isomer mixture whose composition is similar to the one mentioned above. The dehydrogenated product has been distilled for the purpose of separating small quantities of volatile compounds and of moisture as well as small quantities of products with a higher boiling point. The dehydrogenation product thus purified has an iodine number of 63, corresponding to a content of 30% by weight of fluorostyrene isomers, a distillation range of between 120° to 141°C. at 755 mm. and a refractive index n_D^{15} = 1.4982.

The polymerization is conducted for 14 hours at 40°C. The reaction is then stopped by adding methanol in excess. 9.20 g. of solid polymer are separated, corresponding to a conversion of about 50% of the monomer initially introduced; 58 g. of a mixture consisting of fluoro-ethyl-benzene and fluorostyrene isomers are recovered from the residual liquid after elimination of the excess methanol through washing with water and distillation.

The polymer thus produced is then subjected to extraction with acetone. It comprises a smaller fraction soluble in acetone which is amorphous on x-ray examination even after prolonged annealing at temperatures ranging between 100° and 200°C. The intrinsic viscosity in toluene of this polymer fraction is 1.3 at 30°C. The main fraction which is insoluble in acetone, can not be dissolved in methyl ethyl ketone either. It swells in benzene and is soluble in hot tetralin. The intrinsic viscosity in tetralin at 100°C. is 2.94.

This latter fraction appears partially crystalline under the x-rays. The crystallinity increases after annealing for 1 1/2 hours at 200°C. The temperature at which complete disappearance of crystallinity takes place under the polarizing microscope, is about 260°C. By x-ray examination of the polymer, an identity period along the fiber axis of 6.5 ±0.05 A. will be found. The distances between lattice planes, corresponding to the reflexes appearing in a spectrum taken with the aid of a Geiger counter, are d_{hkl} = 11.1; 5.6; 5.0; 4.2 and 3.7 A.

Hydroxyfluoroalkyl-Substituted Styrenes

A process described by W.J. Middleton; U.S. Patent 3,179,640; April 20, 1965; assigned to E.I. du Pont de Nemours and Company provides (A) styrenes containing on the benzene ring a hydroxy[di(polyfluoroalkyl)]methyl substituent, the styrene having the formula

(I)

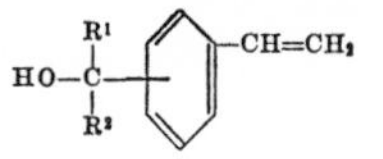

where R^1 and R^2 are, individually, the same or different monovalent polyfluoroalkyl, including perfluoroalkyl, ω-hydroperfluoroalkyl and ω-chloroperfluoroalkyl, radicals, or jointly, a divalent perfluoroalkylene radical; (B) polymers of such compounds; and (C) the preparation of the new monomers by pyrolysis of hydroxydi(polyfluoroalkyl)methyl-substituted ethylbenzenes [X being hydrogen in Formula (II)] or certain α-substituted ethylbenzenes (X being other than hydrogen) of the formula

(II)

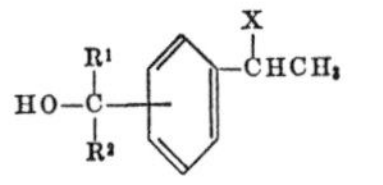

where R^1 and R^2 are as previously defined and X is hydrogen, halogen (chlorine or bromine) or a hydrocarbonoyloxy group

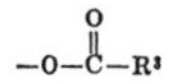

R^3 being hydrocarbon of up to 18 carbons free from aliphatic or nonaromatic unsaturation, preferably alkyl.

The pyrolytic process for synthesizing the styrenes can be represented by the equation:

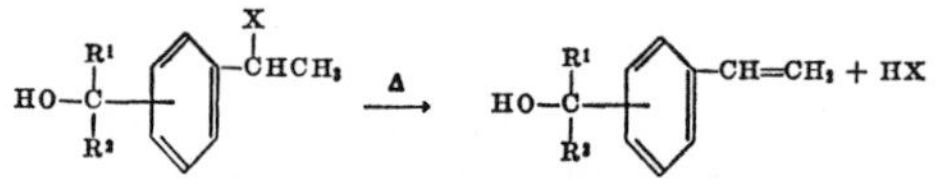

(II)   (I)

where R^1, R^2 and X are as above. When X is hydrogen, the product HX is molecular hydrogen; and the process is a pyrolytic dehydrogenation which can be effected at an appropriately high temperature (about 500° to over 800°C.) in the presence of an oxide catalyst such as chromia-alumina(Cr_2O_3/Al_2O_3) or Fe/Cu oxide. When X is halogen, HX is a hydrogen halide; and the process is a pyrolytic dehydrohalogenation which can be effected at a temperature in the range of about 200° to about 800°C., suitably in a continuous vapor phase flow procedure.

When X is a hydrocarbonoyloxy group, HX is a hydrocarboncarboxylic acid; and the process is a pyrolytic deacylation which can be effected by heating the carboxylic ester at its boiling point or higher, i.e., in the range of about 200° to about 800°C., suitably in a distillation procedure or in a continuous vapor phase flow operation. The pyrolysis products are recovered by condensation of the vapors. The desired hydroxydi(fluoroalkyl)methylstyrene is separated and purified by conventional methods, especially by distillation.

The hydroxydi(fluoroalkyl)methyl-substituted ethylbenzenes that are pyrolyzed to the corresponding styrenes are obtainable by processes described in U.S. Patent 3,148,220. The following examples illustrate the process.

Example 1:   p-(2-Hydroxyhexafluoro-2-Propyl)Styrene —

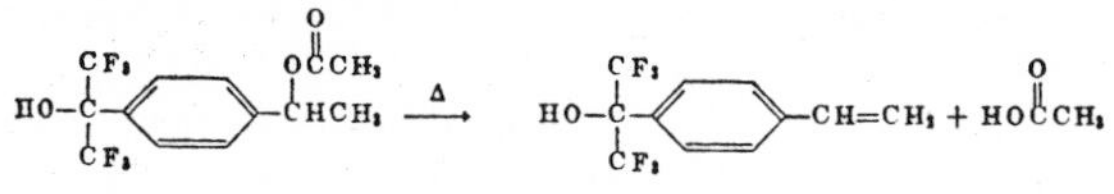

α,α-bis(trifluoromethyl)-4-(α'-acetoxyethyl) benzyl alcohol was prepared in the following way: A mixture of 20 g. (0.12 g. mol) of silver acetate and 35 g. (0.1 g. mol) of α,α-bis(trifluoromethyl)-4-(α'-bromoethyl)benzyl alcohol in 100 ml. of acetonitrile was stirred until the exothermic reaction subsided. The reaction mixture was then filtered, and the filtrate was distilled under reduced pressure. There was obtained 23.1 g. of viscous, liquid product, BP 141° to 142°C./2.5 mm., which solidified on cooling. The solid product was identified as α,α-bis(trifluoromethyl)-4-(α'-acetoxyethyl)benzyl alcohol, MP 75°C.

Analysis for $C_{13}H_{12}F_6O_3$	C	H	F
Calculated	47.28	3.67	34.52
Found	47.68	3.67	32.88

In a distillation flask connected to a fractionating column was placed 100 g. of α,α-bis(trifluoromethyl)-4-(α'-acetoxyethyl)benzyl alcohol. The alcohol was heated strongly at atmospheric pressure, and 16 g. of acetic acid, BP 100° to 120°C., was fractionally separated at a reflux ratio of about 2:1 at the still head. Continued distillation at atmospheric pressure yielded material, BP 120° to 240°C., which was redistilled under reduced pressure. There was obtained 42 g. of product which was identified as p-(2-hydroxyhexafluoro-2-propyl)styrene, BP 77° to 78°C./5 mm., n_D^{25} 1.4520.

Analysis for $C_{11}H_8F_6O$	C	H	F
Calculated	48.90	2.99	42.19
Found	49.29	3.16	41.59

The product was characterized by its infrared spectrum, containing the following bands: 2.78 and 2.83μ, for OH; 3.23, 3.27 and 3.31μ, for $\equiv$CH; 6.11μ, for conjugated olefinic C=C; 6.17, 6.37, and 6.58μ, for aromatic C=C; 11.9μ, for para-disubstituted aromatic group. The proton nuclear magnetic resonance (NMR) spectrum contains the following bands, all shifted to lower field from tetramethylsilane reference (60 megacycles): 7.24, 7.39, 7.64, and 7.78 ppm, for aromatic hydrogens; 5.10, 5.29, 5.51, 5.80, 6.37, 6.56, 6.67, and 6.86 ppm, for vinyl hydrogens; 3.55 ppm, for hydroxyl hydrogen.

Example 2: Free Radical Polymerization of p-(2-Hydroxyhexafluoro-2-Propyl)Styrene — A solution of 20 g. (0.074 g. mol) of freshly prepared p-(2-hydroxyhexafluoro-2-propyl)styrene and 0.2 g. (0.00147 g. mol) of α,α'-azodiisobutyronitrile in 50 ml. of benzene was heated at reflux for 5 hours. The mixture was cooled, and 200 ml. of pentane was added. The solid that formed was broken up, collected on a filter, and dried in a vacuum; yield, 18 g. The polymer was a white powder that softened at 150°C. and was soluble in acetone, benzene and dilute (5%) aqueous sodium hydroxide; η_{inh}=0.08, at 1% in acetone at 25°C.

A soluble polymer prepared in the above manner was deposited on paper and on nylon fabric from a solution in toluene. The coated paper and fabric showed improved water and oil repellencies in comparison with uncoated controls.

A mixture of 5.0 g. (0.0185 g. mol) of freshly redistilled p-2-hydroxyhexafluoro-2-propyl)styrene and 0.05 g. (0.00023 g. mol) of 1,1'-azobis(cyclopropylpropionitrile) was placed under nitrogen in a closed glass vessel and allowed to stand at room temperature. After 30 minutes, the mixture was a viscous syrup, at 70 minutes it was a soft gel, and at 95 minutes it was a hard, slightly yellow solid. The solid was shaken with 100 ml. of acetone for 2.5 days, at which time part was dissolved and the remainder was highly swollen. Homogenization of the swollen acetone mixture in a high-speed, high-shear mixer produced a fluid mixture containing only a small amount of undissolved material. The homogenized mixture was filtered and the filtrate was diluted with an excess of distilled water, whereby a gummy coagulate was obtained which changed into a mass of stiff, coherent particles on prolonged stirring. The particles were isolated by filtration and dried in a vacuum oven at 50°C. for 90 minutes. The resultant slightly yellow product was readily broken up and reduced to a powder by grinding. It was soluble in acetone, with inherent viscosity (0.1% conc., 25°C.) of 2.02. Films of 5 to 6 mils thickness, molded at 165°C. and 500 pounds pressure, were clear, stiff and self-supporting.

Polymers prepared in the above manner were codissolved with polyhexamethylene adipamide (6,6-nylon) in perfluoroacetone hydrate, and films cast from the resultant solutions were clear and homogeneous.

Poly(Perfluoroalkyl)Methylene

A process described by R.N. Haszeldine and R. Fields; U.S. Patent 3,234,149; February 8, 1966; involves highly crystalline polymers with high melting points derived from fluorinated diazo compounds, and particularly to a poly(trifluoromethyl)methylene with a melting point in excess of 300°C. This polymer is particularly desirable because its melting point is close to that of the commercially important polytetrafluoroethylene and is higher than the melting point of any other known fluoropolymer. While 2,2,2-trifluorodiazoethane is the preferred starting material and the corresponding polymer, poly(trifluoromethyl)methylene having a melting point above 300°C. is the preferred product of the process. The method is applicable to the use of starting materials having the general formula:

$$R-CHN_2$$

where R is a radical selected from the group consisting of perfluoroalkyl, perfluorochloroalkyl, perfluoroaryl and perfluorochloroaryl radical. The most preferred starting materials are those compounds where R is a perfluoroalkyl group containing from 1 to about 4 carbon atoms such as 2,2,2-trifluorodiazoethane, 2,2,3,3,3-pentafluorodiazopropane, 2,2,3,3,4,4,4-heptafluorodiazobutane, and 2,2,3,3,4,4,5,5,5-nonafluorodiazopentane. The polymer obtained has the formula

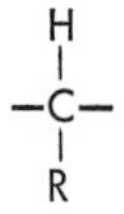

$$-\underset{\underset{R}{|}}{\overset{\overset{H}{|}}{C}}-$$

where R is a radical selected from the group consisting of perfluoroalkyl, perfluorochloroalkyl, perfluoroaryl and perfluorochloroaryl radical. Their high melting points and the sharpness of their infrared spectra compared with those of other polymers show that they are highly crystalline polymers.

The starting materials for the process can be prepared by conventional means such as the method employed by Gilman and Jones, J.A.C.S. 65, pp. 1458 to 1460, for the preparation of 2,2,2-trifluorodiazoethane. The following examples illustrate the process.

Example 1: 1.23 g. by weight of 2,2,2-trifluorodiazoethane is sealed in vacuo in a vessel constructed of "dreadnought" glass. The vessel is shielded from light and is maintained at from 18° to 26°C. for about 2 1/2 years after which time the pressure in the vessel is between 2 and 3 atmospheres. At the end of this period, the vessel is opened and the following products are obtained:

 0.137 g. by weight of nitrogen;
 0.39 g. of unchanged 2,2,2-trifluorodiazoethane equivalent to 32% of that contained in
   the vessel initially;
 0.41 g. by weight of a complex liquid mixture;
 0.234 g. of poly(trifluoromethyl)methylene, representing a yield of 38% calculated on a
   molar basis.

The poly(trifluoromethyl)methylene is recovered and analyzed as follows.

 Calculated: C, 29.3%
              H, 1.2%
 Found: C, 29.2%
              H, 1.1%

The melting point of the polymer was in excess of 300°C.

Example 2: 0.0478 g. of the poly(trifluoromethyl)methylene prepared in Example 1 is heated in vacuo in a sealed tube for 2 hours at 360°C. Analysis of the contents of the tube at the end of this period indicates 0.029 g. of trans-1,1,1,-4,4,4-hexafluorobutene-2 representing 61% yield based on the original polymer and calculated on a molar basis. A small quantity of charred residue remains in the tube.

Example 3: If 2,2,3,3,4,4,4-heptafluorodiazobutane is substituted for the 2,2,2-trifluorodiazoethane of Example 1, a solid polymeric material is obtained which is similar to poly(trifluoromethyl)methylene.

Allyl Fluoride

In a process described by S. Margulis, K. Haim and L.M. Shorr; U.S. Patent 3,524,839; August 18, 1970; assigned to Israel Mining Industries-Institute for Research and Development, Israel polymers may be prepared by polymerization or copolymerization of allyl fluoride in the presence of an initiator selected from the group of azo-compounds of the general formula

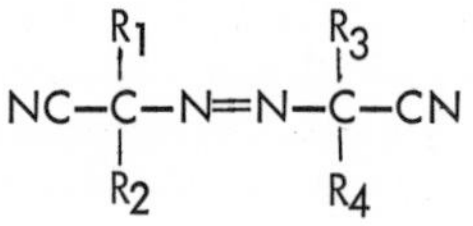

$$NC-\underset{\underset{R_2}{|}}{\overset{\overset{R_1}{|}}{C}}-N=N-\underset{\underset{R_4}{|}}{\overset{\overset{R_3}{|}}{C}}-CN$$

In the formula shown on the previous page R_1, R_2, R_3 and R_4 are each a hydrocarbon radical. It has also been found that in some instances it is advantageous to employ in addition photo excitation. Wavelengths of the order of 3000 to 5000 A. have been found suitable for this purpose. The process is illustrated by the following examples.

Example 1: Preparation of Allyl Fluoride Monomer — The allyl fluoride was prepared by the method described by F.W. Hoffman, J. Org. Chem. 14, 105-10 (1949) from allyl chloride and potassium fluoride in diethylene glycol.

Example 2: Polymerization of Allyl Fluoride — The polymerization runs were performed in glass pressure tubes of total volume 60 ml. sealed by a pressure valve. The reaction tube in which 2% by weight of azo-isobutryronitrile (AIBN) was introduced after previous sweeping with dry nitrogen, was cooled in CO_2-acetone and 5 ml. of allyl fluoride was then distilled into it at atmospheric pressure and exclusion of moisture. Before sealing the tube, dry nitrogen was again passed through it. The tubes were immersed in a constant water bath at 65° to 70°C. for 100 hours. At the end of the reaction time, the reaction tube was opened and the unreacted monomer recovered by distillation of the gases into a CO_2-acetone cooled trap.

The polymeric product contained 28.9% F. No acidic vapors were detected in the reaction mixture, which indicated that dehydrofluorination did not occur. The infrared analysis proved the existence of the polyallyl fluoride structure as well as the presence of nitrile groups. This proves that some of the initiator goes over into the product and remains there as contaminant, which accounts for the fact that fluorine content is somewhat below the theoretical value.

Example 3: Copolymerization of Allyl Fluoride with Vinyl Acetate — 0.1 g. of AIBN and 1 g. of freshly distilled vinyl acetate were introduced into a glass pressure tube of 60 ml. volume. The tube was swept with dry nitrogen and cooled in a CO_2-acetone bath. Approximately 5 ml. (4 g.) of pure allyl fluoride were distilled into the tube. After again sweeping with nitrogen the tube was sealed and heated at 75° to 80°C. for 100 hours. After cooling, the unreacted allyl fluoride was distilled out, first at room temperature, then by heating the reaction product up to 80°C. under vacuum and trapping the volatiles in CO_2-acetone. The remaining yellow viscous resin was analyzed by infrared spectroscopy in a chloroform solution which confirmed the presence of both polymerized allyl fluoride and vinyl acetate moieties. Its fluorine content was 26.0% thus corresponding to a copolymer composition of 82% by weight allyl fluoride and 18% by weight vinyl acetate.

POLYMER MODIFICATION WITH POLAR GROUPS

Carboxy Functional Elastomers — Alkylenimine Curatives

In a process described by N.L. Watkins, Jr.; U.S. Patent 3,198,770; August 3, 1965; assigned to Minnesota Mining and Manufacturing Company a highly fluorinated polymer having at least about two active hydrogens per molecule, particularly in the form of carboxyl or hydroxyl groups, is mixed with an alkylenimine derivative. This composition is then applied to a surface or is fabricated, such as by injecting the composition into a mold cavity, and is cured to form a highly fluorinated solid polymer. The curing reaction can be effected at room or ambient temperature, although higher temperatures may be employed, if desired, to accelerate the rate of cure.

The highly fluorinated polymers of this process have between about 30 and about 70% fluorine substitution and preferably are flowable (no volatiles present) at about room temperature. Although various highly fluorinated polymers can be employed, the vinyl addition type and the polyester type are preferred. It has been found that those highly fluorinated polymers with active hydrogen atoms in the form of hydroxyl or carboxyl groups, preferably carboxyl groups, are capable of being cured with the polyalkylenamides.

The presence of carboxyl groups on the fluorinated polymer chain tends to provide the more rapid cure rate and the most effectively cured product. Such carboxyl groups may be provided by introducing a monoolefinic hydrocarbon acid, such as acrylic, methacrylic, etc.; saturated hydrocarbon polycarboxylic acid, such as adipic, succinic, diglycolic, isophthalic, sebacic, azelaic, thiadipropionic, etc.; or their fluorohalocarbon counterparts, such as the trifluorochlorodicarboxylic acids of U.S. Patents 2,806,865 and 2,806,866, into the polymer chain.

Those fluorinated polymers with only terminal carboxylic or hydroxyl groups are cured by a chain extension mechanism, whereas fluorinated polymers with carboxyl or hydroxyl groups appended elsewhere on the central carbon chain are cured by a crosslinking mechanism. In those instances in which the highly fluorinated polymer has both terminal and nonterminal carboxyl or hydroxyl groups, chain extension and crosslinking may be effected simultaneously in the presence of the polyalkylenamides of this process.

The polyester type, highly fluorinated polymers are produced by reacting a dicarboxylic acid or the corresponding acid halide having up to about 33 carbon atoms, e.g., adipic acid, adipyl chloride, succinic acid, glutaric acid, suberic acid, azelayl chloride, 2,4,6-trichlorononafluorosuberic acid, 3,5-bis(perfluoropropyl)-4-thiapimelyl chloride, sebacic acid, isophthalyl chloride, 3-oxaglutaryl chloride, 3-perfluoropropylglutaryl chloride, 3-perfluoroheptylglutaryl chloride, 3,6-dithia-octanedioic acid, 3,5-dithia-heptanedioic acid, perfluoroadipic acid, etc., preferably an aliphatic hydrocarbon dicarboxylic acid, with an aliphatic diol, preferably a fluorinated aliphatic diol, having from 2 to about 12 carbon atoms, e.g., 2,2,3,3,4,4-hexafluoropentane diol, 2,2,3,3,4,4,5,5-octafluorohexane diol, 3,3'-oxy-bis-tetrafluoropropanol, N-ethyl perfluorooctane sulfonamido propylene glycol, etc. Fluorinated polyesters containing fluorine in the alcohol moiety are somewhat more stable to hydrolytic attack than those containing fluorine in the acid moiety.

The substantially liquid polyesters generally have an average molecular weight between about 700 and about 10,000. Examples A to C illustrate the preparation of such a polymer. If carboxyl-terminated polyesters are desired, an excess of the diacid, e.g., up to about 50% molar excess as compared to the diol, is used.

Example A: Preparation of the Diester Precursors — Ethylperfluoro-oxydipropionate having the formula

$$O(CF_2CF_2COOC_2H_5)_2$$

is prepared from a mixture of 14.2 grams of perfluoro-oxydipropionic acid (the preparation of which is described in U.S. Patent 2,839,513), 137 grams of ethanol and 260 grams of benzene. This mixture is heated in an apparatus equipped to remove the water-ethanol-benzene azeotrope which forms as the sterification proceeds. After removal of the azeotrope, distillation is continued to isolate the ester product, which has a vacuum boiling point of 75°C. at 3.5 mm. of pressure which is identified as ethylperfluoro-oxydipropionate.

Example B: Preparation of the Fluorinated Diol[3,3'-Oxy-Bis(1,1-Dihydrotetrafluoropropanol)] — About 500 ml. of tetrahydrofuran are charged to a two liter 3-necked flask fitted with a stirrer, dropping funnel and condenser. Thirty-eight grams (1.0 mol) of sodium borohydride are added and the mixture is agitated for several minutes. One hundred and eighty-one grams (0.5 mol) of ethylperfluoro-oxydipropionate are then added slowly, the rate of addition being adjusted to maintain the temperature of the reaction mixture at 45° to 50°C. The mixture is then agitated for approximately 1.5 hours and refluxed for an additional 2 hours (reflux temperature being approximately 67°C.). At the end of this time the mixture is cooled to room temperature and acidified with 60 ml. of 20% sulfuric acid.

The reaction mixture is filtered and the two phases of the resulting filtrate are separated. The lower of these two layers is cooled and filtered and the filtrate is combined with the upper layer. This combined material is then distilled (to remove the tetrahydrofuran) until a reflux temperature of 85° to 87°C. is reached. The remaining material is extracted with diethyl ether which is removed by heating the mixture to 80°C. and placing it under vacuum. The remaining material is filtered to remove a small amount of crystalline precipitate and the resulting filtrate distilled in a metroware Vigreux column. A 72.2 gram middle cut of this distillation (coming off at 84° to 86°C. at 2 mm. of pressure) is re-distilled to form a pure, water-white, viscous liquid product.

This purified 3,3'-oxy-bis(tetrafluoropropanol) which is obtained in 52% of theoretical yield from the diester is found to contain 26.0% of carbon (as compared to a calculated value of 25.9%) and 54.1% of fluorine (as compared to a calculated value of 54.7%).

Example C: Preparation of a Polyester of Adipic Acid and 3,3'-Oxy-Bis(Tetrafluoropropanol) — The apparatus for this polyester preparation consists of a 100 ml. flask fitted with a thermometer, a gas addition tube and a water cooled condenser. 16.1803 grams (0.08842 mol) of redistilled adipoyl dichloride, a mobile liquid, and 24.5752 grams (0.08842 mol) of redistilled 3,3'-oxy-bis(tetrafluoropropanol), a viscous, water-white liquid, are charged to the flask, the latter being washed into the flask with diethyl ether. A nitrogen sweep of the reaction mixture is initiated along with a gradual warming of the reaction mixture from ambient temperature. The flow of cooling water through the condenser is not begun until all of the diethyl ether added to wash the fluorinated diol into the reaction flask has evaporated and been swept from the reaction flask. After the diethyl ether has been removed and the flow of condenser water initiated, the gradual increase in the temperature of the reaction mixture is continued, the rate being approximately 100°C. per hour. The exit stream of the nitrogen sweep is checked periodically for hydrogen chloride gas which is formed in the reaction (and the presence of which indicates that the reaction is not as yet complete).

Approximately 70 minutes after initiation of the heating, the temperature of the reaction mixture has reached 150°C. and the temperature is maintained approximately at that point throughout the remainder of the reaction period. After a total reaction time of approximately 3 1/2 hours it is observed that the exit nitrogen stream is free of hydrogen chloride. The pressure on the reaction mixture is reduced to approximately 100 mm. for a short time in order to eliminate all of the dissolved gases and the reaction is terminated. The resulting polyester is an extremely viscous liquid which does not solidify upon cooling. This material may be characterized as follows:

Number of active hydrogen atoms per molecule	2
Active hydrogen content (milliequivalents per gram)	.408
Equivalent weight per active hydrogen	2,450

Other Fluorinated Monomers and Polymers

The preparation of adipic acid-2,2,3,3,4,4-hexafluoropentane diol and adipic acid-2,2,3,3,4,4,5,5-octafluoro-hexanediol polyesters is achieved in a similar fashion. The polyvinyl type polymers are produced by reacting an alpha-beta unsaturated carboxylic acid, such as acrylic, crotonic, methacrylic acid, etc., with one or more aliphatic, at least half fluorinated terminally unsaturated mono- or di-olefinic compounds, such as tetrafluoroethylene, trifluorochloroethylene, vinylidene fluoride, perfluoropropene, 1,1-dihydroperfluorobutyl acrylate, 1,1,3-trifluorobutadiene, 3-perfluoromethoxy-1,1-dihydroperfluoropropyl acrylate, etc., preferably fluorinated monoolefins having less than about 10 carbon atoms.

These polymers may be cured by a crosslinking reaction. The substantially liquid vinyl type polymers have a molecular weight range below about 50,000. Illustrative of these polyvinyl type polymers are the copolymer of 1,1-dihydroperfluorobutyl acrylate and acrylic acid, the copolymer of 3-perfluoromethoxy-1,1-dihydroperfluoropropyl acrylate and acrylic acid, and the terpolymer of perfluoropropene, acrylic acid and vinylidene fluoride. U.S. Patents 2,642,416, 2,803,615, 2,826,564 and 2,839,513 describe a method for preparing the above and other fluorinated acrylates and polymers thereof. U.S. Patent 3,080,347 shows a method for preparing the terpolymer of perfluoropropene, vinylidene fluoride, and an alpha-beta unsaturated carboxylic acid, which includes the polymerization of between about 15 and about 50 mol percent of perfluoropropene, between about 50 and about 85 mol percent of vinylidene fluoride, and between about 0.1 and about 10 mol percent of an alpha-beta unsaturated monocarboxylic acid at a temperature between about 50°C. and 200°C. in the presence of a chain transfer agent, such as carbon tetrachloride or dodecyl mercaptan.

The curing agents employed in this process are alkylenimine derivatives of the formula

$$Q\left[N\underset{CR'R''}{\overset{CH_2}{<}}\right]_n$$

where Q is an n valent radical,
(n is 2 or more, preferably 2 or 3),
N is linked to an atom having a valence of 4 or 5, and
R' and R'' are hydrogen or an alkyl group preferably having
    from 1 to 4 carbon atoms.

Q may be an aliphatic, aromatic or alicyclic organic radical which does not contain an active hydrogen, but which may contain atoms other than carbon, such as oxygen, sulfur, etc. Q may also be an inorganic radical, such as

$$-\overset{|}{\underset{|}{P}}=O \quad and \quad -\overset{|}{\underset{|}{P}}=S$$

In the preferred curing agents Q is selected from the groups consisting of

$$-\overset{|}{\underset{|}{P}}=O, \quad -\overset{|}{\underset{|}{P}}=S \qquad and \qquad -\overset{O}{\overset{\|}{C}}-(Y)_x-R-(Y)_x-\overset{O}{\overset{\|}{C}}-$$

where Y is either O or —NH—,
x is either 0 or 1, and
R is a divalent aliphatic, aromatic or alicyclic radical, which
    may contain atoms other than carbon, e.g., oxygen, sulfur, etc.

The phosphorus-containing alkylenimine derivatives include, for example, tris(1-aziridinyl)phosphine oxide, tris(1-aziridinyl)phosphine sulfide, N,N-diethyl-N',N''-diethylenethiophosphoramide and N,N'-diethylbenzene thiophosphondiamide.

The carbonyl-containing curing agents which are particularly preferred have the formula

$$\underset{R'_2C}{\overset{H_2C}{>}}N\overset{O}{\overset{\|}{C}}-(Y)_x-R-(Y)_x-\overset{O}{\overset{\|}{C}}N\underset{CR'_2}{\overset{CH_2}{<}}$$

In the carbonyl-containing curing agent whose formula is shown on the preceding page,

Y is either oxygen or —NH—,
x is either 0 or 1,
R' is hydrogen or a lower alkyl group (i.e., a hydrocarbon radical having
   from 1 to 4 carbon atoms), and
R is a divalent aliphatic, aromatic or alicyclic radical.

R may contain atoms other than carbon, such as oxygen and sulfur, but does not contain an active hydrogen, i.e., a hydrogen which is active to the Zerewitinoff test (inert to Grignard reagents). When x is 0, the compound is a bis-1,2-alkylenamide. When Y is O and x is 1, the compound is a bis-1,2-alkylene carbamate. When Y is —NH— and x is 1, the compound is a bis-1,2-alkylene urea, such as 1,6-hexamethylene N,N'-diethylene urea.

Bis-1,2-alkylene carbamates and their preparation are described in U.S. Serial No. 850,541, filed November 3, 1959. Generally, their preparation involves the reaction of a 1,2-alkylenimine in a water phase with a solution of a chloro-carbonate of a difunctional alcohol in a water immiscible organic solvent, in the presence of an acid acceptor, at a temperature between about -5°C. and 30°C. Illustrative of the bis-carbamates which are useful as curing agents are:

N,N'-bis-1,2-ethylene(1,4-butanediol)carbamate;
N,N'-bis-1,2-propylene(1,4-butanediol)carbamate;
N,N'-bis-1,2-butylene(1,4-butanediol)carbamate;
N,N'-bis-1,2-butylene(diethylene glycol)carbamate;
N,N'-bis-1,2-propylene(triethylene glycol)carbamate;
N,N'-bis-1,2-butylene(triethylene glycol)carbamate;
N,N'-bis-1,2-ethylene(polyethylene glycol-200)carbamate; and
N,N'-bis-1,2-ethylene(polyethylene glycol-1000)carbamate.

Illustrative of the N,N'-bis-1,2-alkylenamides are:

N,N'-bis-1,2-ethylenadipamide;
N,N'-bis-ethylenpentadecyladipamide;
N,N'-bis-1,2-butylenadipamide;
N,N'-bis-1,2-ethylenepimelamide;
N,N'-bis-ethylene thiodipropionamide;
N,N'-bis-ethylene oxydipropionamide;
N,N'-bis-1,2-ethyleneisosebacamide.

When the curable compositions of this process are used to provide a protective coating, they may be coated onto such surfaces as textile fabrics (e.g., nylon, fiberglass, polyacrylonitrile, ethylene glycol-terephthalic acid copolymer, acrylic copolymer, cotton, etc.), metals (e.g., copper, aluminum, steel, etc.), glass or other ceramics, synthetic resins (including the halogenated resins, e.g., polytetrafluoroethylene, etc.), wood, leather, paper, synthetic rubber and cork. It may also be used to impregnate the surface when a porous material is used, such as paper, textile fibers, and leather. Impregnation is aided by the use of a suitable solvent.

These curable compositions are readily adaptable for injection into grooves surrounding the fuel cavity of many aircraft fuel tank designs, to insure a leakproof and temperature resistant seal against hydrocarbon jet fuels as well as vapor pressure developed in the tank resulting from aerodynamic heating during flight. The cured sealant compositions are inert and resist degradation by hydrocarbon fuels and long term high temperature exposure. Other uses as aircraft seal-ants include use in pressurized cabins, and as gaskets, valve diaphragms and O-rings, and as an interlining or edge sealer for laminated windshields. They provide an excellent adhesive for highly fluorinated polymers, such as Teflon.

The following examples illustrate the process. In the examples [η] is equal to

$$\frac{\ln \frac{\eta \text{ solution}}{\eta \text{ solvent}}}{C}$$

where C is the concentration of polymer in grams per 100 ml. of
   solution, and
η solution and η solvent are viscosities in consistent units.

Example 1: A linear, tacky, viscous, liquid fluorinated polymer was prepared by reacting equimolar proportions of 3,3'-oxy-bis-tetrafluoropropanol and adipic acid, the polymer having a neutralization equivalent of 4900 and 41%

fluorine content. A portion of this fluorinated polymer (0.980 gram) was dissolved in 6 ml. of chloroform, and 0.0672 g. of N,N'-bisethylene isosebacamide (25% excess over theoretical) was added at room temperature in an aluminum cup. The solution was evaporated slowly, and the polymer deposited as a film in the cup. This film was cured for 1 3/4 hours at 120°C. to produce a tack-free solid rubber (0.8 gram). This rubber was milled with 0.04 gram of dicumyl peroxide and cured at 310°F. for 30 minutes to form a snappy rubber with good solvent resistance.

Isosebacic acid, of which the above isosebacamide is a derivative, consists of 72 to 80% of 2-ethylsuberic acid, 12 to 18% of 2,4-diethyladipic acid and 6 to 10% of n-sebacic acid.

Example 2: Using the fluorinated polyester of Example 1, 5.16 grams of polymer were combined with 0.282 gram of N,N'-bisethylene isosebacamide, and the dried film was cured at 120°C. for 75 minutes. A crosslinked rubber was produced which could not be dissolved in chloroform even after a 10 day immersion at room temperature. Physical properties were as follows:

Tensile strength, psi	66
Elongation, %	180
Set at break, %	0
Gehman T_{10} (ASTM D1053-54T), °F.	40
Brittle point, °F.	72
Volume swell at 25°C. in Type B fuel (ASTM D471-55T), %	30

Example 3: One hundred parts of a liquid copolymer $[\eta]$ or intrinsic viscosity of 0.07 (in xylene hexafluoride) of 99% by weight 1,1-dihydroperfluorobutylacrylate and 1% by weight of acrylic acid (51.8% fluorine, prepared in bulk using dodecyl mercaptan as a regulator) were mixed with 6 parts by weight of N,N'-bisethylene sebacamide. The mixture was poured into an open steel mold and allowed to cure for 24 hours at room temperature. A tacky, rubbery solid was formed which could be removed from the mold and exhibited no cold flow on standing. The solid is not rendered thermoplastic by heating to 350°F. and is insoluble in xylene hexafluoride.

Example 4: A terpolymer of 58 parts by weight of perfluoropropene, 4 parts by weight of acrylic acid and 38 parts by weight of vinylidene fluoride (63.4% fluorine) was prepared. The intrinsic viscosity (taken using acetone as the solvent) was 0.030. Nine parts by weight of N,N'-bisethylene isosebacamide were mixed thoroughly into the terpolymer and the mixture was poured onto 17-7 stainless steel panels and cured for 4 hours at 250°F. The elastomer sheet was divided into 3 portions, A, B, and C. Evaluation:

A. After additional cure cycle of 119 hours at 400°F. —

Weight loss	1.1%
Appearance	Little change
Adherence to stainless steel	Good

B. After additional cure cycle of 117 hours at 500°F. —

Weight loss	5.0%
Appearance	Darkened, bubbled, flexible
Adherence to stainless steel	Adhesion lost

C. After additional cure cycle of 72 hours at 180°F. submerged in 70% isooctane:30% toluene —

Percent volume swelling	1.6%
Adherence to stainless steel	Good

Alkylene Oxide Copolymers

M. Hauptschein; U.S. Patent 3,004,961; October 17, 1961; assigned to Pennsalt Chemicals Corporation has found that certain fluorinated unsaturated compounds can be copolymerized with certain alkylene oxides to give new polymeric products useful as plastics, lubricants, elastomers and protective coatings. The polymers containing a number of oxygen atoms in the molecule are, in effect, internally plasticized. Moreover, by copolymerization with alkylene oxide, it has been found possible to obtain with relative ease polymeric products from fluoroolefins and diolefins which had previously been considered extremely difficult to polymerize.

Specific fluorinated olefins and diolefins which have been found especially satisfactory for polymerization include, for example, trifluorochloroethylene, (CF_2=CFCl); perfluoropropene; (CF_2=CFCF$_3$); 1,1-difluoro-2-chloroethene, (CF_2=CHCl); and perfluorobutadiene, (CF_2=CFCF=CF$_2$). Examples of suitable alkylene oxides are ethylene oxide and propylene oxide.

It should be pointed out that products obtained in accordance with the process are true copolymers; i.e., the molecules are chains comprising a number of alternating olefin or diolefin and alkylene oxide units. They are therefore distinguished from polymeric materials formed by telomerization of olefins or diolefins, where the molecule is a chain of repeating olefin units with telogen terminal units. They are also distinguished from homopolymers where the molecule is a chain having only one repeating unit. The process is illustrated by the following examples.

Example 1: Chlorotrifluoroethylene (9.5 g. or 0.082 mol) and ethylene oxide (13.4 g. or 0.3 mol) were sealed in a Pyrex tube and exposed to a Hanovia ultraviolet lamp, 5 cm. from the tube, for 63 hours. 16.5 g. of a colorless viscous liquid was collected. No unreacted chlorotrifluoroethylene was detected. The liquid was dissolved in dichloromethane and extracted exhaustively with water. The lower layer obtained by extraction was separated, dried with anhydrous sodium sulfate and methylene chloride was evaporated off. The resulting jelly-like solid was heated at 100° in vacuo in an Abderhalden pistol. Upon analysis, the composition of the jelly was found to be (percent by weight) carbon 30.82; hydrogen 2.95. This corresponds to 46.8 mol percent of chlorotrifluoroethylene in the copolymer.

Example 2: Perfluoropropene (62 g.), ethylene oxide (47 g.) and di-tertiary-butyl peroxide (6 g.) were heated at 145°C. for 72 hours. Approximately 54 g. of an oily polymer having a boiling point above 45°C. at below 0.5 mm. Hg were collected. Upon fractionation, it was found that about 35% of this material boiled at between 45° and 100°C. at 0.5 mm. Hg. This fraction had a refractive index of $n_D^{24} = 1.3534$ and upon analysis was found to comprise (percent by weight) carbon 33.09; hydrogen 2.82, corresponding to 41.0 mol percent of perfluoropropene in the copolymer. The viscosity of this material at various temperatures was determined. These viscosities were 2.22 centistokes (cs.) at 212°F., 61.15 cs. at 62.0°F. and 279.3 cs. at 32.0°F.

These values lie on a nearly straight line when plotted on an ASTM (D341–43) viscosity chart. The slope of this line, showing the temperature dependence of viscosity, i.e., the so-called ASTM slope, was 1.1. This shows improvement over other fluorocarbon oils.

A second fraction of the crude polymer, comprising about 50% of the crude, was found to boil at between 100° and 145°C. at 0.5 mm. The refractive index of this fraction was $n_D^{24} = 1.3779$. Its densities were $d_4^{22} = 1.466$, $d_4^0 = 1.494$. Upon analysis it was found to comprise (percent by weight) carbon 35.82; hydrogen 3.27; fluorine 45.6. This corresponds to 31.8 mol percent of perfluoropropene in the copolymer.

Example 3: In a dry nitrogen atmosphere 11 g. (0.25 mol) of ethylene oxide and 37 g. (0.23 mol) of perfluorobutadiene were introduced into a heavy-walled Pyrex ampoule which was then sealed and exposed to a Hanovia ultraviolet lamp at about 5 cm. distance for one week. In addition to some unreacted starting materials, there was collected 8 g. of a polymeric residue. Upon analysis this was found to comprise carbon 33.83%; hydrogen 1.85%; fluorine 56.96%, these percentages being by weight. This corresponds to 57 mol percent of perfluorobutadiene.

The product was a pale buff paste. Upon heating at 0.1 mm. pressure with a free Bunsen flame, it remained stable, neither coloring nor melting to a clear liquid. Infrared examination indicated the presence of CF_2 and CH_2 groups, with substantial amounts of unsaturation. The polymer is suitable for vulcanization to form a highly stable rubber.

Polymeric Peroxides

J.L. Anderson and R.E. Putnam; U.S. Patent 2,971,949; February 14, 1961; assigned to E.I. du Pont de Nemours and Company describe the preparation of copolymers of oxygen and 1,1,4,4-tetrafluoro-1,3-butadienes having at least one hydrogen bonded to the 2 and 3 carbons. The copolymers are prepared by contacting a 1,1,4,4-tetrafluoro-1,3-butadiene of the formula

$$\underset{\substack{| \\ F}}{\overset{\substack{F \\ |}}{C}}=\underset{\substack{| \\ Y}}{C}-\underset{\substack{| \\ H}}{C}=\underset{\substack{| \\ F}}{\overset{\substack{F \\ |}}{C}}$$

where Y is hydrogen, a halogen, carboxyl, or a hydrocarbon or halohydrocarbon group, preferably of 1 to 8 carbons in which the halogen has an atomic number of 9 to 35, inclusive, with at least 10 mol percent of oxygen, at a temperature in the range of –40° to 30°C. The following examples illustrate the process.

Example 1: A stainless steel shaker tube of about 80 parts water capacity, cooled to –50°C., is charged with 7 parts of 1,1,4,4-tetrafluoro-1,3-butadiene, and oxygen is added to a total pressure of 2,410 lb./sq. in. in the tube. The tube is then shaken for 16 hours at room temperature (about 30°C.). At the end of this period the tube is vented to release excess oxygen and any unpolymerized tetrafluorobutadiene. There is obtained 7 parts of white solid polymer as a tough film of the walls of the shaker tube. This polymer is soluble in acetone; it explodes violently when heated to 122°C. and it readily liberates iodine from potassium iodide in aqueous acetone. Elemental analysis and chemical reactivity show this polymer to be a 1:1 copolymer of 1,1,4,4-tetrafluoro-1,3-butadiene and oxygen of predominantly peroxidic structure. Analysis — Calculated for $C_4F_4H_2O_2$: C, 30.4%; H, 1.3%; F, 48.1. Found: C, 30.5%; H, 1.6%; F, 48.4%.

Example 2: A stainless steel shaker tube of about 80 parts water capacity is charged with 4.5 parts of 1,1,4,4-tetra-fluoro-1,3-butadiene and is then pressured to 495 lb./sq. in. with oxygen. The closed tube is then shaken at 28°C. for a period of 16 hours. After the tube is opened there is obtained 5 parts of a white polymer as a tough film on the walls of the tube. This polymer is soluble in acetone. A sample of the polymer is dissolved in acetone and acetic acid, and is digested with hydrogen iodide. The liberated iodine is then titrated with standard sodium thiosulfate solution. The results of this treatment indicate that the polymer contains 1.1 mols of peroxide per mol of diene. The high peroxide content indicates that the polymer is cross-linked to some extent by peroxide bridges.

Polyfluorocarbon Oxides

N. Kowanko; U.S. Patent 3,423,364; January 21, 1969; assigned to Minnesota Mining and Manufacturing Company has found that fluorocarbon carbonyl compounds in combination with perfluoroolefins react with ozone giving rise to polyhalocarbon oxides having significantly modified structure and properties. The mechanism for entry by the perfluoro-carbon carbonyl compound into this polymerization process is unknown. This process provides a convenient one-step synthesis for making polyfluorocarbon oxides and does not require the use of light or relatively expensive starting materials.

In the process, three classes of starting material are necessarily employed. One of the three classes of starting material are perfluoroolefins. Preferably each perfluoroolefin molecule contains from about 2 through 15 carbon atoms. A more preferred class of perfluoroinated olefins are those containing from 3 through 6 carbon atoms. A most preferred perhalo-olefin is perfluoropropene.

A second class of starting material used in the process are perfluorocarbon carbonyl compounds. A preferred class of perfluorocarbon carbonyl compounds are polyfluorocarbon ketones containing from 3 to 15 carbon atoms. Another pre-ferred class of perfluorocarbon carbonyl compounds are carboxylic acid fluorides. A preferred acid fluoride is perfluoro-butyryl fluoride.

A third class of starting material used in the process is a reagent mixture containing as the essential active (reactive) component at least about 0.001 weight percent ozone (based on total reagent mixture weight). This reagent mixture prior to being used in the process can be in the gas phase, in the liquid phase, or can constitute a combination thereof. The ozone can be derived from any conventional source. Oxygen can be optionally present in this reagent mixture.

In general, the process proceeds by first blending the starting materials of both the first and second class together and then contacting the resulting mixture (e.g., ozone) so that a polymerization reaction results. The starting materials of the first and second class are generally present in a molar ratio ranging between 1 and 0. Preferred molar ratio ranges between 0.2 and 0.05.

This reaction is highly exothermic, one consequence of which is that it is desirable to limit the maximum quantity of unreacted ozone present in a reaction zone at any given time relative to the total amount of perfluoroolefin and per-fluorocarbon carbonyl compound present to a level which will prevent excessive, even explosive, rates of reaction from occurring. The process can be practiced either batchwise or continuously.

The process is illustrated by the following example. It will be noted that in the examples, on the basis of the available F^{19} NMR and infrared spectroscopic data, and except for a change both in the functional end groups and in the labile end sequences, a backbone polymer structure is substantially unaltered by hydrolysis.

Example: A gaseous mixture comprising O_2 (95 to 98%) and O_3 (2 to 5%), prepared by passing oxygen through an electric discharge, is bubbled slowly (0.01 to 0.02 cu. ft./min.) through a refluxing mixture of n-perfluorobutyryl fluoride (1.2 ml.) and perfluoropropene (6 ml.) in a borosilicate glass vessel provided with an efficient Dry Ice (solid CO_2) condenser. The reaction mixture is thus maintained at a temperature of about -30°C. A vigorous reaction is observed in this instance within two minutes of the time of the introduction of ozone into the n-perfluorobutyryl fluor-ide-perfluoropropene mixture.

The onset of this reaction is evidenced by evolution of heat and of gases which fume in air (analysis shows the evolved gases to contain COF_2, CF_3COF, and other unidentified compounds), and by rapid uptake of the incoming gases by the reaction mixture. After about 50 minutes the reaction subsides and reflux stops. A product (oil, 0.1 to 0.2 ml.) remains after vacuum stripping at 10^{-4} mm. Hg to remove and separate the more volatile components which are largely of low molecular weight and examined separately.

The product oil is washed by shaking with 1 ml. distilled water. After separating the water is found to contain fluoride ion. The F^{19} NMR spectrum of the water-washed polymer in $CFCl_3$ shows the presence of the following bands:

11-14 (V.W.)	80 (S)
53 (W.)	86-89 (W., broad)
55 (M.)	134-137 (V.W. broad)
57-59 (W.)	145 (W.)
75 (W.)	

where (V.W.) stands for very weak; (W.) stands for weak; (M) stands for medium; (S) stands for strong and (VS) stands for very strong bands of the particular chemical shift.

In the above case the spectrum is complex with small unassigned peaks. Typically, the fluorine atoms of (—OCOF) end groups give rise to bands in the region 11-14 $\phi*$. The fluorine atoms of (—CF$_2$—O—), (CF$_3$—O—), and like groups, give rise to bands in the region 53-59 $\phi*$. The NMR spectrum clearly indicates the pressure of a complex polyfluoro-carbon oxide chain structure incorporating perfluorobutyryl fluoride.

The infrared spectrum give absorption maxima at 5.6, 7.6-9.5, 10.2, and 13.4μ, indicating the presence of COOH, C—F, and C—CF$_3$ groups. The overall appearance of the spectrum, especially in the intense 7.6-9.5μ region, further demonstrates that the structure of the polymer has been modified by the presence of perfluorobutyryl fluoride.

In general, the polyfluorocarbon oxides and the hydrolyzed products which are prepared according to the process are useful as intermediates for the manufacture of the following two classes of products:

(1) Stabilized materials, especially liquids, useful, for instance, as heat transfer media or lubricants, and

(2) Cured materials and curable formulations, useful, for instance, in sealant systems and adhesive systems, and as elastomeric materials of construction, particularly where high temperature serviceability in corrosive environments is required.

Thiocarbonyl Fluorides

A process described by W.J. Middleton; U.S. Patent 3,240,765; March 15, 1966; assigned to E.I. du Pont de Nemours and Company involves polymers of thiocarbonyl fluoride and thiocarbonyl chloride fluoride, including copolymers of these monomers with other copolymerizable monomers. These polymers have recurring structural units of the formula

$$-S-\overset{\displaystyle F}{\underset{\displaystyle X}{C}}-$$

where X is fluorine or chlorine.

These polymers range from low molecular weight, i.e., about 2,000, waxy solids melting at 30° to 40°C. to high molecular weight, tough, solid polymers softening at temperatures up to 230°C. These polymers have inherent viscosities, measured at 0.1% concentration in chloroform at 25°C., of up to about 5. The solid polymers can be pressed at elevated temperatures and pressures into rubbery films. Some of these pressed films crystallize on standing at ordinary temperatures to opaque plastic films.

In brief, a method of preparing the polymers is as follows: Thiophosgene dimer (tetrachloro-1,3-dithietane) is fluorinated and the resulting polyfluoro-1,3-dithietane is pyrolyzed. The resulting thiocarbonyl fluoride is then polymerized by any of a variety of methods. The preparation of the polyfluoro-1,3-dithietane involves the reaction of tetrachloro-1,3-dithietane with antimony trifluoride.

The proportions should be such that there is an amount of antimony trifluoride at least stoichiometrically equivalent to the tetrachlorodithietane present. Preferably an excess of antimony trifluoride, e.g., up to 100% excess or more can be present. A reaction medium which is a solvent for the antimony trifluoride, such as tetramethylene sulfone, is ordinarily present. Both tetrafluoro-1,3-dithietane and monochlorotrifluoro-1,3-dithietane are produced.

Thiocarbonyl fluorides are produced by the pyrolysis of the polyfluoro-1,3-dithietanes. Pyrolysis temperatures range from about 400° to 900°C., with temperatures of 450° to 500°C. being especially suitable. A convenient way of carrying out the pyrolysis is by passing the polyfluoro-1,3-dithietane through a reaction tube constructed of an inert metal such as nickel or platinum which has been heated to the reaction temperature. Preferably an inert diluent such as helium or nitrogen is passed through the reaction tube concurrently with the polyfluoro-1,3-dithietane as this improves the yield of the desired product. When tetrafluoro-1,3-dithietane is pyrolyzed as described above, thiocarbonyl fluoride, $S=CF_2$, containing less than 5% of sulfur- or fluorine-bearing non-polymerizable impurities is formed. This high degree of purity renders the product especially useful for polymerization. The process is illustrated by the following examples.

Example 1: Preparation of Thiocarbonylfluoride by Pyrolysis of Tetrafluoro-1,3-Dithietane — 40 g. of tetrafluoro-1,3-dithietane, whose physical properties are as follows,

$$BP\ 49°C,\ MP\ -6°C.,\ n_D^{25}\ 1.3908,\ d_4^{20}\ 1.6036,$$

having been purified by vigorous agitation with a mixture of 5 parts of 10% aqueous sodium hydroxide solution and 1 part of 30% aqueous hydrogen peroxide until the yellow color disappears, followed by drying of the lower organic layer over silica gel and distillation, is added dropwise over a period of 2 hours to the top of a platinum tube 1/2 inch in diameter and 25 inches long inclined at an angle of 30° to the horizontal and heated to 500°C. over a length of 12 inches.

A slow stream of helium (20 ml. per minute) is passed through the tube during the pyrolysis. The effluent gases are condensed in successive traps cooled by a mixture of acetone and carbon dioxide and liquid nitrogen, respectively. The material in the traps is combined and distilled through a 16-inch column packed with glass helixes. There is obtained 34.0 g. (85% yield) of a colorless liquid, boiling at -54°C. Analysis of this product by the mass spectrometer indicates it to be thiocarbonyl fluoride, CSF_2, of 98% purity.

Example 2: The preparation of thiocarbonyl chloride fluoride by pyrolysis of chlorotrifluoro-1,3-dithietane is as follows. Using the apparatus and procedure of Example 1, 9.03 g. of chlorotrifluoro-1,3-dithietane is pyrolyzed during a period of 1 hour at 500°C. There is obtained by distillation of the condensates in the two cold traps 4.3 g. of nearly colorless thiocarbonyl fluoride, boiling at -54°C., 1.18 g. of yellow thiocarbonyl chloride fluoride, CSFCl, boiling at 6°C. and 1.38 g. of red thiophosgene, $CSCl_2$, boiling at 72° to 74°C. Thiocarbonyl fluoride and thiocarbonyl chloride fluoride are identified by their nuclear magnetic resonance spectra and their mass spectrometer patterns.

Example 3: Poly(thiocarbonyl fluoride) is prepared in the following manner. Thiocarbonyl fluoride of 95.5% purity is placed in a stainless steel container and stored at room temperature for 6 weeks. The container is then opened and a clear syrupy residue is removed. After storage in a vacuum desiccator for 2 days, this residue solidifies to a white wax melting at 30° to 35°C. The polymer is not visibly affected by boiling concentrated nitric acid or by boiling 10% sodium hydroxide solution. The poly(thiocarbonyl fluoride) is soluble in chloroform and it can be reprecipitated from the chloroform solution by addition of methyl alcohol.

Analysis — Calculated for $(CF_2S)_n$: C, 14.63%; F, 46.30%; S, 39.06%. Found: C, 14.79%; F, 46.62%; S, 39.11%. The molecular weight determination was made by dissolving a weighed amount of polymer in tetrachlorodifluoroethane, measuring the freezing point depression and calculating the molecular weight by standard procedures. In this example, 0.176 g. of polymer dissolved in 47.8 g. of tetrachlorodifluoroethane depressed the freezing point 0.077°C. This corresponds to a molecular weight of 2150.

N,N-Bis(Perfluoroalkyl)Aminoethylenes

F.S. Fawcett; U.S. Patent 3,311,599; March 28, 1967; assigned to E.I. du Pont de Nemours and Company describes the preparation and polymerization of N,N-bis(perfluoroalkyl)aminoethylenes having the general formula

$$(R_f)_2NCX'=CXX'$$

Each R_f, which can be the same or different, is perfluoroalkyl of up to
6 carbon atoms,
X is hydrogen or fluorine,
X' is hydrogen, fluorine, chlorine or $(R_f)_2N-$,

with the proviso that the two R_f substituents and the two X' substituents need not be, respectively, the same, and that not more than one X' is $(R_f)_2N-$.

Among the unusual properties possessed by the N,N-bis(perfluoroalkyl)aminoethylenes is their hydrolytic stability. These compounds are not hydrolyzed readily by water as are the known N,N-bis(alkyl)aminoethylenes having halogens on the vinyl group. The products of this process can also be polymerized to homopolymers and to copolymers with one or more other polymerizable ethylenic compounds.

The N,N-bis(perfluoroalkyl)aminoethylenes can be polymerized to colorless, solid polymers by bulk or solution polymerization, alone or in the presence of one or more polymerizable ethylenic compounds, at moderately elevated temperatures in the presence of free radical liberating initiators. For example, N-vinylbis(trifluoromethyl)amine polymerizers in the presence of dinitrogen difluoride at 50° to 60°C. under 3,000 atm. pressure to a colorless, solid polymer which is moldable at 100° to 150°C. into stiff, clear, transparent, tough films. Copolymers are prepared by copolymerizing a mixture of the N,N-bis(perfluoroalkyl)aminoethylene with one or more other polymerizable ethylenic compounds, e.g., ethylene, under similar conditions. The N-(trifluorovinyl)bis(trifluoromethyl)amine and ethylene form a copolymer when treated with dinitrogen difluoride at 60°C. under 3,000 atm. pressure. The following examples illustrate the process.

Example 1:   Preparation of N-Vinylbis(Trifluoromethyl)Amine —   A. A mixture of 20 g. of N-chlorobis(trifluoromethyl) amine and 6 g. of ethylene is placed in a 240 ml. shaker tube lined with the corrosion-resistant alloy known as "Hastelloy" and is heated at 100°C. for 2 hours, 125°C. for 2 hours and 150°C. for 4 hours. The volatile reaction product that is formed is collected in an evacuated cylinder and it amounts to 20 g. Distillation of this product gives 11.4 g. (50% conversion) of colorless 1:1 adduct, N-(2-chloroethyl)bis(trifluoromethyl)amine, BP 28° to 30°C./143 mm., n_D^{24} 1.3203. The fluorine nuclear magnetic resonance spectrum shows a single resonance in the CF_3N— region and the proton magnetic resonance spectrum shows a peak for CH_2.

Analysis — Calculated for $C_4F_6H_4NCl$: C, 22.3%; H, 1.87%; F, 53.0%; N, 6.52%; Cl, 16.5%

Found: C, 22.83%; H, 2.25%; F, 52.49%; N, 6.36%; Cl, 17.09%.

B. A pyrolysis tube, consisting of an "Inconel" tube 3/8 in. in inside diameter and 6 inches long packed with nickel gauze and with the packed section heated by means of an electric heater, is connected directly to a gas chromatographic column (1/4 in. diameter and 12 feet long, packed with the ethyl ester of the acid known commercially as Kel-F Acid 8114 on firebrick and operated at 50°C.) Provision is made for collecting samples of the gas stream at the times when material is being eluted from the chromatographic column. With the pyrolysis tube heated at 650°C., a stream of helium at a rate of 60 ml./min. is passed through the tube and then into the gas chromatographic column. N-(2-chloroethyl)bis(trifluoromethyl)amine (20 microliters of liquid) is introduced by means of a hypodermic needle into the gas stream which carries the reactant through the heated zone and the pyrolysis products are delivered to the gas chromatographic column.

A sample of the material eluted at 4.2 min. after injection time is collected and found by mass spectrometric analysis to be N-vinylbis(trifluoromethyl)amine, $(CF_3)_2N$—CH=CH_2.

Example 2:   Polymerization of N-Vinylbis(Trifluoromethyl)Amine — A mixture of 2 g. of N-vinylbis(trifluoromethyl) amine and 5.8 milligrams of dinitrogen difluoride in a sealed platinum tube is heated for 6 hours at 50° to 60°C. under 3,000 atm. pressure. There is obtained 1.83 g. of colorless, solid polymer. This polymer is insoluble in hot N,N-dimethylformamide or hot toluene and is not affected by cold concentrated sulfuric or nitric acid, or by n-butylamine at its boiling point. The polymer is molded at 100° to 150°C. to a stiff, clear, transparent, tough, self-supporting film.

Example 3:   Preparation of a Copolymer of N-(Trifluorovinyl)Bis-(Trifluoromethyl)Amine and Ethylene — A mixture of approximately 2.8 g. of N-(trifluorovinyl)bis(trifluoromethyl)amine, 0.3 g. of ethylene and 6 mg. of dinitrogen difluoride (all measured as gases at 1 atm., 300 cc, 300 cc, and 2 cc volumes, respectively) is placed in a platinum tube at liquid nitrogen temperature. The tube is sealed and the mixture is then heated with shaking at 60°C. for four hours under 3,000 atm. pressure. On opening the tube a volatile liquid is evolved and there remains 0.4 g. of white, solid copolymer of ethylene and N-(trifluorovinyl)bis(trifluoromethyl)amine.

This polymer is pressed at 50° to 90°C. into films that are transparent and elastic. The film shows infrared absorption bands at 3.4μ (CH) and strong absorption at 7 to 9μ (CF). The polymer is insoluble in hot xylene. The copolymer contains 53.57% F and 4.74% N.

Example 4:   Preparation of a Copolymer of N-Vinylbis(Trifluoromethyl)Amine and Tetrafluoroethylene — By the procedure of Example 3, a mixture of 2.0 ml. of liquid perfluorodimethylcyclobutane and 2 g. of N-vinylbis(trifluoromethyl)amine, 1.3 g. of tetrafluoroethylene, and 6 mg. of dinitrogen difluoride is heated at 60°C. and 3,000 atm. pressure for 4 hours. There is obtained a white, tacky polymer which on drying in vacuum, cutting into small pieces and further drying at 100°C. in vacuum to constant weight gives 3.32 g. of white, tough, horny solid copolymer of N-vinylbis(trifluoromethyl)amine and tetrafluoroethylene.

The copolymer contains 5.11% N. The copolymer is pressed into clear, colorless, flexible, transparent films at 140° to 180°C. The copolymer film shows infrared absorption at 3.3μ (CH) and strong absorption at 7 to 10μ (CF). Films are manually drawn cold, or when warm, to the extent of several times the original length.

GENERAL PROCESSING TECHNIQUES

<u>CATALYSTS</u>

Bis(Trifluoromethyl) Peroxide

In a process described by <u>R.A. Darby and E.K. Ellingboe; U.S. Patent 3,069,404; December 18, 1962; assigned to E.I. du Pont de Nemours and Company</u> the polymerization of polymerizable ethylenically unsaturated compounds is carried out in contact with bis(trifluoromethyl) peroxide as the polymerization initiator. The polymerization is usually run at temperatures of 100° to 250°C. in order to obtain a reasonable rate of polymerization.

Bis(trifluoromethyl) peroxide, CF_3OOCF_3, has been obtained in small amounts by electrolysis of trifluoroacetic acid [Swarts, Bull. Soc. Chim., Belg. 42, 102 (1933)]. It is prepared much more conveniently and in satisfactory yields by the methods described by Porter and Cady in J. Am. Chem. Soc. 79, 5628 (1957), especially by the silver fluoride catalyzed reaction of carbon monoxide with fluorine at about 180°C. It is a stable, nonexplosive, colorless gas boiling at about -37°C. The process is illustrated by the following examples.

Example 1: A pressure vessel of 80 ml. capacity, constructed of a corrosion-resistant nickel-iron-molybdenum alloy, was evacuated, cooled to -80°C., and charged while cold with 30 g. of vinyl fluoride and 20 mg. of bis(trifluoromethyl) peroxide. The peroxide was measured by displacement as a gas at known volume and pressure, according to calculation based on the perfect gas laws. The reaction vessel was heated to 175°C. over a period of 45 minutes and then held at this temperature for 6 hours while being agitated.

The internal pressure during this period dropped from 2,650 to 1,210 psi. The reaction vessel was then cooled to room temperature and opened. There was obtained 16 g. of polyvinyl fluoride. Films of this polymer, formed by pressure at 10,000 psi pressure at temperatures in the range of 185° to 230°C., were tough and transparent.

Example 2: In an 80 ml. capacity corrosion-resistant pressure vessel was placed 20 g. of perfluoro(dimethylcyclobutane) [Hauptschein et al., J. Am. Chem. Soc. 80, 842 (1958)] to serve as solvent and diluent in the polymerization reaction. The pressure vessel was cooled to -80°C., evacuated and charged with 12 g. of vinylidene fluoride, 12 g. of hexafluoropropylene and 20 mg. of bis(trifluoromethyl) peroxide. After heating at 175°C. for 6 hours with agitation, during which time the internal pressure fell from 2,520 to 1,930 psi, there was obtained 6.3 g. of a rubbery, acetone-soluble copolymer of vinylidene fluoride and hexafluoropropylene. Under the same conditions, except that the heating period was 12.5 hours, there was obtained 10.2 g. of a similar copolymer.

Example 3: Into a 100 ml. stainless steel reactor was charged 30 ml. of perfluoro(dimethylcyclobutane), 2.5 g. of hexafluoropropylene and 15 g. of tetrafluoroethylene. Bis(trifluoromethyl) peroxide was injected into the reactor until the pressure had increased by one atmosphere. The vessel was heated to 150°C. and agitated at a temperature of 150° to 175°C. under autogenous pressure for one hour. On workup, there was obtained 5.6 g. of a tetrafluoroethylene/hexafluoropropylene copolymer containing approximately 20% of hexafluoropropylene and having a specific melt viscosity of 20.4×10^4 poises. The polymer could be melt extruded into tough, clear beading.

Bis(trifluoromethyl) peroxide has been found to possess certain advantages as a polymerization initiator in comparison with the commonly used organic peroxides or inorganic persulfates. The principal advantage is that polymers prepared with bis(trifluoromethyl) peroxide as the initiator are free of end groups which are known to impart chemical and thermal instability. For example, tetrafluoroethylene/hexafluoropropylene copolymers are prepared as descrbied in Example 3 were analyzed for end groups by their characteristic infrared absorption, a sensitive method capable of quantitative determination of carboxyl, acyl fluoride or vinyl end groups. The analyses indicated that none of these groups were present in detectable amounts. These end groups are found to be present in appreciable amounts in polymers prepared with inorganic persulfates or organic peroxides, and their presence has been an impediment to the development of maximum stability in polymeric products.

Fluoroazoalkanes

In a process described by D.D. Coffman; U.S. Patent 3,047,553; July 31, 1962; assigned to E.I. du Pont de Nemours and Company the polymerization of ethylenically unsaturated compounds under the usual polymerization conditions is carried out in the presence, as the polymerization initiator, of a polyfluoroazoalkane of the formula R—CF₂—N=N—CF₂—R, where R stands for fluorine, a perfluorocarbon radical or an omega-hydroperfluorocarbon radical. The polyfluoroazoalkanes of the above formula include hexafluoroazomethane, $CF_3N=NCF_3$, and the higher perfluoroazoalkanes and omega-hydroperfluoroazoalkanes which may be represented by the formula X—R₁—CF₂—N=N—CF₂—R₁—X, where R₁ is a divalent aliphatic perfluorocarbon radical, i.e., a radical consisting only of carbon and fluorine atoms, and X is fluorine or hydrogen.

Hexafluoroazomethane, a gas boiling at –32°C., has been reported in the literature. It can be prepared by the described methods or, much more conveniently, by reacting cyanogen chloride with chlorine and a fluoride of an alkali metal of atomic number 11 to 19 under substantially anhydrous conditions, and at a temperature of at least 150°C. The process is illustrated by the following examples.

Example 1: A 100 ml. glass tube was cooled in liquid nitrogen, and charged with 8.4 g. of tetrafluoroethylene and 0.14 g. of hexafluoroazomethane, after which the tube was evacuated to remove the air and sealed. The tube was irradiated with the light from a 110 volt, 275 watt sun lamp at 0° to 30°C. for 1.5 hours. There was obtained 7.2 g. of white polytetrafluoroethylene, which retained very good color after hot pressing. The polymer had a molecular weight of approximately 2,000,000, as determined by melt viscosity. No end groups were apparent in its infrared spectrum.

Similar results were obtained under similar conditions except that only 0.017 g. of hexafluoroazomethane was used for 10 g. of tetrafluoroethylene. Conversion to the polymer was 84% after one hour's exposure to the light of the sun lamp. When the hexafluoroazomethane was omitted, no polymerization took place under the same conditions.

Example 2: A 100 ml. glass tube cooled in liquid nitrogen was charged with 8 g. of tetrafluoroethylene, 6 g. of hexafluoropropylene and 0.2 g. of hexafluoroazomethane. After evacuating and sealing, the tube was irradiated with the light of a sun lamp for 4 hours at a temperature of 20° to 60°C. There was obtained 5.7 g. of a white copolymer of tetrafluoroethylene and hexafluoropropylene containing approximately 3 mol percent of polymerized hexafluoropropylene, as indicated by its infrared spectrum. The molecular weight of this copolymer was approximately 300,000 to 500,000. No end groups were detectable in the infrared spectrum of the copolymer. Copolymerization also took place in a similar system placed in a steel bomb and irradiated through a small quartz window by a mercury vapor lamp for 4 hours at 75°C.

Titanium Tetrachloride, Triethyl Aluminum and Tetrahydrofuran

G.F. Helfrich and E.J. Rothermel, Jr.; U.S. Patent 3,380,977; April 30, 1968; assigned to The Dow Chemical Company describe a process for making fluorinated polymers which comprises effecting reaction between a mixture of ingredients comprising one or more polymerizable fluorinated ethylenes and a catalytic mixture of titanium tetrachloride, triethyl aluminum and tetrahydrofuran. The following examples illustrate the process.

Example 1: In each of a series of experiments 100 parts by weight of dry, inhibitor-free vinyl fluoride monomer were individually sealed under nitrogen in a dry glass bottle containing a catalyst mixture composed of titanium tetrachloride in amount sufficient to provide from 200 to 300 ppm titanium based on the weight of the monomer; triethyl aluminum in amount sufficient to provide from 1 to 5 mols of triethyl aluminum per mol of titanium tetrachloride, and tetrahydrofuran in amount sufficient to provide from 10 to 20 mols of tetrahydrofuran per mol of triethyl aluminum; such catalyst mixture being prepared by first admixing the triethyl aluminum and tetrahydrofuran for a period of about 30 minutes at a temperature of about 25°C. in the reaction bottle followed by the addition of the titanium tetrachloride.

The bottles containing the mixtures were then individually capped and shaken to insure uniform mixing, and the contents subsequently reacted for a period of about 6 hours at 40°C. in a constant temperature bath. At the end of the reaction period, each bottle was opened, allowed to cool and the polymeric material washed with methanol containing 3 weight percent of hydrochloric acid, and then dried. The individually recovered polymeric products were each fine white powders having a melting point of from 120° to 140°C. These polymers did not discolor or otherwise decompose when subjected to temperatures up to about 150°C. and were each capable of being formed into tough film materials having excellent clarity.

Example 2: In each of a series of experiments 100 parts by weight of dry, inhibitor-free chlorotrifluoroethylene monomer were polymerized as described in Example 1, utilizing a polymerization temperature of about 55°C. and a reaction time of about 17 hours. The recovered polymeric products were each fine white powders having a melting point of from 140° to 150°C. These polymeric materials did not discolor or otherwise decompose when subjected to temperatures of up to about 270°C. and were capable of being formed into film materials having excellent clarity.

Transition Metal Halide Treated Silica

A process described by J.C. MacKenzie and A. Orzechowski; U.S. Patent 3,285,898; November 15, 1966; assigned to Cabot Corporation involves the polymerization and copolymerization of halogenated, conjugated and unconjugated, α-mono- and diolefins such as tetrafluoroethylene, vinyl chloride, chlorotrifluoroethylene, 2-bromopropylene, 3-bromo-1,4-hexadiene and 2-chloro-1,3-butadiene.

In the process, halogenated α-mono- and diolefins are polymerized and/or copolymerized by catalysts comprising (a) the product of the reaction carried out under certain conditions between a halide-type compound of a metal of Group IVa, Va, VIa, VIIa or period 4 of Group VIII and hydroxyl groups on the surface of a finely divided particulate inorganic solid, and (b) an organometallic compound.

The polymerization or copolymerization reaction can be effected at suitable temperatures within the range of from about -80° to 250°C., and pressures ranging from below atmospheric upwardly to any desired maximum pressure. Inorganic solids suitable for the process generally include any inorganic compound which is available in finely divided particulate form with hydroxyl groups on the surface thereof. For example, oxides such as alumina, zirconia and silica, carbon blacks such as channel black and furnace black, and aluminates such as corundum are generally suitable for the purposes of the process. In particular, inorganic solids having an average particle diameter of less than 0.1 micron and having at least 1×10^{-4} equivalents per gram of hydroxyl groups chemically bound to the surface are preferred.

Examples of suitable compounds are halides such as zirconium tetrachloride, vanadium tetrachloride, ferric chloride, manganese dichloride and titanium tetraiodide, and oxyhalides such as chromium oxychloride and vanadium oxychloride. The conditions under which reaction between the transition metal halide and the finely divided, inorganic solid can be accomplished are subject to considerable variation.

However, in order to obtain a catalyst component with exceptionally high activity and reproducible character and performance, it has been found to be all important that the finely divided, inorganic solid be essentially dry and anhydrous (i.e. free of molecular water in any form) at the time, it is brought into contact with the transition metal halide. In addition, it is recommended that the reaction of the inorganic solid and the transition metal halide be accomplished so as to allow by-products of the reaction to be eliminated from the reaction zone in order to insure that the reaction goes to completion.

Generally, the reaction can be carried out by contacting the inorganic solid with the transition metal-halide, preferably in an inert hydrocarbon medium, and maintaining the two reactants in intimate contact for a period of time sufficient to effect the desired chemical reaction resulting in the chemical bonding of the transition metal to the inorganic solid. The length of time required to effect a given amount of such reaction and chemical bonding is largely dependent upon the temperature of the reaction mixture.

Generally speaking, almost any temperature between 0° and 300°C. and even higher temperatures can be used satisfactorily, but room temperature to about 105°C. is generally preferred. Assuming provision is made for intimate contact of the dry inorganic solid and the transition metal halide, the minimum time required to accomplish the chemical reaction will vary from 1 hour at room temperature to about 15 minutes at temperatures of 100°C. or over. Temperatures substantially higher than 300°C., e.g., 500°C. are completely needless and therefore of little or no interest. The following examples illustrate the process.

Example 1: To a 2,000 ml., three-neck, glass reaction vessel there is added 25 g. of "Cab-O-Sil," a pyrogenic silica produced by Cabot Corporation, which has an average particle diameter of about 10 millimicrons and a hydroxyl group content on the surface of about 1.5 milliequivalents per gram. The reaction vessel is then dried in a vacuum oven for 24 hours at a temperature of about 120°C. Subsequently, the vessel is sealed without exposing the silica to the atmosphere and there is charged to the vessel 25 millimols of titanium tetrafluoride and 1,000 milliliters of anhydrous hexane.

The vessel is then continuously stirred and maintained at a refluxing temperature for a period of 24 hours while the contents are swept by a stream of dry nitrogen. Subsequently, the extent of the reaction between titanium tetrafluoride and the silica is determined by measuring the quantity of HF removed from the vessel by the nitrogen stream, and by testing the liquid contents of the vessel for the absence therein of titanium tetrafluoride. The silica is found to have 25 milliatoms of titanium chemically bound to the surface.

80 milliliters of this slurry containing about 2 milliatoms of titanium bound to the surface of about 2 g. of silica, is then transferred from this reaction vessel to an 8 oz. bottle which has been flushed with dry nitrogen. Next, without exposure to the atmosphere, 2 millimols of diisobutylaluminum chloride followed by 200 millimols of purified 20 chloro-1,3-butadiene (chloroprene) are charged into the bottle. The bottle is then continuously agitated at ambient temperatures for about 48 hours. The reaction products are analyzed and it is found that about 1.5 g. of solid 2-chloro-1,3-butadiene

(neoprene) polymer has been produced.

Example 2: To a 500 milliliter reaction bomb there is added about 80 milliliters of the catalyst slurry produced in Example 1 which contains about 2 milliatoms of titanium chemically bound to the surface of about 2 g. of silica. Next, 2 millimols of triisobutylaluminum are added to the bomb, followed by pressurization with 200 millimols of chlorotrifluoroethylene gas. The reaction vessel is then heated to, and maintained at 80°C. with continuous agitation for 24 hours. The reaction products are analyzed and it is found that solid chlorotrifluoroethylene polymer has been produced. When, under the same conditions, the triisobutylaluminum or the silica bearing chemically combined titanium on the surface is utilized alone as the catalyst, no solid polymer is produced.

Redox Catalysts

G. Borsini, M. Modena, C. Nicora and M. Ragazzini; U.S. Patent 3,401,155; September 10, 1968; assigned to Montecatini Edison SpA, Italy describe a process for the production of highly linear fluorinated polymers of improved mechanical characteristics, comprising polymerizing, at low temperature, fluorine-substituted ethylene monomers in the presence of a redox catalyst comprised of (A) at least one organometallic compound of the formula:

$$
\begin{array}{ccc}
R' & & R'' \\
 & \diagdown \diagup & \\
 & Me & \\
 & \diagup \diagdown & \\
R' & & R'''
\end{array}
$$

Me is a metal selected from the group consisting of germanium, tin and lead.
R' is a member selected from the group consisting of lower alkyl of up to 8 carbon atoms, aryl of up to 10 carbon atoms, cyclo-lower alkyl of up to 8 carbon atoms and lower alkaryl of up to 18 carbon atoms.
R'' and R''' which may be the same or different are members selected from the group consisting of R', halo, nitrate, lower alkoxy of up to 8 carbon atoms, carboxy and divalent cations of a strong inorganic acid.

(B) at least one tetravalent cerium salt of a strong inorganic acid, and in the further presence of an inert solvent for the active catalyst. Particularly favorable results may be obtained by employing ethylenically unsaturated fluorinated compounds such as, for example, tetrafluoroethylene, monochlorotrifluoroethylene, vinyl fluoride, vinylidene fluoride, trifluoroethylene and vinylene fluoride ($FHC=CHF$).

These fluorinated monomers can be homopolymerized and copolymerized among themselves or with ethylene and α-olefins and other ethylenically unsaturated compounds such as, for example, propylene, isobutene, butadiene, styrene and their analogs. Particularly good results have been obtained by using a redox type catalytic system comprised of tetraalkyl lead and by an ammonium and tetravalent cerium salt of a strong inorganic acid such as sulfuric acid, nitric acid and pyrophosphoric acid.

The catalytic system, according to this process permits one to operate, with appreciable economic yields, at temperatures lower by from 20° to 60°C. than those generally used with standard catalytic systems; therefore, polymers and copolymers are obtained having enhanced mechanical characteristics that can only be related to the low polymerization temperature. The process for the preparation of polymeric materials may be conducted at temperatures varying from -100° to +50°C., but preferably from -60° to +20°C. The following examples illustrate the process.

Example 1: A stainless steel autoclave, having a capacity of 2 liters, equipped with stirrer and thermoregulation jacket, was scavenged, cooled to -5°C. and then charged with a solution of 0.975 ml. of tetraethyl lead in 900 ml. of a mixture of tertiary butanol and water in the ratio of about 8.5 to 1.5 by volume. A mixture of ethylene and tetrafluoroethylene, containing 14 mol percent tetrafluoroethylene, was then introduced until a pressure of 20 atmospheres was reached. A solution of 2.74 g. of $(NH_4)_2Ce(NO_3)_6$ and of 2 ml. HNO_3 65% in 200 cc of a mixture formed from tertiary butanol and water (in the previously mentioned ratios) was subsequently fed therein.

Pressure and temperature were kept constant for 3 hours and 40 minutes. Thereafter a mixture formed of 50 cc methanol, 3 cc of 30% hydrogen peroxide and 5 cc of 65% HNO_3, was fed into the autoclave. The excess gas was eliminated and the reaction mass discharged. The polymer was separated by filtering, washed with methanol and dried. A quantity of 39 g. of ethylenetetrafluoroethylene copolymer was obtained and its elementary analysis showed 48.5% fluorine, corresponding to 33% by mols of tetrafluoroethylene.

This product contained no fraction soluble in boiling acetone and, submitted to fractional extraction in boiling xylene, 99% thereof remained insoluble. The melting point, determined via the disappearance of birefringence in a polarized light microscope, was 237°C. This product was workable according to techniques suitable for thermoplastics and did not exhibit any appreciable degradation. The coefficient of rigidity (G') determined by a damping test, was 6 x 10^8 at 100°C. and 1 x 10^8 at 214°C. On the contrary, an ethylenetetrafluoroethylene copolymer having the same composition, and prepared in the same solvent but at 70°C. using an ammonium persulfate initiator, was partially soluble in boiling acetone (5%) and xylene (80%), and underwent severe degradation whenever submitted to press or extruder molding, and had a coefficient of rigidity equal to 3 x 10^8 at 100°C. and 1 x 10^8 at 174°C.

Example 2: A stainless steel autoclave, having a capacity of 1/2 liter, equipped with stirrer and thermoregulation jacket, was scavenged, cooled to –15°C. and charged with a solution of 0.24 ml. tetraethyl lead in 100 ml. of a mixture formed from tertiary butanol, water and methanol with ratios 8 : 1 : 1 by volume. A solution of 0.69 g. of $(NH_4)_2Ce(NO_3)_6$ and of 0.5 ml. of HNO_3 at 65% in 200 ml. of the above specified solvent mixture was then fed therein. A mixture of tetrafluoroethylene and ethylene containing 12.5 mol % of tetrafluoroethylene was subsequently introduced until a pressure of 14.51 atmospheres was attained in the autoclave.

After approximately 3 hours of reaction at constant temperature and pressure, the catalyst was quenched by means of a mixture consisting of methanol (50 ml.), H_2O_2 at 30% (3 ml.) and HNO_3 at 65% (5 ml.): the excess gas was expelled and the reaction mass discharged. A quantity of 7 g. polymer was obtained, melting at 235° to 236°C. and having 47.5% of fluorine, corresponding to 32 mol % of C_2F_4.

Example 3: Solvent and catalysts in the quantity specified in Example 2 were fed into an autoclave having a capacity of 1/2 liter, and cooled to –15°C. A mixture of ethylene and monochlorotrifluoroethylene with 23 mol % of monochlorotrifluoroethylene was introduced until a pressure of 13.54 atmospheres was attained. The reaction was stopped after 3 hours and there was obtained a quantity of 9 g. of copolymer with 22.4% chlorine, corresponding to 39.5 mol % of C_2ClF_3, and which melts at 181° to 182°C.

Boron Alkyls Activated with Oxygen

M. Ragazzini and D. Carcano; U.S. Patent 3,371,076; February 27, 1968; assigned to Montecatini Edison SpA, Italy have found that the process of polymerizing ethylene and monochlorotrifluoroethylene proceeds with higher speed and with better yields when the boron alkyl catalyst is activated with oxygen or, at any rate, with oxygen-containing substances capable of liberating under suitable conditions oxygen such as for instance the peroxides. It is particularly advantageous to use, as a catalyst, complexes formed by boron alkyls with substances having electron-donor character; the complex in fact is much more stable than boron alkyl alone and can more easily be activated with oxygen. The following examples illustrate the process.

Example 1: Into a shaking autoclave of 1 liter capacity there are charged under nitrogen atmosphere in orderly succession 2 ml. of boron trimethyl previously activated by bubbling of oxygen, 5 ml. of ether, 250 ml. of heptane, 40 g. of monochlorotrifluoroethylene and 40 atmospheres of ethylene. The temperature is kept at 25°C. and the autoclave is put under agitation for 5 hours. After that period, 30 g. of copolymer are discharged from the autoclave with a percentage of chlorine of 24% corresponding to a molar ratio:

$$\frac{ethylene}{monochlorotrifluoroethylene} = 1$$

Example 2: The test is conducted as in Example 1, replacing boron triethyl by boron hydride. After equal duration of reaction, 15 g. of polymer are obtained.

Titanium Trifluoride and Aluminum Triethyl

A.T. Polishuk; U.S. Patent 3,098,844; July 23, 1963; assigned to Sun Oil Company describe a method for preparing relatively high molecular weight, predominantly crystalline polymers which are relatively inert to oxidation. According to the process, halogen-substituted olefins having from 3 to 8 carbon atoms are polymerized to relatively high molecular weight, predominantly crystalline solid polymers by contacting the monomer with a metal halide catalyst and an activator for the catalyst.

The preferred halogen-substituted alpha-olefins are those having halogen atoms attached to carbon atoms which are tertiary in a polymer of the material, such as the beta unsaturated carbon atom. For example, in 2-fluoropropene $(CH_2=CF–CH_2)$ the beta unsaturated carbon becomes a tertiary carbon after polymerization. Branched monomers can be used, but it is preferred that such monomers have no hydrogen atoms attached to the tertiary carbon atoms. For example, 3-methyl-2,3-dichlorobutene-1 can be used with good results, since in the polymer product all tertiary carbon atoms are attached to chlorine atoms.

The preferred monomer is accordingly one which has no hydrogen atoms which would be attached to a tertiary carbon in a polymer of the material. However, other halogen-substituted alpha-olefins may be used in the process with good results, although some of the polymers produced are less resistant to oxidation than those which have no tertiary hydrogen atoms. Suitable monomers for the process include allyl chloride, allyl fluoride, allyl bromide, allyl iodide, 1,2,3-trifluoropropene, 2-chlorobutadiene-1,3, 2-bromohexene-1,2,3-dibromopropene, 2,3-dichloropropene, 2-fluorobutene-1,3-bromo-3,3-difluoropropene and 2-bromo-1,1-difluoropropene. The catalysts suitable for the process are subhalides of metals of groups IV, V and VI of the periodic table. The subhalides of titanium and zirconium, such as titanium trichloride, titanium difluoride, zirconium tribromide and zirconium diiodide are the preferred catalysts, however the subhalides of other metals of groups IV, V and VI are also effective catalysts.

The catalysts used in this process must be activated by a suitable activator, such as an aluminum trialkyl. The aluminum trialkyls, such as aluminum triethyl, and aluminum triisobutyl are the preferred activators. The following example illustrates the process.

Example: A reactor fitted with a stirrer is partially filled with 1,000 parts of a mixture of saturated hydrocarbons, mostly octanes, in an atmosphere of nitrogen. With the temperature of this reaction medium maintained at 80°C. about 250 parts of 2-fluoropropene are injected therein at a pressure of 250 psig. One part of aluminum triethyl and 1.6 parts of titanium trifluoride are then added, and polymerization starts immediately, as evidenced by a drop in the pressure in the reactor.

Additional 2-fluoropropene is added from time to time to maintain the pressure in the reactor at approximately 250 psig and the temperature is maintained at about 80°C. After 5.0 hours, reaction has virtually stopped, due to the coating of the catalyst particles by solid polymer. The reaction medium is then drained off and the solid polymer transferred to a ball mill where 1,000 parts of methanol are added. The ball mill is started and run for one hour. The methanol is then removed by filtration, and the polymer washed with dilute nitric acid. The polymer is then washed with water and dried. About 175 parts of white crystalline polymer are obtained, having an average molecular weight of about 210,000. This polymer may be formed into useful articles by the usual methods, e.g., molding or extruding.

CROSS-LINKING

Nitrogen Fluorides

T.L. Cairns and C.S. Cleaver; U.S. Patent 3,162,623; December 22, 1964; assigned to E.I. du Pont de Nemours and Company describe the cross-linking of polymers with nitrogen fluorides. Specific examples of nitrogen fluorides which are operative in the process include cis- and trans-dinitrogen difluoride (N_2F_2), nitrogen trifluoride (NF_3), dinitrogen tetrafluoride (N_2F_2), trifluoromethyldifluoronitride and bis(trifluoromethyl)fluoronitride.

Example 1: Several 1/2" x 3" strips of 3-mil polyethylene film, each weighing 60 mg. were placed in a 17 cc glass vessel closed on the top with a stopcock and ball joint. The apparatus was evacuated and filled to 640 mm. pressure with 36 to 40 mg. of cis-N_2F_2. The vessel was suspended for 45 minutes in a water bath whose temperature was 80° to 85°C. During this time, the films became slightly tan. Gas chromatography of the gas mixture at this time showed that the mixture was 29% N_2/air and 68% cis-N_2F_2.

The tube was placed in a bath for another 30 minutes at 80° to 85°C. Gas chromatography now showed that the gas was 26% air/N_2 and 73% cis-N_2F_2, suggesting that very little reaction was occurring after the original 45 minutes. During the experiment 9 to 10 mg. of N_2F_2 was consumed. The film strips were removed. A piece of the treated polyethylene failed to dissolve in boiling xylene, but swelled slightly. The zero-strength temperature of the treated film was 213°C. versus 104°C. for the untreated control. The control polymer was completely soluble in boiling xylene.

Example 2: Several 1/2" x 2 1/2" x 3 mil films were made from a 60/40 (weight ratio) 1,1-difluoroethylene/hexafluoropropylene copolymer. This copolymer was soluble in acetone and of inherent viscosity (η_{inh}) 0.95. Four of the films, each weighing 0.10 g., were treated as in Example 1 for one hour at 80°C. with cis-N_2F_2 at a pressure of 600 mm. During the experiment, 11 mg. of the N_2F_2 was consumed. The treated films were unchanged in color but became very elastic and much stronger than untreated controls. The treated films were insoluble in acetone.

Example 3: A copolymer composed of 60 parts of 1,1-difluoroethylene and 40 parts of hexafluoropropylene was formulated as indicated in the following table and pressed into films 5 to 10 mils thick.

Formulation	A	B	C
Parts copolymer	100	100	100
Parts carbon black	-	18	-
Parts MgO	15	15	-

Portions of the films were exposed to N_2F_2 at 600 mm. pressure and 100°C. for 0, 1 and 3 hours and tested, both with and without additional curing, as shown in the table on the following page.

Formulation	A		B		C		
Hours Exposure to N_2F_2 at 100°C.*	1	3	1	3	0	1	3
Tensile strength at break, psi	1,020	1,050	1,840	1,750	200-300	1,000	850
Elongation at break, percent	640	535	500	400	500-600	700	620
Modulus at 100% elongation, psi	160	125	260	290	150	120	100
Additional Oven Cure, 18 Hours at 200°C.							
Tensile strength at break, psi	1,570	1,600	1,950	1,920	–	800	–
Elongation at break, percent	545	430	410	355	–	620	–
Modulus at 100% elongation, psi	130	160	225	290	–	100	–

*0.8% N_2F_2 based on weight of compounded polymer was consumed in the 1 hour exposures.
1.6% N_2F_2 based on weight of compounded polymer was consumed in the 3 hour exposures.

Metal Coordination Complexes

C.D. Dipner; U.S. Patent 3,245,968; April 12, 1966; assigned to Minnesota Mining and Manufacturing Company describe the curing of polymers by the use of metal coordination complexes. The addition of the chromium compound to the polymer apparently causes a cross-linking of the polymer chain thereby increasing thermal stability and decreasing the solubility of the polymer.

As such, the improved polymer compositions contain a small amount of irreversibly bound microscopic chromium-containing particles. In contradistinction to fillers, the chromium is dispersed in such a manner that no particles or discrete units are detectable even using the electron microscope. Regardless of the method of incorporating the chromium the chromium must become molecularly dispersed and intimately associated with the individual polymer chains in order for the reinforcing effect to manifest itself.

Chromium coordination complexes may be prepared by reacting a chromium metal salt with a 3 to 20 carbon aliphatic carboxylic acid. Because these complexes are intended to function at elevated temperatures, those acids which have maximum thermal stability are preferred. Maximum thermal stability is attained by the use of perfluorohalo carboxylic acids in the preparation of the chromium complexes. In preparing a complex the chromium in the form of a salt is reacted with the acid. Suitable chromium salts are the acetates, nitrates, chlorides, oxychlorides, etc.

Blending of the metal coordination complex with the polymer can be achieved in any manner which insures homogeneity. Thus, the polymer complex can be blended in a ball mill, pebble mill, or in other suitable blending equipment. The amount of chromium coordination complex employed will vary to a large extent depending upon the desired degree of cross-linking but will generally range between 0.1 and 20 weight percent based on the weight of polymer and preferably between 0.5 and 5%.

Curing or cross-linking of the treated polymer is effected by maintaining the polymer at a temperature above about 225°C. and preferably above 250°C. for a period of time above 1 hour. The heating period can be continued indefinitely although 15 days is a practical maximum. Preferably, the heating period is between 1 and 48 hours.

The following examples illustrate the process. In the examples, the term "ZST" is used. This term refers to an empirical test for determining the physical characteristics of polymeric materials and is a reflection of the molecular weight of the polymer involved, i.e., the higher the ZST, the higher the molecular weight. In the case of compositions containing the chromium compounds of this process, the increase in the ZST is due to the cross-linking action. The ZST test is described in "Modern Plastics," October 1954, page 146.

Example 1: Steel panels which had been blasted with BB No. 25 grit were coated with a chrome complex of

$$Cl(CF_2CFCl)_6CF_2COOH$$

by immersion in a 5 weight percent solution of the complex in acetone. One panel was dip coated twice and the other was dip coated 4 times. The panels were baked 5 minutes at 150°C. after each dip. The complex coated panels were then coated with a dispersion of a homopolymer of trifluorochloroethylene in xylene-diisobutyl ketone (80/20) dispersant and were baked for one-half hour at 250°C. (480°F.). There was no fusion of the polymer particles. The panels were baked again for one hour with no evidence of fusion. The homopolymer coating stripped easily from the complex treated panel.

After an additional 4 hours of baking at about 250°C. there was still no fusion. The melt viscosity of the applied homopolymer had increased, indicating a cross-linking reaction. When the previously described dispersion was applied to a grit-blasted but untreated steel panel, fusion of the particles occurred after baking for one-half hour at about 250°C.

Example 2: A blend of 5% by weight of the chromium complex of $Cl(CF_2CFCl)_8CF_2COOH$ was prepared by mixing in a mortar and pestle. A molded sheet was prepared from this material by pressing at 2,500 psi for 3 minutes at 260°C. The ZST of this pressing was the same as that of the original of the homopolymer, i.e., about 710 seconds. The pressed sheet was bluish-purple in color. A piece of this sheet was heated at 250°C. for 7 days. The resulting green material has a ZST in excess of 5,000 seconds.

Example 3: A blend of about 1% of the chromium complex of $Cl(CF_2CFCl)_8CF_2COOH$ and a homopolymer of trifluoro-chloroethylene was prepared by ball milling for 2 hours. This blend was pressed for 3 minutes at 260°C. and 2,500 psi. These pressings were slightly colored and had essentially the same ZST as the starting material, i.e., about 710 seconds. No marked increase in ZST was observed after heating at 190°C. for 16 hours. However, heating at 250°C. for 12 hours gave a material with a ZST in excess of 5,000 seconds. It should be noted, that these sheets were cured at 250°C. by suspending them in a 250°C. oven for the 12 hour period. Dimensionally similar sheets which did not contain chromium complex did not support their own weight at 250°C.

Example 4: This example illustrates the use of metal coordination complex with a liquid polymer. Approximately 54 g. of a homotelomer of trifluorochloroethylene having the general formula $Cl(CF_2CFCl)_nCl$ and boiling between 130° and 170°C. at 0.5 mm. was admixed with 6 g. of the chromium complex of $Cl(CF_2CFCl)_6CF_2COOH$ and heated to 200°C. until bubbling stopped (due to water in either the oil or the complex). The mixture was then heated to 250°C. for 15 minutes after which it was cooled to room temperature. The oil was gelled or thickened.

Polyamines

In a process described by F.W. West; U.S. Patent 2,979,490; April 11, 1961; assigned to Minnesota Mining and Manufacturing Company a hydrofluorinated polymeric material is reacted with a polyamine to produce a cross-linked polymer. The polyamine which serves as a cross-linking or vulcanizing agent, which is preferably an acyclic hydrocarbon polyamine, is uniformly mixed with the hydrofluorinated polymeric material and is permitted to cross-link therewith.

The linear polymers which are cross-linked in the process are those which are produced by the polymerization of hydrofluorinated olefins, that is, olefins containing only hydrogen and fluorine substituents on the carbon atoms and preferably having at least one fluorine atom per carbon atom. The hydrofluorinated olefins are preferably hydrofluoroethylenes such as, for example, trifluoroethylene, 1,2-difluoroethylene and vinyl fluoride.

The preferred copolymers are those of the above hydrofluorinated ethylenes and hexafluoropropene or tetrafluoroethylene. Most preferred of the hydrofluorinated olefin polymers which are cross-linked by the process are the polymers where at least half of the carbon atoms are bonded to fluorine atoms. Examples of preferred cross-linking agents which are employed in the process are ethylenediamine, diethylenediamine, diethylenetriamine, triethylenetetramine, tetraethylenepentamine, hexamethylenediamine, pentamethylenediamine, decamethylenediamine and undecamethylenediamine.

The amount of vulcanizing agent employed in the reaction can be varied over a wide range depending upon the number of the cross-links desired in the polymer and the particular polymer employed. The amount of polyamine employed as a cross-linking agent may be varied between 0.1 and 15 parts by weight per 100 parts by weight of polymer. Generally, an amount between 1 and 10 parts by weight per 100 parts by weight of polymer is used in the cross-linking process.

In carrying out the process, the desired amount of polyamine vulcanizing agent is added and is uniformly and intimately mixed with the unvulcanized hydrofluorinated polymer or with the unvulcanized hydrofluorinated polymeric mixture at a temperature not exceeding about 100°C. The resulting mixture is then pressed at a temperature between about 100° and 250°C., preferably between 150° and 180°C. for a period of from 10 to 15 hours, preferably from 0.5 to 2 hours under a pressure of between 100 and 2,000 psig, preferably between 500 and 1,000 psig. The vulcanizable admixture may be pressed into sheets or pressed in a mold.

The molds may be coated with a silicone emulsion or silicone oil, for example, DC-200, and prebaked for about 4 hours at a temperature of about 250°C. Certain elastomers have been found to have better release properties when molded under these conditions, however, this coating step may be omitted if desired. The resulting pressed vulcanizate is then baked in an oven at a temperature of between 100° and 200°C., preferably at 175°C. for a period of from 2 to 25 hours depending upon the cross-sectional thickness of the sample. The molds are usually baked at atmospheric pressure, however, pressures up to 15 or 20 atmospheres can be applied if so desired. The following example illustrates the process.

Example: A clear tetrahydrofuran solution of polytrifluoroethylene was prepared by shaking 100 parts by weight of the polymer in 1,900 parts by weight of tetrahydrofuran at room temperature and then were added 6 parts by weight of tetraethylenepentamine. After shaking for a few minutes, or until the solution was complete, a film was cast on a glass plate. The solvent was allowed to evaporate from the film at room temperature using no special conditions to obtain the film, such as protecting the film from atmospheric moisture or light, or applying irrigation or moisture to the film. At this time the film was readily dissolved by tetrahydrofuran. The film was then heated to 130°C. for 4 hours and allowed to cool to room temperature after which the film was again tested with tetrahydrofuran. It was found that the film was completely insoluble in tetrahydrofuran and also in any of the other common organic solvents such as esters, ketones, ether, alcohols, aromatic and aliphatic hydrocarbons and halocarbons. The film was then reheated at 170°C. and at 200°C., and the solubility behavior determined after each heating treatment and the results are shown in the table below where the cross-linked polymer is compared with polytrifluoroethylene plastic which was prepared without the addition of a cross-linking agent.

Solubility Properties* of Cast Films of Polytrifluoroethylene Plastic
at Several Stages in their Cure with Tetraethylenepentamine

Polymer	Parts of Tetraethylene-pentamine Present per 100 Parts of Polymer	Before Heating	Curing Temperature (time)		
			130°C. (4 hrs.)	170°C. (16 hrs.)	200°C. (16 hrs.)
Polytrifluoroethylene plastic	None (blank run).	Soluble	Soluble	Soluble	Substantially soluble.
	6	Substantially soluble.	Insoluble, tough.	Insoluble, tough.	Insoluble, tough.

*The solubility properties of this sample were determined by completely immersing in the solvent, tetrahydrofuran, at room temperature and allowing the mixture to stand overnight without agitation. When a rating of "soluble" or "substantially soluble" is given, little if any cure is indicated. The rating of "insoluble, tough" indicates a good cure and is the best rating possible in this test.

The same general procedure was repeated using p,p'-diphenylmethanediisocyanate and also using toluenetriisocyanate instead of tetraethylenepentamine. Polyisocyanates give inferior cures compared with that of the polyamine. Any of the other polyamines previously cited and particularly ethylenediamine, diethylenetriamine, triethylenetetramine and hexamethylenediamine may be substituted in this example to give equally satisfactory cures of any of the previously mentioned hydrofluorinated polymers, particularly the elastomeric copolymers of hydrofluorinated ethylenes containing one fluorine atom per carbon atom and hexafluoropropene or tetrafluoroethylene.

Curing Agents

F.J. Honn and W.M. Sims; U.S. Patent 2,965,619; December 20, 1960; U.S. Patent 2,999,854; September 12, 1961 and U.S. Patent 3,318,854; May 9, 1967; all assigned to Minnesota Mining and Manufacturing Company describe the vulcanization of fluorinated, linear, saturated elastomers. The cross-linking agents used to produce the cross-linked elastomers may be of various types and may depend on any of several reactions to produce their cross-linking. In general, the cross-linking agents react to remove a hydrogen or halogen atom from a carbon atom on the polymer chain and thereby produce a free radical spot on the chain which is capable of linking to a similar free radical spot on another chain, either directly or indirectly. Among the cross-linking agents which may be used in this process are the peroxy-type compounds, basic metal oxides and inorganic polysulfides.

The peroxy-type compounds include both organic and inorganic compounds which contain oxygen atoms directly linked to oxygen atoms. Among the organic compounds are the acyl and acoyl peroxides and hydroperoxides, such as ditertiary butyl peroxide, dilauryl peroxide, dibenzoyl peroxide, and ditertiary butyl hydroperoxide. The organic peroxy-type compounds also include peresters having either organic or inorganic peroxy oxygen. The former would include such compounds as alkyl and aryl perbenzoates. The latter would include alkyl and aryl persulfates. Among the inorganic peroxy compounds are hydrogen peroxide and metal peroxides, such as lead peroxide, barium peroxide and zinc peroxide.

Among the basic oxides which may be used as linking agents are magnesium oxide, zinc oxide and lead oxide (PbO). It is believed that these basic acting compounds react with and remove a halogen atom from a carbon atom on the linear chain and produce a free radical spot on the chain. This free radical spot links directly to a similar spot on another chain and thereby produces a cross-linked polymer.

Among the inorganic polysulfides which may be used as linking agents are the alkali metal polysulfides and ammonium polysulfide. The generally accepted formula for a sodium polysulfide, for example, is $Na(S)_nNa$. It is believed that this compound breaks up into two sodium ions and a bivalent chain of sulfur atoms. Each of the sodium ions reacts with and removes a halogen atom from a linear polymer chain and leaves a free radical spot on the chain. Each end of the chain of sulfur atoms links to a free spot on a different polymer chain and thereby links the chains together. Cross-linking agents may be easily incorporated into the elastomers by mechanical mixing, either with or without plasticizers. Such mechanical mixing involves shearing forces and is carried out in equipment such as 2-roll mills, Banbury (internal) mixers and screw-type plasticators, which resemble extruders. Somewhat elevated temperatures of the order of from 50° to 100°C. ordinarily prevail in the mixing operation due to the mixing action itself and to the exothermic nature of the linking reaction. Articles to be molded are then heated in the mold with additional heat, as by hot air, steam or hot press platens, thereby shaping and cross-linking simultaneously. The temperature in the mold may range from 120° to 200°C. The following examples illustrate the process.

Example 1: 100 parts by weight of equimolar copolymer of chlorotrifluoroethylene and vinylidene fluoride, 15 parts by weight of a mixture of low molecular weight oily polymers of chlorotrifluoroethylene, as a plasticizer, and 15 parts by weight of benzoyl peroxide were thoroughly blended together on the cold rolls of a 2-roll mill. The blend was then press-cured for about 1 hour at 149°C. The cure was found to have converted the normally acetone-soluble copolymer into an insoluble product.

Example 2: 100 parts by weight of a copolymer of chlorotrifluoroethylene and vinylidene fluoride, containing 62 mol percent of the former and 38 mol percent of the latter, 30 parts by weight of carbon black, 2 parts by weight of stearic acid and 3 parts by weight of sodium polysulfide were thoroughly blended together on the cold rolls of a 2-roll mill. The blend was then press-cured for about 1 hour at 171°C. The cure was found to have converted the normally acetone-soluble copolymer into an insoluble product.

MISCELLANEOUS

Polymerization in the Presence of Polymeric Perfluorinated Polyepoxide

D. Sianesi and G.C. Bernardi; U.S. Patent 3,493,530; February 3, 1970; assigned to Montecatini Edison SpA, Italy have found that by carrying out the polymerization or copolymerization of perfluorinated olefins in the presence of macromolecular perfluorinated polyperoxides, consisting essentially of C, F and O atoms, very stable polymeric mixtures are obtained. Furthermore, one of the components, the olefin homopolymer or copolymer formed during the reaction, has a higher thermal stability than the corresponding homopolymer or copolymers obtained by the methods previously used.

The process for the polymerization and copolymerization of halogenated olefins is characterized in that one or more halogenated olefins are subjected to temperatures between -100° and +350°C. under a pressure from 0.01 to 200 atmospheres in the presence of macromolecular perfluorinated polyperoxides having the general formula $(C_3F_6O_x)_n$, in which n is a whole number from 5 to 100 and x is higher than 1 and lower than 2. There are also present terminal acid fluoride groups of the —COF type, or transformation products, obtained by hydrolysis, salt-formation, amidation and decarboxylation.

The macromolecular perfluorinated polyperoxides used in the process are described in Italian Patent 802,089. They are the reaction products obtained by photo-chemical reaction of oxygen with a perfluoroolefin in the liquid phase. Their preparation in the presence of ultraviolet radiations is described in U.S. Patent 3,442,942. These polyperoxides have the appearance and consistency of an oil or a viscous syrup. The following example illustrates the process.

Example: As the polyperoxide, a product is used which is derived from a photo-chemical reaction, carried out up to a transformation of about 10%, of liquid perfluoropropylene at -30°C., with molecular oxygen, under atmospheric pressure, in the presence of ultraviolet light, coming from a high pressure Hg-vapor generator of the Nanau Q81 type.

By evaporation of unreacted perfluoropropylene and after further treatment under vacuum of 0.5 mm. Hg at room temperature, prolonged for 12 hours in order to eliminate any trace of volatile products, the peroxide polymer was obtained in the form of a highly viscous colorless liquid. The composition of the product corresponds to an average formula $(C_3F_6O_{1.17})_n$. The acidimetric equivalent weight was evaluated at around 1.2×10^3 by titration at 20°C. with a normal NaOH solution of the terminal groups —COF and —COOH, and determination with thorium nitrate of the hydrolyzable fluorine ions. By titration of iodine freed from an initiator sample after prolonged contact with NaI at room temperature in a 1:4 solution of CF_2Cl—$CFCl_2$ and acetic anhydride, an active oxygen content corresponding to about 0.1 oxygen atoms per each combined C_3F_6 unit was found.

In a stainless steel autoclave having an inner volume of 100 cc, 56.0 g. of this polyperoxide are introduced. The autoclave is then closed and vacuum is applied up to 0.5 mm. Hg, nitrogen is introduced into the autoclave and vacuum is applied again so as to completely remove air. The autoclave is then immersed in a cooling bath at -80°C. and 20.0 g. of tetrafluoroethylene monomer are introduced into the autoclave by vacuum distillation. The closed autoclave is introduced into an oscillating furnace and then heated to the constant temperature of 100°C. This temperature is maintained for 16 hours under agitation. The autoclave is then heated to 120°C. for 4 hours, while maintaining the agitation.

After this time, the autoclave is cooled to room temperature, the residual gases are let out, and the product contained in the autoclave is collected. 64 g. of a seemingly homogeneous, white solid having the consistency of a grease are obtained, which has a high resistnace to the action of heat, and of chemical reactants. A sample of this material is kept for 5 hours at the temperature of 200°C. without noticing any evidence of demixing between the oil and the polyolefin forming the intimate mixture. Similar comparison tests were carried out relating to the stability with respect to the separation of the liquid and solid phases in mixtures obtained by mechanical dispersion of various types of polytetrafluoroethylene as fine powder, separately obtained by conventional methods, in fluorooxygenated oils having a composition similar to that of this example. In each case a rapid demixing effect was observed after only short periods of heating of the mixtures at temperatures between 50° and 100°C.

Samples of the product obtained in the above described reaction are subjected to extraction treatments with various solvents such as benzene, n-heptane, carbon tetrachloride, methylene chloride and ethyl acetate at their boiling temperature. In no case is even partial solubilization of the product observed. A sample of 20 g. of this product is subjected to continuous extraction with boiling 1,1,2-trifluorotrichloroethane. Initially, the formation of a translucid, seemingly homogeneous gel, is observed, which, as the treatment goes on, dissolves, leaving as the residue, after extraction for 72 hours, 3.1 g. of a high molecular weight polymer having a crystalline melting point of 320° to 325°C., and exceptional characteristics of thermal and chemical resistance. The IR absorption spectrum of this polymer corresponds to that of a polytetrafluoroethylene in which nonperfluorinated terminal groups are absent.

Monomer Purification Techniques

Y. Kometani, T. Sueyoshi and M. Tatemoto; U.S. Patent 3,300,538; January 24, 1967; assigned to Thiokol Chemical Corporation have found that by treating the fluoroolefin monomer prior to polymerization with a sulfuric anhydride selected from the group consisting of sulfur trioxide, fuming sulfuric acid and the alkali metal pyrosulfates (or any such compound capable of generating sulfur trioxide), it is possible to remove the olefinic impurities from the monomer and improve the thermal stability of the polyfluoroolefin produced by the radical-induced polymerization of the treated monomer.

Treatment of the fluoroolefin monomer with the sulfuric anhydride may be accomplished in the gaseous or liquid state, using either sulfur trioxide (in any of its polymorphic forms), fuming sulfuric acid, or the alkali metal pyrosulfates, or any compound capable of generating SO_3. Although a wide range of process conditions may be employed to purify the fluoroolefin monomer, particularly satisfactory results have been obtained by treating the fluoroolefin monomer prior to polymerization with from 2 to 10% by weight of the sulfuric anhydride at a temperature in the range from ambient room temperatures to about 150°C. for a period of time in the range from 10 seconds to 30 minutes, and preferably using about 5% by weight of the sulfuric anhydride at temperatures ranging from room temperature to about 100°C. for periods of time in the range from 30 seconds to 10 minutes. The following examples are illustrative of the process.

Example 1: 1,1,2-trifluoro-2-chloroethylene, which was prepared by dechlorinating 1,1,2-trifluoro-1,2,2-trichloroethane in methanol at reflux temperatures with zinc powder, and initially purified by distilling the monomer and drying it over silica gel, was passed through a reaction tube (5 cm. in diameter and 150 cm. in length) packed with Raschig rings (5 mm. in diameter and 10 mm. in length) together with 5% by weight of sulfur trioxide, over a period of 5 minutes at ambient room temperature and atmospheric pressure. Upon recovery, the treated monomer was washed with water and then dried.

To illustrate the differences in the physical properties of polytrifluorochloroethylene prepared from treated and untreated monomer, two polymer samples (designated as "A" and "B") were prepared under identical reaction conditions by separately polymerizing in vacuo 500 g. of SO_3-treated monomer (Sample A) and an equal weight of untreated monomer (Sample B) in a 500-ml. autoclave, using from 0.05 to 0.3 g. of trichloroacetyl peroxide at a temperature of -15° ± 1°C. Each polymer sample was then heat aged at 300°C. and atmospheric pressure and its NST value (i.e., "no strength temperature") measured before and after the heat treatment, using the method described in the article by W.T. Miller, Modern Plastics, p, 146, October 1954.

In addition to the measurement of the NST value in each test, the absorption in the infrared spectrum at 5.31 microns and at 4.43 microns were determined on films (0.1 mm. thickness) of each polymer sample before and after the heat treatment. Since the absorption band at 5.31 microns characteristic of the

$$-\underset{\underset{O}{\|}}{C}-F$$

radical, while the band at 4.43 microns is characteristic of polytrifluorochloroethylene, the ratio of absorptions at 5.31 microns with respect to that at 4.43 microns is an index of the degree of pyrolysis of the polymer. An increase in the NST value and a decrease in the ratio of absorption $[D_{5.31}/D_{4.43}]$ both are indicative of deterioration of the polymer upon heat treatment.

The results of these tests, which are summarized in the table on the following page, clearly demonstrate the improved thermal stability which is obtained in polytrifluorochloroethylene (Sample A) prepared from SO_3-treated monomer when compared to an identical polymer (Sample B) prepared from untreated trifluoroethylene.

Thermal Properties of Polytrifluorochloroethylene Prepared from Treated and Untreated Monomer

Sample	Test No.	Treatment of monomer	N.S.T.		Ratio of absorption spectra $D_{5.31}/D_{4.43}$	Physical appearance
			Before heat treatment	After heat treatment		
A	1	Treated with SO₃	310	317	0.069	Unblistered.
	2	do	275	283	0.057	Do.
B	3	Untreated	314	277	0.199	Do.
	4	do	290	275	0.121	Do.
	5	do	268	243	0.141	Blistered.

Example 2: Tetrafluoroethylene, which was prepared by the pyrolysis of chlorodifluoromethane and initially purified by distillation, was passed through a reaction tube packed with Raschig rings together with 5% by weight (based on the weight of tetrafluoroethylene) of sulfur trioxide, using the same reaction conditions described in Example 1. The yield of treated monomer was 98% of theory, based on the weight of untreated monomer passed into the reaction tube.

Two polymer samples (again designated as "A" and "B") were prepared under identical reaction conditions by separately polymerizing in vacuo 125 g. of SO₃- treated tetrafluoroethylene (polymer Sample A) and an equal weight of untreated tetrafluoroethylene (polymer Sample B) in a 500-ml. stainless steel autoclave, using a solution in 250-ml. of water of 0.003 g. of potassium persulfate and 0.001 g. of sodium bisulfate to initiate polymerization. The polymerization reaction, which was carried out at temperatures in the range from 0° to 5°C. for 3 hours, yielded in each instance 100 g. of polytetrafluoroethylene.

Each polymer sample was then preformed under the pressure of 300 kg./cm.² and subjected to heat treatment at 380°C. for periods of time ranging from 0.5 hour to 9 hours. Measurements were made on each sample at the end of the particular heat treatment of its tensile strength at the yielding point and at the breaking point, as well as is percent elongation at the breaking point. The results of these physical tests are summarized in the table below. Analysis of these results show the markedly improved physical properties of polytetrafluoroethylene (Sample A) prepared from SO₃-treated monomer when compared to polymer (Sample B) prepared from untreated tetrafluoroethylene.

Physical Properties of Polytetrafluoroethylene Prepared from Treated and Untreated Monomer

Period of thermal treatment (hrs.)	Tensile strength at yielding point (kg./mm.²)		Tensile strength at break (kg./mm.²)		Elongation at break (percent)		Molecular weight (in ten thousands)	
	A	B	A	B	A	B	A	B
0.5	1.75	1.53	1.95	2.55	270	380	1,100	800
1	1.71	1.40	2.05	1.89	260	415	900	600
3	1.70	1.27	2.05	1.44	295	330	900	500
6	1.70	1.26	2.02	1.26	320	115	900	400
9	1.70	1.26	2.05	1.26	305	115	900	400

Constant Pressure Process

A process described by A.N. Bolstad; U.S.Patent 3,163,628; December 29, 1964; assigned to Minnesota Mining and Manufacturing Company involves a method of polymerizing one or more monomers at a constant pressure which is below the saturation pressure of the monomer or at least of one of the comonomers being polymerized.

The constant pressure method of the process is easily adapted to continuous operation in equipment having no moving parts. A large safety factor is also involved, since homopolymerization of a single monomer or polymerization between two or more monomers is controlled so that there is never a large amount of monomer or monomers present in the reaction zone, whereby a sudden and uncontrollable reaction of large amounts of material is avoided. The safety of the operation is linked to the control factor of the process, the temperature and pressure being easily regulated and maintained constant.

Monomers which may be copolymerized according to the process may be divided into the following groups, the division being made on the basis of the boiling points of the monomers. The maximum pressures given are those below which the monomers will be in the vapor phase at room temperature, i.e., 25°C., and above which the monomers will begin to condense, i.e., the saturation pressure. The minimum pressures given are those which are necessary to obtain substantial amounts of polymer product.

Group: Relative Volatility		Boiling Point of Monomers	Max. Pressure at Room Temp.	Minimum Pressure
1	Most	-180° to -40°C.	500 psig	50
2	Less	-40° to -10°C.	75 psig	25
3	Least	-10° to +25°C.	40 psig	Atmospheric

Illustrative examples of monomers in each of these three groups are as follows:

Group 1 — Vinylidene fluoride, trifluoroethylene, tetrafluoroethylene, vinyl fluoride, ethylene and propene.

Group 2 — Chlorotrifluoroethylene, vinyl chloride, 1,1-chlorofluoroethylene, 2,3,3,3-tetrafluoropropene-1, perfluoropropene, 3,3,3-trifluoro-propene-1 and 1,1,3,3,3-pentafluoropropene.

Group 3 — 1,1-dichloro-2,2-difluoroethylene, vinylidene chloride, bromotri-fluoroethylene, 2-chloropentafluoropropene, vinyl bromide, isobutene, 1-chloro-2,2-difluoroethylene, perfluoroacrylonitrile, 2-chloropenta-fluoropropene, perfluoroisobutene, butadiene and fluoro-1,3-dienes such as perfluorobutadiene, 1,1,2,4,4-pentafluorobutadiene; 1,1,2,4-tetrafluorobutadiene; 1,1-difluorobutadiene; and fluoroprene.

In one method of producing polymers according to the process, a polymerization bomb is charged with a catalyst solution, evacuated and connected to a cylinder containing a monomer or mixture of monomers, the cylinder being connected to the bomb by means of a conduit having a needle valve. The feed is then introduced into the bomb through the needle valve at a controlled rate sufficient to maintain the pressure of polymerization at a desired constant value. Agitation of the contents of the bomb is provided by rocking the system, and the bomb may be heated, if desired. At the end of the poly-merization period, the feed of the monomer or monomers to the bomb is discontinued, and the product is removed in the form of a latex which is then coagulated with an electrolyte or by freezing. The product is then washed and dried to obtain the desired polymer product.

Referring to Figure 9.1, the apparatus consists of a vertically elongated vessel 2, which may be fabricated from any non-corrosive material such as stainless steel, having a jacket 4. Any suitable heat exchange medium may be passed through the jacket for the purpose of controlling the temperature of polymerization within the vessel 2. A conduit 6 connects to a catalyst solution storage and with a proportioning gear pump 8 which pumps the aqueous solution through the conduit 10 the valve 12 and the spray head 14 mounted in the top of the elongated vessel 2. The spray head uniformly distributes a fine spray of the aqueous catalyst solution downwardly through the elongated vessel or polymerization reactor, where it contacts rising monomer vapor or vapors introduced through the conduit 16 and the pressure regulator 18 into the bottom of the elongated reaction vessel 2 from a monomer storage.

The monomer or monomers are flashed as vapors from the pressure regulator 18 below the foraminous plate 20 mounted in the bottom of the vessel 2, the plate uniformly distributing the monomer vapors throughout the cylindrical reaction vessel 2. A conduit 22 connects the vapor space, or upper portion of the reaction vessel 2, with the suction side of a recircu-lating compressor 24 and a conduit 26 is connected to the discharge side of the recirculating compressor, terminating at a point 28 below the foraminous plate 20. A conduit 30 having a valve 32 therein is provided for withdrawal of polymer latex product from the bottom of the vertically elongated reaction vessel 2. If desired, the tower 2 may be packed with any conventional packing such as broken stone, clay spheres, Carborundum, glass rings, porcelain saddles and porcelain rings.

In the operation of this apparatus, the reaction vessel 2 is brought to the desired temperature by means of a heat transfer medium circulating through the jacket 4, and an aqueous catalyst solution is introduced through the spray head 14 by means of the gear pump 8. As the spray of aqueous catalyst system descends through the reaction vessel 2, it is met by an upflowing stream of monomer or monomers introduced through the conduit 16, the pressure regulator 18, and the dis-persing foraminous plate 20. As the aqueous catalyst solution collects in the bottom of the vessel 2, the level of the liquid rises until the foraminous plate 20 is covered, and additional catalyst may be introduced continuously or inter-mittently.

The monomers are introduced at a rate sufficient to maintain the pressure within the vessel constant, and monomers from the vapor space are recirculated through the conduit 22, the recirculating compressor 24, and the conduit 26 in order to maintain the monomer system at a constant composition. As the latex polymer product forms, it may be withdrawn from the system through the conduit 30 and valve 32 at the bottom of the reaction vessel. Thus, by correlating the feed of catalyst solution, the feed of monomer or monomers, and the withdrawal of the polymer product, the reaction may be operated continuously in the apparatus shown with sufficient monomer or monomers being added at all times to maintain the pressure within the reactor 2 constant and by adding sufficient catalyst, preferably continuously, to make up for that

FIGURE 9.1: CONSTANT PRESSURE PROCESS, POLYMERIZATION BOMB

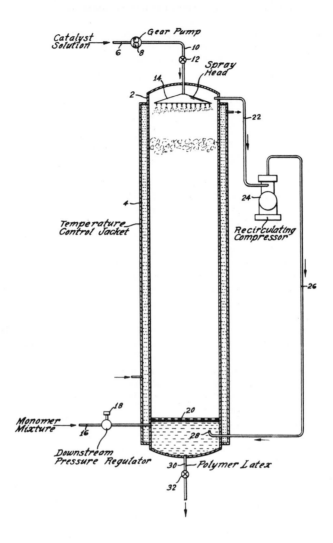

Source: A.N. Bolstad; U.S. Patent 3,163,628; December 29, 1964

lost in the withdrawal of the polymer latex product. The pressure within the vessel 2 is maintained at all times below the saturation pressure of at least one of the monomers being polymerized so that at least one of the monomers exists substantially completely in the vapor phase, although the monomer is being recirculated through the aqueous catalyst solution and some monomer is dissolved at all times. The process is illustrated by the following examples.

Example 1:   In the preparation of a copolymer containing 50 mol % of each of 1,1-chlorofluoroethylene and tetrafluoroethylene employing the constant pressure technique, the following emulsion redox polymerization system was used.

	Parts by Weight
Water, deionized	200.0
Potassium persulfate	1.0
Sodium metabisulfite	0.4
Ferrous sulfate	0.1
Perfluorooctanoic acid	1.0

A stainless steel polymerization bomb was charged with the following catalyst solutions, freezing the contents in a solid carbon dioxide-acetone bath after each addition.

(1) 150 parts of water containing 1 part of dissolved perfluorooctanoic acid
(2) 30 parts of water containing 1 part of dissolved potassium persulfate
(3) 10 parts of water containing 0.1 part of ferrous sulfate heptahydrate
(4) 10 parts of water containing 0.4 part of sodium metabisulfite.

The bomb was then evacuated and connected to a steel cylinder equipped with a pressure gauge and a needle valve located between the bomb and the steel cylinder. The steel cylinder contained a mixture of monomers consisting of a quantity of 1,1-chlorofluoroethylene and tetrafluoroethylene, in the liquid phase, calculated to provide a feed containing 50 mol percent of each monomer, or a feed containing 22.3 parts of 1,1-chlorofluoroethylene and 27.7 parts of tetrafluoroethylene.

The needle valve between the steel cylinder and the polymerization bomb was opened and the feed was introduced into the bomb in the gaseous phase at a rate sufficient to maintain the pressure of polymerization at 50 psig. The entire system was rocked at ambient temperature which was about 25°C. At the end of 21 hours, 50 parts of feed had been introduced into the bomb and the polymerization was then stopped. The product in the bomb was in the form of a white latex. The latex was coagulated with a hot dilute sulfuric acid-sodium chloride solution.

The coagulated product was collected, thoroughly washed with warm and cold water, and dried to constant weight in vacuo at a temperature of 35°C. A rubbery polymeric product was obtained in 90% conversion, and was found to contain 20.1% chlorine, or 50 mol percent of combined 1,1-chlorofluoroethylene and 50 mol percent of combined tetrafluoroethylene.

The above procedure was repeated with the exception that the feed, containing 50 mol percent of 1,1-chlorofluoroethylene and 50 mol percent of tetrafluoroethylene was introduced into the polymerization bomb at a rate sufficient to maintain the polymerization pressure between 75 and 100 psig. After 6.2 hours of polymerization at a temperature of 25°C., the polymerization was stopped. The latex was dried in the same manner as that set forth above and a rubbery product was obtained having approximately the same composition as that of the polymer produced above.

Example 2:   In the preparation of a copolymer of 1,1-chlorofluoroethylene and tetrafluoroethylene under autogenous conditions of pressure, three runs were conducted in heavy walled glass polymerization tubes using the same recipe of Example 1, except that no perfluorooctanoic acid was employed. After the tubes were charged with 200 parts of water, 1.0 part of potassium persulfate, 0.4 part of sodium metabisulfite and 0.1 part of ferrous sulfate heptahydrate, the tubes were evacuated and further charged by flash distillation at liquid nitrogen temperature with 44.6 parts of 1,1-chlorofluoroethylene and 55.4 parts of tetrafluoroethylene, corresponding to a charge containing 50 mol percent of each monomer.

After sealing the tubes under vacuum at the temperature of liquid nitrogen, they were rotated end over end in a water bath, the temperature of which was automatically controlled at 20°C. The polymerizations were conducted under autogeneous conditions of pressure. At the end of the period indicated in the table below, the polymerizations were stopped and the tubes were placed in a liquid nitrogen bath to coagulate the products. The coagulated polymeric products were collected, washed several times with water and dried to constant weight in vacuo at a temperature of 35°C. The results of these three runs are given below.

			Composition of the Product		
Run Nos.	Percent Conversion	Reaction Time (hrs.)	Mol Percent $CH_2=CFCl$	Mol Percent $CF_2=CF_2$	Characteristics
1	54	24	82	18	Rubbery
2	97	23	51	49	Powdery (very slightly rubbery)
3	8	0.3	80	20	Rubber

Comparison of the results obtained in these three runs with those obtained in Example 1 points up several of the advantages of using the constant pressure technique over the ordinary method of contact polymerizations under autogenous conditions of pressure. The reactivities of 1,1-chlorofluoroethylene and tetrafluoroethylene are quite different, the reactivity ratio of 1,1-chlorofluoroethylene being 2.8 + 0.3 and the reactivity of tetrafluoroethylene being 0.1 + 0.1. Thus, no feed containing both of these monomers will yield a copolymer of the same composition as the feed when the copolymerization is conducted under autogenous pressure or when the monomers are reacted in the liquid phase.

In Run No. 2 above, a 51:49 copolymer of 1,1-chlorofluoroethylene:tetrafluoroethylene was obtained from a 50:50 charge. Although this particular run yielded a copolymer of approximately the same composition as the feed, the product was very heterogeneous. The product was also a powder, although a rubbery product was expected and desired. The heterogeneity and powdery nature of this product is accounted for by the fact that as the relative concentration of the less reactive monomer, $CF_2=CF_2$, to the more reactive monomer, $CH_2=CFCl$, increases, the $CF_2=CF_2$ is drawn into the reaction to a greater extent. Therefore, the resulting polymer possessed a preponderance of the tetrafluoroethylene monomer unit at either end of the polymer chain, the product thereby assuming the powdery nature of polytetrafluoroethylene.

This unevenness of reaction of the two monomers produced the excessive spread of monomer ratios found in the copolymers listed in the previous table. The use of constant pressure or feeding the monomers into the reaction zone in the vapor phase and maintaining the polymerization pressure below that which causes condensation of the monomers does not necessitate the calculation, based on reactivity ratios, of a specific molar charge to obtain a product of desired composition.

Also, in order to produce a more homogeneous copolymer when autogenous pressures are used, the conversion must be kept below about 30% by weight. On the other hand, the composition and homogeneity of the copolymer produced using the constant pressure technique is independent of conversion and, in general, is dependent only on the composition of the feed, a specific feed of monomer producing a copolymer having the same composition as the feed.

Example 3: In the preparation of a copolymer containing about 80 mol % of vinylidene fluoride and about 20 mol % of 1,1-chlorofluoroethylene, the following emulsion polymerization recipe was used.

	Parts by Weight
Water, deionized	200.0
Potassium persulfate	1.0
Sodium metabisulfite	0.4
$Cl(CF_2-CFCl)_3CF_2COOH$	1.0

A stainless steel polymerization bomb was charged with the following catalyst solutions, freezing the contents in a solid carbon dioxide-acetone bath after each addition:

(1) 150 parts of water containing 1.0 part of dissolved $Cl(CF_2-CFCl)_3-CF_2COOH$.
(2) 30 parts of water containing 1.0 part of dissolved potassium persulfate.
(3) 20 parts of water containing 0.4 part of dissolved sodium metabisulfite.

The bomb was then evacuated and connected to a steel cylinder equipped with a pressure gauge and a needle valve located between the bomb and the steel cylinder. The steel cylinder contained a mixture of monomers consisting of a quantity of vinylidene fluoride and 1,1-chlorofluoroethylene in the liquid phase, calculated to provide a feed containing 80 mol percent of vinylidene fluoride and 20 mol percent of 1,1-chlorofluoroethylene, or a feed containing 76.1 parts of vinylidene fluoride to 23.9 parts of 1,1-chlorofluoroethylene. The needle valve between the steel cylinder and the polymerization bomb was opened and the feed was introduced into the bomb in the gaseous phase a a rate sufficient to maintain the pressure of polymerization at 140 psig. The entire system was rocked at a temperature of 50°C. for a period of 21 hours. The polymer latex was coagulated, collected, washed and dried as described in Example 1. A solid product was obtained and was found to contain 8.9% fluorine corresponding to a copolymer containing 82 mol percent of vinylidene fluoride and 18 mol percent of 1,1-chlorofluoroethylene.

COMPANY INDEX

INVENTOR INDEX

U.S. PATENT NUMBER INDEX

NOTICE

Nothing contained in this Review shall be
construed to constitute a permission or recom-
mendation to practice any invention covered
by any patent without a license from the patent
owners. Further, neither the author nor the
publisher assumes any liability with respect to
the use of, or for damages resulting from the
use of, any information, apparatus, method or
process described in this Review.

ABS RESIN MANUFACTURE 1970

by C. Placek

Chemical Processing Review No. 46

ABS (acrylonitrile-butadiene-styrene) resins make up one of the most rapidly growing segments of the polymer industry. There are 39 producers of ABS throughout the world. U.S. consumption in 1969 was about 500 million pounds, with a total world production of about 1,200 million pounds. It is conceivable that U.S. production alone could reach 1,000 million pounds by 1975. This is a market to seek out for new and enlarged business ventures. This book is designed with that in mind.

This book offers detailed practical process information based on the patent literature for manufacture of ABS resins. The Table of Contents below indicates the type of information provided by this comprehensive survey.

1. Straight ABS Materials
   Basic Process
   Molded Vulcanized Mixture
   Thermoplastic Molding Composition
   Thermoplastic Casting Composition
   Ternary Polymer Composition
   Polycomponent Blends
   Nonblooming ABS Composition
   Graft Copolymers
   Acrylonitrile Styrene (AS) Graft on ABS
     Copolymer
   AS with Polybutadiene
   ABS with Butadiene Styrene
   AS Polymerized with Polybutadiene
   Interpolymer with Rubbery
     Polybutadiene
   ABS Polyblends
   ABS Polymerized in Presence of
     Polystyrene
2. ABS Modified with Acrylic Derivatives
   ABS with Acrylate Type Rubber
   Acrylic Acid Esters and ABS
   Blends with Poly (Methyl Methacrylate)
   Rubbery Methacrylic Acid Terpolymer
   Acrylate and Vinyl Alkyl Ether with ABS
   Methyl Methacrylate with ABS
   Use of n-tert-Butylacrylamide
3. ABS from Alpha-Methylstyrene (AMS)
   Polymerization of Acrylonitrile/AMS
   AMS/Acrylonitrile Blend with
     AMS/Diolefin
   ABS with AMS/Acrylonitrile Copolymer

   Interpolymer of Rubber Acrylonitrile
     and tert-Alkylstyrene
   AMS and Methacrylate in ABS
     Composition
4. Miscellaneous Modifiers
   Ethylenically Unsaturated Dicarboxylic
     Acid Anhydride
   Use of Vinyl Chloride
   Use of Divinyl Benzene
   Copolymerization with Silane Diols
   Epsilon-Caprolactam/ABS Blend
   Sulfone/ABS Blend
   Amine Modifiers
   Itaconate Compounds
   Carbonyls, Hydroxyls and Carboxyls
   Ethylene/Vinylene Carbonate
   Polycarbonate/ABS Blend
   Dibutyl Fumarate
   Mineral Oil
   Hevea Rubber
5. Modification of Properties
   Flame Resistance
   Light and Temperature Stable
   Antistatic
   Improved Adhesion
6. Process Variations
   Process Sequence
   Cell Products
   Monomer Addition to Latex
   Elimination of Coagulation Problems
   Specific Initiators
   Two Stage Process
   Biphase Plastics
   Stable Aqueous Suspension
   Coagulating ABS Latex
   Continuous Monomer Addition
   Stable Polymerization Solutions
   Suspending Agents
   Polymerizable Solvent Adhesive
     Composition
   Incremental Addition of Styrene
   Peroxide-Cured Blends
   Solvents to Control Polymerization
   Shock Resistant Polymers
   Polymerization Along a Flow Path
   Decomposition of Peroxy Oxygen

   233 pages

   $35

SOUNDPROOF BUILDING MATERIALS 1970

by M. Ranney

U.S. sales of noise abatement equipment are rising at an annual rate of 15 to 25%. This can be attributed in large measure to an increasing awareness of the health menace of excess noise and to the attendant interest in noise pollution abatement and also to some law enforcement action. The Walsh-Healey Act, which took effect in 1969, now regulates noise in factories for companies with government contracts in excess of $1,000. These factors signify a rapidly growing market.

The current market for noise and vibration control, a relatively untapped market, is placed at $170 million in annual sales, and this figure is expected to double within five years. Building materials for shelter construction are considered one of the areas with the most promising growth potential. A number of processes have been patented, largely relating to the use of fibrous glass, mineral wool and fiberboard; but gypsum, perlite, ceramics, and most recently, plastics have been studied in detail for this use. This book summarizes pertinent U.S. patent literature through March 1970 relating to soundproofing processes and techniques for this increasingly important industry.

The following shortened Table of Contents will provide an indication of the scope and content of this volume. The book's usefulness as a tool for making the most of this new market will become apparent from a study of the contents where the data contained is indicated by the numbers of processes included by the number in ( ).

1. Fibrous Glass and Mineral Wool
   Fibrous Glass (14)
   Surface Coating Techniques for Fibrous Glass (5)
   Mineral Wool (10)
   Flake Glass Panels (1)
2. Fiberboard
   Perforation Techniques (6)
   Panel Design (6)
   Surface Coatings (3)
   Miscellaneous (5)
3. Gypsum—Ceramic—Perlite
   Gypsum (4)
   Ceramic (3)
   Perlite, Vermiculite and Silicate (5)
   Miscellaneous (3)
4. Plastic Products
   Polystyrene (3)
   Polyurethane (4)
   Miscellaneous (8)
5. General Processes
   Flexible Partitions (3)
   General Construction (7)

   217 pages

   $35

MEAT PRODUCT MANUFACTURE 1970

by E. Karmas

Food Processing Review No. 14

This most recent addition to the Food Processing Review series is concerned with the latest technology in the field of preparing packaged meats in both ready-to-cook and ready-to-eat forms. The information contained in this volume has been derived primarily from the U.S. patent literature since 1960, which provides you with comprehensive process information. This information has been divided into two sections—general processing methods, and specific products—to insure the greatest usefulness of the available material.

Meat has usually been eaten only at mealtimes because of the amount of time necessary for its preparation. However, with the aid of modern technology, meat is becoming increasingly available as both a snack food and a convenience item. Much of this book pertains to just these new types of meat foods.

The Table of Contents below indicates the many additional areas covered in this survey. The numbers in ( ) indicate the number of processes described under that heading.

A. General Processing

1. Curing Methods and Ingredients
   Accelerated Curing and Color Stability (13)
   Dietetic Curing Compositions (3)

2. Increased Water Binding and Yield
   Phosphates, Hydroxides and Other Chemicals (7)
   Various Yield Increasing Agents (5)

3. Improved Curing Formulations
   Nitrite Stability (6)
   Handling and Stability of Ascorbates (2)
   Homogenity of Curing Compositions (7)

4. Integral Meats
   Mechanical Methods and Apparatus (5)
   Binding Agents (5)

5. Smoking
   Production of Smoke or Smoke Flavor Concentrates (5)
   Smoking Methods and Apparatus (4)

6. Thermal Processing and Sterilization
   Methods and Apparatus (2)
   Commercial Sterilization (2)
   Thermal Treatment (2)

7. Miscellaneous Processing Methods
   Auxiliary Curing (2)
   Product Releasing Agents (3)
   Meat Conditioning (2)

B. Products

8. Bacon Production
   Bacon Curing (5)
   Bacon Processing (3)
   Bacon Cooking (2)

9. Patty Type Products
   Fresh Patties (4)
   Dehydrated Patties with Binder (3)

10. Dehydrated Convenience and Snack Products
    Meat Tidbits (5)
    Pork Rind Snacks (3)

11. Modified and Novel Products
    Conventional Products (2)
    Filled and Shaped Products (2)
    Liquid or Powdered Meat Protein (2)
    Shelf-Stability with Antioxidants and other Additives (3)

273 pages

$35

EUROPEAN FOOD MARKET RESEARCH SOURCES 1970

by Noyes Data S.A.

This book is designed to serve as a guide to the mounting volume of statistical and market intelligence material currently available on the European food industry. It will enable the researcher to pinpoint, with the minimum of delay, those publications most likely to be of assistance, as well as indicate a wide range of sources which might not previously have been consulted.

The book contains an international section, listing sources with a world or European coverage and summarizing their contents. The book also deals with sixteen countries of Western Europe. In each case, references are classified under the following headings: (A) Government statistics and reports, (B) Other statistics and reports, (C) Trade associations, (D) Food trade journals, (E) Other newspapers and periodicals, (F) Directories, (G) Advertising statistics, (H) Bank reviews.

Although principally concentrating on the food processing industry, many sources have been included which also contain information on related sectors, such as fresh foods or self-service distribution. Entries are accompanied, where appropriate, by a synopsis of their contents and the name and address of the issuing authority. Since many statistics are collected and never published, these addresses will often be the key to further important sources of primary data.

European Food Market Research Sources' has been compiled with a view to providing a balanced and comprehensive selection of essential sources. It will prove an invaluable aid to those undertaking research into the food industries of Europe and a useful addition to the reference sections of commercial libraries.

This book was prepared by our subsidiary (Noyes Data S.A.) with offices in Zurich and London. On-the-spot coverage by our experienced European editors brings you a valuable, up-to-date marketing guide to this important industry.

111 pages

$19

PHARMACEUTICAL AND COSMETIC FIRMS U.S.A.

This publication describes the 600 leading pharmaceutical and cosmetic firms in the United States with as much information as possible to assist you in your sales, market research, acquisition, divestiture, or employment efforts. Includes (1) ethical, (2) proprietary, (3) veterinary, (4) private formula, (5) cosmetics, and (6) toiletries firms.

Contains this information (where available) for United States firms:

Name, Address, and Telephone Number
Ownership
Annual Sales Figure
Number of Employees
Names of Executives
Subsidiaries and Affiliates
Plant Locations
Products

This guide puts much information that is difficult to obtain, at your fingertips. Considerable effort was undertaken to include only those firms that have some significance. The pharmaceutical and cosmetic industry is difficult to assess because of the large number of very small companies; these companies have not been included. This enables you to concentrate your efforts on the larger firms.

This book has two indexes—a subsidiary and division index and a zip code index. The zip code index is an extremely important sales tool. This index lists the companies in numerical order by zip code, thereby providing you with an easy-to-use, invaluable, geographical index.

Pharmaceutical industry sales in the United States are over $6 billion, and cosmetics sales are over $4 billion. These are two large industries, that are growing rapidly.

This guide can help you:

Concentrate on the big buyers
Prepare market reports
Increase sales effectiveness
Research potential acquisitions and divestitures
Employment and personnel guide

212 pages

$20

3

MODERN BREAKFAST CEREAL PROCESSES 1970
by R. Daniels
Food Processing Review No. 13

Describes in detail production processes and equipment for the manufacture of modern breakfast cereals. These include both ready-to-eat and quick-cooking products.

Offers detailed practical information for the manufacture and production of these cereal products based on the U.S. patent literature. 61 processes included. Abbreviated Table of Contents follows.

Dough Cooking and Extrusion Processes
Treatment Prior To Puffing
Puffing Processes
Processes For Whole Cereal Grains
Cereal Shaping Processes
Sugarcoating Process
Fruit Incorporation and Nutritional Enrichment
Quick Cooking Cereal Products

217 pages. $35

FRESH MEAT PROCESSING 1970
by Dr. E. Karmas
Food Processing Review No. 12

This Food Processing Review, deals with 106 detailed processes covering essential developments in the fresh meat processing industry since 1960. The book provides a well-organized tour through the field; the processes included are well researched and presented as an easy-to-use guide to what is being done in this vital field today.

The material has been divided into two parts; processes for enhancing palatability, and preservation processes. The numbers in ( ) after each heading indicate the number of processes for each entry.

A. Palatability: Tenderness (33), Flavor and Tenderness (8), Flavoring (12), Color (13), Integral Texture (6). B. Preservation: Moisture Retention (9), Antimicrobial Treatment (10), Ionizing Radiation (7), Other Methods of Preservation (8). 236 pages. $35

SOLUBLE TEA PRODUCTION PROCESSES 1970
by Dr. N. Pintauro
Food Processing Review No. 11

This book describes production processes for producing soluble tea and offers a wealth of detailed practical information based primarily on the U.S. patent literature. Describes 73 specific processes in this field with substantial background information. The Table of Contents is listed below. The numbers in ( ) indicate the number of processes in that category.

Withering and Rolling (4)
Fermentation, Firing and Sorting (8)
Extraction (13)
Recovery of Aroma (10)
Tannin-Caffeine Precipitate (Cream) (15)
Filtration and Concentration (8)
Dehydration Process (6)
Agglomeration and Aromatization (9)

Illustrations. 183 pages. $35

ALCOHOLIC MALT BEVERAGES 1969
by M. Gutcho
Food Processing Review No. 7

The traditional brewing process is a batch operation, costly and time consuming. There would be economic advantages to improved continuous processes which would require less capital investment for plant and equipment, give savings in labor, better use of raw materials, shorter processing time, and a more uniform product.

Detailed descriptive process information is found in this review, based on 157 U.S. Patents in the brewing field, issued since 1960. The 157 processes are organized in 7 chapters which tend to follow the steps in the brewing process.

Contents: Malting, Wort, Hops, Fermentation, Freeze Concentration and Reconstitution of Beer, Chillproofing, Preservation against Microbiological Spoilage, Foam, Indexes. Illustrations. 333 pages. $35

CONFECTIONARY PRODUCTS MANUFACTURING PROCESSES 1969
by M. Gutterson
Food Processing Review No. 6

This book is of technological significance in that it details over 200 processes for producing confections, based on the U.S. patent literature since 1960.

Based solely on new technology, this book offers substantial manufacturing information relating to this field. The wide scope of detailed data can be seen by the chapter headings indicated below:

Candy
Chocolate Products
Whipped Products
Icings
Gels
Coatings and Glazes
Gums and Stabilizers
Egg Products
Marshmallows and Meringues
Puddings
Frozen Confections
Chewing Gum
Other Confections
Indexes

Illustrations. 321 pages. $35

EDIBLE OILS AND FATS 1969
by Dr. N. E. Bednarcyk
Food Processing Review No. 5

This book describes in detail 225 recent process developments.

Shortenings; Fluid, Plastic, Miscellaneous: Margarine and Spreads; Margarine Oils, Highly Nutritional Oil Blends, Antispattering Agents, Fluid and Whipped Margarines, Flavor, Color, and Texture Modifications, Low Calorie Spreads: Salad Oils, Mayonnaise and Emulsified Dressings; Crystallization Inhibitors, Emulsified Dressings, Flavored Salad Oils, Low Calorie Dressings: Frying and Cooking Oils; Equipment, Breakdown Inhibitors, Antispattering Additives, Other Additives: Hard Butters; Preparation by Fractional Crystallization, Preparation by Ester Exchange, Miscellaneous: Oil Processing; Antioxidants and Stabilizers; Emulsifiers and Emulsions; Mixed Ester Emulsifiers, Dried Emulsion, Miscellaneous: Peanut Butter and Spreads; Chocolate Products; Indexes. Illustrations. 404 pages. $35

ANIMAL FEEDS 1970
by M. Gutcho
Food Processing Review No. 10

This is a significant work, based on the patent literature, that shows you how to prepare a wide variety of modern feed products from numerous sources. It discusses the use of many chemicals, additives, and supplements used for animal feeds.

The 278 processes covered in this book are listed in the Table of Contents below. The numbers in ( ) indicate the number of processes described under that heading.

Introduction; Forage and Fodder (32), Fats and Oils (21), Molasses and Flavoring (15), Estrogens as Growth Stimulators (17), Antibiotics as Anabolic Stimulators (33), Antioxidants in Feeds (6), Minerals and Vitamins (13), Growth-Promoting Chemical Additives (24), Poultry Feeds (40), Ruminant Feeds (23), Feed for Swine (10), Pet and Other Feeds (19), Feed Products from Industrial Waste and By-Products (25). Indexes. 350 pages. $35

BAKED GOODS PRODUCTION PROCESSES 1969
by M. Gutterson
Food Processing Review No. 9

This book describes 201 recent processes for the production of baked goods. Based on the patent literature, it offers an up-to-date comprehensive publication of manufacturing processes.

There is a substantial amount of information in this book relating to the use of various chemicals and related additives.

Contents: Bread, Yeast Leavened Products, Chemically Leavened Products, Leavening Non-Leavened Products, Air Leavened Products, Refrigerated Doughs, Emulsifiers and Dough Improvers, Miscellaneous, Indexes. Illustrations. 353 pages. $35

SOLUBLE COFFEE MANUFACTURING PROCESSES 1969
by Dr. N. Pintauro
Food Processing Review No. 8

This book describes significant manufacturing processes for producing soluble coffee, and offers a wealth of detailed practical information based primarily on the U.S. patent literature. Describes 114 specific processes in this field with substantial background information.

Introduction: Roasting, Extraction, Filtration and Concentration, Recovery of Aromatic Volatiles, Spray Drying and Other Dehydration Processes, Freeze Drying Processes, Aromatization of Soluble Coffee Powder, Agglomeration Techniques for Soluble Coffee, Decaffeinated Soluble Coffee, Packaging of Soluble Coffee. Illustrations, Indexes, 254 pages. $35

SNACKS AND FRIED PRODUCTS 1969
by Dr. A. Lachmann
Food Processing Review No. 4

The sales of snack foods in the U.S. may reach the two billion dollar mark in 1969. Many companies are actively working on new snack foods or on improved processes. The patent literature on french fried potatoes, potato chips, corn chips and other crisps is continually growing and it is the purpose of this book to present this literature in easy readable form.

French fried potatoes and their methods of production are described in the second chapter. The next chapter deals with potato chips, still the most popular product of the snack food industry. The U.S. market for potato chips is estimated to be approximately 600 million dollars in 1969. In Chapter Four the processes for corn chips are covered; in Chapter Five, apple crisps. Chapter Six describes processes for expanded chips and some specialty items; and the last chapter deals with batter mixes. Many illustrations. 181 pages. $35

PROTEIN FOOD SUPPLEMENTS 1969
by R. Noyes
Food Processing Review No. 3

The 126 Processes in this book are organized in 8 chapters by raw material source including the important newer processes for producing protein by fermentation of hydrocarbons. Another chapter on textured foods describes in detail a number of processes for producing these products that simulate meat. Indexes by company, inventors and patent number help in providing easily obtainable information.

This book is based upon the patent literature and serves a double purpose in that it supplies detailed technical information and can be used as a guide to the U.S. Patent literature on processes to obtain protein materials.

Contents: Hydrocarbon Fermentation, Fish-Based Protein, Soybeans, Cottonseed, Other Oilseeds and Legumes, Wheat and Gluten, Milk-Based Protein, Textured Foods, Miscellaneous, Indexes. Many illustrations. 412 pages. $35

DEHYDRATION PROCESSES FOR CONVENIENCE FOODS 1969
by R. Noyes
Food Processing Review No. 2

Describes 236 up-to-date dehydration processes for producing specific foods. Most detailed body of information ever published.

The detailed, descriptive process information in this book is based on 236 U.S. patents in the food dehydration field—issued between January 1960 and May 1968. This book serves a double purpose in that it supplies detailed technical information, and can be used as a guide to the U.S. patent literature on dehydration of foods. By indicating only information that is significant, and eliminating much of the legal jargon in the patents; this book then becomes an advanced commercially oriented review of food dehydration processes.

Dry Milk Products, Cheese and Yoghurt, Eggs, Fruit and Vegetable Juices, Fruits, Potatoes, Vegetables, Coffee, Tea, Miscellaneous. Many illustrations. 367 pages. $35

FREEZE DRYING OF FOODS AND BIOLOGICALS 1968
by R. Noyes
Food Processing Review No. 1

The detailed, descriptive process information in this book is based on 105 U.S. patents in the freeze drying field—issued between January 1960 and May 1968. Serves a double purpose in that it supplies detailed technical information, and can be used as a guide to the U.S. patent literature on freeze drying.

Higher costs of freeze drying are due to high capital costs for equipment, as well as high operating costs due to high energy consumption and more limited output. This book contains many cost-cutting ideas.

Since the products obtained by freeze drying are considerably superior in most cases than those produced by other processes, considerable research and development work has been carried out attempting to lower costs. This book will give you the latest developments and recent advances. Numerous illustrations. 313 pages. $35

Noyes Data Corporation is the world's leading publisher of books relating to actual production techniques for producing specific products. They are down-to-earth practical books; theoretical considerations are only mentioned where pertinent to a basic understanding of the subject.

SYNTHETIC PERFUMERY MATERIALS 1970
by M. Gutcho
Chemical Processing Review No. 45

This Review shows you how to produce synthetic perfumery materials. It contains a valuable odor index.

The 152 U.S. patents included in this book are distributed among the 11 areas as shown below:

From Terpenic Materials (28)
Alcohols (11)
Esters (18)
Ethers (19)
Aldehydes (10)
Ketones (18)
Lactones, Pyrones, Substituted Phenols and Guinones (12)
Other Structures (7)
Naphthalene and Indene Derivatives (7)
Compounds with Scent of Ambergris or Irone (9)
Product Application (13)

273 pages. $35

PHTHALOCYANINE TECHNOLOGY 1970
by Y. L. Meltzer
Chemical Processing Review No. 42

Advances in phthalocyanine technology have been truly explosive during the past few years. New phthalocyanine products, processes and applications have poured forth from industrial, governmental and academic laboratories at a rapid pace. These advances in technology have made themselves felt in the market place and in government programs, and have contributed to corporate sales and profits. At the same time, however, competition has become more intense in the phthalocyanine field making it imperative to keep up with the latest technological advances.

Examines recent developments in phthalocyanine technology as reflected in U.S. patents and other literature. The first 23 chapters discuss up-to-date manufacturing processes for phthalocyanine pigments and dyes. Chapters 24 through 31 discuss unusual new applications for phthalocyanines. 390 pages. $35

ION EXCHANGE RESINS 1970
by C. Placek
Chemical Processing Review No. 44

This report on ion exchange resins provides detailed information on 126 U.S. patents issued since 1960 concerning the composition and manufacture of ion exchange materials. This Review, by its organization, also provides a guide to these ion exchange resins by grouping them according to physical form, behavior characteristics, etc.

1. Anion Exchange Resins
2. Cation Exchange Resins
3. Resins For Removing Metals
4. Resins Having Mixed Properties
5. Specific Use Resins
6. Unconventional Materials
7. Process Emphasis
8. Properties of Ion Exchange
9. Ion Exchange Membranes
10. Emphasis on Shapes

329 pages. $35

RADIATION CHEMICAL PROCESSING 1969
by R. Whiting
Chemical Processing Review No. 41

A number of radiation induced chemical processes are already operating commercially. The radiation processing of chemicals has been growing at an annual rate of about 25% per year. Currently, $100 to 150 million worth of irradiated products are produced in the United States per year, however, it has been forecast that by 1980, the value of products receiving radiation treatment will be close to $1,000 million per year.

This book surveys the radiation processing field and is based on the U.S. patent literature since 1960. Over 250 separate processes are described in detail in the chemical, polymer, rubber, petroleum, textile and other fields.

Contents: Polyolefins, Other Polymers, Elastomers, Hydrocarbons, Organic Chemicals, Inorganic and Organo-Metallic Compounds, Other Processes, Indexes. 377 pages. $35

PHOTOCHEMICAL PROCESSES 1969
by B. Albertson
Chemical Processing Review No. 36

The purpose of this Review is to provide an up-to-date description of industrial photochemical technology as recorded in the U.S. Patent literature since 1960. Describes in detail 210 photochemical production processes.

Photochemistry, long known as a selective powerful tool, has excited the imagination of chemical engineers as a method of manufacturing various chemicals. From the economic point of view, photochemistry has the following advantages: 1. Many materials cannot be obtained in any other way. 2. Specificity of effects cannot be obtained by other methods. 3. Less need for high pressures and temperatures can lower costs.

Introduction, Photohalogenation, Photonitrosation, Organic Photochemical Reactions, Inorganic Photochemical Reactions, Photopolymerization, Indexes. Illustrations. 185 pages. $35

CHLORINE AND CAUSTIC SODA MANUFACTURE RECENT DEVELOPMENTS 1969
by Dr. R. Powell
Chemical Processing Review No. 33

Contents:
General Considerations
Preparation of Brine for Electrolysis
Brine Electrolysis in Diaphragm Cells
Brine Electrolysis in Mercury Cells
Combination Mercury Cell and Amalgam Fuel Cell System
Recovery of Mercury from Brine
Production of Caustic Soda in the Decomposer Unit
Coordinated Operation of Diaphragm and Mercury Cells
Platinum-Coated Titanium Anodes
Electrolysis of Sea Water
Electrolysis of Hydrogen Chloride
Catalytic Oxidation of Chlorides
Cracking of Hydrogen Chloride
Nitrosyl-Chloride Route
Cooling, Drying and Purification of Chlorine
Purification of Caustic Soda

Numerous illustrations. 265 pages. $35

ALKALI METAL PHOSPHATES 1969
by Dr. M. W. Ranney
Chemical Processing Review No. 34

The 97 processes described are based on U.S. patents issued since 1960 and offer a comprehensive treatment of up-to-date technical information.

Contents: Orthophosphates, Metaphosphates, Pyrophosphates, Tripolyphosphates, Phosphites/Hypophosphites. Numerous illustrations. 344 pages. $35

CITRIC ACID PRODUCTION PROCESSES 1969
by R. Noyes
Chemical Processing Review No. 37

Detailed descriptions of production processes for citric acid, based on the patent literature. The Table of Contents is indicated below:

Processing, Iron Impurities, Other Microorganisms, Recovery and Purification, Other Processes. Indexes. 157 pages. $24

AMINES, NITRILES AND ISOCYANATES PROCESSES AND PRODUCTS 1969
by M. Sittig
Chemical Process Review No. 31

Material covered includes: Manufacture of Amines, Manufacture of Mono-Nitriles, Acrylonitrile Derivatives, Isocyanate Manufacture, Future Trends. 62 illustrations. 201 pages. $35

ALCOHOLS, POLYOLS, AND PHENOLS MANUFACTURE AND DERIVATIVES 1968
by M. Sittig
Chemical Process Review No. 23

Contents: Introduction, Manufacture of Alcohols, Manufacture of Glycols, Manufacture of Polyols, Manufacture of Phenol, Reactions of Alcohols, Reactions of Glycols and Polyols, Phenol Derivatives. 201 illustrations. 344 pages. $35

MONOSODIUM GLUTAMATE AND GLUTAMIC ACID 1968
by Dr. R. Powell
Chemical Process Review No. 25

This technical Review with numerous flow diagrams and equipment designs, describes in detail processes for manufacture of MSG.

U.S. demand in 1970 is expected to be 80 million pounds. World production of MSG was 240 million pounds in 1967. Numerous illustrations. 256 pages. $35

PRACTICAL DETERGENT MANUFACTURE 1968
by M. Sittig
Chemical Process Review No. 27

Contents: Manufacture of Branched-Chain Olefins, Linear Alpha Olefins, and Linear Paraffins; Competitive Routes to Straight-Chain Alcohols, Alkylaromatics and other Detergent Raw Materials; Sulfation and Sulfonation Processes; Detergent Formulation. 76 illustrations. 212 pages. $35

POLYACETAL RESINS, ALDEHYDES, AND KETONES 1968
by M. Sittig
Chemical Process Review No. 26

Polyformaldehyde or polyoxymethylene is assuming increased importance.

Contents: Introduction, Production of Aldehydes, Production of Ketones, Reactions of Aldehydes, Reactions of Ketones, Aldehyde Polymers, Future Trends. 81 illustrations. 237 pages. $35

HYDRAZINE MANUFACTURING PROCESSES 1968
by Dr. R. Powell
Chemical Process Review No. 28

General Considerations, Raschig Process, Bayer Process, Processes Based on Urea, Electrical Glow Discharge Processes, Chemo-Nuclear Processes, Direct Production of Anhydrous Hydrazine, Other Processes, Recovery, Concentration, Stabilization. Numerous illustrations. 252 pages. $35

5

CARBON BLACK TECHNOLOGY RECENT DEVELOPMENTS 1968
by Dr. R. Powell
Chemical Process Review No. 21

Introduction; Feedstocks, Channel Blacks, Furnace Blacks, Thermal Blacks, Acetylene Blacks, High Structure Carbon Blacks, Low Structure Carbon Blacks, Unconventional Processes, Carbon Black Pelletizing, Other Finishing Treatments. Numerous illustrations. 242 pages. $35

HYDROGEN PEROXIDE MANUFACTURE 1968
by Dr. R. Powell
Chemical Process Review No. 20

This Review is concerned with chemical processes for manufacturing hydrogen peroxide. Major emphasis is placed on the anthraquinone processes. General Considerations, The Anthraquinone Process, Other Processes, Purification, Concentration, Stabilization. Numerous illustrations. 221 pages. $35

ELECTRO-ORGANIC CHEMICAL PROCESSING 1968
by Dr. C. Mantell
Chemical Process Review No. 14

This volume has been written from the viewpoint of the chemical engineer, with emphasis on plant processes, operating data, and plant design. The commercial successes in this field have been attained by the chemical engineering approach. 186 pages. $35

AROMATICS MANUFACTURE AND DERIVATIVES 1968
by M. Sittig
Chemical Process Review No. 17

Contents: Introduction, Production of Aromatics, Separation of Aromatics, Purification of Aromatics, Reactions giving Hydrocarbon Products, Other Reactions, Phenol Production, Styrene Manufacture and Derivatives, Future Trends. 73 illustrations. 232 pages. $35

CATALYSTS AND CATALYTIC PROCESSES 1967
by M Sittig
Chemical Process Review No. 7

Contents: Hydrocarbon Conversion Processes, Hydrocarbon Polymerization Processes, Hydrocarbon Oxidation Processes, Other Processes. Future Trends. 109 illustrations. 303 pages. $35

TITANIUM DIOXIDE AND TITANIUM TETRACHLORIDE 1968
by Dr. R. Powell
Chemical Process Review No. 18

Describes titanium dioxide processing steps (a) chlorination of rutile ore, (b) separation of sublimated solids, (c) purification of crude titanium tetrachloride, and (d) conversion of titanium tetrachloride to titanium dioxide. Numerous illustrations. 306 pages. $35

INDUSTRIAL GASES MANUFACTURE AND APPLICATIONS 1967
by M. Sittig
Chemical Process Review No. 4

This book discusses conventional cryogenic air separation and purification techniques in considerable detail.

This book also discusses newer techniques such as adsorption using molecular sieves, and permeation using various membrane materials. 313 pages. 103 illustrations. $35

PLASTICIZER EVALUATION AND PERFORMANCE 1967
by I. Mellan

This book will help you evaluate: 1. A new plasticizer with a known resin; 2. A known plasticizer with an unknown resin; 3. A new plasticizer with a new resin.

The 110 tables of performance data, and the standard methods of testing described in this book, provide data from which one can estimate what a particular plasticizer will do in a specific resin.

The plasticizers were chosen for inclusion, primarily by their greatest demand in the industry. Chapters: 1. Introduction; 2. Testing; 3. Comparative Performance Data; 4. Performance Data of Individual Plasticizers; 5. Brand Names and Their Manufacturers; 6. Abbreviations for Coding Plasticizers; 7. Chemical Names of Plasticizers and Their Brand Names; 8. Brand Names of Plasticizers and Their Chemical Names; 9. Bibliography. 178 pages $20

SUGAR ESTERS 1968
by Research Corporation

Already approved for use in foods in a number of countries, use of sugar esters in food in the United States awaits FDA clearance.

Sugar esters are an important new raw material for the food industry.

Contains papers presented at California Symposium 1967. 134 pages. $15

FLAME RETARDANT POLYMERS 1970
by M. Ranney

Summarizes selected process technology for the use of fire retardant imparting additives and reactive intermediates for major polymeric plastic materials, with particular emphasis on recent technology in the areas of polymer esters, polystyrene and polyurethane foam. There are 144 separate processes included, all based on the U.S. patent literature.

An abbreviated Table of Contents is listed. The numbers in parentheses indicate the number of processes included for each entry.

Polyethylene and
Polypropylene (15)
Polystyrene (19)
Polyurethanes (50)
Polyesters (13)
Other Polymer Systems (26)
General Utility Additives (21)

263 pages. $35

COMPATIBILITY AND SOLUBILITY 1968
by I. Mellan

Normally, it requires laborious testing to determine compatibility of polymers, resins, elastomers, plasticizers, and solvents. Predictions made without testing or literature searching, are usually unreliable. We all know of the many commercial products that were failures, when compatibility and solubility considerations were ignored.

This book helps you evaluate proper materials by the use of 224 tables. The tables were originally published by manufacturers of various materials, and reproduced in this book, to give you a reference work to this important subject. The tables are organized in three sections — polymers and resins, plasticizers and esters, and solvents. The tables in each section indicate the solubility and compatibility of the particular material with a wide range of other materials. 304 pages. $20

CELLULAR PLASTICS RECENT DEVELOPMENTS 1970
by K. Johnson

Recent accomplishments in the production of cellular plastics includes among the 190 total processes, 71 polyurethane processes and a number of techniques for polyolefin and polystyrene foam. Included information can be seen in the abbreviated Table of Contents where the numbers in ( ) indicate how many processes are covered in each area. Also contains company, author, and patent number indexes which will help you make the greatest use of the information.

Polyolefins (15)
Polyvinyl Chloride (15)
Polystyrene (22)
Rubber (15)
Polyurethanes (71)
Polyesters and Epoxides (15)
Urea-Formaldehyde and
Phenolic Resins (9)
Other Cellular Products (27)

280 pages. $35

SYNTHETIC LEATHER FROM PETROLEUM 1969
by M. Sittig
Chemical Process Review No. 29

The newer synthetic leathers offer permeability to water vapor and air, as does natural leather. In addition, the newer products offer good tear strength and softness. It is expected that these newer synthetic leathers will make substantial inroads in shoe and other markets.

These modern synthetic leathers consist, in general, of nonwoven mats of such fibers as polyesters impregnated with such binder resins as polyurethanes. The fiber-binder combination is then rendered porous by one of a number of different processes. Introduction: Manufacture of Fiber-Forming Polymers, Types of Structural Fibers Used, Mat Formation, Mat Treatment, Binder Polymer Formulation, Other Ingredients, Fiber-Binder Combination, Consolidation, Coating, Introduction of Porosity, Surface Modification, Product Properties, Future Trends. 84 illustrations. 214 pages. $35

ELECTRODEPOSITION AND RADIATION CURING OF COATINGS 1970
by Dr. M. W. Ranney

The advantages of electrodeposition; pinhole free coating, automated operation, elimination of fire hazards and air pollution problems, low operating costs and fast throughput make this an attractive method.

Radiation curing will also reach significant commercial status, due to its advantages of rapid curing, elimination of ovens, and use of solvent-free vehicles.

1. Electrodeposition—Techniques—Control of Bath Stability, Electrical Parameters, Miscellaneous Feed Control, 2. Electrodeposition—Formulation—Acid-Containing Resins, Cathode Deposition, Miscellaneous Emulsions, Nonaqueous Systems, 3. Electrodeposition—General—Metal Treatment, Fillers and Pigments, Miscellaneous, 4. Radiation Curing—Formulations—Polymeric Vehicles, Wood Impregnation 5. Radiation Curing—Equipment 170 pages. $35

SYNTHETIC FIBERS FROM PETROLEUM 1967
by M. Sittig
Chemical Process Review No. 1

Production processes for the four major synthetic fibers: nylon, polyesters, acrylics, and polyolefins. Contains a wealth of processing data with particular emphasis on process conditions. A technical evaluation of the synthetic fiber industry. 116 illustrations. 275 pages. $35

PLASTIC FILMS FROM PETROLEUM RAW MATERIALS 1967
by M. Sittig
Chemical Process Review No. 6

This Review concerns production of synthetic polymer films. Contents: Raw Materials, Polymer Manufacture, Film Production Processes, Film Treating Processes, Future Trends. 118 illustrations. 276 pages. $35

POLYOLEFIN PROCESSES 1967
by M. Sittig
Chemical Process Review No. 2

Manufacturing techniques and processes for polyolefin resins. Includes: Current Status of Various Polyolefins, Processes Using Non-Metallic Catalysts, Metal Oxide Catalysts, Metal Alkyl-Reducible Metal Halide Catalysts; Polymer After-Treatment, and Fabrication. 96 illustrations. 234 pages. $35

STEREO-RUBBER AND OTHER ELASTOMER PROCESSES 1967
by M. Sittig
Chemical Process Review No. 3

Synthesis of new polymers with rubber-like elasticity is a vital growing industry. Modification of molecular architecture of elastomers for specific properties is subject of worldwide research. Review summarizes state of the art in elastomer industry; new routes to monomers and polymerization techniques. 215 pages. 77 illustrations. $35

THERMOELECTRIC MATERIALS 1970
by M. Sittig
Electronics Materials Review No. 7

Manufacturing processes for:

Alkali Mn Tellurides, Sn and Bi Selenides and Tellurides, Bi Telluride, Boron Materials, BP, Cd-Sn Arsenide, Ce Sulfide Matrices, Ce Titanate, CsCl Melt, Cr and Co Silicide, GeTe, Au-Ni Alloys, Au-Ag Combinations, Graphite, In-Ga Arsenide, Pb-Pb Dioxide, PbTe, Mn-Ge Telluride, Mn Silicide, HgTe, Organics, Rhodium Arsenide Antimonide, Ruthenium Arsenide Antimonide, Sm Sulfide, Si-C Matrices, Si-Ge Alloys, Ag-Sb Telluride, Ag Selenide, Sr Titanate, Thallium Telluride, Tin Oxide, Tin Telluride, Ti-Va Ceramics, Uranium Dioxide, Uranium Monosulfide, Uranium Nitrate, Zn Antimonide, Zr in Zirconia Matrix.

73 illustrations. 235 pages. $35

PHOTOCONDUCTIVE MATERIALS 1970
by M. Sittig
Electronics Materials Review No. 8

Photoconductors have a number of important applications, such as; television camera tubes, solar cells, photoelectric cells, solid state light amplifiers, electrophotographic copying processes.

They offer the promise of a proliferation of applications in the near future such as new forms of computer circuitry, optical carrier communications systems using laser light sources, and in color copying processes. There are 86 processes relevant to the manufacture of these materials. They cover the three areas indicated in the abridged Table of Contents.

Electrophotographic Materials
Vidicon Tube Materials
Photocell Materials

288 pages. $35

DOPING AND SEMICONDUCTOR JUNCTION FORMATION 1970
by M. Sittig
Electronics Materials Review No. 4

Donor or acceptor impurities may be added to the liquid phase from which semiconductor crystals are grown or they may later be deposited on the crystal surface and diffused inward. In either case the addition of such impurities is called doping. Where there is formed a transition between a P-type and an N-type material as a consequence of one of these operations, a semiconductor junction is formed. This book describes 115 processes for doping and semiconductor junction formation.

Production of Alloyed Junctions, Diffusion Processes, Melt Grown Junctions, Doping During Melting, Simultaneous Dopant and Substrate Deposition, Spark Doping Processes, Doping by Particle Bombardment, Hydrothermally Grown Junctions, Doping Epitaxially Grown Layers, Future Trends. Indexes. 318 illustrations. 318 pages. $35

FLAME RETARDANT TEXTILES 1970
by Dr. M. W. Ranney
Textile Processing Review No. 3

Describes 177 commercial processes to produce flame retardant textiles and fabrics.

Most activity is based on chemical modification of cellulose through hydroxyl groups. Use of phosphoric acid, urea-phosphates, and other phosphorylating agents all confer flame retardant properties to cellulose. A significant portion of this book is devoted to the latest in application of phosphorous containing materials.

Numbers in ( ) indicate the number of processes described. Ammonium Salts, Borates (12); Antimony, Titanium Metal Oxides (25); Amine-Phosphorus Products (21); Aziridines, APO, APS (21); Methylol-Phosphorus Polymers, THPC (27); Phosphonitrilic Chlorides (9); Trialkyl Phosphates and Phosphonates (26); Silicones Isocyanates, Miscellaneous (10); Nylon, Acrylics (18). 373 pages. $35

MAGNETIC MATERIALS 1970
by M. Sittig
Electronics Materials Review No. 9

This latest review in the Electronics Materials series surveys the empirical state of the art with regard to magnetic materials as revealed primarily in recent U.S. patents.

In arranging the sections of the text with number three as can be seen from the Table of Contents below, the author has listed the various compositions alphabetically according to principal component. Ferrites containing both magnesium and manganese may be covered twice—once as magnesium manganese ferrites where the magnesium is the principal constituent, and once as manganese magnesium ferrites where manganese is the principal constituent.

Simple Magnetic Oxide Materials, Complex Magnetic Oxide Materials, Magnetic Metallic Materials. 286 pages. $35

ELECTROLUMINESCENT MATERIALS 1970
by M. Sittig
Electronics Materials Review No. 6

The phenomenon of electroluminescence offers a variety of proven and potential applications such as illuminated display panels and color TV tubes. Much research and development is being put into the evolution of this new breed of luminescent materials.

This Electronics Materials Review is concerned with manufacturing processes for producing electroluminescent materials. The numbers in ( ) indicate the number of processes given for producing each of the indicated electroluminescent materials. This book, based on the latest U.S. patent literature, serves as a guide to a productive new industry.

Introduction, Cathode Ray Tube Phosphors (20), Color Television Phosphors (25), Electroluminescent Lamp Materials (20), Electroluminescent Diode Materials (14), Future Trends. 306 pages. $35

PRODUCING FILMS OF ELECTRONIC MATERIALS 1970
by M. Sittig
Electronics Materials Review No. 5

Films of varying thickness have a multitude of applications in the fabrication of modern electronic devices. This book describes numerous processes for film production as indicated in the Contents below. The numbers in ( ) indicate the number of processes described under that heading.

Cathode Sputtering (10)
Vacuum Evaporation (18)
Gas Plating (2)
Electroplating (6)
Electroless Deposition (6)
Film Deposition From Reacting Vapors (32)
Explosive Evaporation (1)
Squeezing on Optical Flats (1)
Electron Deposition (2)
Ion Beam Deposition (1)
Reaction & Deposition From Solution (5)
Oxide Films by Reaction at Solid Surfaces (7)
Carrier Transport in Vapor Phase (6)
97 Illustrations. 295 pages. $35

SOIL RESISTANT TEXTILES 1970
by Dr. M. W. Ranney
Textile Processing Review No. 5

Ideal soil release finishes must be capable of releasing stains readily and preventing redeposition of soil during laundering. Treatments should render manmade fibers and durable press reactants less attractive to oily stains and should be more easily wetted.

This report summarizes the developments in soil retardant and soil release finishes in both the carpet industry and in textile manufacture. It includes the newest technology associated with the use of acrylates and fluorochemical treatments. The numbers in ( ) following each treating agent indicate the number of processes covered for that particular compound.

Introduction: Metal Oxides and Salts For Carpet Treatment (15), Acrylic and Vinyl Polymers (10), Silicones (6), Fluorochemical Compounds (72), General Treatments (14). 216 pages. $35

WATERPROOFING TEXTILES 1970
by Dr. M. W. Ranney
Textile Processing Review No. 4

This Textile Processing Review summarizes the technology of water resistant treatments for textiles and fabrics as described in the U.S. patent literature since the early 1950's. 246 waterproofing processes are included—64 relate to use of fluorochemicals.

The numbers in ( ) after each entry in the Table of Contents, where the treatment processes are organized by the agent used, indicates the number of production processes for each agent.

Production Processes for Waterproofing Textiles using the following Agents: Metal Salts and Wax Containing Formulations (44); Silicones and Alkyl Polysiloxanes (53), Organofunctional Silanes and Fluorosilanes (20), Acrylics (8), Nitrogen Containing Compounds (30), Fluorochemical Compounds (64), Elastomer, Vinyl, Polyolefin Vapor Permeable Fabrics (27), Miscellaneous Treatments (9). 353 pages. $35

CREASEPROOFING TEXTILES 1970
by Dr. M. W. Ranney
Textile Processing Review No. 2

Summarizes with detailed process information relating to textile creaseproofing agents used to obtain wash and wear, or permanent press fabrics. Over 300,000 words, describes 343 processes in this field. Shows you what chemical agents used, and processes by which they are applied.

Dimethyllolethylene Urea and Related Compounds, Aldehyde-Urea Condensates, Uron Resins, Aminoplasts—Catalyst Performance, Melamine Derivatives, Triazones, Carbamates, Other Nitrogen-Containing Compounds, Phosphorus-Amino Compounds—Aziridines, Aminoplast-Thermoplastic Resin Compositions, General Processing Techniques, General Formulations, Aldehydes, Acetals, Epoxies, Epihalohydrins, Sulfones, Sulfonium Salts, Cross-Linking Agents, Miscellaneous, Polymeric Coatings—Rubber, Vinyl, Silicones, Radiation Curing, Wool, Nylon, and Others. Indexes. 460 pages. $35

DYEING OF SYNTHETIC FIBERS 1969
by C. Whiting
Textile Processing Review No. 1

The dyeing of synthetic fibers continues to be a challenge. Successful methods however, have been developed. Chapters 2 through 7 of this Review are divided into two sections. The first presents various new dyes and dyeing processes which are applicable to synthetic fibers. The second section concerns many auxiliary products available to aid in production of an acceptably dyed product. These include dye improving agents, leveling and retarding agents, agents used in after treatment to achieve optimum fastness properties.

Introduction: Dyeing Polyolefin and Polypropylene Fibers, Dyeing Polyamide Fibers, Dyeing Polyester Fibers, Dyeing Acrylic Fibers, Dyeing Hydrophobic Fibers, Dyeing Glass Fibers, Dyeing Miscellaneous Fibers, Dyes Applicable to More than One Type of Fiber, Dyeing Fiber Blends. 257 pages. $35